T0392688

Imidazole-Based Drug Discovery

Heterocyclic Drug Discovery Series

Imidazole-Based Drug Discovery

Shikha Agarwal
Department of Chemistry, Mohanlal Sukhadia University, Udaipur, Rajasthan, India

Series editor

Ruben Vardanyan

ELSEVIER

Elsevier
Radarweg 29, PO Box 211, 1000 AE Amsterdam, Netherlands
The Boulevard, Langford Lane, Kidlington, Oxford OX5 1GB, United Kingdom
50 Hampshire Street, 5th Floor, Cambridge, MA 02139, United States

Copyright © 2022 Elsevier Inc. All rights reserved.

No part of this publication may be reproduced or transmitted in any form or by any means, electronic or mechanical, including photocopying, recording, or any information storage and retrieval system, without permission in writing from the publisher. Details on how to seek permission, further information about the Publisher's permissions policies and our arrangements with organizations such as the Copyright Clearance Center and the Copyright Licensing Agency, can be found at our website: www.elsevier.com/permissions.

This book and the individual contributions contained in it are protected under copyright by the Publisher (other than as may be noted herein).

Notices
Knowledge and best practice in this field are constantly changing. As new research and experience broaden our understanding, changes in research methods, professional practices, or medical treatment may become necessary.

Practitioners and researchers must always rely on their own experience and knowledge in evaluating and using any information, methods, compounds, or experiments described herein. In using such information or methods they should be mindful of their own safety and the safety of others, including parties for whom they have a professional responsibility.

To the fullest extent of the law, neither the Publisher nor the authors, contributors, or editors, assume any liability for any injury and/or damage to persons or property as a matter of products liability, negligence or otherwise, or from any use or operation of any methods, products, instructions, or ideas contained in the material herein.

Library of Congress Cataloging-in-Publication Data
A catalog record for this book is available from the Library of Congress

British Library Cataloguing-in-Publication Data
A catalogue record for this book is available from the British Library

ISBN: 978-0-323-85479-5

For information on all Elsevier publications
visit our website at https://www.elsevier.com/books-and-journals

Publisher: Susan Dennis
Acquisitions Editor: Emily McCloskey
Editorial Project Manager: Ivy Dawn Torre
Production Project Manager: Sruthi Satheesh
Cover Designer: Mark Rogers

Typeset by STRAIVE, India

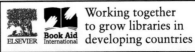

ELSEVIER Book Aid International
Working together to grow libraries in developing countries
www.elsevier.com • www.bookaid.org

Dedication

This book is dedicated to my family,
friends, and my research group for their encouragement
and continuous support.

Contents

2. Biological profile of imidazole-based compounds as anticancer agents

Ayushi Sethiya, Jay Soni, Nusrat Sahiba, Pankaj Teli, Dinesh K. Agarwal, and Shikha Agarwal

3. Recent advancements on imidazole containing heterocycles as antitubercular agents

Dinesh K. Agarwal, Jay Soni, Ayushi Sethiya, Nusrat Sahiba, Pankaj Teli, and Shikha Agarwal

4. Imidazole derivatives: Impact and prospects in antiviral drug discovery

Pankaj Teli, Nusrat Sahiba, Ayushi Sethiya, Jay Soni, and Shikha Agarwal

Contributors

Numbers in parenthesis indicate the pages on which the authors' contributions begin.

Dinesh K. Agarwal (35, 133, 291, 323), Department of Pharmacy, PAHER University, Udaipur, Rajasthan, India

Shikha Agarwal (1, 35, 133, 167, 195, 263, 291, 323), Synthetic Organic Chemistry Laboratory, Department of Chemistry, Mohanlal Sukhadia University, Udaipur, Rajasthan, India

Nusrat Sahiba (35, 133, 167, 195, 263, 291), Synthetic Organic Chemistry Laboratory, Department of Chemistry, Mohanlal Sukhadia University, Udaipur, Rajasthan, India

Ayushi Sethiya (1, 35, 133, 167, 195, 263, 323), Synthetic Organic Chemistry Laboratory, Department of Chemistry, Mohanlal Sukhadia University, Udaipur, Rajasthan, India

Jay Soni (1, 35, 133, 167, 323), Synthetic Organic Chemistry Laboratory, Department of Chemistry, Mohanlal Sukhadia University, Udaipur, Rajasthan, India

Pankaj Teli (35, 133, 167, 263, 291), Synthetic Organic Chemistry Laboratory, Department of Chemistry, Mohanlal Sukhadia University, Udaipur, Rajasthan, India

About the authors

Dr. Shikha Agarwal has been working as Assistant Professor in Department of Chemistry, Mohanlal Sukhadia University, Udaipur, since 2012. She received a Gold Medal in MSc Chemistry in 2006 from the University of Rajasthan, Jaipur. She was a recipient of JRF and SRF from CSIR, New Delhi and qualified GATE—2006 with 98th percentile. She was awarded a PhD from the University of Rajasthan, Jaipur in 2011. She has published more than 50 research and review articles in various national and international journals of NISCAIR, Wiley, Elsevier, Springer, Bentham Science, and Taylor & Francis. She has published several books and contributed chapters. Her research interests are in synthetic organic chemistry, green chemistry, catalysis, and combinatorial and medicinal chemistry. She has four major research projects (one UGC and three RUSA-MHRD) to her credit.

Ayushi Sethiya has done her postgraduation (MSc) from University College of Science, Mohanlal Sukhadia University (India) and received a Gold Medal in MSc Chemistry in 2015. She is NET and GATE qualified and is the recipient of the UGC-MANF fellowship. Presently, she is pursuing her PhD under the supervision of Dr. Shikha Agarwal as Senior Research Fellow. Her research interests focus on synthesizing biologically active heterocyclic scaffolds via eco-benign pathways. She has published more than 20 articles in renowned journals.

Nusrat Sahiba received her BSc (2013) and MSc (2015) in organic chemistry from the Mohanlal Sukhadia University, Udaipur. She qualified in the CSIR-UGC national eligibility test in December 2017 and was awarded with a research fellowship. Currently she is pursuing her doctoral research as a Senior Research Fellow under the supervision of Dr. Shikha Agarwal in Mohanlal Sukhadia University. Her research work focuses on the chemistry of heterocyclic scaffolds, their synthesis using different catalytic systems, and eco-friendly pathways. She has published more than 18 articles in journals of international repute.

Dinesh K. Agarwal is working as Associate Professor in the Department of Pharmacy, PAHER University, Udaipur. He submitted his PhD thesis in March, 2021. He completed his M. Pharma (Pharmocology) in 2012 from RUHS, Rajasthan. His research interests are medicinal chemistry and pharmacology. He has published more than 20 articles in journals of international repute.

Jay Soni received his BSc and MSc (Industrial Chemistry) from University College of Science, Mohanlal Sukhadia University, Udaipur, in 2017. Presently, he is pursuing a PhD in the Department of Chemistry, MLSU under the supervision of Dr. Shikha Agarwal. His research is focused on synthetic organic chemistry, heterocyclic synthesis, green chemistry, and medicinal chemistry. He has published more than 15 articles in journals of international repute.

Pankaj Teli received his BSc and MSc (Industrial Chemistry) from University College of Science, Mohanlal Sukhadia University, Udaipur, in 2017. He qualified the CSIR-UGC national eligibility test in December 2017 and was awarded with a research fellowship. Presently, he is working as a Senior Research Fellow in the Department of Chemistry, MLSU under the supervision of Dr. Shikha Agarwal. His research is focused on synthetic organic chemistry, heterocyclic synthesis, green chemistry, and medicinal chemistry. He has published more than 10 articles in reputed and high-quality journals.

Preface

In past decades, interdisciplinary research has been of great interest to scientists. Imidazole scaffolds act as a viaduct between organic synthesis and medicinal chemistry, and compel researchers to explore new drugs. This text aims to allow researchers/students to fabricate the foundation of introductory organic chemistry and attain a deep level of knowledge and understanding about drug discovery. There have been major developments in synthetic organic chemistry in recent years, and efforts have been made in shaping the new edition of this book to make it more useful to students, instructors, researchers, and other readers. The expanding applications of imidazole-based drug discovery are reflected by amplified discussion of this area. This book includes a study of past decades on the synthesis and biological applications of imidazole derivatives. The book has been divided into eight chapters to offer a sustainable approach for readers.

In Chapter 1, synthetic aspects of imidazole derivatives are considered. In Chapters 2–7, the different biological profiles, viz., anti-cancer, anti-tubercular, anti-viral, anti-microbial, anti-oxidant, and miscellaneous activities of imidazole analogs with their mechanism of action are discussed. Finally, Chapter 8 includes imidazole-based drugs and drug discovery, with their future possibilities. Biological activities are discussed in separate chapters for better understanding of the readers. Significant information is provided in the form of tables, figures, and schemes.

Our goal is to present a broad and fairly detailed view of the core area of imidazole-based drug discovery. We have approached this goal by extensive use of both the primary and review literature, and the sources are referenced. This book tries to review the most current developments and future perspectives on different disease therapies to achieve the ultimate goal of disease eradication. The role of imidazole in contemporary science and technological innovations to expand the rational drugs, critical challenges, and future research directions are discussed. The overriding aim of the book is to summarize the recent advancements achieved in the synthesis and biological applications of imidazole to date, which will lead to better approaches for future compounds.

Acknowledgments

First of all, I want to express my gratitude to the Omnipresent, Omniscient, and Almighty God, the glorious fountain and continuous source of inspirations.

I thank Professor D.C. Gautam Department of Chemistry, University of Rajasthan, Jaipur, Rajasthan, India, for her encouragement, blessings, and academic support.

I owe my deepest gratitude to my family members and friends for their love and affection. Finally, I thank all those who helped me directly and indirectly during the completion of this work.

Dr. Shikha Agarwal

Chapter 1

Imidazole and its derivatives: Introduction and synthetic aspects

Jay Soni, Ayushi Sethiya, and Shikha Agarwal
Synthetic Organic Chemistry Laboratory, Department of Chemistry, Mohanlal Sukhadia University, Udaipur, Rajasthan, India

1. Introduction

In the 1800s, the chronicle of heterocyclic chemistry was initiated with the progress of organic chemistry. Presently, more than 70% of organic chemistry journals and other publications are based on heterocyclic chemistry [1–3]. Heterocyclic moieties play a prestigious role in science. Among the heterocyclic functionalities, nitrogen-based heterocyclic compounds, i.e., imidazoles, play an important role for mankind. They have immense importance in chemical sciences, biological sciences, and material science for nonlinear optical applications and catalysts in synthesis [4–7]. Several imidazole nuclei containing naturally occurring compounds are found in nature, viz. histamine, biotin, pilocarpine, naamidine A, cyclooroidin, agelastatin, etc. (Fig. 1.1).

Imidazole was first synthesized in 1858 by Debus, who isolated derivatives from three-component compounds like dicarbonyl, aldehyde, ammonia, or its salts [8, 9]. After this reaction, many syntheses were demonstrated after 1858 allowing access to various imidazole derivatives. Imidazole has specific characteristics, i.e., it is planar, a five-membered heterocyclic molecule which constitutes three carbon atoms and two nitrogen atoms at the first and third positions. It is soluble in water and a highly polar compound due to the presence of dipole moment. It has an amphoteric nature, i.e., it works as an acid as well as a base. This property is explained by resonance interactions, which increase the basicity of two nitrogen atoms and make electrophilic and nucleophilic attacks feasible. The resonance structures of imidazole are shown in Fig. 1.2 [10].

Imidazole comprises a huge area of research in pharmaceuticals due to its ability to treat numerous harmful and dreadful diseases. It can be antibacterial

Imidazole-Based Drug Discovery. https://doi.org/10.1016/B978-0-323-85479-5.00003-4
Copyright © 2022 Elsevier Inc. All rights reserved.

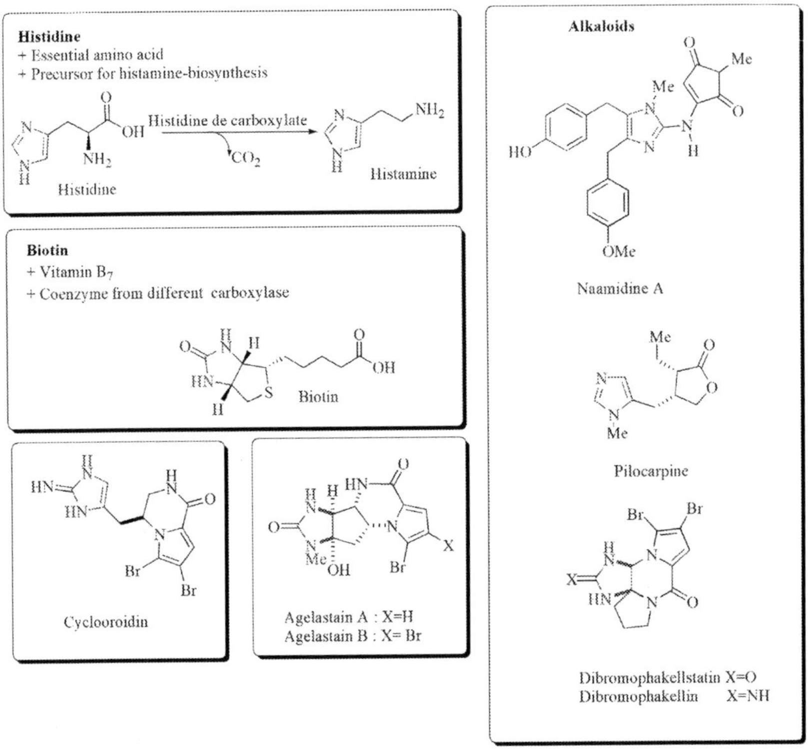

FIG. 1.1 Natural product containing imidazole scaffold.

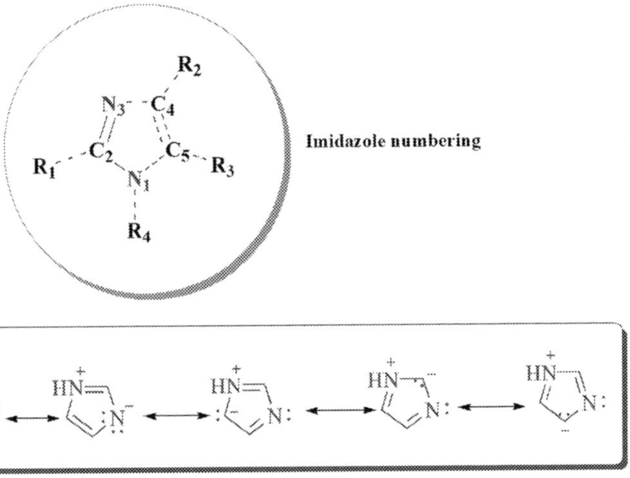

FIG. 1.2 Numbering and resonance structures of imidazole.

[11–14], antifungal [15–17], antidiabetic [18], antiparasitic, antioxidant [19], antitumor [20–22], antituberculosis [23], antiinflammatory [24–26], antimalarial, anticancer [27–29], antihypertensive, antidepressant, anxiolytic, etc. (Fig. 1.3). Some biologically important compounds containing imidazole are shown in Fig. 1.4. In other usage, imidazolium salts are used in several ways such as to extract metal ions from aqueous solutions and metal nanoparticles coating, offering antimicrobial action and creating oriented liquid crystals [30]. In bioactive applications, imidazole is used as imidazolium hydrogels, antiarrhythmics, and antimetastatic agents. Imidazolium salts have potential power of antimicrobial activity [31] and are also used in liquid crystals for molecular self-assembly [32]. Nishide and coauthors used poly(1-vinyl imidazole) for oxygen transport by coordinating cobalt as oxygen carriers to form reversible oxygen-binding polymer membranes [33].

Nowadays, many researchers have reported several methods for the synthesis of mono, di-, tri-, and tetra-substituted imidazole derivatives, fused imidazoles, and hybrid imidazoles using different moieties and reaction conditions involving the use of nontoxic, more efficient, inexpensive, reusable, nanoparticles, catalysts, as well as different eco-friendly methodologies such as microwave irradiation (MWI) and ultrasonication, to improve atom-economy, yield, etc. A plethora of review articles and monographs are available which emphasize the various synthetic strategies of imidazoles. In recent times, Rani and coauthors [34] reviewed synthesis of tri-substituted imidazoles. Kumar et al. [35] investigated imidazole and its derivatives with remarkable studies on their chemical/biological uses. Gupta et al. [36] represented the synthesis

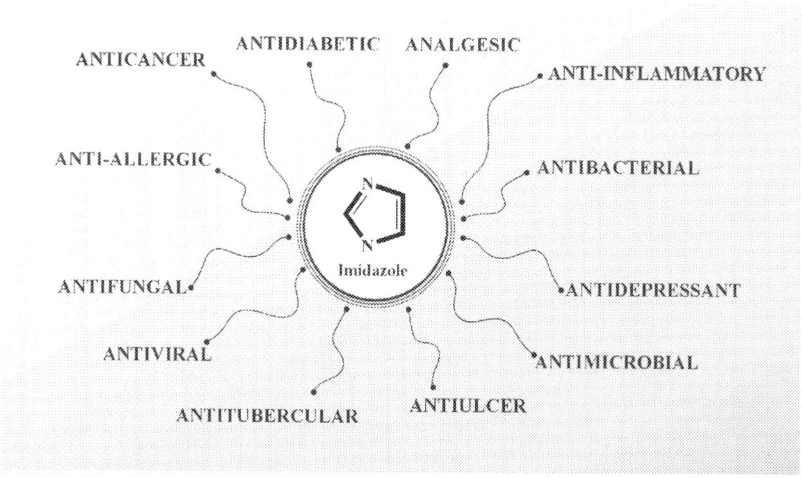

FIG. 1.3 Biological profile of imidazole derivatives.

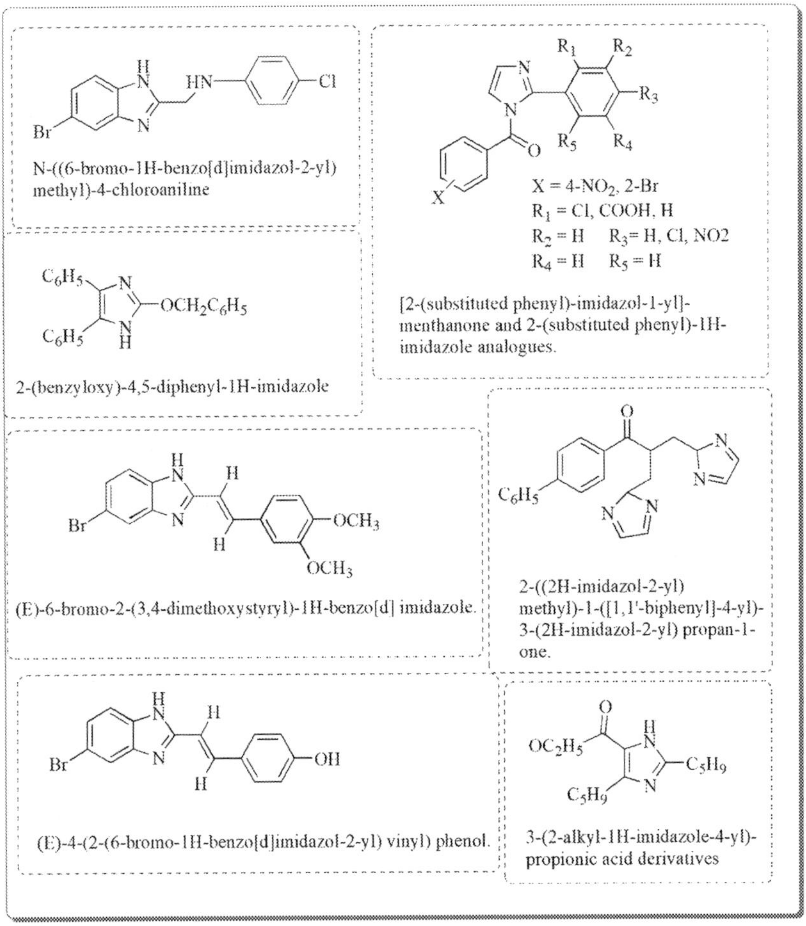

FIG. 1.4 Biologically active imidazole scaffolds.

of bioactive imidazoles. A review by Fan et al. [37] was focused mainly on the tubercular activity of imidazole. Jay and coauthors represented the synthesis and biological profile of imidazoles [38]. Pardeshi et al. prepared a review on different synthetic approaches of benzimidazoles [39]. Prajapat et al. provided a complete outline of recent progress of benzimidazole-based compounds in the whole range of medicinal chemistry [40]. Due to a wide range of applicability, there has been increasing interest in the development of efficient methodologies for the synthesis of imidazole derivatives. This chapter covers the literature from 2005 to 2020, presenting the diverse synthetic strategies of substituted imidazoles.

2. Synthesis of substituted imidazoles

Substituted imidazoles are useful substrates for the synthesis of molecules of biological or pharmaceutical interest. The activity of heterogeneous catalysts has been remarkably explored in achieving high selectivity and conversion rates in the synthesis of imidazoles. Imidazoles are classified on the basis of their structural units (Fig. 1.5).

2.1 Mono-substituted imidazole derivatives

Mono-substituted imidazoles have been synthesized via bisfunctionalization of 1,2-di-substituted acetylenes by ruthenium carbonyl to form cis-enediol diacetates, followed by subsequent reaction with ammonium carbonate (Scheme 1.1). Interestingly, this is one of the few recent examples in the literature in which mono-substituted NH-imidazoles were synthesized. Both electron-rich and electron-deficient aromatics were tolerated under the reaction conditions [41].

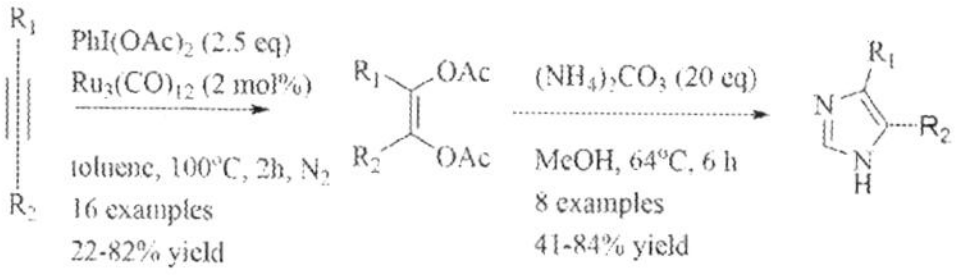

SCHEME 1.1 Synthesis of mono-substituted imidazole via ruthenium-catalyzed oxidation of alkynes.

Imidazole ketone and analogs were synthesized using 1-aryl-2-bromoethanones with imidazole. The other dioxolane derivatives were synthesized from substituted acetophenones (Scheme 1.2) [42].

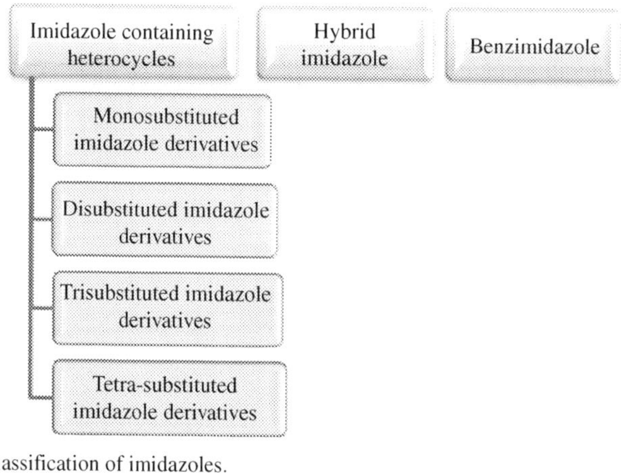

FIG. 1.5 Classification of imidazoles.

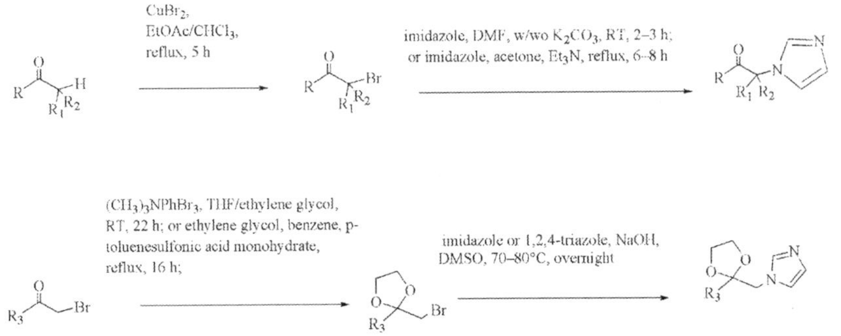

SCHEME 1.2 Synthetic route of mono-substituted imidazole based ketone and dioxolane derivatives.

N-arylimidazoles were synthesized by a one-pot reaction of iodobenzene and imidazole as model arylating agents with K_3PO_4 as a base in the presence of CuI as a catalyst and DMF as a solvent system at 35–40 °C for 40 h with excellent yields (98%) (Scheme 1.3) [43].

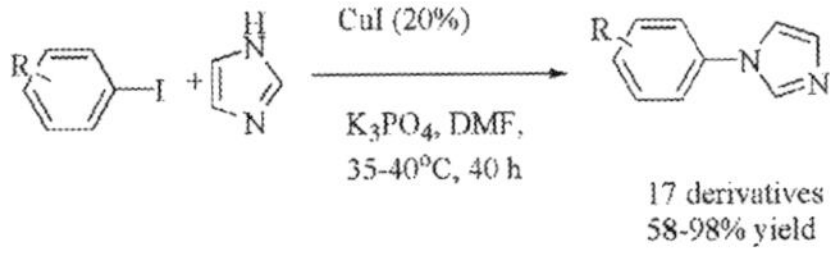

SCHEME 1.3 Catalytic N-arylation of imidazole with aryl iodides by CuI.

2.2 Di-substituted imidazole derivatives

Di-substituted imidazoles were synthesized by a one-pot, two-component reaction of aryl methyl ketone and 1,4-dioxane with selenium dioxide (SeO_2) at reflux overnight in an oil bath. The product was separated from solid black SeO_2 and cooled at RT. Afterward, the mixture of ammonium acetate and ethanol was added in ice water. The synthesized derivatives were further used as imidazole-based fluorogenic sensors to detect selectivity and sensitivity of Fe^{3+} ions (Scheme 1.4) [44].

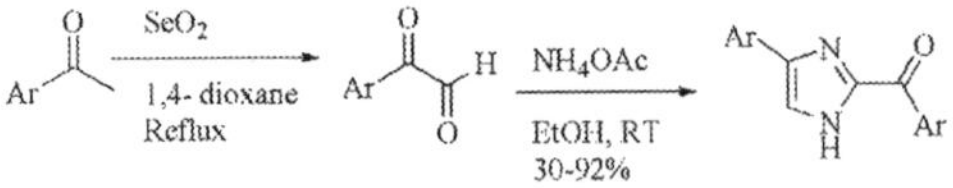

SCHEME 1.4 Synthesis of di-substituted imidazole.

Di-substituted imidazoles were synthesized using a single-step reaction of N,N-dimethoxyamide and were treated with phosphorus pentachloride in the presence of hydroiodic acid as a reductive agent (Scheme 1.5) [45].

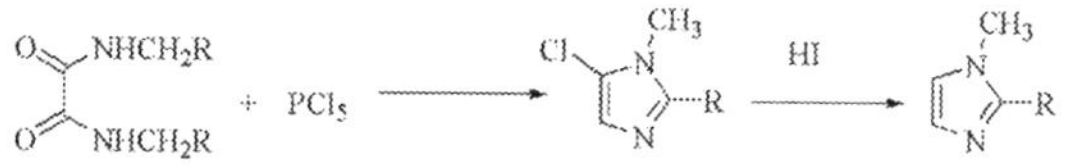

SCHEME 1.5 Synthesis of 1-ethyl-2methyl imidazole.

In 2010, Chen and his coworkers devised an efficient methodology for the synthesis of benzoylimidazole by structural modification of imidazole-4-carboxylic amide (AICA) derivatives and 4-substituted methoxylbenzoyl-aryl-thiazoles (SMART). The reaction was done in multiple steps using different reagents (Scheme 1.6) [46].

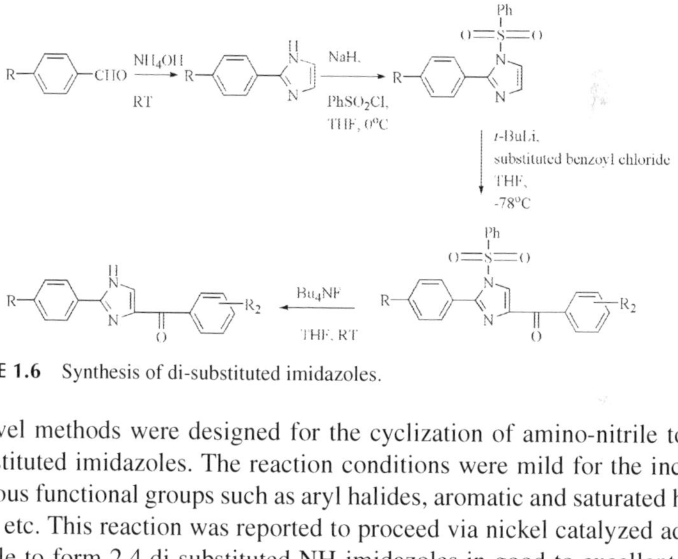

SCHEME 1.6 Synthesis of di-substituted imidazoles.

Novel methods were designed for the cyclization of amino-nitrile to form di-substituted imidazoles. The reaction conditions were mild for the inclusion of various functional groups such as aryl halides, aromatic and saturated heterocycles, etc. This reaction was reported to proceed via nickel catalyzed addition to nitrile to form 2,4 di-substituted NH-imidazoles in good to excellent yields (Scheme 1.7) [47].

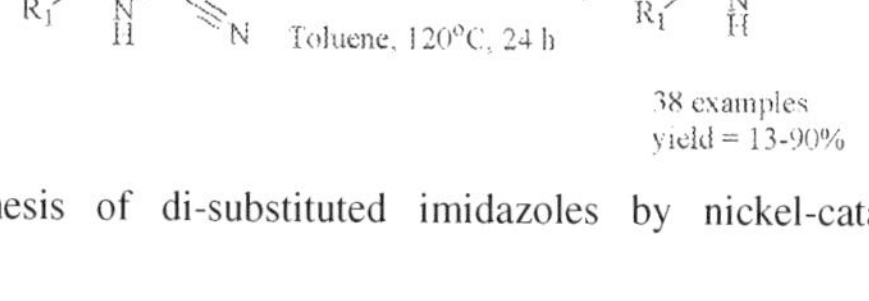

SCHEME 1.7 Synthesis of di-substituted imidazoles by nickel-catalyzed cyclization of amido-nitriles.

Methyl propiolate and substituted amidoximes reacted in the presence of catalytic amount of 1,4-diazabicyclo [2.2.]octane (DABCO) under microwave irradiation to furnish NH-imidazole in moderate yields. The C-4 position was substituted by an ester moiety and the reaction conditions were also tolerant to aryl halides and heterocycles (Scheme 1.8) [48].

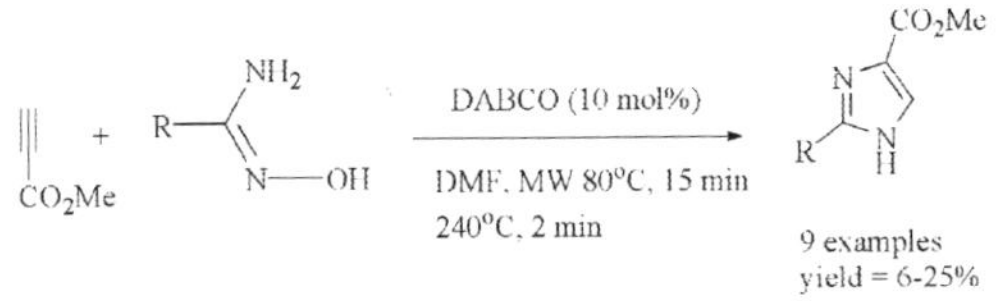

SCHEME 1.8 Synthesis of di-substituted imidazole by nucleophilic catalyst-based protocol.

Recently, imidazole derivatives were synthesized by a metal-free, one-pot process from ketones via oxidation and coupling with aldehydes and ammonium acetate (Scheme 1.9) [49].

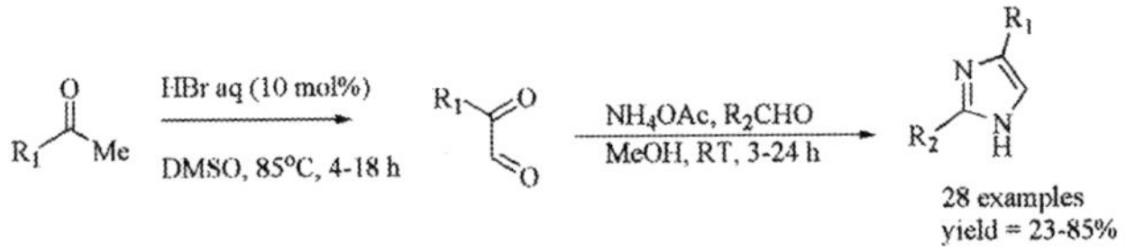

SCHEME 1.9 Synthesis of di-substituted imidazole by acid-catalyzed oxidation of ketones.

2.3 Tri-substituted imidazole derivatives

2.3.1 Using 2,3-diaminomaleonitrile

A novel and eco-friendly protocol was developed for one-step synthesis of 2-aryl-4,5-dicarbonitrile imidazole derivatives from aromatic aldehydes and 2,3-diaminomaleonitrile in the presence of CAN/NA (0.05/0.4 equiv.) via stirring at 120 °C under solvent-free conditions for 15–30 min with good yields (68%–88%) (Scheme 1.10) [50].

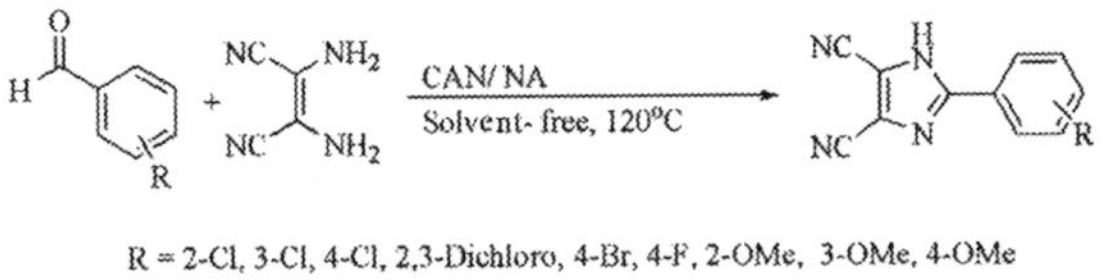

SCHEME 1.10 Synthesis of dicyano imidazole derivatives.

2.3.2 Using the Van Leusen method

A novel process for the synthesis of 1,4,5-tri-substituted imidazoles-containing trifluoromethyl group was developed using N-aryltrifluoroacetimidoyl chlorides, TosMIC, and sodium hydride in dry THF at room temperature, under an argon atmosphere (Scheme 1.11) [51].

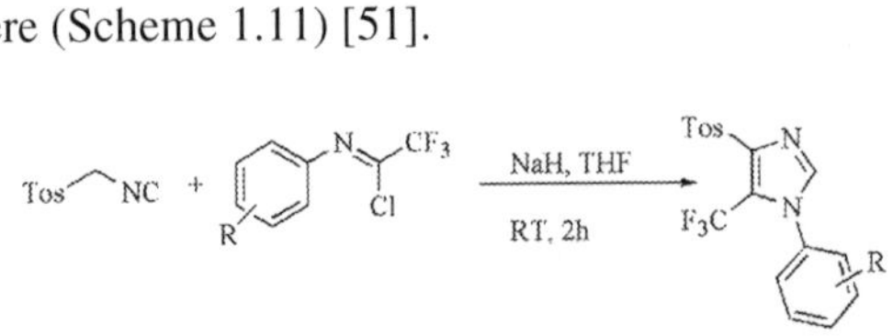

SCHEME 1.11 Synthesis of imidazole using the Van Leusen method.

2.3.3 *Using 2,3-dioxo-3-substituted propanoates*

The synthesis of 2,4,5-tri-substituted imidazole derivatives was reported using 2,3-dioxo-3-substituted propanoates, ammonium acetate, and various aldehydes in EtOH and AcOH as catalysts at room temperature for 3 days and moderate yields (27%–65%) were obtained [52,53] (Scheme 1.12).

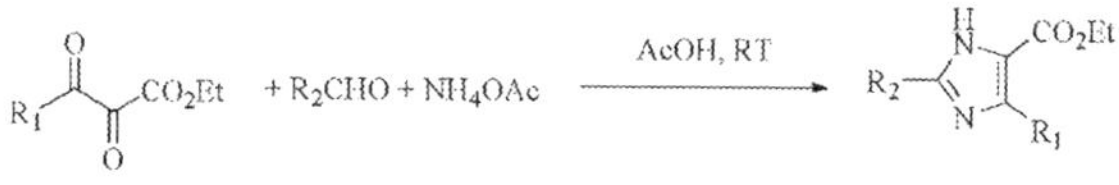

R_1 = Ph, n-Pr
R_2= Ph, 4-MeOC_6H_4, 4-ClC_6H_4, 4-(O_2N)C_6H_4, 2-(O_2N)C_6H_4, Et

SCHEME 1.12 Synthesis of 2,4,5-tri-substituted imidazole derivatives.

2.3.4 *Using α-aminoketones*

A simple and fast protocol was used for the synthesis of 1,4,5-tri-substituted imidazoles from α-amino ketones in the presence of NH_2CHO (0.02 mL) and THF at 180 °C for 8 h, and good yields (50%–83%) were obtained [54] (Scheme 1.13).

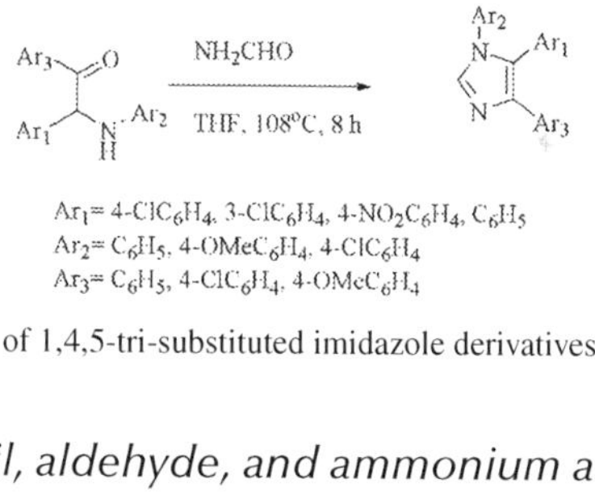

Ar_1= 4-ClC_6H_4, 3-ClC_6H_4, 4-$NO_2C_6H_4$, C_6H_5
Ar_2= C_6H_5, 4-OMeC_6H_4, 4-ClC_6H_4
Ar_3= C_6H_5, 4-ClC_6H_4, 4-OMeC_6H_4

SCHEME 1.13 Synthesis of 1,4,5-tri-substituted imidazole derivatives.

2.3.5 *Using benzil, aldehyde, and ammonium acetate*

The multicomponent one-pot synthesis of tri-substituted imidazole derivatives involved a combination of benzil, substituted aldehydes (R-CHO), and ammonium acetate (NH_4OAc) under various reaction conditions (Fig. 1.6).

Simple, fast and safe synthesis of 2,4,5-tri-substituted imidazole derivatives was developed via a one-pot multicomponent condensation reaction of benzil, various aldehydes, and ammonium acetate with ethanol as a solvent system in the presence of A-MFGO (0.15 g) as a catalyst at room temperature. To optimize the reaction conditions, several solvents were tested such as acetonitrile, THF, EtOH, DMF, DCM, etc. EtOH was chosen for the reaction, and 0.15 g of catalyst

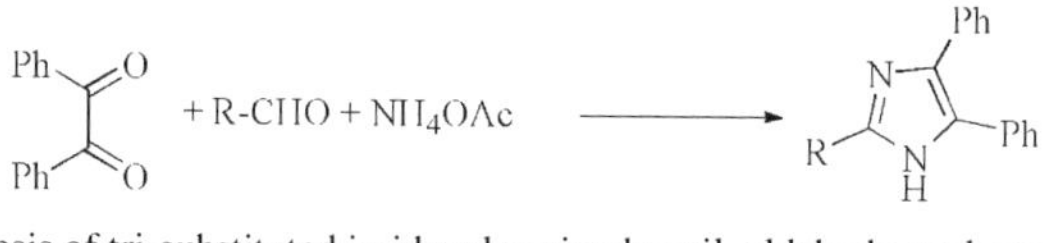

FIG. 1.6 Synthesis of tri-substituted imidazole using benzil, aldehyde, and ammonium acetate.

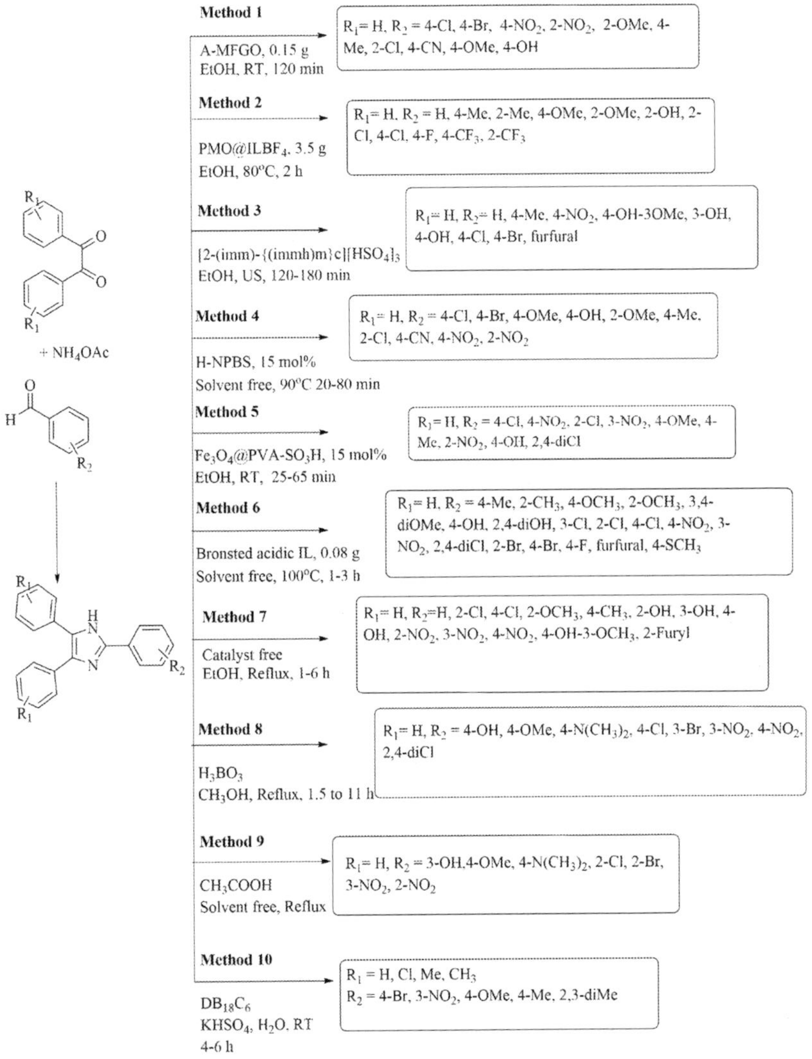

FIG. 1.7 Synthesis of 2,4,5-tri-substituted imidazole derivatives.

produced high yields (81%–95%) in 120 min. The catalyst was reused up to five times without any loss in catalytic efficiency [55] (Fig. 1.7, Method 1).

Afterward, synthesis of 2,4,5-tri-substituted imidazole derivatives was done in the presence of PMO@ILBF$_4$ (1.0) (3.5 g) as a catalyst at 80 °C for 2 h [56]. Various catalysts were used for the comparison with PMO@ILBF$_4$(1.0) such as MIP nanoreactors, *N*-acetyl glycin, urea-ZnCl$_2$, DABCO, rGO-NiO-NC, etc. The catalyst was reused for up to six runs with excellent yield, 99% [57] (Fig. 1.7, Method 2).

A simple, fast and environmental benign method was reported for the preparation of 2,4,5-tri-substituted imidazole derivatives in the presence of

$[2\text{-(imm)-}\{(\text{immh})\text{m}\}\text{c}][HSO_4]_3$ as an inexpensive and efficient catalyst under ultrasound irradiation with moderate to high yields (60%–86%) in 120–180 min [57,58]. To optimize the reaction conditions, several solvents were examined such as methanol, EtOH, DCM, acetonitrile, H_2O, chloroform, acetone, THF, etc. Various acid catalysts were compared with the reported catalysts such as H_2SO_4, $H_2SO_4.SiO_2$, etc. The catalyst was reused until the third run with minor changes in catalytic efficiency (Fig. 1.7, Method 3).

In 2019, 2,4,5-tri-substituted imidazole derivatives were synthesized in the presence of H-NPBS (15 mol%) at 90 °C in an oil bath without any solvent [58]. High yields (85%–95%) were produced in a short time, 20–80 min. The catalyst was reused until the third run (Fig. 1.7, Method 4).

The catalyst (Fe_3O_4@PVA-SO_3H) was prepared for the synthesis of 2,4,5-tri-substituted imidazole derivatives at room temperature for 25–65 min using EtOH and good to excellent yields, 84%–98% were obtained [59]. To optimize the reaction conditions, several catalysts were compared with Fe_3O_4@PVA-SO_3H such as ZrO_2-Al_2O_3 (99.2%), HNO_3@nanoSiO_2 (91%), ZnO nanorods (83%), Al_2O_3 (78%), TBAB (65%), etc. The recyclability and reusability of the catalyst without any treatment was also reported. The catalyst was reused up to 10 times with few changes in the catalytic efficiency (Fig. 1.7, Method 5).

The preparation of 2,4,5-tri-substituted imidazole derivatives was also reported in the presence of *N*-methyl 2-pyrrolidone hydrogen sulfate as Brønsted acidic ionic liquid (0.08 g) under solvent-free conditions at 100 °C in 1–3 h and 78%–98% yields were obtained [60]. The catalyst was recovered and reused until the seventh run without loss of its activity. The substrate conversion percentage to the product was 100% and TOF was obtained at 5 h^{-1}. Further, 40 derivatives of 1,2,4,5-tetra-substituted imidazoles under solvent-free conditions in 15–60 min were also produced in high yields, 80%–97% (Fig. 1.7, Method 6).

The preparation of tri-substituted imidazole derivatives without any catalyst was reported with good yields, 78%–98% in 1–6 h [61]. Tetra-substituted imidazole derivatives were also synthesized without any catalyst. The advantages of this reaction are that it is greener, one-pot, needs no catalyst, enables reusability of catalyst, and is not time-consuming (Fig. 1.7, Method 7).

Synthesis of tri-substituted imidazole derivatives was performed in the presence of H_3BO_3 as a catalyst and CH_3OH as a solvent under reflux conditions with 72%–92% yields in 1.5–11 h [62]. The catalyst was reused several times (Fig. 1.7, Method 8). Triphenyl imidazole derivatives were also synthesized in the presence of acetic acid at reflux (Fig. 1.7, Method 9).

A facile synthesis of tri-substituted imidazoles using DB18C6 and $KHSO_4$ in the presence of water as a solvent at room temperature for 4–6 h was reported and good to excellent yields (52%–92%) were obtained. DB18C6 and $KHSO_4$ could be reused at least six times. The authors also reported the synthesis of tetra-substituted imidazole derivatives (Fig. 1.7, Method 10) [63].

Various synthetic pathways have also been designed for the synthesis of tri-substituted imidazole derivatives, and these are depicted in Fig. 1.8.

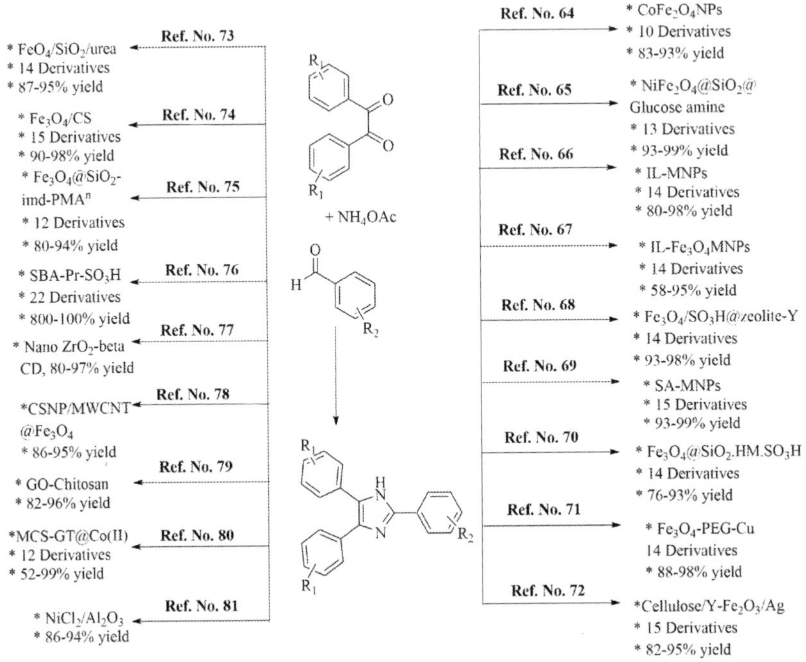

FIG. 1.8 Synthesis of tri-substituted imidazole derivatives under various reaction conditions [64–81].

2.4 Tetra-substituted imidazole derivatives

2.4.1 *Using aminoethylpiperazine*

A one-pot multicomponent condensation reaction of benzil, aminoethylpiperazine, several aldehydes, and ammonium acetate with ethanol in the presence of SO_4^{2-}/Y_2O_3 at 80 °C for 10h was reported for the synthesis of 1,2,4,5-tetra-substituted imidazole derivatives and high yields (83%–93%) were obtained. No reaction happened in water. Several catalysts were tested such as PTSA, CAN, AcOH, L-Proline, etc. The catalyst was reused up to five times without significant change in catalytic efficiency [82] (Scheme 1.14).

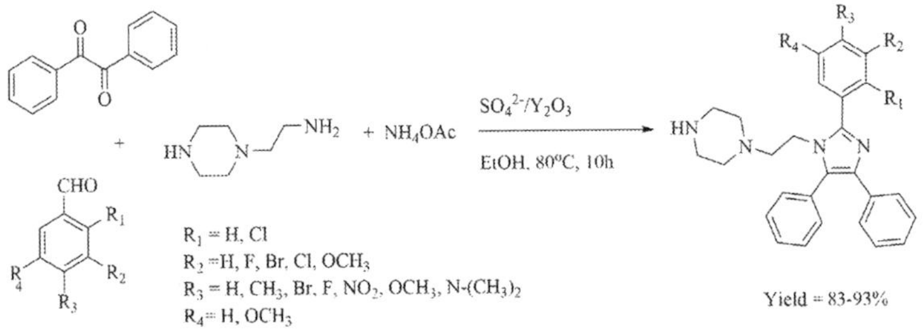

SCHEME 1.14 Synthesis of 1,2,4,5-tetra-substituted imidazole derivatives.

2.4.2 Using prop-2-ynylamine

A one-pot reaction of benzil, several aldehydes, ammonium acetate, and prop-2-ynylamine in the presence of $CuFe_2O_4$NPs (10 mol%) as a catalyst in H_2O:EtOH (1:1) under reflux for 30–50 min was developed. Sixteen derivatives were synthesized in excellent yields (84%–95%). The catalyst was reused up to the sixth run (Scheme 1.15) [83].

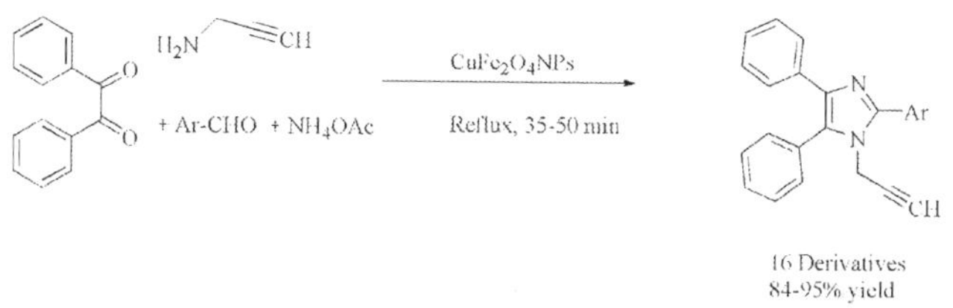

SCHEME 1.15 Synthesis of tetra-substituted imidazole derivatives.

2.4.3 Using benzil, aldehydes, ammonium acetate, and aniline

The multicomponent, one-pot synthesis of tri-substituted imidazole derivatives involved combination of benzil, substituted aldehydes (R-CHO), ammonium acetate (NH_4OAc), and aniline under various reaction conditions (Fig. 1.9).

1,2,4,5-tetra-substituted imidazole derivatives were synthesized from a one-pot, multicomponent condensation reaction of benzil, various aldehydes, aniline, and ammonium acetate in the presence of $Fe_3O_4@SiO_2@Si$-$(CH_2)_3$@N-Ligand@Co (0.05 g) as a catalyst at 100 °C without any solvent. The yield was found to be low (53%) when no catalyst was used [84]. The catalyst was reused up to the fifth run without any loss in catalytic activity (Fig. 1.10, Method 1).

Synthesis of 1,2,4,5-tetra-substituted imidazole derivatives was reported in the presence of $Fe_3O_4@SiO_2$/BNC as a highly effective and recyclable catalyst at 120 °C under solvent-free conditions for 19–65 min and produced high yields, 85%–95%. The catalyst was reused until the fifth run without much change in catalytic efficiency [85]. The authors have also synthesized 2,4,5-tri-substituted imidazoles using the same catalyst without any solvent at 120 °C and found high to excellent yields (75%–95%) (Fig. 1.10, Method 2).

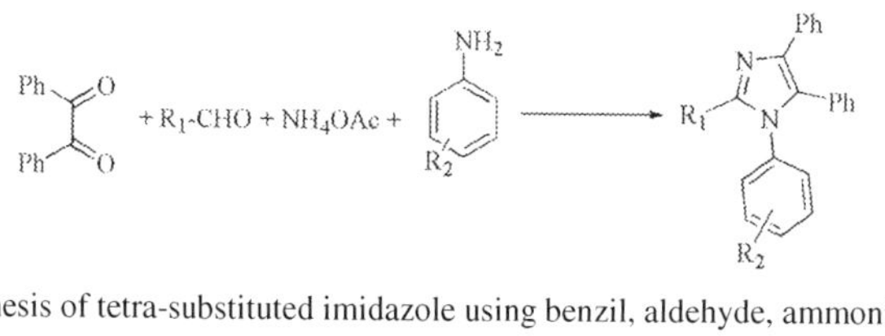

FIG. 1.9 Synthesis of tetra-substituted imidazole using benzil, aldehyde, ammonium acetate, and aniline.

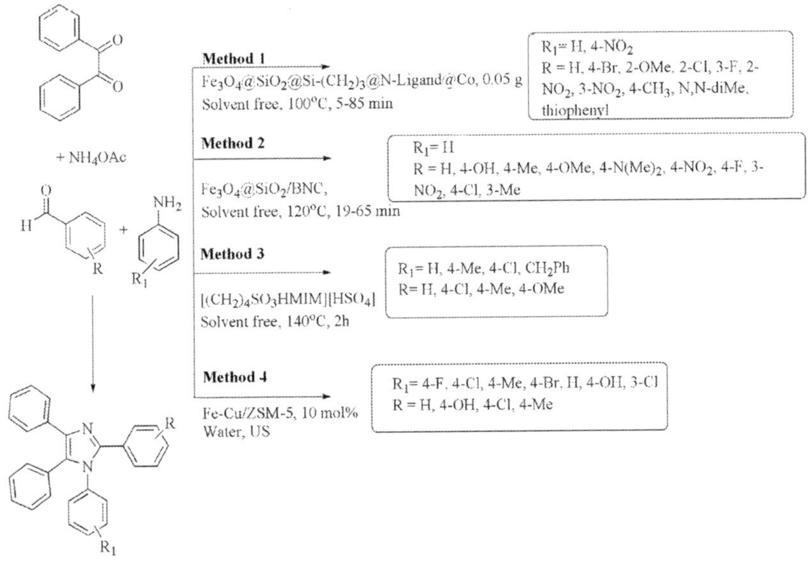

FIG. 1.10 Synthesis of 2,3,4,5-tetra-substituted imidazole derivatives.

Similarly, synthesis of 1,2,4,5-tetra-substituted imidazole derivatives was done in the presence of $[(CH_2)_4SO_3HMIM][HSO_4]$ as a catalyst without any solvent at 140°C for 2h [86]. To optimize the reaction conditions, 15mol% catalyst was selected for the high yield production, i.e., 85%–94%. Reusability, good yields, simple, and easy handling are the advantages of the reaction (Fig. 1.10, Method 3).

A one-pot, multicomponent condensation reaction for the synthesis of tetra-substituted imidazole from benzil, aldehydes, aniline, and ammonium acetate with water in the presence of Fe-Cu/ZSM-5 under ultrasound irradiation was reported and high yields (99%) were obtained (Fig. 1.10, Method 4) [87].

Several synthetic pathways have also been designed for the synthesis of tetra-substituted derivatives and these are depicted in Fig. 1.11.

3. Imidazole hybrids

Over the past years, five- and six-membered azaheterocyclic compounds have received considerable attention due to their important applications from pharmacological, industrial, and synthetic points of view. Imidazole (and its benzo-derivative, benzimidazole) and other heterocyclic derivatives are core scaffolds, widely present in many classes of drugs (of natural or synthetic origin), displaying a large variety of interesting biological activities (antimicrobial, antifungal, antiinflammatory, antihypertensive, antineuropathic, antihistaminic, etc.). Synthesis of highly substituted fused imidazole-containing 5,5 and 5,6 fused bicyclic heterocycles has been described.

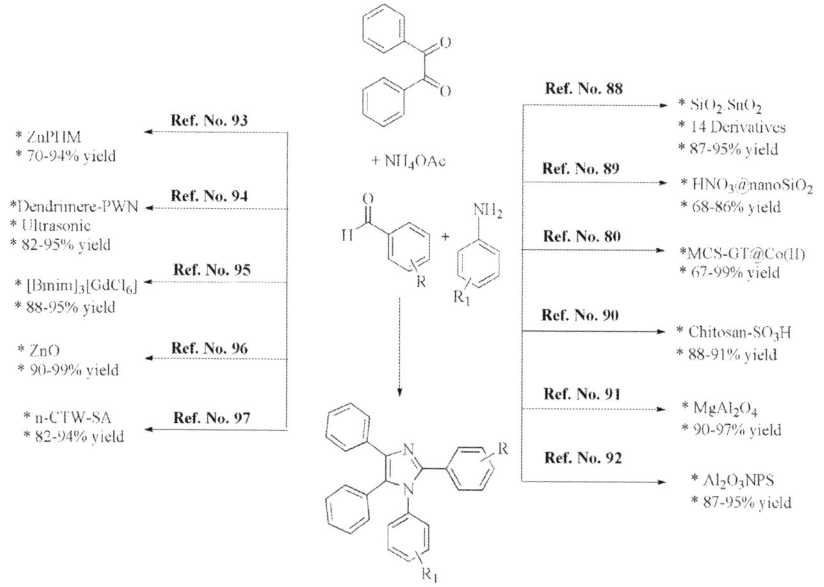

FIG. 1.11 Synthesis of tetra-substituted imidazole derivatives using various reaction conditions [88–97].

This section is subdivided into two different parts:

a. Imidazole fused heterocycles and
b. Benzimidazole fused heterocycles.

3.1 Imidazole fused heterocycles

The bis- and poly(imidazoles) were synthesized by a one-pot reaction of 1,2-diketone with the corresponding bis- and poly(aldehydes) and ammonium acetate using ZnO as a nanocatalyst under microwave irradiation [98] (Scheme 1.16).

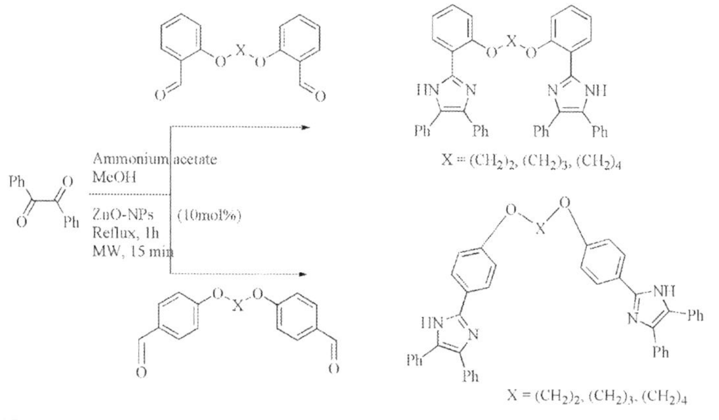

SCHEME 1.16 Synthesis of a series of bis((4,5-diphenyl-1H-imidazol-2-yl)phenpoxy)alkenes.

The structurally diverse imidazole hybrids were synthesized by the decarboxylative annulation of α-azidochalcones and L-proline using Ru^{+3}-P4VP complex as a catalyst under visible light irradiation at ambient temperature. The reaction was carried out in batch as well as continuous-flow conditions. The continuous-flow conditions gave high yields, 70%–94% in just 2 min compared to batch conditions, which took up to 16 h with slightly lower yields (63%–73%) [99].

4-Methyl-1-*N*-(*p*-tolyl)imidazoles were synthesized by *N*-arylation of 4-methylimidazole, 4-bromotoluene, and potassium *tert*-butoxide catalyzed by Cu_2O at 160–170 °C and established as effective inhibitors of atmospheric and acidic copper corrosion (Scheme 1.17) [100].

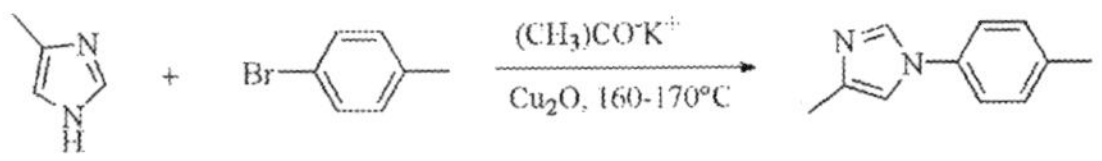

SCHEME 1.17 Synthesis of hybrid imidazole.

4-((4-(4,5-diaryl-1*H*-imidazol-2-yl)phenoxy)methyl)-1-substituted-1*H*-1,2,3-triazoles were synthesized from the reaction of triazole-substituted aryl aldehydes and 1,2-dicarbonyl compounds in the presence of ammonium acetate and acetic acid under reflux conditions. The compounds exhibited good antibacterial and antifungal activity (Scheme 1.18) [101].

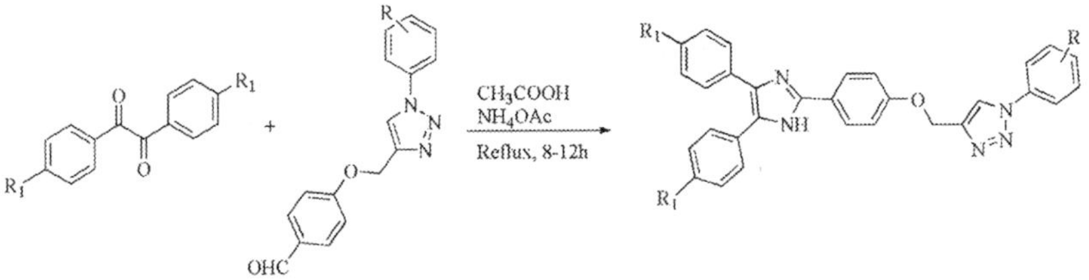

SCHEME 1.18 Synthesis of 4-((4-(4,5-diaryl-1H-imidazol-2-yl)phenoxy)methyl)-1-substituted-1H-1,2,3-triazoles.

A simple, efficient, and one-pot reaction has been described for the synthesis of benzo[*d*]imidazo [2,1-b]thiazoles by the condensation of aromatic ketones, NBS (N-bromosuccinimide) and 5-(biphenyl-4-yl)-1,3,4-thiadiazol-2-amine in the solvent system of PEG-400 and water under microwave irradiations at 80–85 °C, and it gave excellent yields (94%–98%) of products in short reaction times (11–14 min) (Scheme 1.19) [102].

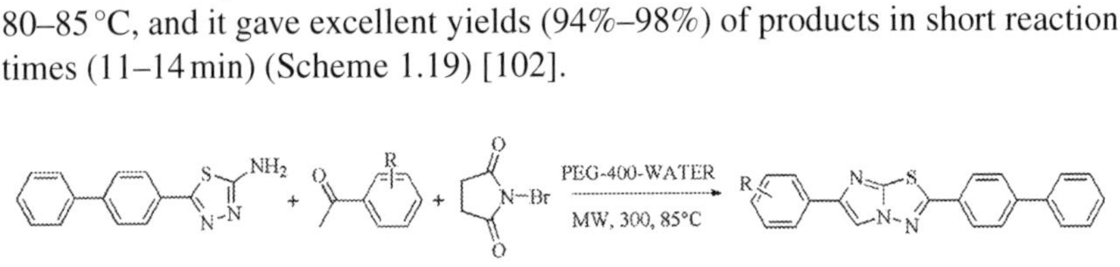

(**R** = H, 4-Cl, 4-F, 2,4-(Cl)$_2$, 4-Br)

SCHEME 1.19 Synthesis of benzo[*d*]imidazo [2,1-b]thiazoles using PEG-400.

An effective procedure has been developed for the synthesis of a series of imidazole derivatives through condensation of substituted guanylhydrazone and phenylglyoxal monohydrate in ethanol solvent at 78 °C and moderate yields (51%–78%) were obtained (Scheme 1.20) [103].

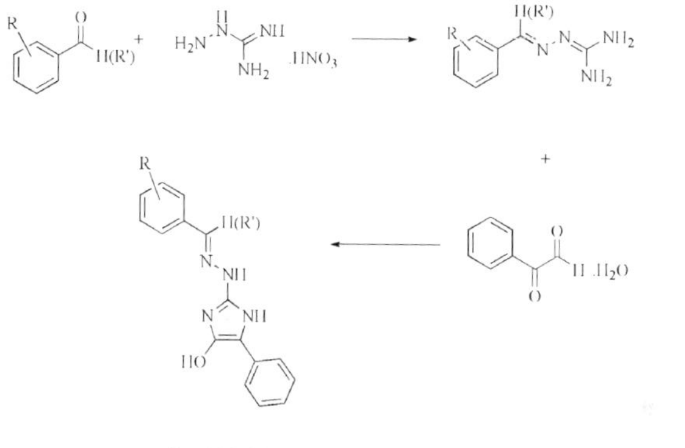

SCHEME 1.20 Synthesis of highly substituted imidazole.

A practical and easy two-step microwave-assisted approach was described for the synthesis of N-(4-((2-chloroquinolin-3-yl)methylene)-5-oxo-2-phenyl-4,5-dihydro-1*H*-imidazol-1-yl)(aryl)amides. In the first step, hippuric acid, 2-chloroquinoline-3-carbaldehyde, and acetic anhydride underwent Perkin condensation in the presence of anhydrous sodium acetate and microwave irradiation at 400 W and produced compounds, 4-((2-chloroquinolin-3-yl)methylene)-2-phenyloxazol-5(4H)-ones. Further, 4-((2-chloroquinolin-3-yl)methylene)-2-phenyloxazol-5(4H)-ones reacted with N-aminoarylcarboxamides in pyridine under reflux conditions and yielded a final product. Most of the derivatives showed good to excellent antimicrobial properties (Scheme 1.21) [104].

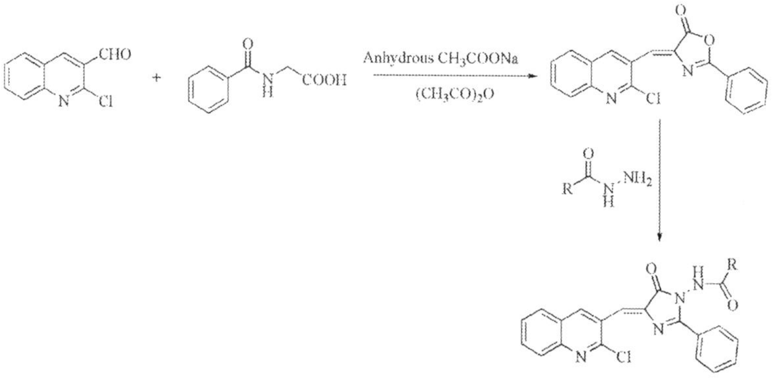

SCHEME 1.21 Synthesis of N-(4-((2-chloroquinolin-3-yl)methylene)-5-oxo-2-phenyl-4,5-dihydro-1H-imidazol-1-yl)(aryl)amides.

3.2 Benzimidazole fused heterocycles

Novel naphthalene and perylene imide imidazoles were synthesized via condensation of 4,5-diaminophthalonitrile and N-(2-ethylhexyl)-naphthalenetetracarboxylic-monoimide-monoanhydride in the presence of anhydrous $Zn(OAc)_2$ and dry quinoline at 150 °C in an argon atmosphere (Scheme 1.22) [105].

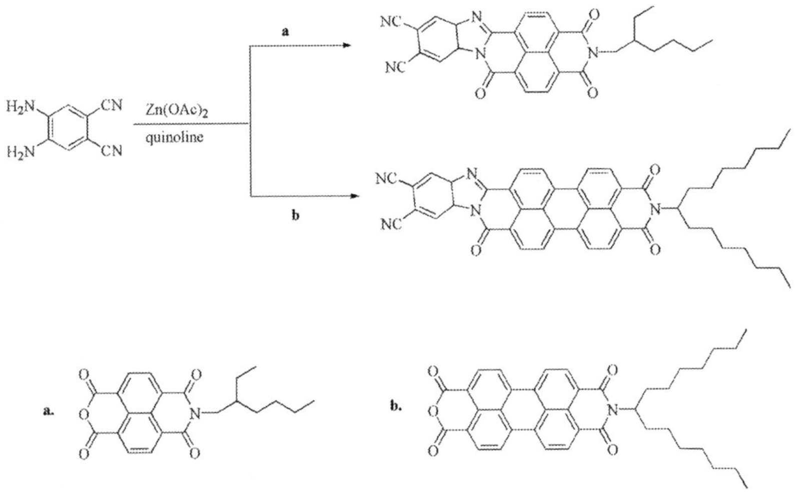

SCHEME 1.22 Preparation of perylene imide imidazoles.

Various kinds of N-heterocyclic benzimidazole derivatives were synthesized by the reaction of benzimidazole derivatives and arynes as precursors using CsF and K_2CO_3 as cocatalysts (Scheme 1.23) [106].

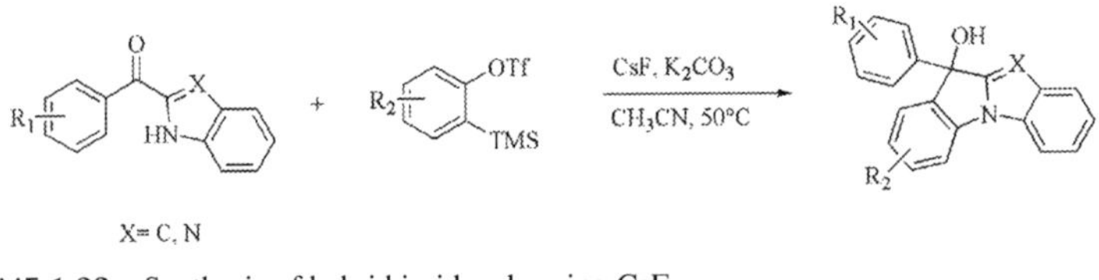

SCHEME 1.23 Synthesis of hybrid imidazole using CsF.

A range of benzo[4,5]-imidazo[1,2-c]-pyrimidines was developed by the reaction of 2-(2-bromovinyl)- and 2-(2-bromoaryl)-benzimidazoles with primary amides in the presence of CuI as a catalyst and DMF as a solvent under microwave irradiation (Scheme 1.24) [107].

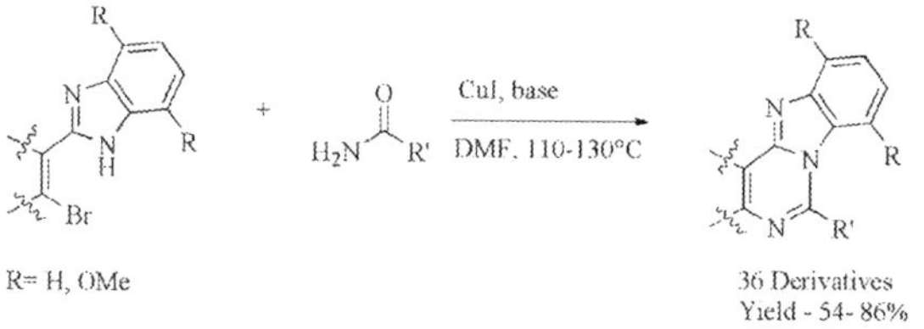

SCHEME 1.24 Preparation of benzo[4,5]-imidazo[1,2-c]-pyrimidines.

A facile pathway has been introduced for the synthesis of fused heterocyclic polycyclic compounds, pyrrole[1,2- a]benzimidazoles, piperidine[1,2-*a*] benzimidazoles, and oxa-fused benzimidazoles using silver nitrate and iodine at room temperature by *exo*-dig and *endo*-dig cyclization (Scheme 1.25) [108].

Several hybrid imidazole derivatives [109–119] synthesized under several reaction conditions are summarized in Fig. 1.12.

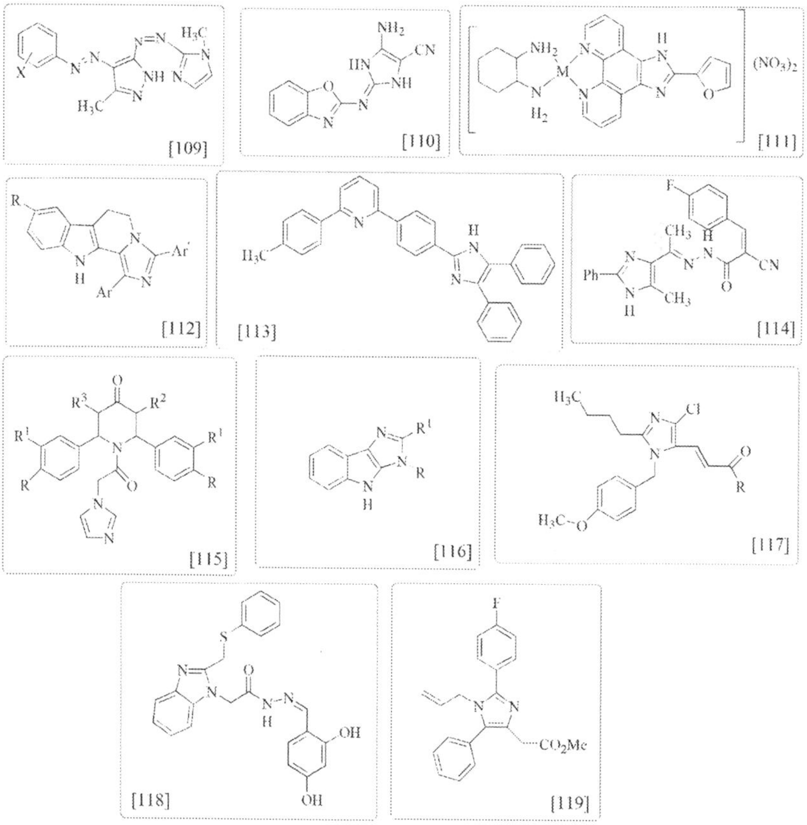

FIG. 1.12 Synthesis of hybrid imidazole derivatives.

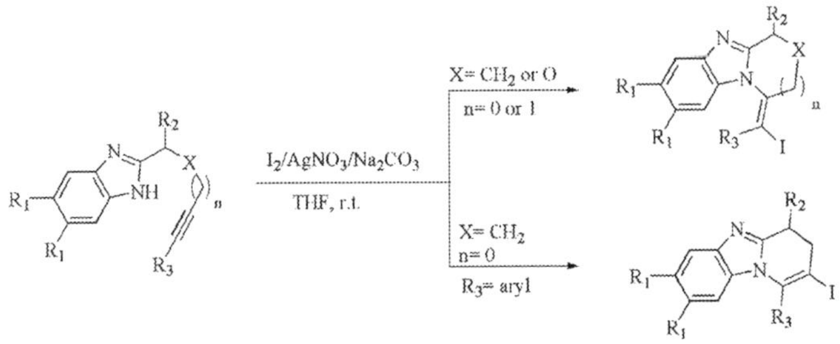

SCHEME 1.25 Synthesis of hybrid imidazole derivatives.

4. Benzimidazole

Benzimidazole, a bicyclic hetero-aromatic organic compound, consists of a benzene ring and imidazole ring fused at the 4- and 5-positions. It shows potent activity in different pharmaceutical fields [120, 121]. Benzimidazole derivatives have been used in treatment of hypertension, as a tyrosine kinase inhibitor, in the treatment of gastric ulcer, etc., as shown in Fig. 1.13. Various pathways for the synthesis of benzimidazole have been discussed.

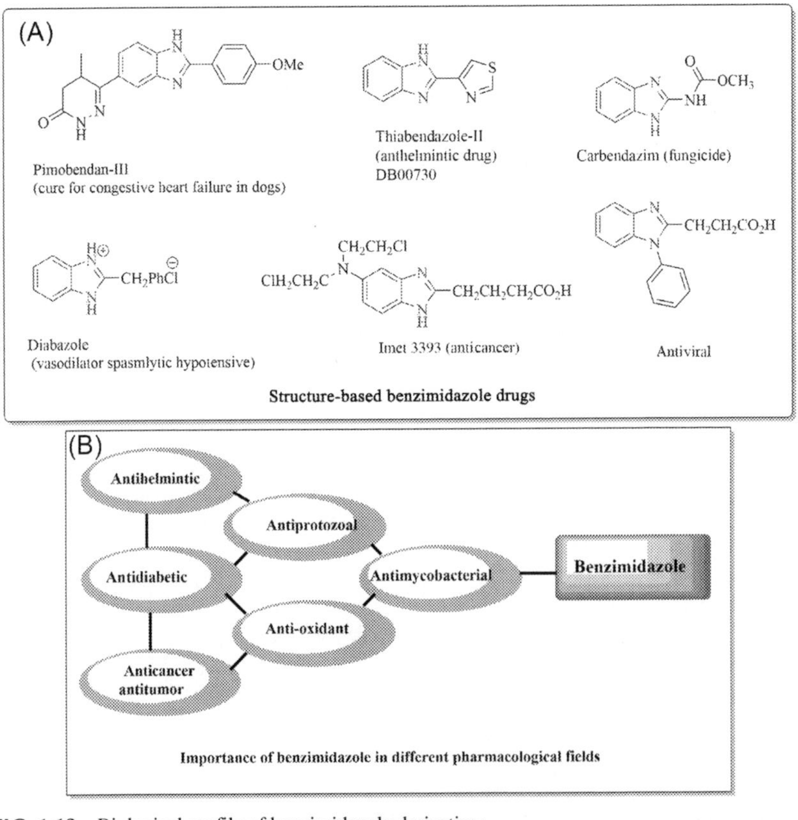

FIG. 1.13 Biological profile of benzimidazole derivatives.

4.1 Hoebrecker synthesis

The first synthesis of benzimidazole was done by Hoebrecker in 1872, who reduced 2-nitro-4-methylacetanilide and obtained 2,5-dimethylbenzimidazole [122] (Scheme 1.26).

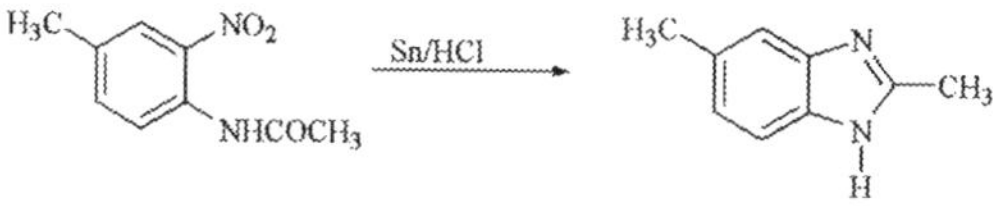

SCHEME 1.26 Synthesis of benzimidazole using 2-nitro-4-methylacetanilide.

Several methods have been reported for benzimidazole synthesis via condensation of ortho phenylenediamine and aldehyde, nitriles, carboxylic acid, alcohols, and amines using various catalysts.

4.2 Synthesis of benzimidazoles using *o*-phenylenediamine and acids

2-Substituted benzimidazoles were obtained by the reaction of *o*-phenylenediamines and carboxylic acids in high yields by heating under reflux, in the presence of conc. HCl, NH_4Cl, alumina, silica gel, zeolite, etc. [123–126] (Scheme 1.27).

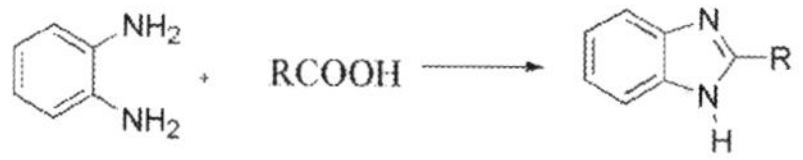

SCHEME 1.27 Synthesis of benzimidazoles using *o*-phenylenediamine and carboxylic acids.

4.3 Synthesis of benzimidazoles using *o*-phenylenediamine and acid anhydride

o-Phenylenediamines, when heated under reflux for several hours with acetic anhydride, got completely converted to 2-methylbenzimidazole [123] (Scheme 1.28).

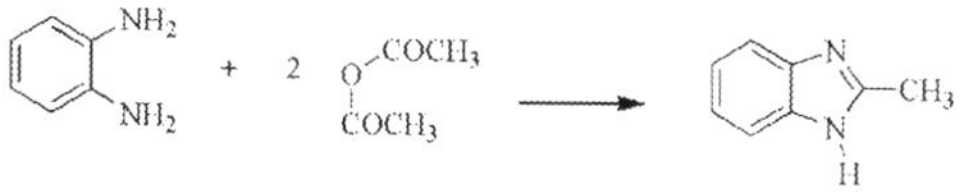

SCHEME 1.28 Synthesis of benzimidazoles using *o*-phenylenediamine and acid anhydride.

4.4 Synthesis of benzimidazoles using *o*-phenylenediamine and alcohols

Benzyl alcohol reacted with *o*-phenylenediamines to yield 2-aminobenzimidazoles in high yields via ultrasonication and photoradiation for 5 min in the presence of NiO@Anatase/rutile-TiO_2 nanoparticles [127] (Scheme 1.29).

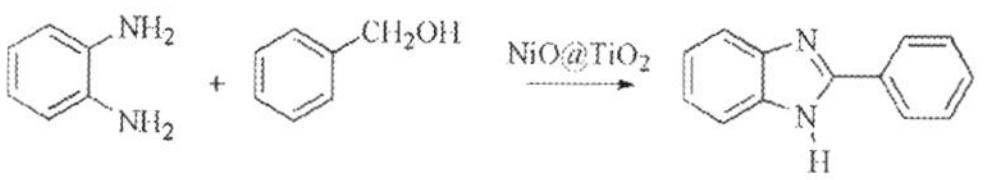

SCHEME 1.29 Synthesis of benzimidazoles using *o*-phenylenediamine and alcohols.

4.5 Synthesis of benzimidazoles using *o*-phenylenediamine and alkyl halides

Qiu and coauthors synthesized 2-arylbenzimidazoles using arylmethyl halides and *o*-phenylenediamine in high yields using CuBr through aerobic oxidation reaction [128] (Scheme 1.30).

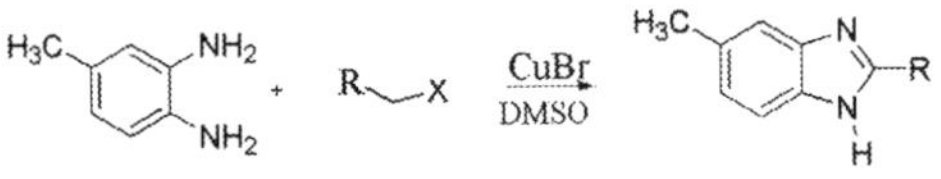

SCHEME 1.30 Synthesis of benzimidazoles using *o*-phenylenediamine and alkyl halides.

4.6 Synthesis of benzimidazoles using ketones

Reaction of benzene-1,2-diamine hydrochloride salts with several ketones has been investigated, which resulted to N-alkyl substituted benzimidazole derivatives [121] (Scheme 1.31).

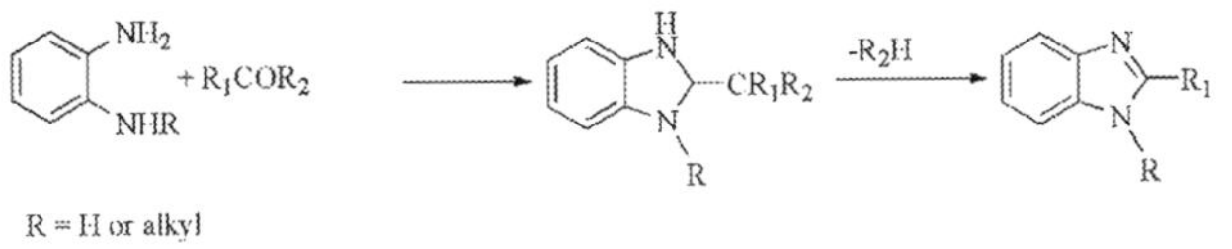

SCHEME 1.31 Synthesis of benzimidazole using ketones.

2-Phenyl-5 (or 6)-methyl benzimidazole derivatives were synthesized by heating benzene-1,2-diamine with various acetophenones at 180 °C for 15 min [129] (Scheme 1.32).

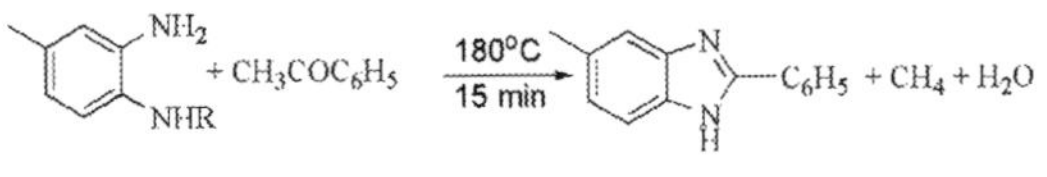

SCHEME 1.32 Synthesis of benzimidazole using ketones.

The cyclization reaction of 2-iodoaniline with benzonitriles was carried out in the presence of K^{t}OBu and DMAc, and the reaction mixture was heated at 120 °C for 24 h to afford different derivatives of benzimidazole. The scope of the reaction was expanded with a variety of substituents of benzonitrile, where the desired product was obtained in excellent yields [130] (Scheme 1.33) irrespective of substitution on benzonitrile scaffold.

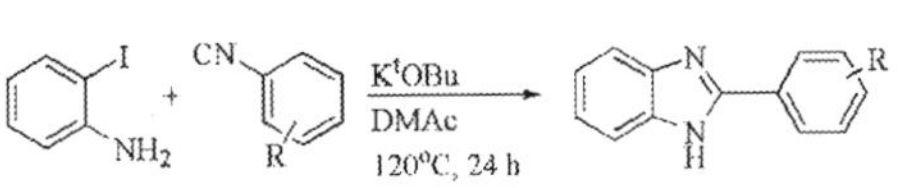

SCHEME 1.33 Synthesis of benzimidazole using benzonitrile/KtOBu.

Various substituted urea and thiourea or ammonium thiocyanate was treated with benzene-1,2-diamine using ball milling technique under solvent-free conditions for certain time periods and produced benzimidazol-2-ones (Scheme 1.34). In addition, studies showed that the ball milling technique offered some benefits over conventional methods like full conversion rate, short time for reaction completion, less toxic, and clean profile, compared to other conditions with short and easy experimental protocols [131–133].

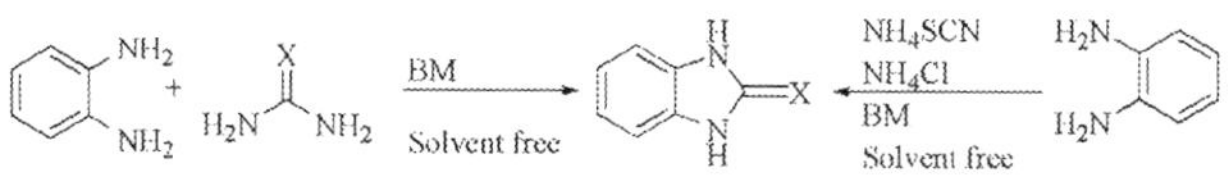

SCHEME 1.34 Synthesis of benzimidazol-2-one using urea and benzimidazol-2-thione.

4.7 Synthesis of benzimidazoles using *o*-phenylenediamine and aldehydes

Aldehydes and *o*-phenylenediamines reacted with each other to form 2-substituted benzimidazoles in the presence of different reaction conditions, viz. cupric acetate, nitrobenzene, $LaCl_3$, air, indium triflate, zinc triflate, Na_3AlF_6, molecular oxygen and visible light, transition metals, $BiWO_6$(CTAB) assisted tungsten lamp 35 W, etc. [123, 134–138] (Scheme 1.35). Some more protocols used for the synthesis of benzimidazoles are presented in Table 1.1.

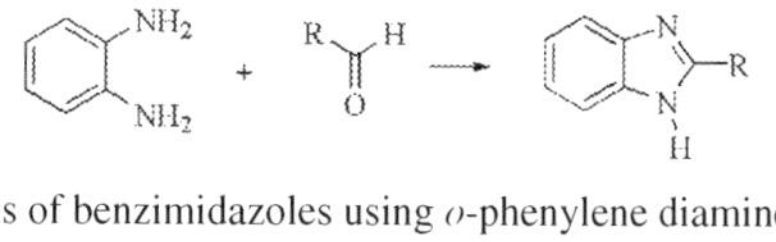

SCHEME 1.35 Synthesis of benzimidazoles using *o*-phenylene diamine and aldehydes.

5. Conclusion

In the preceding years, a lot of research work has been done on the synthesis, chemistry, and biological characteristics of imidazole-based heterocycles and, as demonstrated in this chapter, much interest has been shown in the last 15 years in the synthesis and evaluation of the biological and pharmacological properties of imidazole analogs. The exhaustive literature study carried out in this period involved designing and development of facile, novel, and efficient synthetic methodologies for various imidazole-based compounds that exhibited potential applications in several streams of medical science. These imidazole-based compounds possessed anticancer, antimicrobial, antiinflammatory, antioxidant, antitubercular, antinociceptive, and antiobesity agents as well as showing beneficial effects in neurodegenerative disorders including Alzheimer's disease and many more. This chapter has provided plentiful and clear information on imidazole-based compounds using various classification

TABLE 1.1 Synthesis of benzimidazoles under various reaction conditions.

S. no.	Reactant	Catalyst	Reaction condition	Ref.
1	*o*-Diaminobenzenes + carboxylic acid	Conc. HCl	Philip's method, reflux	[124]
2	Trichloroacetimidate + *o*-phenylenediamine	AcOH	Room temp.	[139]
3	*o*-Phenylenediamine + different aromatic acid	$EtOH.NH_4Cl$	80–90 °C	[125]
4	*o*-Phenylenediamine + aromatic, aliphatic and heterocyclic carboxylic acid	Alumina, silica gel and zeolite HY	Microwave 160–560 W for 5–9 min, solvent-free	[126]
5	*o*-Phenylenediamine + various aldehydes	Sodium metabisulfite ($Na_2S_2O_5$)	DMF, room temp.	[140]
6	Unsubstituted and substituted phenylenediamine + aldehyde	Nitrobenzene	Water, 140 °C	[134]
7	*o*-Phenylenediamine + several aldehydes	Indium triflate [$In(OTf)_3$]	Solvent-free	[137]
8	*o*-Phenylenediamine + aromatic aldehydes	Sodium hexafluroaluminate (Na_3AlF_6)	50 °C, EtOH	[141]
9	*o*-Phenylenediamine + different aldehydes	Dioxane dibromide	Room temp., acetonitrile, 30–60 min	[142]
10	*o*-Phenylenediamine + aldehyde	Zinc triflate	Ethanol, reflux, 8 h	[143]
11	*o*-Phenylenediamine + benzaldehydes	Nickel acetate ($Ni(OAc)_2$)	$CHCl_3$, room temp.	[144]
12	*o*-Phenylenediamine + aldehydes	Iodine	Water, 80–90 °C, 1.2–1.5 h	[145]
13	*o*-Phenylenediamine + various aldehydes	Sodium dodecyl sulfate	Water, reflux, room temp.	[146]
14	*o*-Phenylenediamine + various aldehydes	GO/Co/Mn	Water, US irritation	[147]
15	*o*-Phenylenediamine + different aldehydes	DBU	DCM, room temp., 10 h	[148]
16	2-Nitrophenol + aldehydes	Ag/Pd WO_2	HCOOH, 80 °C	[149]

approaches like substituted imidazoles, fused imidazoles, imidazole hybrids, and benzimidazole-based compounds for the better understanding of readers.

Although numerous pathways have been designed and developed for the synthesis of imidazole-based heterocycles, there is still a wide gap in the application of these heterocycles and greener routes for their synthesis. There should be focus on novel, safe, cost-effective, and green synthetic pathways to develop future drugs considering purity and stereochemistry as the most acute parameters. Additional efforts for the fabrication and development of new, safe, and less resistant drugs are ongoing, and advancements are fervently sought.

Conflict of interest

The authors declare no conflict of interest, financial or otherwise.

Acknowledgments

The authors are thankful to Department of Chemistry, Mohan Lal Sukhadia University, Udaipur, India for providing the necessary library facilities for carrying out the work. The author A. Sethiya is also thankful to UGC-MANF (201819-MANF-RAJ-2018-19-91971) for providing Senior Research Fellowship to carry out this work.

Funding source

This work was supported by UGC-MANF (201819MANF-2018-19-RAJ-91971).

References

[1] Gupta RR, Kumar M, Gupta V. Heterocyclic chemistry: volume II. In: Five-membered heterocycles. Springer Science & Business Media; 2013.

[2] Del Vecchio A, Destro G, Taran F, Audisio D. Recent developments in heterocycle labeling with carbon isotopes. J Labelled Comp Radiopharm 2018;61(13):988–1007.

[3] Kalal P, Gandhi D, Prajapat P, Agarwal S. Biological and synthetic studies of four, five and six membered heterocycles. Heterocycl Lett 2017;7(2):513–40.

[4] Mohammad A. A mini review: biological significances of nitrogen hetero atom containing heterocyclic compounds. Int J Bioorg Chem 2017;2:146–52.

[5] Zhao C, Qiao X, Yi Z, Guan Q, Li W. Active centre and reactivity descriptor of a green single component imidazole catalyst for acetylene hydrochlorination. Phys Chem Chem Phys 2020;22(5):2849–57. https://doi.org/10.1039/c9cp06005g.

[6] Olofson A, Yakushijin K, Horne DA. Synthesis of marine sponge alkaloids oroidin, clathrodin, and dispacamides. Preparation and transformation of 2-amino-4,5-dialkoxy-4,5-dihydroimidazolines from 2-aminoimidazoles. J Org Chem 1998;63:1248.

[7] Wan Y, Hur W, Cho CY, Liu Y, Adrian FJ, Lozach O, Bach S, Mayer T, Fabbro D, Meijer L, Gray NS. Synthesis and target identification of hymenialdisine analogs. Chem Biol 2004;11:247–59.

[8] Kudelko A, Zieliński W, Jasiak K. Synthesis of novel 1-[(1-ethoxymethylene)amino]imidazol-5(4H)-ones and 1,2,4-triazin-6(5H)-ones from optically active α-aminocarboxylic acid hydrazides. Tetrahedron Lett 2013;54:4637–40.

[9] Debus H. Ueber die einwirkung des ammoniaks auf glyoxal. Justus Liebigs Ann Chem 1858;107(2):199–208.

[10] Verma A, Joshi S, Singh D. Imidazole: having versatile biological activities. J Chem 2013;2013:1.

[11] Khabnadideh S, Rezaei Z, Khalafi-Nezhad A, Bahrinajafi R, Mohamadia R, Farrokhroz AA. Synthesis of N-alkylated derivatives of imidazole as antibacterial agents. Bioorg Med Chem Lett 2003;13(17):2863–5.

[12] Kalaria PN, Satasia SP, Avalani JR, Raval DK. Ultrasound-assisted one-pot four-component synthesis of novel 2-amino-3-cyanopyridine derivatives bearing 5-imidazopyrazole scaffold and their biological broadcast. Eur J Med Chem 2014;83:655.

[13] Lu B, Lu F, Ran L, Yu K, Xiao Y, Li Z, Dai F, Wu D, Lan G. Self-assembly of natural protein and imidazole molecules on gold nanoparticles: applications in wound healing against multidrug resistant bacteria. Int J Biol Macromol 2018;119:505–16.

[14] Moraski GC, Thanassi JA, Podos SD, Pucci MJ, Miller MJ. Onestep syntheses of nitrofuranyl benzimidazoles that are active against multidrug-resistant bacteria. J Antibiot 2011;64(10):667–71.

[15] Jeanmart S, Gagnepain J, Maity P, Lamberth C, Cederbaum F, Rajan R, Jacob O, Blum M, Bieri S. Synthesis and fungicidal activity of novel imidazole-based ketene dithioacetals. Bioorg Med Chem 2018;26(8):2009–16.

[16] Park NH, Shin KH, Kang MK. Antifungal and antiviral agents. Pharmacol Ther Dent 2017;488–503. https://doi.org/10.1016/B978-0-323-39307-2.00034-5.

[17] Bhatt HB, Sharam S. Sythesis, characterization and biological evalution of some trisubsituted imidazole/thiazole derivatives. J Heterocycl Chem 2014;52(4):1126–31. https://doi.org/10.1002/jhet.1992.

[18] Koh A, Molinaro A, Ståhlman M, Khan MT, Schmidt C, Mannerås-Holm L, Wu H, Carreras A, Jeong H, Olofsson LE, Bergh PO. Microbially produced imidazole propionate impairs insulin signaling through mTORC1. Cell 2018;175(4):947–61.

[19] Hu DC, Chen LW, Yang YX, Liu JC. Syntheses, structures and antioxidant activities of two new cu(II) complexes with a benzimidazole schiff base ligand. Inorg Nano-Met Chem 2018. https://doi.org/10.1080/15533174.2013.843562.

[20] Keppler BK, Wehe D, Endres E, Rupp W. Synthesis, antitumor activity, and x-ray structure of bis(imidazo l ium) (imidazole) pentachlororuthenate(111), (ImH) (RuImCl5). Inorg Chem 1987;26(6):844–6. https://doi.org/10.1021/ic00253a014.

[21] Antonini I, Claudi F, Cristalli G, Franchetti P, Grifantini M, Martelli S. Heterocyclic quinones with potential antitumor activity. 2. Synthesis and antitumor activity of some benzimidazole-4,7-dione derivatives. J Med Chem 1988;31(1):260–4. https://doi.org/10.1021/jm00396a041. 3336025.

[22] Inoue T, Shimozato O, Matsuo N, Mori Y, Shinozaki Y, Lin J, Watanabe T, Takatori A, Koshikawa N, Ozaki T, Nagase H. Hydrophobic structure of hairpin ten-ring pyrrole-imidazole polyamides enhances tumor tissue accumulation/retention in vivo. Bioorg Med Chem 2018;26(9):2337–44. https://doi.org/10.1016/j.bmc.2018.03.029. 29622411.

[23] Alegaon SG, Alagawadi KR, Sonkusare PV, Chaudhary SM, Dadwe DH, Shah AS. Novel imidazo[2,1-b][1,3,4]thiadiazole carrying rhodanine-3-acetic acid as potential antitubercular agents. Bioorg Med Chem Lett 2012;22(5):1917–21. https://doi.org/10.1016/j.bmcl.2012.01.052. 22325950.

[24] Toja E, Selva D, Schiatti P. 3-Alkyl-2-aryl-3H-naphth[1,2-d]imidazoles, a novel class of nonacidic antiinflammatory agents. J Med Chem 1984;27(5):610–6. https://doi.org/10.1021/jm00371a010. 6609233.

[25] Bender PE, Hill DT, Offen PH, Razgaitis K, Lavanchy P, Stringer OD, Sutton BM, Griswold DE, DiMartino M, Donald T, Walz Lantos OI, Lad CB. 5,6-Diaryl-2,3-dihydroimidazo [2,l-b] thiazoles: a new class of immunoregulatory anti inflammatory agents. J Med Chem 1985;28(9):1169–77. https://doi.org/10.1021/jm00147a008. 4032421.

[26] de Gaetano M, Butler E, Gahan K, Zanetti A, Marai M, Chen J, Cacace A, Hams E, Maingot C, McLoughlin A, Brennan E. Asymmetric synthesis and biological evaluation of imidazole-and oxazole-containing synthetic lipoxin A4 mimetics (sLXms). Eur J Med Chem 2019;162:80–108.

[27] James DA, Koya K, Li H, Chen S, Xia Z, Ying W, Wu Y, Sun L. Conjugated indole-imidazole derivatives displaying cytotoxicity against multidrug resistant cancer cell lines. Bioorg Med Chem Lett 2006;16:5164–8.

[28] Ozkay Y, Isikdag I, Incesu Z, Akalın G. Synthesis of 2-substituted-N-[4-(1-methyl-4,5-diphenyl-1H-imidazole-2-yl)phenyl]acetamide derivatives and evaluation of their anticancer activity. Eur J Med Chem 2010;45:3320–8.

[29] Bellina F, Guazzelli N, Lessi M, Manzini C. Imidazole analogues of resveratrol: synthesis and cancer cell growth evaluation. Tetrahedron 2015;71(15):2298–305.

[30] Anderson EB, Long TE. Imidazole- and imidazolium-containing polymers for biology and material science applications. Polymer 2010;51:2447–54.

[31] Demberelnyamba D, Kim KS, Choi S, Park SY, Lee H, Kim CJ. Synthesis and antimicrobial properties of imidazolium and pyrrolidinonium salt. Bioorg Med Chem 2004;12:853–8.

[32] Batra D, Seifert S, Firestone MA. The effect of cation structure on the mesophase architecture of self-assembled and polymerized imidazolium-based ionic liquids. Macromol Chem Phys 2007;208:1416–27.

[33] Preethi N, Shinohara H, Nishide H. Reversible oxygen-binding and facilitated oxygen transport in membranes of polyvinylimidazole complexed with cobalt-phthalocyanine. React Funct Polym 2006;66:851–5.

[34] Rani N, Sharma A, Singh R. Trisubstituted imidazole synthesis: a review. Mini Rev Org Chem 2015;12(1):34–64. https://doi.org/10.2174/1570193X11666141028235010.

[35] Kumar M, Kumar D, Raj V. Studies on imidazole and its derivatives with particular emphasis on their chemical/biological applications as bioactive molecules/intermediated to bioactive molecule. Curr Synth Syst Biol 2017;5(1):135–45. https://doi.org/10.4172/2332-0737.1000135.

[36] Gupta P, Gupta JK. Synthesis of bioactive imidazoles: a review. J Chem Sci 2017;27(1):1–6.

[37] Fan YL, Jin XH, Huang ZP, Yu HF, Zeng ZG, Gao T, Feng LS. Recent advances of imidazole-containing derivatives as anti-tubercular agents. Eur J Med Chem 2018;150:347–65. https://doi.org/10.1016/j.ejmech.2018.03.016. 29544148.

[38] Soni J, Sethiya A, Sahiba N, Agarwal DK, Agarwal S. Contemporary progress in the synthetic strategies of imidazole and its biological activities. Curr Org Synth 2019;16:1086–112.

[39] Pardeshi VAS, Chundawat NS, Pathan SI, Sukhwal P, Chundawat TPS, Singh GP. A review on synthetic approaches of benzimidazoles. Synth Commun 2021;51(4):485–513.

[40] Prajapat P, Kumawat M, Talesara GL, Kalal P, Agarwal S, Kapoor CS. Benzimidazole scaffold as a versatile biophore in drug discovery: a review. Chem Biol Interact 2018;8(1):1–10.

[41] Ruan Y, Chen Y, Gu L, Luo Y, Yang Z, He L. Preparation of imidazole derivatives via bisfunctionalization of alkynes catalyzed by ruthenium carbonyl. Synthesis 2019;51:3520–8.

[42] Roman G, Vlahakis JZ, Vukomanovic D, Nakatsu K, Szarek WA. Heme oxygenase inhibition by 1-Aryl-2-(1H-imidazol-1-yl/1H-1,2,4-triazol-1yl) ethanones and their derivatives. Chem Med Chem 2010;5:1541–55.

[43] Zhu L, Li G, Luo L, Guo P, Lan J, You J. Highly functional group tolerance in copper-catalyzedn-arylation of nitrogen-containing heterocycles under mild conditions. J Org Chem 2009;74(5):2200–2. https://doi.org/10.1021/jo802669b.

[44] Kuzua B, Tana M, Ekmekcib Z, Menges N. A novel fluorescent sensor based on imidazole derivative for Fe^{3+} ions. JOL 2017;1192:1096–103.

[45] Wallach O. Imidazole and its derivatives. Ber 1876;184:33–5.

[46] Chen J, Wang Z, Li C, Lu Y, Vaddady PK, Meibohm B, et al. Discovery of novel 2-aryl-4-benzoyl-imidazoles targeting the colchicines binding site in tubulin as potential anticancer agents. J Med Chem 2010;53:7414–27.

[47] Fang S, Yu H, Yang X, Li J, Shao L. Nickel-catalyzed construction of 2,4-disubstituted imidazoles via C-C coupling and C-N condensation cascade reactions. Adv Synth Catal 2019;361:3312–7.

[48] Shabalin DA, Dunsford JJ, Ngwerume S, Saunders AR, Gill DM, Camp JE. Synthesis of 2,4-disubstituted imidazoles via nucleophilic catalysis. Synlett 2020;31:797–800.

[49] de Toledo I, Grigolo TA, Bennett JM, Elkins JM, Pilli RA. Modular synthesis of di-and tri-substituted imidazoles from ketones and aldehydes: a route to kinase inhibitors. J Org Chem 2019;84(21):14187–201.

[50] Kalhor M, Seyedzade Z, Zarnegar Z. $(NH_4)_2Ce(NO_3)_6/HNO_3$ as a high-performance oxidation catalyst for the one-step, solvent-free synthesis of dicyano imidazoles. Polycyclc Aromat Compd 2019. https://doi.org/10.1080/10406638.2019.1686402.

[51] Bunev AS, Vasiliev MA, Statsyuk VE, Ostapenko GI, Peregudov AS. Synthesis of 1-aryl-4-tosyl-5-(trifluoromethyl)-1*H*-imidazoles. J Fluor Chem 2014;163:34–7.

[52] Wong LC, Gehre A, Stanforth SP, Tarbit B. Convenient synthesis of highly substituted imidazole derivatives. Synth Commun 2013;43(1):80–4.

[53] Noriega-Iribe E, Díaz-Rubio L, Estolano-Cobián A, Barajas-Carrillo VW, Padrón JW, Salazar-Aranda R, Díaz-Molina R, García-González V, Chávez-Santoscoy RA, Chávez D, Córdova-Guerrero I. In vitro and in silico screening of 2,4,5-trisubstituted imidazole derivatives as potential xanthine oxidase and acetylcholinesterase inhibitors, antioxidant, and antiproliferative agents. Appl Sci 2020;10:2889–94. https://doi.org/10.3390/app10082889.

[54] Takashima R, Tsunekawa K, Shinozaki M, Suzuki Y. Selective synthesis of 1,4,5-trisubstituted imidazoles from a-imino ketones prepared by N-heterocyclic-carbene-catalyzed aroylation. Tetrahedron 2018;74:2261–7.

[55] Hanoon HD, Kowsari E, Abdouss M, Ghasemi MH, Zandi H. Highly efficient and simple protocol for synthesis of 2,4,5-triarylimidazole derivatives from benzil using fluorinated graphene oxide as effective and reusable catalyst. Res Chem Intermed 2017;43:4023–41.

[56] Tan J, Li JR, Hu Y. Novel and efficient multifunctional periodic mesoporous organosilica supported benzotriazolium ionic liquids for reusable synthesis of 2,4,5-trisubstituted imidazoles. J Saudi Chem Soc 2020;24(10):777–84.

[57] Hilal DA, Hanoon HD. Brønsted acidic ionic liquid catalyzed an eco-friendly and efficient procedure for synthesis of 2,4,5-trisubstituted imidazole derivatives under ultrasound irradiation and optimal conditions. Res Chem Intermed 2020;46:1521–38. https://doi.org/10.1007/s11164-019-04048-z.

[58] Alinezhad H, Tajbakhsh M, Maleki B, Oushibi FP. Acidic ionic liquid [H-NP]HSO_4 promoted one-pot synthesis of dihydro-1H-indeno[1,2-b]pyridines and polysubstituted imidazoles. Polycycl Aromat Compd 2020;40(5):1485–500. https://doi.org/10.1080/10406638.2018.1557707.

[59] Maleki A, Rahimi J, Valadi K. Sulfonated Fe_3O_4@PVA superparamagnetic nanostructure: design, in-situ preparation, characterization and application in the synthesis of imidazoles as a highly efficient organic–inorganic Brønsted acid catalyst. Nano-Struct Nano-Objects 2019;18:100264–70.

[60] Shaterian HR, Ranjbar M. An environmental friendly approach for the synthesis of highly substituted imidazoles using Brønsted acidic ionic liquid, N-methyl-2-pyrrolidonium hydrogen sulfate, as reusable catalyst. J Mol Liq 2011;160:40–9.

[61] Wang M, Gao J, Song Z. A practical and green approach toward synthesis of 2,4,5-trisubstituted imidazoles without adding catalyst. Prep Biochem Biotechnol 2010;40:347–53.

[62] Zhang F, Gao Q, Chen B, Bai Y, Sun W, Lv D, Ge M. A practical and green approach towards synthesis of multi-substituted imidazoles using boric acid as efficient catalyst. Phosphorus Sulfur 2016;191(5):786–9. https://doi.org/10.1080/10426507.2015.1100184.

[63] Mukhopadhyay C, Tapaswi PK. A facile and efficient synthesis of tri- and tetrasubstituted imidazoles with potassium hydrogen sulfate and DB18C6 in an aqueous medium. Green Chem Lett Rev 2012;5(2):109–20.

[64] Gupta S, Lakshman M. Magnetic nano cobalt ferrite: an efficient recoverable catalyst for synthesis of 2,4,5-trisubstituted imidazoles. J Med Chem Sci 2019;2(2):51–4.

[65] Fekri LZ, Nikpassand M, Shariati AB, Zarkeshvari R, Norouz Pour NJ. Synthesis and characterization of amino glucose-functionalized silica-coated NiFe2O4 nanoparticles: A heterogeneous, new and magnetically separable catalyst for the solvent-free synthesis of 2,4,5-trisubstituted imidazoles, benzo[d]imidazoles, benzo[d] oxazoles and azo-linked benzo[d]oxazoles. Organomet Chem 2018;871:60–73. https://doi.org/10.1016/j.jorganchem.2018.07.008.

[66] Safari J, Zarnegar Z, Rahimi F. Immobilized ionic liquid on superparamagnetic nanoparticles as an effective catalyst for the synthesis of tetrasubstituted imidazoles under solvent-free conditions and microwave irradiation. C R Chim 2013;16(10):920–8. https://doi.org/10.1016/j.crci.2013.01.019.

[67] Safari J, Zarnegar Z. Magnetic nanoparticle supported ionic liquid as novel and effective heterogeneous catalyst for synthesis of substituted imidazoles under ultrasonic irradiation. Monatsh Chem 2013;144:1389–96. https://doi.org/10.1007/s00706-013-1015-6.

[68] Kalhor M, Zarnegar Z. Fe_3O_4/SO_3H@zeolite-Y as a novel multi-functional and magnetic nanocatalyst for clean and soft synthesis of imidazole and perimidine derivatives. RSC Adv 2019;9:19333–46. https://doi.org/10.1039/C9RA02910A.

[69] Safari J, Zarnegar Z. Sulphamic acid-functionalized magnetic Fe_3O_4 nanoparticles as recyclable catalyst for synthesis of imidazoles under microwave irradiation. J Chem Sci 2013;125:835–41. https://doi.org/10.1007/s12039-013-0462-2.

[70] Naeimi H, Aghaseyedkarimi D. $Fe_3O_4@SiO_2{\cdot}HM{\cdot}SO_3H$ as a recyclable heterogeneous nanocatalyst for the microwave-promoted synthesis of 2,4,5-trisubstituted imidazoles under solvent free conditions. New J Chem 2015;3:9415–21. doi: 10.10/C5NJ01273B.

[71] Zarnegar Z, Safari J. Catalytic activity of Cu nanoparticles supported on Fe_3O_4–polyethylene glycol nanocomposites for the synthesis of substituted imidazoles. New J Chem 2014;38:4555–61. https://doi.org/10.1039/C4NJ00645C.

[72] Maleki A, Movahed H, Paydar R. Design and development of a novel cellulose/γ-Fe_2O_3/Ag nanocomposite: a potential green catalyst and antibacterial agent. RSC Adv 2016;6:13657–65. https://doi.org/10.1039/C5RA21350A.

[73] Maleki A, Alrezvani Z, Maleki S. Design, preparation and characterization of urea-functionalized Fe_3O_4/SiO_2 magnetic nanocatalyst and application for the one-pot multicomponent synthesis of substituted imidazole derivatives. Cat Commun 2015;69:29–33. https://doi.org/10.1016/j.catcom.2015.05.014.

[74] Zarnegar Z, Safari J. Fe_3O_4@chitosan nanoparticles: a valuable heterogeneous nanocatalyst for the synthesis of 2,4,5-trisubstituted imidazoles. RSC Adv 2014;4:20932–9. https://doi.org/10.1039/C4RA03176H.

[75] Esmaeilpour M, Javidi J, Zandi M. One-pot synthesis of multisubstituted imidazoles under solvent-free conditions and microwave irradiation using $Fe_3O_4@SiO_2$-imid-PMAn magnetic porous nanospheres as a recyclable catalyst. New J Chem 2015;39:3388–98. https://doi.org/10.1039/C5NJ00050E.

[76] Ziarani GM, Badiei A, Lashgari N, Farahani Z. Efficient one-pot synthesis of 2,4,5-trisubstituted and 1,2,4,5-tetrasubstituted imidazoles using SBA-Pr-SO_3H as a green nano catalyst. J Saudi Chem Soc 2016;20:419–27. https://doi.org/10.1016/j.jscs.2013.01.005.

[77] Girish YR, Kumar KSS, Thimmaiah KN, Rangappa KS, Shashikanth S. ZrO_2-b-cyclodextrin catalyzed synthesis of 2,4,5-trisubstituted imidazoles and 1,2-disubstituted benzimidazoles under solvent free conditions and evaluation of their antibacterial study. RSC Adv 2015;5:75533–46. https://doi.org/10.1039/C5RA13891D.

[78] Shaabani A, Afshari R, Hooshmand SE. Crosslinked chitosan nanoparticle-anchored magnetic multi-wall carbon nanotubes: a bio-nanoreactor with extremely high activity toward click-multi-component reactions. New J Chem 2017;41:8469–81. https://doi.org/10.1039/C7NJ01150D.

[79] Maleki A, Paydar R. Graphene oxide-chitosan bionanocomposite: a highly efficient nanocatalyst for the one-pot threecomponent synthesis of trisubstituted imidazoles under solvent-free conditions. RSC Adv 2015;5:33177–84. https://doi.org/10.1039/C5RA03355A.

[80] Singh H, Rajput JK. Co(II) anchored glutaraldehyde crosslinked magnetic chitosan nanoparticles (MCS) for synthesis of 2,4,5-trisubstituted and 1,2,4,5-tetrasubstituted imidazoles. Appl Organomet Chem 2018;32. https://doi.org/10.1002/aoc.3989, e3989.

[81] Heravi MM, Bakhtiari K, Oskooie HA, Taheri S. Synthesis of 2,4,5-triaryl-imidazoles catalyzed by $NiCl_2 \cdot 6H_2O$ under heterogeneous system. J Mol Cat A: Chem 2007;263(1–2):279–81.

[82] Rajkumar R, Kamaraj A, Krishnasamy K. Multicomponent, one-pot synthesis and spectroscopic studies of 1-(2-(2,4,5-triphenyl-1H-imidazol-1-yl)ethyl)piperazine derivatives. J Taibah Univ Sci 2015;9(4):498–507. https://doi.org/10.1016/j.jtusci.2014.12.001.

[83] El-Remaily MAEAAA, Abu-Dief AM. $CuFe_2O_4$ nanoparticles: an efficient heterogeneous magnetically separable catalyst for synthesis of some novel propynyl-1H-imidazoles derivatives. Tetrahedron 2015;71:2579–84. https://doi.org/10.1016/j.tet.2015.02.057.

[84] Gilan MM, Khazaei A, Seyf JY, Sarmasti N, Keypour H, Mahmoudabadi M. Synthesis of magnetic Fe_3O_4@SiO_2@$Si(CH_2)_3$@N-ligand@co with application in the synthesis of 1,2,4,5 substituted imidazole derivatives. Polycycl Aromat Compd 2019. https://doi.org/10.1080/10406638.2019.1666886.

[85] Hosseini S, Kiasat AR, Farhadi A. Fe_3O_4@SiO_2/bipyridinium nanocomposite as a magnetic and recyclable heterogeneous catalyst for the synthesis of highly substituted imidazoles via multi-component condensation strategy. Polycycl Aromat Compd 2019;43:1–11. https://doi.org/10.1080/10406638.2019.1616306.

[86] Davoodnia A, Heravi MM, Safavi-Rad Z, Tavakoli-Hoseini N. Green, one-pot, solvent-free synthesis of 1,2,4,5-tetrasubstituted imidazoles using a Brønsted acidic ionic liquid as novel and reusable catalyst. Synth Commun 2010;40(17):2588–97.

[87] Safa KD, Feyzi A, Allahvirdinesbat M, Sarchami L, Panahi PN. Synthesis of novel organosilicon compounds possessing fully substituted imidazole nucleus sonocatalyzed by Fe-Cu/ZSM-5 bimetallic oxides. Synth Commun 2015;45(3):382–90. https://doi.org/10.1080/00397911.2014.962056.

[88] Borhade AV, Tope DR, Gite SG. Synthesis, characterization and catalytic application of silica supported tin oxide nanoparticles for synthesis of 2,4,5-tri and 1,2,4,5-tetrasubstituted imidazoles under solvent-free conditions. Arab J Chem 2017;10:559–67. https://doi.org/10.1016/j.arabjc.2012.11.001.

[89] Nikoofar K, Dizgarani SM. HNO_3@Nano SiO_2: an efficient catalytic system for the synthesis of multi-substituted imidazoles under solvent-free conditions. J Saudi Chem Soc 2017;21:787–94. https://doi.org/10.1016/j.jscs.2015.11.006.

[90] Khan K, Siddiqui ZN. An efficient synthesis of tri- and tetrasubstituted imidazoles from benzils using functionalized chitosan as biodegradable solid acid catalyst. Ind Eng Chem Res 2015;54:6611–8. https://doi.org/10.1021/acs.iecr.5b00511.

[91] Safari J, Akbari Z, Naseh S. Nanocrystalline $MgAl_2O_4$ as an efficient catalyst for one-pot synthesis of multisubstituted imidazoles under solvent-free conditions. J Saudi Chem Soc 2016;20:250–5. https://doi.org/10.1016/j.jscs.2012.10.012.

[92] Reddy BP, Vijayakumar V, Arasu MV, Al-Dhabi NA. c-Alumina nanoparticle catalyzed efficient synthesis of highly substituted imidazoles. Molecules 2015;20:19221–35. https://doi.org/10.3390/molecules201019221.

[93] Agarwal S, Kidwai M, Poddar R, Nath MA. Facile and green approach for the one-pot multicomponent synthesis of 2,4,5-triaryl- and 1,2,4,5-tetraarylimidazoles by using zinc-proline hybrid material as a catalyst. Chem Select 2017;2:10360–4. https://doi.org/10.1002/slct.201702222.

[94] Esmaeilpour M, Javidi J, Dehghani F, Zahmatkesh S. One-pot synthesis of multisubstituted imidazoles catalyzed by dendrimer-pwan nanoparticles under solvent-free conditions and ultrasonic irradiation. Res Chem Intermed 2017;43:163–85. https://doi.org/10.1007/s11164-016-2613-9.

[95] Akbari A. Tri(1-butyl-3-methylimidazolium) gadolinium hexachloride, ($[Bmim]_3[GdCl_6]$), a magnetic ionic liquid as a green salt and reusable catalyst for the synthesis of tetrasubstituted imidazoles. Tetrahedron Lett 2016;57:431–4. https://doi.org/10.1016/j.tetlet.2015.12.053.

[96] Kiumars B, Mohammad MK, Akbar N. One-pot synthesis of 1,2,4,5-tetrasubstituted and 2,4,5-trisubstituted imidazoles by zinc oxide as efficient and reusable catalyst. Monatsh Chem 2011;142:159–62. https://doi.org/10.1007/s00706-010-0428-8.

[97] Kolvari E, Zolfagharinia S. A waste to wealth approach through utilization of nano-ceramic tile waste as an accessible and inexpensive solid support to produce a heterogeneous solid acid nanocatalyst: to kill three birds with one stone. RSC Adv 2016;6:93963–74. https://doi.org/10.1039/C6RA11923A.

[98] Hebishy AM, Abdelfattah MS, Elmorsy A, Elwahy AH. ZnO nanoparticles catalyzed synthesis of bis- and poly(imidazoles) as potential anticancer agents. Synth Commun 2020;50(7):980–96.

[99] Adiyala PR, Jang S, Vishwakarma NK, Hwang YH, Kim DP. Continuous-flow photo-induced decarboxylative annulative access to fused imidazole derivatives via a microreactor containing immobilized ruthenium. Green Chem 2020;22(5):1565–71.

[100] Katava R, Zorko F, Mance AD, Otmačić-Ćurković H, Pavlović G. Synthesis and structure of 4-methyl-1-N-(p-tolyl) imidazole as organic corrosion inhibitor. Mol Cryst Liq Cryst 2017;642(1):29–37.

[101] Subhashini NJ, Kumar EP, Gurrapu N, Yerragunta V. Design and synthesis of imidazolo-1,2,3-triazoles hybrid compounds by microwave-assisted method: evaluation as an antioxidant and antimicrobial agents and molecular docking studies. J Mol Struct 2019;1180:618–28.

[102] Wagare DS, Sonone A, Farooqui M, Durrani A. An efficient and green microwave-assisted one pot synthesis of imidazothiadiazoles in PEG-400 and water. Polycycl Aromat Compd 2019;1–6.

[103] Yavuz SÇ, Akkoc S, Sarıpınar E. The cytotoxic activities of imidazole derivatives prepared from various guanylhydrazone and phenylglyoxal monohydrate. Synth Commun 2019;49(22):3198–209.

[104] Desai NC, Maheta AS, Rajpara KM, Joshi VV, Vaghani HV, Satodiya HM. Green synthesis of novel quinoline based imidazole derivatives and evaluation of their antimicrobial activity. J Saudi Chem Soc 2014;18(6):963–71.

[105] Aksakal NE, Bayar M, Dumrul H, Atilla D, Chumakov Y, Yuksel F. Structural and optical properties of new naphthalene and perylene imide imidazoles. Polycycl Aromat Compd 2019;39(4):363–73.

[106] Li B, Mai S, Song Q. Synthesis of fused benzimidazoles via successive nucleophilic additions of benzimidazole derivatives to arynes under transition metal-free conditions. Org Chem Front 2018;5(10):1639–42.

[107] Dao PD, Lim HJ, Cho CS. Weak Base-Promoted lactamization under microwave irradiation: synthesis of quinolin-2 (1H)-ones and phenanthridin-6 (5H)-ones. ACS Omega 2018;3(9):12114–21.

[108] Zhang X, Zhou Y, Wang H, Guo D, Ye D, Xu Y, Jiang H, Liu H. An effective synthetic entry to fused benzimidazoles via iodocyclization. Adv Synth Catal 2011;353(9):1429–37.

[109] Ataya K, Gokalp MM, Tuncerc BO, Tilk T. Antimicrobial activities and absorption properties of disazo dyes containing imidazole and pyrazole moieties. J Macromol Sci Pure Appl Chem 2017;54(4):236–42.

[110] Remaily MAEAAE, Mohamed SK, Soliman AMM, Ghanya HAE. Synthesis of dihydroimidazole derivatives under solvent free condition and their antibacterial evaluation. Biochem Physiol 2014;3(3):1000139. https://doi.org/10.4172/2168-9652.1000139.

[111] Rasanania SH, Moghadam ME, Soleimania E, Divsalarc A, Ajlood D, Tarlani A, Amiri M. Anticancer activity of new imidazole derivative of 1R, 2R diaminocyclohexane palladium and platinum complexesas DNA fluorescent probes. J Biomol Struct Dyn 2018;36(12):3058–76.

[112] Adib M, Peytam F, Shourgeshty R, Mohammadi-Khanaposhtani M, Jahani M, Imanparast S, Faramarzi MA, Larijani B, Moghadamnia AA, Esfahani EN, Bandarian F, Mahdavi M. Design and synthesis of new fused carbazole-imidazole derivatives as anti-diabetic agents: in vitro α-glucosidase inhibition, kinetic, and in silico studies. Bioorg Med Chem Lett 2019;29(5):713–8.

[113] Nagarajan N, Vanitha J, Ananth DA, Rameshkumar S, Sivasudha T, Renganathan R. Bioimaging, antibacterial and antifungal properties of imidazole-pyridine fluorophores: synthesis, characterization and solvatochromism. J Photochem Photobiol B 2013;127:212–22.

[114] Wahab ABF, Awad GEA, Badria FA. Synthesis, antimicrobial, antioxidant, anti-hemolytic and cytotoxic evaluation of new imidazole-based heterocycles. Eur J Med Chem 2011;461:505–1511.

[115] Ramachandran R, Rani M, Senthan S, Jeong YT, Kabilan S. Synthesis, spectral, crystal structure and in vitro antimicrobial evaluation of imidazole/benzotriazole substituted piperidin-4-one derivatives. Eur J Med Chem 2011;46:1926–34.

[116] Maheta HK, Patel AS, Naliapara YT. Synthesis and microbial study of some novel cyanopyrans and cyanopyridines containing imidazole nucleus. Int J Chem Sci 2012;10:1815–29.

[117] Prabhu M, Radha R. Synthesis, characterization and evaluation of antibacterial and antihelmintic activity of some novel aryl imidazole derivatives. Asian J Pharm Clin Res 2012;5:154–9.

[118] Liu T, Sun C, Xing X, Jing L, Tan R, Luo F, Huang W, Songa H, Li Z, Zhao Y. Synthesis and evaluation of 2-2-(phenylthiomethyl)-1H-benzo[d] imidazol-1-yl-acetohydrazide derivatives as antitumor agents. Bioorg Med Chem Lett 2012;22(9):3122–5.

[119] Nascimento MVPDS, Munhoz ACM, Facchin ABMDC, Fratoni E, Rossa TA, Sá MM, Campa CC, Ciraolo E, Hirsch E, Dalmarco EM. New pre-clinical evidence of anti-inflammatory effect and safety of a substituted fluorophenyl imidazole. Biomed Pharmacother 2019;111:1399–407.

[120] Sharma J, Soni PK, Bansal R, Halve AK. Synthetic approaches towards benzimidazoles by the reaction of *o*-phenylenediamine with aldehydes using a variety of catalysts: a review. Curr Org Chem 2018;22:2276–95.

[121] Alaqeel SI. Synthetic approaches to benzimidazoles from *o*-phenylenediamine: a literature review. J Saudi Chem Soc 2017;21(2):229–37. https://doi.org/10.1016/j.jscs.2016.08.001.

[122] Hoebrecker F. Ber 1872;5:920–6.

[123] Rathod CP, Rajurkar RM, Thonte SS. On benzimidazole synthesis and biological evaluation: a review. Indo American J Pharm Res 2013;3(2):2323–9.

[124] Phillip M. J Chem Soc C 1951;1143–5.

[125] Rithe SR, Jagtap RS, Ubarhande SS. One pot synthesis of substituted benzimidazole derivatives and their characterization. Rasayan J Chem 2015;8(2):213–7.

[126] Saberi A. Efficient synthesis of benzimidazoles using zeolite, alumina and silica gel under microwave irradiation. Iran J Sci Technol 2015;39(1):7–10.

[127] Ziarati A, Badiei A, Ziarani GM, Eskandarloo H. Simultaneous photocatalytic and catalytic activity of p–n junction NiO@anatase/rutile-TiO_2 as a noble-metal free reusable nanoparticle for synthesis of organic compounds. Cat Commun 2017;95:77–82.

[128] Qiu D, Wei H, Zhou L, Zeng Q. Synthesis of benzimidazoles by copper-catalyzed aerobic oxidative domino reaction of 1,2-diaminoarenes and arylmethyl halides. Appl Organomet Chem 2014;28:109–12.

[129] Elderfield RC, McCarthy JR. The reaction of o-phenylenediamines with carbonyl compounds. II. Aliphatic ketones. J Am Chem Soc 1951;73:975–84.

[130] Xiang SK, Tan W, Zhang DX, Tian XL, Feng C, Wang BQ, Zhao KQ, Hu P, Yang H. Synthesis of benzimidazoles by potassium tert-butoxide-promoted intermolecular cyclization reaction of 2-iodoanilines with nitriles. Org Biomol Chem 2013;11:7271–5.

[131] EL-Sayed T, Aboelnaga A, Hagar M. Ball milling assisted solvent and catalyst free synthesis of benzimidazoles and their derivatives. Molecules 2016;21(9):1111.

[132] Schmidt R, Fuhrmann S, Wondraczek L, Stolle A. Influence of reaction parameters on the depolymerization of H_2SO_4-impregnated cellulose in planetary ball mills. Powder Technol 2016;288:123–31.

[133] Rightmire NR, Hanusa TP. Advances in organometallic synthesis with mechanochemical methods. Dalton Trans 2016;45:2352–62.

[134] Mann J, Baron A, Opoku-Boahen Y, Johansson E, Parkinson G, Kelland LR, Neidle S. A new class of symmetric bisbenzimidazole-based DNA minor groove-binding agents showing antitumor activity. J Med Chem 2001;44:138–44.

[135] Venkateswarlu Y, Kumar SR, Leelavathi P. Facile and efficient one-pot synthesis of benzimidazoles using lanthanum chloride. Org Med Chem Lett 2013;3(7):2–8.

[136] Lin S, Yang L. A simple and efficient procedure for the synthesis of benzimidazoles using air as the oxidant. Tetrahedron Lett 2005;46:4315–9.

[137] Trivedi R, De SK, Gibbs RA. A convenient one-pot synthesis of 2-substituted benzimidazoles. J Mol Catal A: Chem 2006;245(1–2):8–11.

[138] Park S, Jung J, Cho JE. Visible-light-promoted synthesis of benzimidazoles. J Org Chem 2014;19:4148–54.

[139] Holan G, Samuel EL, Ennis BC, Hinde RW. 2-Trihalogenomethylbenzazoles. Part I. Formation. J Chem Soc C: Org 1967;20–5.

[140] Özbey S, Kaynak FB, Kuš C, Göker H. 2-(2,4-Dichlorophenyl)-5-fluoro-6-morpholin-4-yl-1H-benzimidazole monohydrate. Acta Crystallogr Sec E: Struct Rep Online 2002;58(10):o1062–4.

[141] Mobinikhaledi A, Hamta A, Kalhor M, Shariatzadeh M. Simple synthesis and biological evaluation of some benzimidazoles using sodium hexafluroaluminate, Na3AlF6, as an efficient catalyst. Iranian J Pharma Res 2014;13(1):95–101.

[142] Birajdar SS, Hatnapure GD, Keche AP, Kamble VM. Synthesis of 2-substituted-1H-benzo[d] imidazoles through oxidative cyclization of O-phenylenediamine and substituted aldehydes using dioxane. Res J Pharm Biosci Res 2014;5(1):487–93.

[143] Srinivasulu R, Kumar KR, Satyanarayana PVV. Facile and efficient method for synthesis of benzimidazole derivatives catalyzed by zinc triflate. Green Sustain Chem 2014;4:33–7.

[144] Vishvanath DP, Ketan PP. Synthesis of benzimidazole and benzoxazole derivatives catalyzed by nickel acetate as organometallic catalyst. Int J Chem Technol Res 2014;8(11):457–65.

[145] Aniket P, Shantanu DS, Anagha OB, Ajinkya PS. Iodine catalyzed convenient synthesis of 2-aryl-1-arylmethyl-1 H-benzimidazoles in aqueous media. Int J ChemTech Res 2015;8:496–500.

[146] Pardeshi SD, Thore SN. Mild and efficient synthesis of 2-aryl benzimidazoles in water using SDS. Int J Chem Phys Sci 2015;4:300–7.

[147] Karami AY, Manafi M, Ghodrati K, Khajavi R, Hojjati M. Nanoparticles supported graphene oxide and its application as an efficient and recyclable nano-catalyst in the synthesis of imidazole derivatives in ultrasound solvent-free condition. Int Nano Lett 2020. https://doi.org/10.1007/s40089-020-00297-8.

[148] John S, Kavya G, Aparna MM, Anuja Krishnan R, Parvathy R, Sreenath TS, Sivan A. Organic base catalysed synthesis of benzimidazole. IOP Conf Ser Mater Sci Eng 2020;872:12142.

[149] Yu C, Guo X, Shen B, Xi Z, Li Q, Yin Z, Liu H, Muzzio M, Shen M, Li J. One-pot formic acid dehydrogenation and synthesis of benzene-fused heterocycles over reusable AgPd/$WO_{2.72}$ nanocatalyst. J Mater Chem A 2018;6:23766–72. https://doi.org/10.1039/c8ta09342c.

Chapter 2

Biological profile of imidazole-based compounds as anticancer agents

Ayushi Sethiya[a], Jay Soni[a], Nusrat Sahiba[a], Pankaj Teli[a], Dinesh K. Agarwal[b], and Shikha Agarwal[a]

[a]*Synthetic Organic Chemistry Laboratory, Department of Chemistry, Mohanlal Sukhadia University, Udaipur, Rajasthan, India,* [b]*Department of Pharmacy, PAHER University, Udaipur, Rajasthan, India*

Abbreviations

A375	human melanoma cell line
A549	adenocarcinomic human alveolar basal epithelial cells
ABI analogs	2-aryl-4-benzoyl imidazole
ALK5 kinase	activin-like kinase 5
AML	acute myeloid leukemia
AR	androgen receptor
ASPC-1	human pancreatic cancer cell line
ATRA	all-trans-retinoic acid
B-Raf	V-RAF murine sarcoma viral oncogene homolog
CA-4	combretastatin A-4
CDK2	cyclin-dependent kinase 2
CML	chronic myeloid leukemia
CNS	central nervous system
COX-2	cyclooxygenase-2
CYP17A1	cytochrome P450 17A1
CYP26A1	cytochrome P450 26A1
DMF	dimethyl formamide
DMSO	dimethyl sulfoxide
DRB	5,6-dichloro-1-β-D-ribofuranosyl benzimidazole
Du 145	human prostate cancer cell line
Dvtm	divinyltetramethyldisiloxane
EBV	Epstein-Barr virus
EGFR	epidermal growth factor receptor
EVI1	human ectopic viral integration site
H1299	human lung carcinoma cell
H2pbic	2-(pyridin-4-yl)-1*H*-benzo[*d*]imidazole-5-carboxylic acid

Imidazole-Based Drug Discovery. https://doi.org/10.1016/B978-0-323-85479-5.00005-8
Copyright © 2022 Elsevier Inc. All rights reserved.

H460	human lung carcinoma cell line
HaCaT cells	aneuploid immortal keratinocyte cell
HCT-15	human colon adenocarcinoma
HEK293	human kidney cancer cell line
HepG2	human liver cancer cell line
HL-60 cell	human leukemia cell line
HMM	hexamethylmelamine
HO-1	heme oxygenase-1
IC_{50}	half maximum inhibitory concentration
IDO1	indoleamine 2,3-dioxygenase
LCL	lymphoblastoid cell line
LNCaP	human prostate cancer cell line
LSC	leukemia stem cell
MAOS	microwave-assisted organic synthesis
MBIC	methyl 2-(5-fluoro-2-hydroxyphenyl)-1*H*-benzo[*d*]imidazole-5-carboxylate
MCF-7	human breast cancer cell line
MDA-MB-231	human breast cancer cell line
MDR	multidrug resistance
MRC-5	human fetal lung fibroblast cell line
MTS	one-step MTT assay
MTT	(3-(4,5-dimethylthiazol-2-yl)-2,5-diphenyltetrazolium bromide)
NCI	National Cancer Institute
NCI-H460	human lung cancer cell line
NSCLC	nonsmall cell lung cancer line
NUGC-3 cells	human gastric cancer cell line
PANC	human pancreatic cancer cell line
PC-3	human prostate cancer cell line
PDGFRβ	platelet-derived growth factor receptor β
PI3K	phosphatidylinositol-3-kinase
PMS	phenazine methosulfate
PPC1	human prostate cancer cell line
PTS	protein thermal shift
RABI analogs	reverse 2-aryl-4-benzoyl imidazole
RAF	rapid accelerated fibrosarcoma
RC_{50}	concentration achieved by half-maximal fluorescence recovery
ROS	reactive oxygen species
SAR	structure activity relationship
SRB	sulforhodamine B assay
Tan-IIA	tanshinone-IIA
VEGF	vascular endothelial growth factor
WM-164	human melanoma cell line
XTT assay	2,3-bis-(2-methoxy-4-nitro-5-sulfophenyl)-2*H*-tetrazolium-5-carboxanilide
ZBG	zinc binding group

1. Introduction

Imidazole is an imperative five-membered heterocycle ring, extensively present in natural products and synthetic compounds [1–3]. Its exclusive structural characteristics with enviable electron-rich features are favorable for imidazole-based fused heterocycles to bind efficiently with an array of enzymes and receptors in biological systems through various weak interactions like hydrogen bonds, ion-dipole, cation-π, π-π stacking, coordination, Van der Waals forces, hydrophobic effects, etc., and therefore they demonstrate widespread bioactivities. Principally, several imidazole-based composites as clinical drugs have been comprehensively used to cure diverse types of diseases with high therapeutic potency [4–6]. Imidazole, being amphoteric, acts as an acid as well as a base and easily interacts with the anions and cations and biological molecules in the body. The associated investigation and development of imidazole derivatives and their therapeutic potential has become an area of intense interest. Furthermore, imidazole has many binding sites that can interact with a range of inorganic metal ions or bind with organic molecules via different types of bonds to generate supra-molecular drugs that bear the bioactivities of imidazole along with the benefits of several supra-molecular drugs, probably imposing a double-action mechanism that is supportive to defeat drug resistance [7–10]. The aforementioned studies showed that these compounds have vast potential in medicinal chemistry and a lot of research has been focused on their practical productive applications in miscellaneous areas. They have a plethora of biological activities like antibacterial, antiviral, antihistaminic, antifungal, antidiabetic, antiobesity, anticancer, and many more. The synthetic strategy and structure activity relationship for the anticancer activity of imidazole derivatives are discussed in this chapter.

Cancer is currently one of the most common diseases and reason behind deaths worldwide. According to the World Health Organization (WHO), cancer deaths are expected to increase, with an estimate of 13.1 million deaths in the year 2030 globally. In the past decades, huge attention has been dedicated to the synthesis and evaluation of potent and biologically active drugs to treat cancer. Several phytochemicals and synthetic organic molecules are used to make prodrugs. These include imidazole-fused heterocycles, which are effective against various type of cancers like breast cancer, lung cancer, leukemia, prostate cancer, etc. [11–13] (Fig. 2.1). This chapter aims to show the potential of functionally substituted imidazole compounds in different types of cancer and discuss

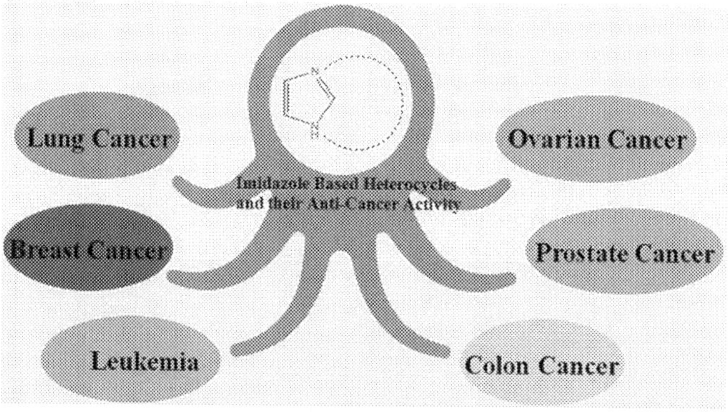

FIG. 2.1 Imidazole-based heterocycles activity toward different types of cancer.

different classes of medicines. Moreover, the pathways reported in the scientific literature in the past decades for the synthesis of imidazole derivatives and their pharmacological and biological significance and properties are considered.

A large number of drugs are clinically available for the treatment of cancer from different categories like natural products (camptothecins, taxols, curcumin), porphyrin (photofrin, visudyne), synthetic alkylating agents (mechlorethaminoxide, chlorambucil), and metal-complexes (cisplatin, carboplatin) [14–19]. Still there is a huge demand for cost-effective drugs to overcome several shortcomings that chiefly involve toxicity, poor curative effect, low solubility and efficacy, high cytotoxicity and side effects, low selectivity, and multidrug resistance. Therefore, extensive investigations have been carried out in attempts to discover ideal noncross-resistant and more tumor-specific therapies to diminish cancer cells with minimal impact on normal cells. A meticulous literature study showed that imidazole derivatives, as anticancer drugs, possess substantial potentiality. Several imidazole-based synthetic compounds are used as anticancer agents [20, 21]. Some clinically important drugs containing imidazole are shown in Fig. 2.2.

2. Imidazole-based heterocycles

Imidazole-based compounds are present in various biologically important molecules like histamine, hemoglobin, vitamin B_{12}, histidine, pilocarpine alkaloids, biotin, nucleic acid bases, etc. They are enormously versatile and present in

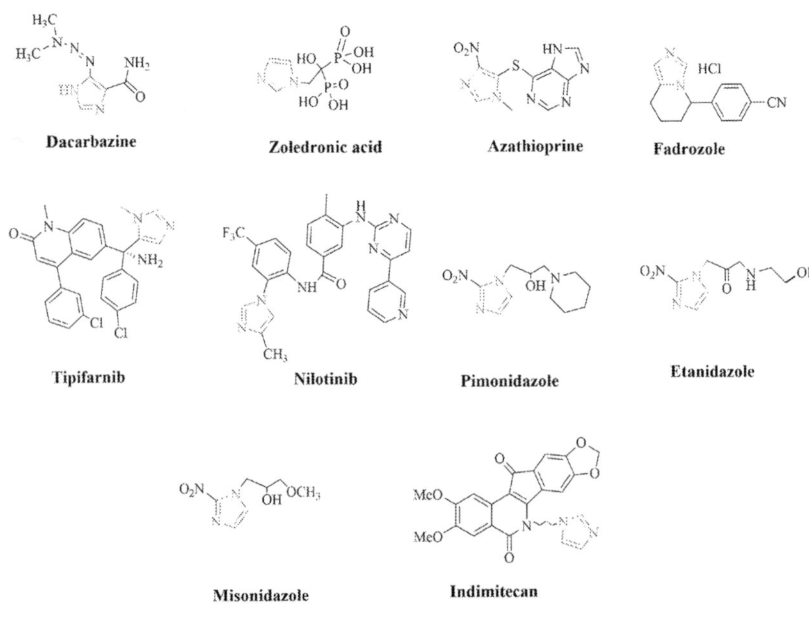

FIG. 2.2 Clinically available drugs on imidazole-based heterocycles.

several marketed drugs to treat different types of diseases. Hence, it is important to keep an eye on the synthetic progress of imidazole-based heterocyclic compounds along with their mechanism of action and SAR characteristics for future accomplishments. Here, we divide the anticancer profiles of imidazole-based heterocycles into various subsections depending on their structural profiles. Several research articles have been published that emphasize the anticancer potency of imidazole-based compounds. The sensation in the research of imidazole-based heterocycles come forward with the use of dacarbazine that prompted importance in the growth of imidazole-based agents. Numerous categories of different structured heterocyclic scaffolds bearing imidazole rings (Fig. 2.3) have been fabricated and developed to treat different types of cancer via various targets such as DNA, mitotic spindle microtubules, VEGF, receptor tyrosine kinases, histone deacetylases, CYP26A1 enzyme, topoisomerases, rapid accelerated fibrosarcoma (RAF) kinases, etc.

2.1 Imidazole as the main nucleus in the anticancer compounds

Imidazole is such an amazing structure that it has several possibilities of substitution (Fig. 2.4).

The position and number of substituents significantly affect the pharmacological characteristics and their biological activities (Fig. 2.5).

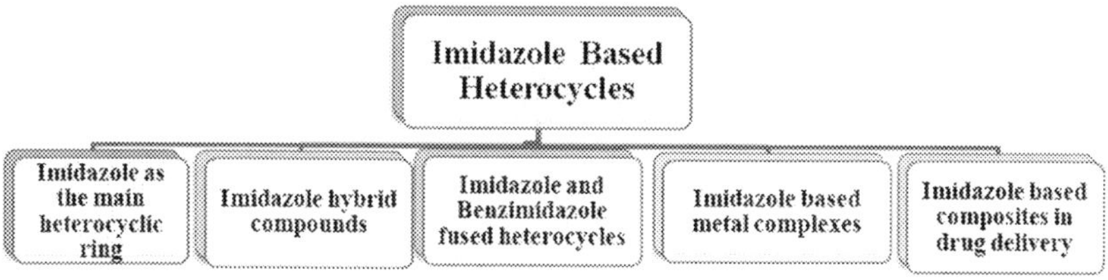

FIG. 2.3 Classification of imidazole-based heterocycles on the basis of their structural unit.

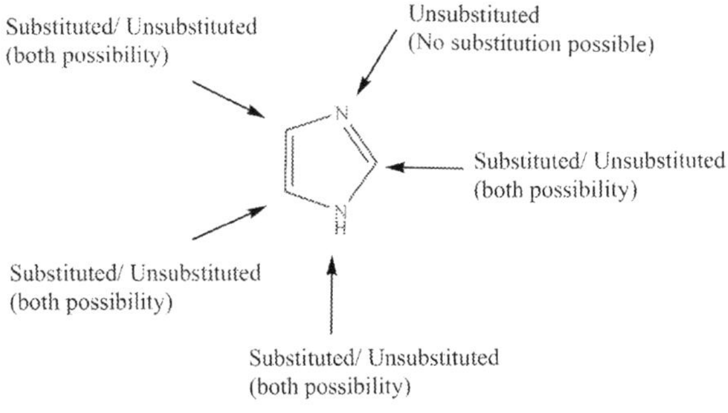

FIG. 2.4 Active sites of imidazole scaffolds.

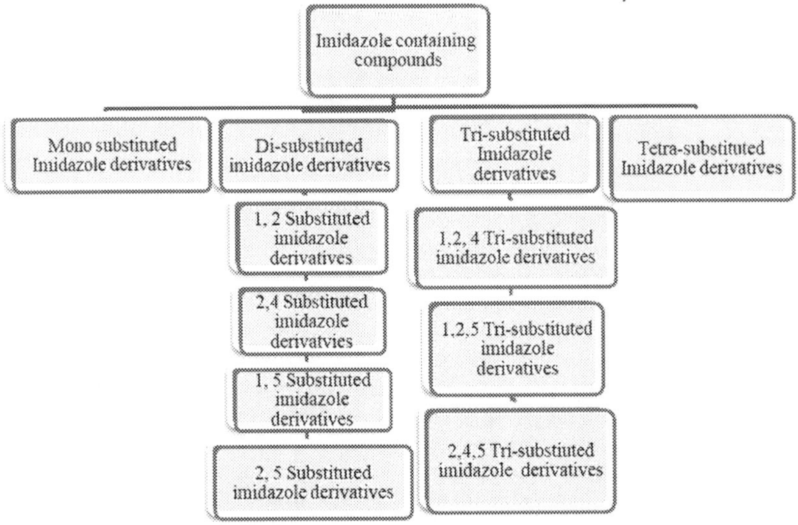

FIG. 2.5 Classification of imidazole containing heterocycles in several subsections.

2.1.1 Mono-substituted imidazole derivatives

This section sums up the various synthetic methodologies of mono-substituted imidazole derivatives and their biological profile as anticancer agents. To fabricate an anticancer drug, various targets need to be hit by the compounds. For the same, compounds are synthesized in such a way that they inhibit or bind the target efficiently. Heme oxygenase-1 (HO-1) is one such target. It is a microsomal enzyme, which gets overexpressed in the case of cancer where it offers a growth advantage and protection over chemotherapy, radiotherapy, and photodynamic healing. HO-1 plays a crucial role in carcinogenesis through its prometastatic, antiapoptotic, and proangiogenic activities. Consequently, targeted inhibition of HO-1 activity might be a novel prospective approach for anticancer treatment. In 2005, numerous N-substituted imidazole derivatives based on the structure of azalanstat (**1**) were prepared and appraised as new blockers of HO-1 (Fig. 2.6). Among the synthesized compounds, compounds (**1a**) and (**1j**) were found to be the most potent toward HO-1 [22].

Later on, structural modifications were done on azalanstat analogs (compounds of series 1) by replacing the aminothiophenols substituent with a hydrogen atom to give (**2a**, **2b**); these derivatives were then examined for their potency against heme oxygenase activity. Additional alteration involved devoid of methyl group in the fourth position of the dioxolane ring and of its dithiolane analog to give (**2c, 2d**). The determined IC_{50} values with their structures are summarized in Fig. 2.7 [23].

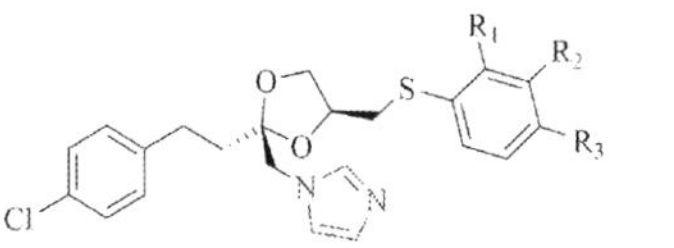

(1) R_1=R_2=H; R_3=NH_2 (Azalanstat) IC_{50} 6 ± 1 μM
(1a) R_1=R_3=H; R_2=NH_2 IC_{50} 1 ± 0.2 μM
(1b) R_3=R_2=H; R_1=NH_2 IC_{50} 5 ± 1 μM

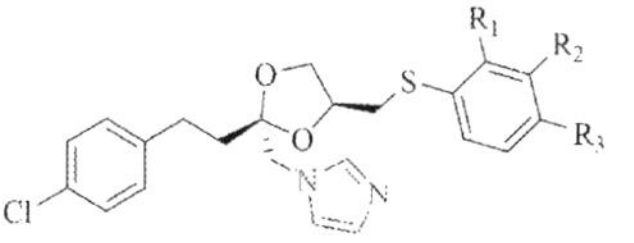

(1c) R_1=R_2=H; R_3=NH_2 IC_{50} 0.52 ± 0.03 μM
(1d) R_1=R_3=H; R_2=NH_2 IC_{50} 1.6 ± 0.7 μM
(1e) R_3=R_2=H; R_1=NH_2 IC_{50} 2.5 ± 0.1 μM

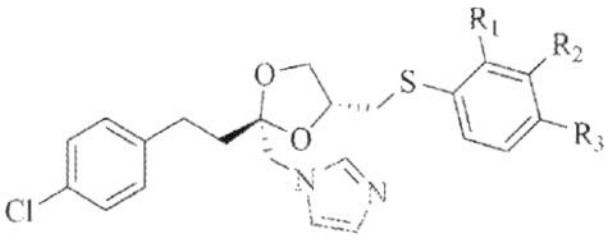

(1f) R_1=R_2=H; R_3=NH_2 IC_{50} 40 ± 8 μM
(1g) R_1=R_3=H; R_2=NH_2 IC_{50} >100 μM
(1h) R_3=R_2=H; R_1=NH_2 IC_{50} >100 μM

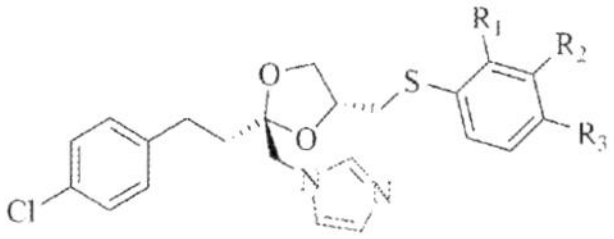

(1i) R_1=R_2=H; R_3=NH_2 IC_{50} 27 ± 9 μM
(1j) R_1=R_3=H; R_2=NH_2 IC_{50} 4 ± 2 μM
(1k) R_3=R_2=H; R_1=NH_2 IC_{50} >100 μM

FIG. 2.6 Inhibitory potency and selectivity of azalanstat (**1**) and various analogs against HO-1.

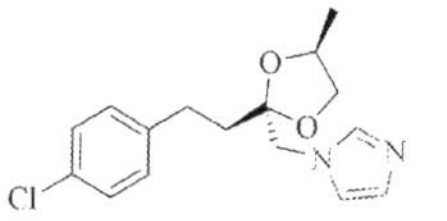

(2a) IC_{50} 0.8 ± 0.2 μM

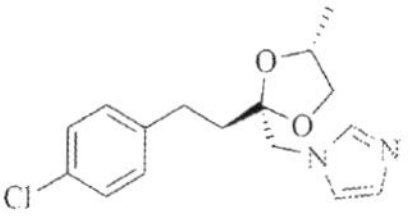

(2b) IC_{50} 2.6 ± 0.4 μM

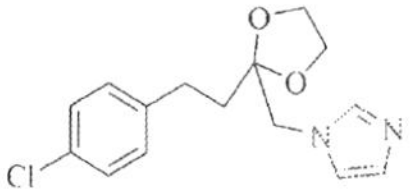

(2c) IC_{50} 4 ± 2 μM

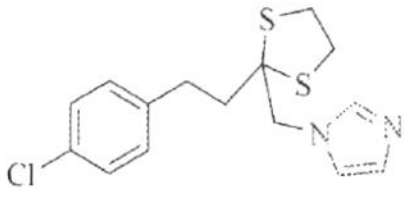

(2d) IC_{50} 4.7 ± 0.6 μM

FIG. 2.7 Selective most potent imidazole-dioxolane derivatives.

In continuation of this, imidazole ketone and analogs were synthesized using 1-aryl-2-bromoethanones with imidazole. The other dioxolane derivatives were synthesized from substituted acetophenone (Scheme 2.1). It was observed that the presence of carbonyl group in the series (**3**) showed good inhibition against HO-1. Among the third series compounds, compounds with R_2 and R_3=H and R_1 with 3,4-dichlorophenyl were found to be the most potent with IC_{50} (1.24 ± 0.05 μM). Among the fourth series compounds (**4**), the naphthalen-2-yl substituted derivative was the most potent, having IC_{50} (2.63 ± 0.04 μM). In addition, compound (**7**) was the most potent one with IC_{50} (1.19 ± 0.02 μM) in comparison to other structural analogs (**5**, **6**, **8**, **9**) (Scheme 2.1) [24].

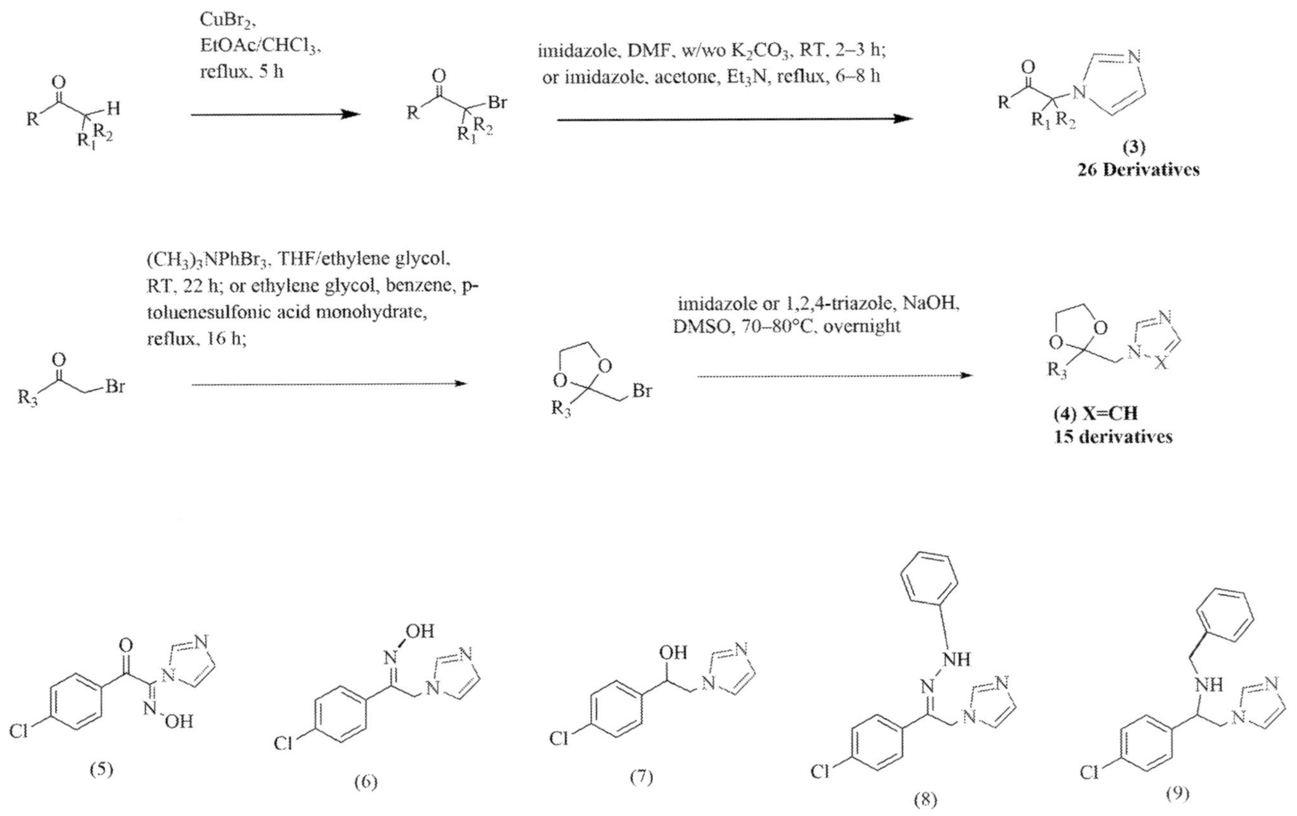

SCHEME 2.1 Synthetic route of imidazole-based ketone and dioxolane derivatives.

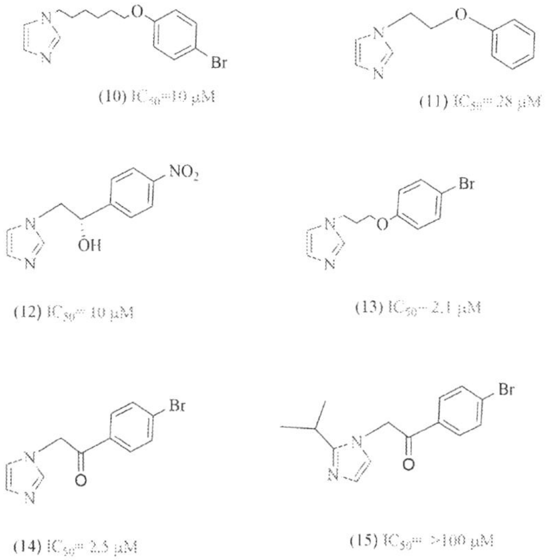

FIG. 2.8 Imidazole derivatives with their IC_{50} values.

In 2012, different imidazole derivatives were examined for their potency against HO-1. For this, an in vitro assay was used and HO-1 was isolated from rat spleen as the microsomal fractions, synthesized by a centrifugation method. The IC_{50} of the synthesized derivatives with their structure is presented in Fig. 2.8. The synthesized derivatives were docked against 3CZY protein and it was interesting to note that the compounds (**10–15**) docked in the active site of Met34, Val50, Phe37, Leu54, Gly139, Arg136, Ser142, Asp140, Leu147, Gly143, Phe166, Phe162, Phe214. Compound (**13**) bearing a 3-bromophenoxybutyl chain was a potent inhibitor (IC_{50}=2.1 μM). Furthermore, product (**13**) formed a complex with heme bound to recombinant HO-1 at low micromolar concentrations [25].

The CYP26A1 enzyme is induced by all-trans-retinoic acid (ATRA) in the liver and other target tissues. Inhibition of the CYP26A1 metabolic enzyme acts as an important approach for finding new anticancer agents. In 2015, amide imidazole derivatives were synthesized and examined for inhibitory activity toward the CYP26A1 enzyme. Liarozole (0.89 μM) was used as a control (Scheme 2.2). All the synthesized compounds exhibited IC_{50} values ranging from 0.22 to 1.11 μM. The in silico studies revealed that the synthesized derivatives showed crucial interactions with the CYP26A1 enzyme. The aromatic main chain created a sequence of interactions with the neighboring amino acid residues (Lys436 and Pro332). The N atom of the imidazole ring made a coordinate bond with Fe^{2+} of the heme group. The flexible ester side chain fitted well into the other pocket. To estimate the in vitro cell growth inhibition and differentiation for CYP26A1, the synthesized compounds (**16–18**) were given with a combination of ATRA in HL60 cells. Among all the derivatives, the increase in inhibition and differentiation inducing actions of compounds (**16a**), (**16d**), (**17a**), (**17b**), (**18a**), and (**18d**) were appreciably superior with respect to liarozole [26].

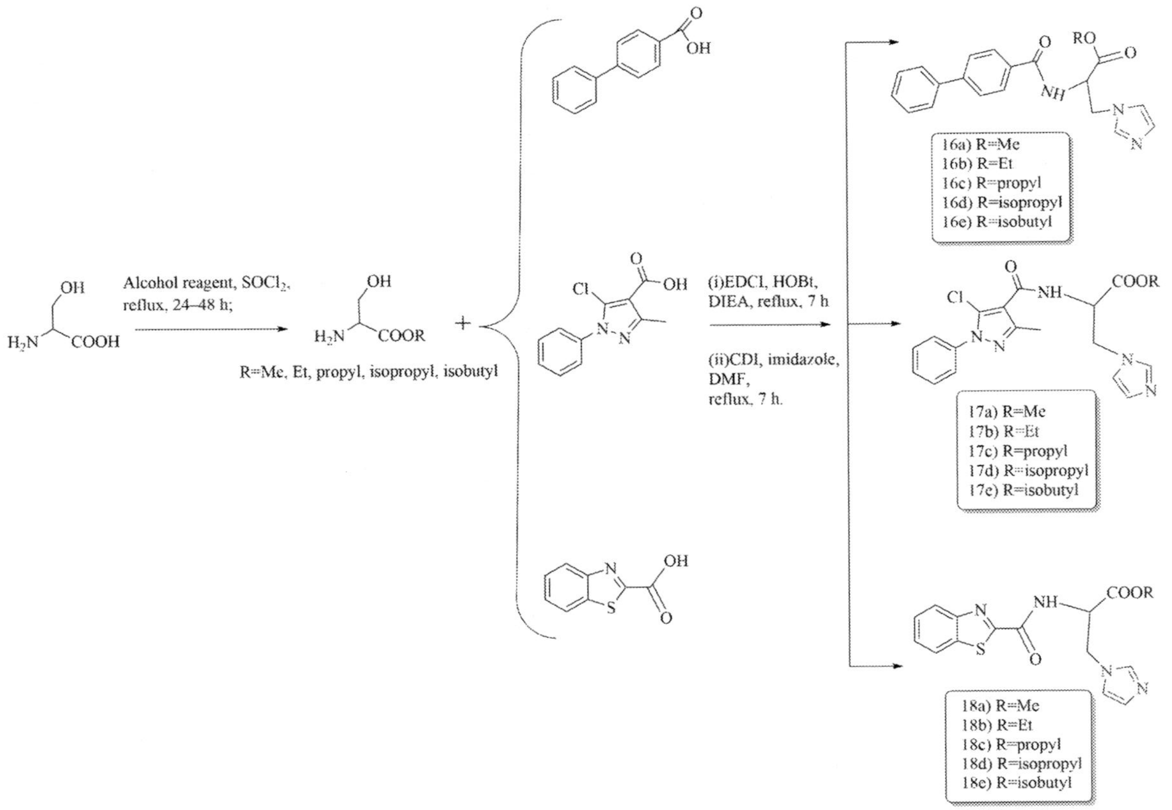

SCHEME 2.2 Synthetic routes of amide imidazole derivatives.

2.1.2 Disubstituted imidazole derivatives

The medicinal use of imidazole-based compounds was introduced in the mid-1990s. Stolinsky et al. [27] demonstrated the effect of 5-(3,3-dimethyl-1-triazeno) imidazole-4-carboxamide (NSC-45388) (**19**) (Fig. 2.9) with a hexamethylmelamine (HMM) drug on 30 cancer patients. They were suffering from lung carcinoma and squamous cervix carcinoma. Among the 30 patients, the responses of only 25 patients were noticed. It was found that 15 patients did not show any positive effect, three patients showed interruption of progression of disease for 1 month or more, without objective improvement, four patients showed objective improvement of 25%–50% for 1 month or more, with subjective improvement, and two patients with objective improvement greater than 50% for 1 month or more, with subjective improvement. Only one patient completed the objective and subjective remission for 1 month or more; regression of all manifestations of disease. Later on, scientists studied the effect of this drug in combination with other drugs for the treatment of solid tumors in children [28, 29].

A series of imidazole-4,5-dicarboxamide derivatives (**20**) (Fig. 2.10) was synthesized and examined for antineoplastic activity against HL-60 cell lines [30]. The fabrication of the imidazole derivatives was done on the basis of the skeleton of tri-substituted purine complexes with cyclin-dependent kinase 2 (**CDK2**). As per the results, among the synthesized compounds, most of them inhibited proliferation of HL-60 cells as measured by MTS mitochondrial functional assay with IC_{50} values in the range of 2.5–25 μM. According to the structure activity relationship, the derivatives were found consistent with predictable hydrogen bonding interactions in the ATP-binding site of **CDK2**. The IC_{50} values for the selected compounds ($\leq$ 10 μM) are illustrated in Table 2.1.

A series of imidazole-based compounds was synthesized by the structural modification in 4-substituted methoxylbenzoyl-aryl-thiazoles and 2-aryl-imidazole-4-carboxylic amide derivatives. In the first, the thiazole ring was replaced with the imidazole ring; in the latter case, the amide group was replaced with carbonyl and sulphonyl groups (**21a–c**) (Fig. 2.11). (This replacement increased water solubility and made the derivatives effective against multidrug-resistant cancer cells. The synthesized compounds were examined against different type of cancer cell lines, viz. human prostate cancer cell lines (PPC1, PC-3, LNCaP, Du 145) and three other cell lines (A375, WM-164, B16-F1) [31]. The IC_{50} values of the selective potent compounds are summarized in Table 2.2.

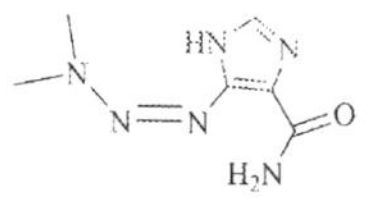

(19)

FIG. 2.9 Structure of 5-(3,3-dimethyl-1-triazeno)imidazole-4-carboxamide.

TABLE 2.1 The IC_{50} values for the selective imidazole-4,5-dicarboxamide derivatives.

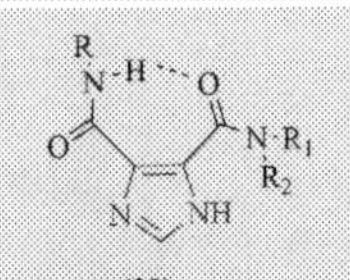

(20)

FIG. 2.10 Imidazole derivatives.

R_1	R_2	R	HL-60 cells (IC_{50}) (μM) Day 2	HL-60 cells (IC_{50}) (μM) Day 4
CH_2-4-Ph-Ph	H	3,4-$Cl_2C_6H_3$	7.0±0.4	2.9±0.3
(*R*)-$CHCH_3Ph$	H	3,4-$Cl_2C_6H_3$	8.8±0.3	5.9±0.6
(*S*)-$CHCH_3Ph$	H	3,4-$Cl_2C_6H_3$	9.2±0.4	5.9±0.6
(*R*)-$CHCH_3CO_2C(CH_3)_3$	H	3,4-$Cl_2C_6H_3$	4.6±0.5	2.5±0.24
(*S*)-$CHCH_3CO_2C(CH_3)_3$	H	3,4-$Cl_2C_6H_3$	4.9±0.3	2.4±0.1
(*R*)-$CHCH_3CO_2C(CH_3)_3$	H	3-ClC_6H_4	8.2±0.7	2.9±0.3
(*R*)-$CHCH_3CO_2C(CH_3)_3$	H	4-ClC_6H_4	6.6±0.5	3.3±0.3
(*R*)-$CHCH_3CO_2C(CH_3)_3$	H	4-$CH_3C_6H_4$	6.3±0.4	3.5±0.3
(*R*)-$CHCH_3CO_2C(CH_3)_3$	H	4-$OCH_3C_6H_4$	6.0±0.5	3.1±0.3
(*R*)-$CHCH_3CO_2C(CH_3)_3$	H	4-$CO_2CH_2CH_3$-C_6H_4	6.3±0.5	3.7±0.4
(*R*)-$CHCH_3CO_2C(CH_3)_3$	H	C_6H_5	4.7±0.4	2.7±0.1

The synthesized compounds possessed good aqueous solubility relative to standard drug combretastatin A4 and paclitaxel. The binding interactions and tubulin polymerization suggested that compounds bound within colchicine binding sites. The molecular docking studies on ABT-751 and DAMA colchicine suggested that the imidazole proton was near and probable to form H-bonds to Thr179 in T5 loop (residues 173–182) of the tubulin α-monomer. The substituent fluorine at the para position of the carbonyl phenyl ring caused interaction Cys241 in T7 and Tyr202 in S6, possibly forming one or two hydrogen bonds. The CO group of some substituents was in adequate nearness to form two H-bond interactions with the NH of Leu-252 in H8 and the side chain of Asp-251 in T7 of the tubulin β-monomer.

In lieu of their previous research, in 2011, Chen and coworkers [32] demonstrated an efficient diversely substituted synthesis of 2-aryl-4-benzoyl-imidazole and also examined their tubulin polymerization potency. The anticancer activity was measured against similar cancer cell lines. The presence of a bulky group in the phenyl ring and modification in N1 decreased the cytotoxicity of the compounds. The IC_{50} values are illustrated in Table 2.3. These compounds showed

TABLE 2.2 The IC_{50} values and structure of selective potent compounds.

		$IC_{50} \pm SEM$ (μM)						
Structure	R	A375	B16-F1	WM164	LnCaP	PC-3	Du145	PPC-1
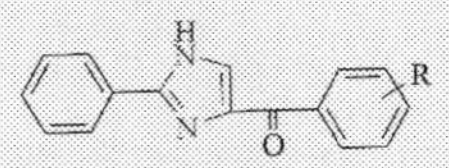 (21a-c) FIG. 2.11 Imidazole derivatives.	3,4,5-$(OMe)_3$	0.16 ± 0.02	0.12 ± 0.01	0.10 ± 0.01	0.15 ± 0.02	0.29 ± 0.04	0.20 ± 0.03	0.13 ± 0.01
	3,5-$(OMe)_2$	2.8 ± 0.5	5.4 ± 0.8	2.1 ± 0.4	3.6 ± 0.3	3.2 ± 0.5	2.6 ± 0.3	2.1 ± 0.2
	4-F	0.58 ± 0.07	0.93 ± 0.1	0.63 ± 0.09	0.61 ± 0.08	2.1 ± 0.03	0.85 ± 0.1	0.57 ± 0.0

TABLE 2.3 The IC_{50} values and structure of selective potent imidazole derivatives.

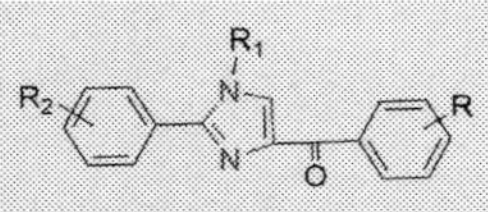

22(a-f)

FIG. 2.12 Imidazole derivatives.

R, R_1, R_2	A375	WM164	LnCaP	PC-3	Du145	PPC-1
IC50 ± SEM (μM)						
$R_1 = H$ $R_2 = 4\text{-}Me$ $R = 3,5\text{-}(OMe)_2\text{-}4\text{-}OH$	0.11 ± 0.02	0.12 ± 0.02	0.13 ± 0.04	0.13 ± 0.08	0.17 ± 0.07	0.11 ± 0.02
$R_1 = H$ $R_2 = 4\text{-}Cl$ $R = 3,5\text{-}(OMe)_2\text{-}4\text{-}OH$	0.68 ± 0.15	0.50 ± 0.06	0.55 ± 0.01	0.30 ± 0.08	1.00 ± 0.21	0.58 ± 0.02
$R_1 = Bn$ $R_2 = 4\text{-}Me$ $R = 3,4,\ 5\text{-}(OMe)_3$	1.06 ± 0.22	1.17 ± 0.35	0.68 ± 0.02	0.47 ± 0.04	1.50 ± 0.52	0.78 ± 0.01
$R_1 = Me$ $R_2 = 4\text{-}Me$ $R = 3,4,5\text{-}(OMe)_3$	0.03 ± 0.01	0.04 ± 0.02	0.03 ± 0.01	0.04 ± 0.02	0.03 ± 0.01	0.03 ± 0.01
$R_1 = 4\text{-}OMePhSO_2$ $R = 4\text{-}F$ $R_2 = 4\text{-}NMe_2$	0.10 ± 0.02	0.10 ± 0.01	0.07 ± 0.02	0.04 ± 0.01	0.13 ± 0.01	0.06 ± 0.01
$R_1 = H$ $R_2 = 4\text{-}Br$ $R = 3,4,5\text{-}(OMe)_3$	0.03 ± 0.01	0.04 ± 0.01	0.03 ± 0.01	0.04 ± 0.01	0.05 ± 0.02	0.03 ± 0.01

improved aqueous solubility, oral bioavailability, and pharmacokinetic properties. Methylation of the imidazole ring increased its oral bioavailability. The SAR study revealed that the previous derivatives [30] were more potent compared to these derivatives (**22a–f**) (Fig. 2.12).

In the subsequent year, the authors designed novel 2-aryl-4-benzoyl-imidazole derivatives (**23a–d**) (Fig. 2.13). These analogs were also targeted on tubulin polymerization and they acted as anticancer agents. An in vitro assessment was carried out to clarify that these compounds are not substrates of P-glycoprotein and therefore may efficiently defeat P-glycoprotein-mediated multidrug resistance. The synthesis of compounds was a multistep process that included indole-3-carboxaldehyde as a precursor. The anticancer activity of these products was estimated in two human metastatic melanoma cell lines (WM164 and A375) and three human prostate cancer cell lines (PC-3, Du 145, and LNCaP) as shown in Table 2.4. The mechanism of action was same as that of previously synthesized compounds [33].

In 2013, Xiao et al. [34] designed and synthesized novel Reverse ABI (RABI) derivatives (**25a–i**) and (**26a–h**) using previously synthesized 2-aryl-4-benzoyl imidazole (ABI) analogs (**24a–e**) (Fig. 2.14). The SAR studies were done by incorporating varied substituents in A, B, and C rings. The antiproliferative activity of the synthesized compounds was studied against eight cancer cell

TABLE 2.4 The IC_{50} values and structure of selective potent compounds.

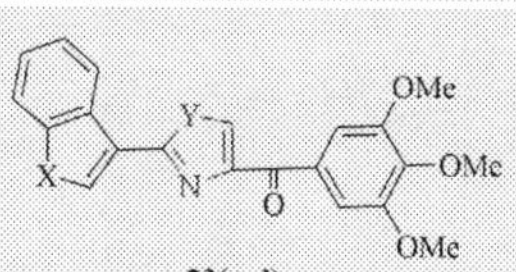

23(a-d)

FIG. 2.13 Imidazole derivatives.

	A375	WM164	LnCaP	PC-3	Du145
X, Y	IC_{50} ± SEM (μM)				
X=O Y=NH	53.4±12.5	73.6±15.2	14.1±6.2	33.3±2.1	45.0±9.6
X=S Y=NH	41.0±5.2	45.7±9.3	29.1±3.9	20.8±0.5	51.2±3.5
X=NH Y=NH	3.2±1.2	5.3±2.0	2.8±0.6	3.7±0.3	3.9±1.0
X=N-CH_3 Y=N-CH_3	18.8±3.6	24.2±4.1	35.9±3.3	40.7±5.1	76.5±4.1
Colchicine	20.6±3.6	29.0±5.2	16.3±4.0	11.5±0.1	11.2±1.1

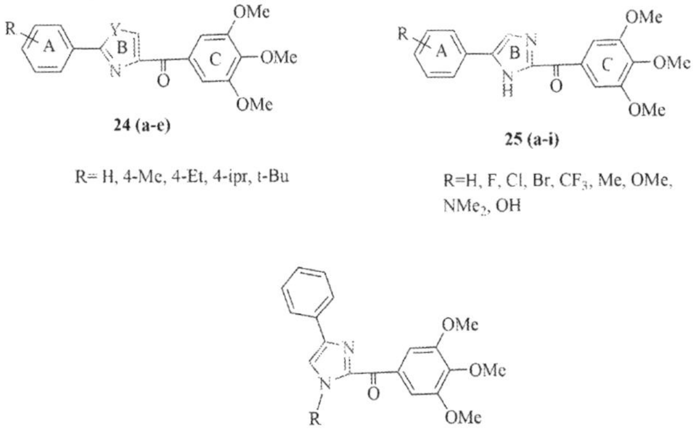

FIG. 2.14 Reverse analogs of 2-aryl-4-benzoyl imidazole.

lines. Furthermore, the cytotoxicity of the synthesized compounds was studied in human melanoma and prostate cancer cell lines. Vinblastine, colchicine, and docetaxel were used as positive controls. The in vitro studies showed that most of the synthesized derivatives were found potent with IC_{50} values in low nM range. Compounds (**25a**) and (**26g**) were found to be the most potent antiproliferative agents among all the synthesized ones. The mechanistic action studies showed the ability of RABI analogs to inhibit tubulin polymerization at the binding site of colchicine and arresting cells in the G_2/M phase.

In recent years, several steroidal imidazole derivatives (Fig. 2.15) were tested as dual AR/CYP17 ligands. Some compounds exhibited good biological activity in both cellular and enzymatic assays. Structure activity relationship studies revealed that commencing the oximino group at the C-3 position of steroidal moiety is helpful for the enrichment of AR antagonistic activity. Most of the derivatives of this series displayed significant CYP17 inhibitory activity in the concentration range of 20 μM compared with the positive control, abiraterone. Among these derivatives, the most effective derivative was (**27a**), which showed the best AR inhibition with $IC_{50}=0.5\,\mu M$; this was 27-fold amplified compared to the similar CYP17 inhibition ($IC_{50}=11\,\mu M$) (Table 2.5). In addition, compound (**27a**) also demonstrated promising anticancer effects on LNCap cell lines with an IC_{50} value of 23 μM that was higher in comparison to positive control flutamide ($IC_{50}=28\,\mu M$) [35].

2.1.3 Trisubstituted imidazole derivatives

A number of studies have been conducted for the synthesis and appraisal of the antitumor characteristics of structural analogs of combretastatin A-4 (CA-4). It is a potent antimitotic agent extracted from the stem wood of the South African tree *Combretum caffrum* [36]. Numerous remarkable review articles

TABLE 2.5 Activity of synthesized compounds against CYP17 inhibitory, AR antagonistic, and antiproliferative effects on LNCap cell lines.

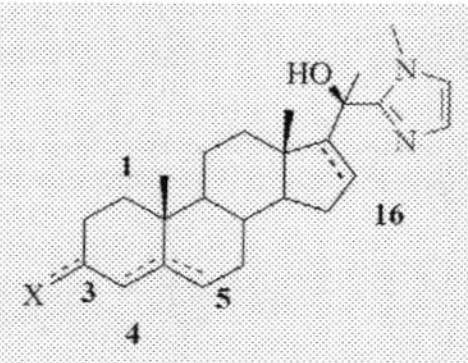

(27)

FIG. 2.15 Steroidal imidazole derivatives.

Compound	X	Double bond in the ring	CYP17 inhibitory activity (%) at μM	IC_{50} values against LNCap cell line (μM)	AR antagonistic activity (%) at 20 μM	IC_{50} (μM)
13a	-OMOM	Δ^{5}	27%	> 80	< 10%	ND
13b	-OMOM	$\Delta^{5,16}$	–	34	< 10%	ND
13c	-OH	Δ^{5}	53%	12	50%	ND
13d	-OH	$\Delta^{5,16}$	61%	> 80	0.2%	ND
13e	= O	Δ^{4}	52%	23	91%	ND
13f	= O	$\Delta^{4,16}$	63%	> 80	37%	ND
13g	= N-OH	Δ^{4}	72%	23	99%	0.5
13h	-OH	Δ^{4}	86%	46	68.3%	3
13i	-OH	$\Delta^{4,16}$	87%	67	20%	ND
Flutamide	–	–	–	28	82%	0.7%

Note: *MOM*, methoxymethyl group; *ND*, not determined.

have confirmed the huge interest shown by researchers in this field [37–39]. Novel 5-membered heterocyclic conjugated combretastatin A-4 derivatives were designed by substituted aromatic aldehydes and substituted aliphatic or anilines using a couple of reagents [40]. The potency of the products was measured against two human cancer cell lines, HCT-15 (MDR +) and NCI-H460 (MDR −), and the inhibition activity on tubulin polymerization was considered. The in vivo anticancer activity was estimated in rats, mice, monkeys, and dogs. The in vivo antitumor activity against murine M5076 reticulum sarcoma was evaluated in mice. The potent derivatives (**28h**, **29f**, **29g**) are illustrated below with their IC_{50} values. The results of the in vivo experiments confirmed that the structural analogs of combretastatin A-4 showed potent antitumor activity. Assessment of compounds (**28h**) and (**29f**) (Fig. 2.16) against murine M5076 reticulum sarcoma in mice indicated that both derivatives were orally efficient with an increasing life span of 38.5% and 40.5%, correspondingly.

Kim and coworkers [41] demonstrated the synthesis of a series of 4(5)-(6-alkylpyridin-2-yl)imidazole compounds (**30**), as shown in Fig. 2.17. To examine the human ALK5 kinase activity, SF9 cells were used. Among the synthesized compounds, the derivatives having quinoxalin-6-yl, methyl and 4-CN substituent showed highest activity and has IC_{50} 0.012 μM. As per the results, the presence of alkyl group at the sixth position of the pyridine ring also affected the potency of compounds. The methyl-substituted derivatives exhibited equal or slightly more cytotoxicity than the related 6-ethylpyridyl derivatives. The position of N atom and bulkiness of substituent significantly affected the influenced ALK5 inhibition. The meta-substituted analogs were more cytotoxic than the para-substituted derivatives. To estimate TGF-β-induced transcriptional activation to ALK5 signaling, HaCaT cell-based luciferase activity was measured using three diverse luciferase receptor genes, viz. p3TP-luciferase, SBE luciferase, and ARE-luciferase, at a concentration of 0.1 μM. The other derivatives, 3- and 4-{[4-(benzo[1,3]dioxol-5-yl)-5-(6-methylpyridin-2-yl)-1*H*-imidazol-2-ylamino]methyl}benzonitrile (**31**) (Fig. 2.18), were also examined for luciferase activity. The IC_{50} values and luciferase activity of all the compounds are provided in Table 2.6.

In 2010, 2-amino-1-arylidenaminoimidazole derivatives were prepared as shown in Scheme 2.3 and (Fig. 2.19). The derivatives were designed from a

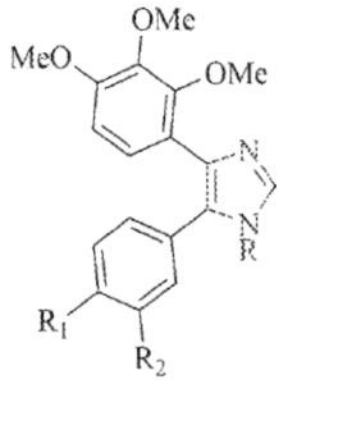

28h R=Me, R_1=R_2=CHCHN(CH_3)
IC_{50} = 20 nM for HCT-15; 32 nM for NCI-H460

29f R=CH_3, R_1= OMe, R_2=NH_2
IC_{50} = 220 nM for HCT-15; 170 nM for NCI-H460

29g R=H, R_1=R_2=CHCHN(CH_3)
IC_{50} = 1.7 nM for HCT-15; 1.8 nM for NCI-H460

FIG. 2.16 Tri-substituted imidazole derivatives.

TABLE 2.6 Inhibitory activity of 4(5)-(6-alkylpyridin-2-yl)imidazoles derivatives on ALK5.

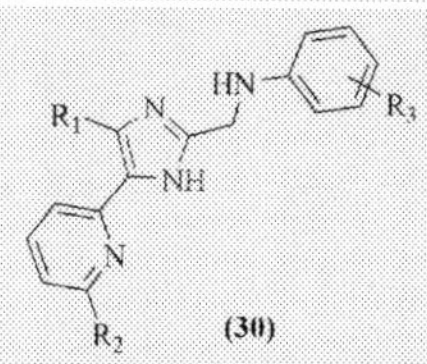

(30)

FIG. 2.17 Imidazole derivatives.

				Luciferase activity (% control)		
R_1	R_2	R_3	IC_{50} (μM)	p3TP-luc	SBE-luc	ARE-luc
[B]	Me	3-CN	0.046	14±6	36±4	62±6
[B]	Me	4-CN	0.174	43±3	35±5	123±26
[B]	Et	3-CN	0.073	21±4	33±6	75±6
[B]	Et	4-CN	0.360	67±12	99±12	136±17
[Q]	Me	3-CN	0.012	7±1	25±9	13±1
[Q]	Me	4-CN	0.042	16±3	23±7	55±6
[Q]	Et	3-CN	0.119	33±7	62±24	79±8
[Q]	Et	4-CN	0.047	32±9	40±7	61±4
[B]	Me	3-$CONH_2$	0.021	28±10	54±18	79±11
[B]	Me	4-$CONH_2$	0.0262	55±28	67±2	119±12
[B]	Et	3-$CONH_2$	0.046	30±2	59±9	64±3
[B]	Et	4-$CONH_2$	0.385	125	96	133±11
[Q]	Me	3-$CONH_2$	0.010	63±14	95±33	99±5
[Q]	Me	4-$CONH_2$	0.075	84±8	101±40	127±12
[Q]	Et	3-$CONH_2$	0.015	67±13	95±23	96±10
[Q]	Et	4-$CONH_2$	0.092	104±7	112±23	98±8

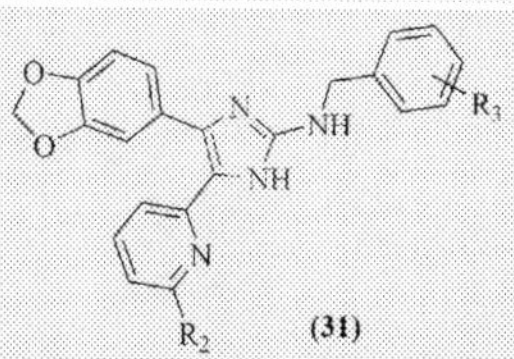

(31)

FIG. 2.18 Imidazole derivatives.

				Luciferase activity (% control)		
R	R_2	R_3	IC_{50} (μM)	p3TP-luc	SBE-luc	ARE-luc
–	Me	4-CN	0.024	33±8	52±18	95±10
–	Me	3-CN	0.153	61±13	66±9	129±15

Notes: [B] represents benzo[1,3]dioxol-5-yl. [Q] represents quinoxalin-6-yl.

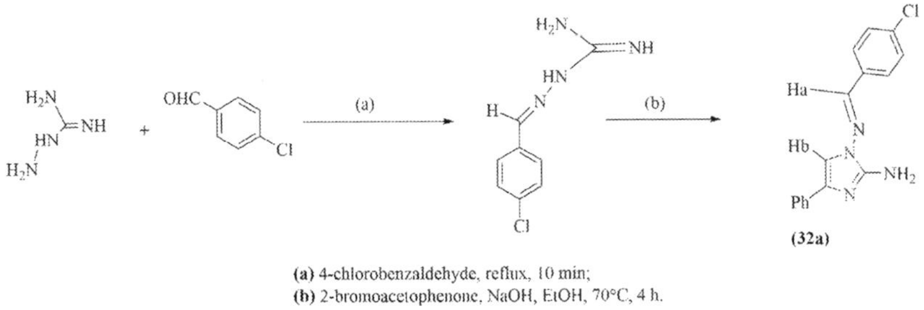

SCHEME 2.3 Synthetic route for the preparation of 2-amino-1-arylidenaminoimidazoles.

hit compound, identified by cancer cell-based high throughput screening assay. Tubulin and microtubule inhibition caused apoptosis of tumor cells, interrupted in various cellular functions, together with cell motility and mitosis. The mechanism of action of these synthesized derivatives showed that it inhibited tubulin polymerization and thus possessed anticancer activity. All the designed compounds were examined for their anticancer activity against NUGC-3 cells (human gastric cancer cells) by MTS and PTS assay system. SAR studies revealed that the presence of electron-withdrawing group increased the cytotoxicity (F > Cl > Br) as well as position of substituents also affected the cytotoxicity. The substituent at the fourth position imparted more cytotoxicity than the second position. Among the synthesized compounds, compounds (**32a**) and (**32e**) were examined for tubulin polymerization and the IC_{50} values are summarized in Table 2.7. Compounds (**32a**) and (**32b**) were examined for antitumor activity

TABLE 2.7 The IC_{50} values of selective potent compounds against human gastric cancer NUGC-3 cells.

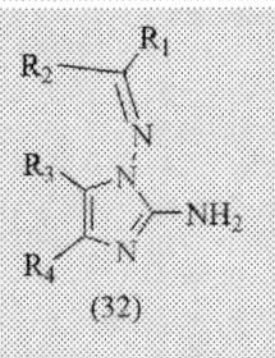

FIG. 2.19 2-Amino-1-arylidenaminoimidazoles.

ID	R_1	R_2	R_3	R_4	IC_{50} (μM)
32a	4-ClC_6H_4	H	H	C_6H_5	0.06
32b	4-FC_6H_4	H	H	C_6H_5	0.05
32c	4-ClC_6H_4	H	H	3-Pyridyl	0.23
32d	4-FC_6H_4	H	H	3-Pyridyl	0.06
32e	4-ClC_6H_4	H	H	2-Thiophenyl	0.08

against MKN-45 and colorectal SW-480 cells by the in vivo method. The compound (**32a**) showed both po and intravenously dose-dependent activities and increased the life spans of the murine leukemic P388 cells-inoculated mice [42].

Also in 2010, imidazole derivatives (**33**) (Fig. 2.20) were devised and their efficacy was examined in the fabrication of DFG-out allosteric B-Raf inhibitor [43]. B-Raf is a serine/threonine-protein involved in the process of cell growth and differentiation. Its mutation occurred in numerous types of cancer, thereby causing cell proliferation due to variation in the ERK/MAPK-signal cascade. It affected the composition of the tumor micro-environment and altered both solid and soluble mediators. The B-Raf V600E mutation corresponds to a frequent oncogenic kinase mutation which was responsible for augmented kinase activity in about 7% of all human cancers, and thus B-Raf was found to be a vital curative target for inhibition. The imidazole derivatives were synthesized by a [3+2] cycloaddition reaction followed by conjugation with quinazoline (Scheme 2.4).

The DFG loop has four interactions that are important to bind it. The first interaction was related to the central core that linked the hydrogen-bond network between a conserved glutamate side chain and the amide N–H from the aspartate included in the movement of the DFG loop. The second approach of interaction was at the selectivity site that was formed when the phenylalanine of the DFG loop vacated its lipophilic pocket and this was occupied by the CF_3 group. The next interaction was the gatekeeper region, which was near to a preserved lysine side chain that was usually included in triphosphate binding. The last interaction site was the "hinge region," which in general bonded with the flat aromatic adenine ring of ATP and established crucial hydrogen bonds with both "ATP site" and "allosteric site" inhibitors. The different binding sites are shown in Scheme 2.4, and IC_{50} values for selective potent compounds are summarized in Table 2.8.

COX-2 inhibitors play an important role in apoptosis and carcinogenesis. In 2012, 4,5-bisaryl imidazolyl imidazole derivatives (**34**) were designed and screened for anticancer activity on Caco-2 colorectal cancerous cell lines by MTT assay [44]. The effect of synthesized compounds was evaluated on gene expression pattern of 112 genes using DNA microarray and clustering analysis. Bcl-2, Bax, and caspase-3 mRNA expression and their relationship were studied by quantitative real-time PCR. The derivatives were synthesized using microwave-assisted organic synthesis (MAOS) at 360W and 85°C with a mixture of 1-(4-fluorobenzyl)-2-(methylthio)-1*H*-imidazol-5-carbaldehyde, benzyl, and ammonium acetate (Scheme 2.5). The studies showed that the effect of 4,5-bisaryl imidazolyl imidazole derivatives on Caco-2 cells were time- and concentration-dependent. Increase in levels of Bax and caspase-3 mRNA and no change in the levels of *Bcl-2* mRNA expression showed that a direct relationship exists between Bax/Bcl-2 ratio and caspase-3 gene expression in Caco-2 cells when treated with COX-2 inhibitors. In vitro growth inhibitory concentration (GI_{50}) of Caco-2 cells is summarized in Table 2.9.

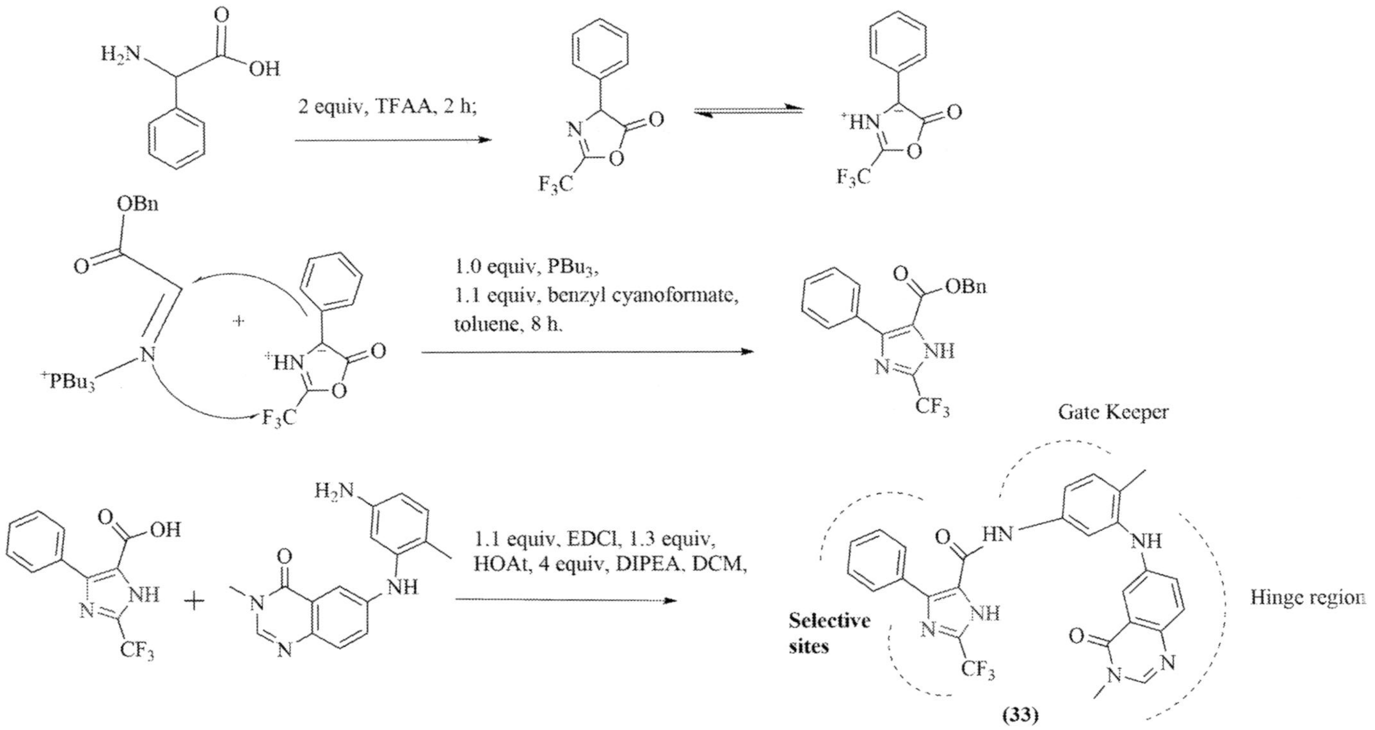

SCHEME 2.4 [3+2] Cycloaddition reaction for the synthesis of quinoline imidazole derivatives.

TABLE 2.8 The IC_{50} values of selective quinoline imidazole derivatives.

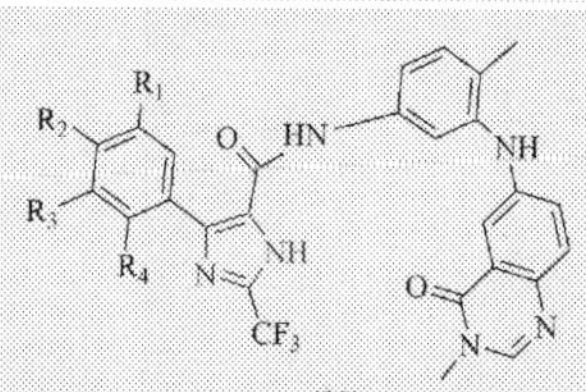

(33)

FIG. 2.20 Imidazole derivatives.

ID	R_1	R_2	R_3	R_4	IC_{50} B-Raf V600E (nM)	IC_{50} P38a (nM)
33a	H	H	H	H	21.4	405.2
33b	H	Me	H	H	27.9	1648.0
33c	H	F	H	H	0.4	16.0
33d	Cl	F	H	H	53.2	48.8
33e	CH_3	F	H	H	19.7	60.1
33f	F	Me	H	H	0.3	100.2
33g	CF_3	H	H	H	49.6	105.4
33h	F	F	H	H	0.3	11.0
Sorafenib	–	–	–	–	120.1	84.2

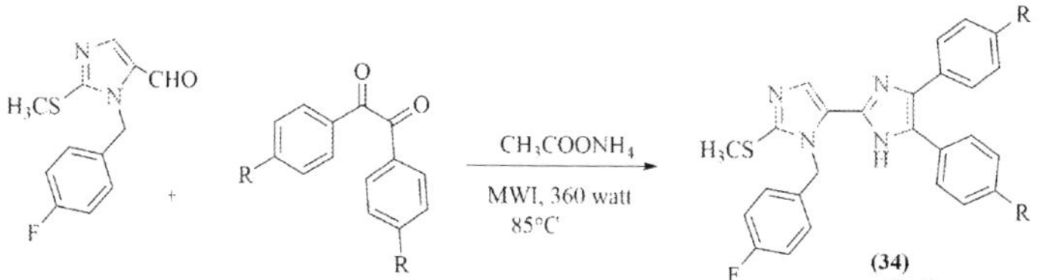

(34)
a) R=F
b) R=H
c) R=OMe

SCHEME 2.5 Synthesis of 2,4,5 tri-substituted imidazole.

Also in 2012, several 2,4,5-triaryl-1*H*-imidazole derivatives were synthesized and evaluated for their antiproliferative activity profile against a panel of five cell lines, namely three nonsmall cell lung cancers (NSCLC) (H460, H1299, and A549), one breast cancer (MCF-7), and one normal diploid embryonic lung cell line (MRC-5), using XTT assay [45]. The activity results

TABLE 2.9 In vitro growth inhibitory concentration (GI_{50}) of Caco-2 cells, caspase-3 mRNA fold change, and Bax/Bcl-2 mRNA expression.

Compound	GI_{50} (μM)	Caspase-3 mRNA fold change	Bax/Bcl-2 mRNA expression
34a	79.61 ± 1.870	5.2 ± 0.35	2.21
34b	29.83 ± 1.465	1.96 ± 0.27	2.17
34c	21.20 ± 0.729	8.17 ± 0.81	4.42
Celecoxib	23.85 ± 1.107	7.64 ± 0.93	4.09

showed that compounds (**35f**) and (**35g**) were highly potent against H1299 with < 0.1 μM IC_{50} value compared to topotecan (IC_{50} > 10.0 μM). Moreover, compounds (**35f**) and (**35g**) showed only marginal cytotoxicity against the cell lines, H460, A549, MCF-7, and MRC-5. It was also found by flow cytometric analysis that compound (**35f**) induced H1299 cell cycle arrest at G0/G1 through the inactivation of p38 MAPK, JNK, ERK kinases, SIRT1, and survivin pathways. The compounds (**36c**) and (**36d**) were active against the four cancer cell lines (H1299, A549, MCF7, and MRC-5) with IC_{50} value < 1 μM. The SAR studies proved that the presence of flexible alkylamine moiety on 2,4,5-triaryl-1*H*-imidazole pharmacophore developed novel compounds and showed selective activity against the growth of NSCLC cell lines (H1299) (Fig. 2.21).

In the following year, 4-aryl-5-(4-(methylsulfonyl)phenyl)-2-alkylthio and 2-alkylsulfonyl-1*H*-imidazole derivatives were prepared (Scheme 2.6) and their COX-1 and COX-2 inhibitory potency and selectivity in human whole blood assay was determined [46]. 2-Sulfonylalkyl derivatives (**39**, **40**) were found to be less potent and selective than 2-alkylthio derivatives (**37**, **38**). The potencies (IC_{50} values) of the synthesized compounds were determined and compared to those of the reference molecules SC-560 as a selective COX-1 inhibitor and DuP-697 and celecoxib as selective COX-2 inhibitors. Compounds

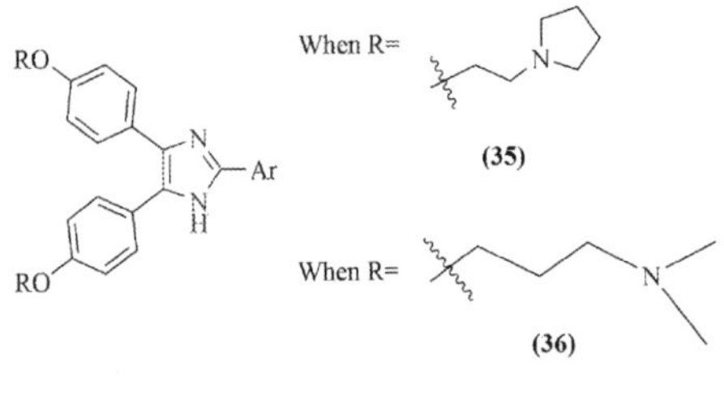

Ar = a) Phenyl b) 4-OMePhenyl c) 4-BrPhenyl d) 4-CF_3Phenyl e) furfural f) 5-BrFurfural g) 5-NO_2 Furfural

FIG. 2.21 Imidazole derivatives.

(**37a** $IC_{50}=0.06\,\mu M$) and (**38a** $IC_{50}=0.05\,\mu M$) were found to be the most potent and comparable to the standard compounds. Molecular modeling studies showed that the residence time of ligands in the active site of 2-sulfonylalkyl and 2-alkylthio analogs increased due to the interaction of the 2-sulfonylalkyl group with Arg120 in COX-1 and an extra hydrogen bond with Tyr341 in COX-2, respectively.

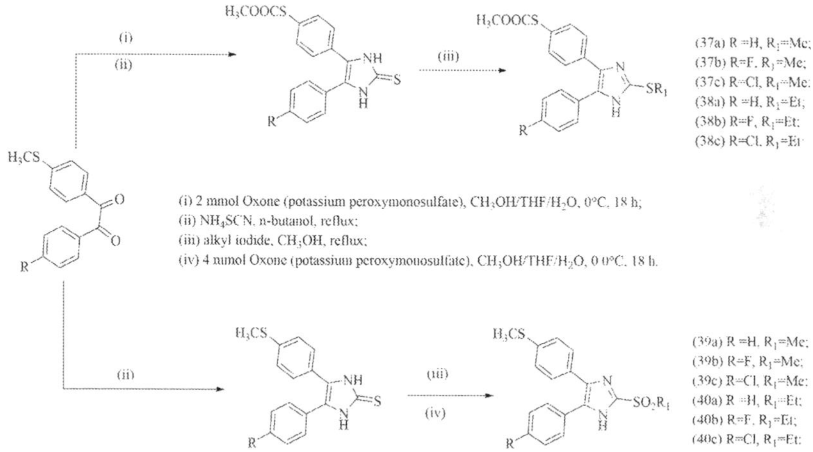

SCHEME 2.6 Synthetic route of tri-substituted imidazole derivatives.

For the preparation of derivatives, two methods were used. Method A involved oxidation of benzoin to corresponding sulfoxides using oxone and further treatment with excess ammonium thiocyanate in n-butanol to prepare the compound, phenyl-substituted carbonothioate. This compound was further treated with alkyl iodide to produce 4-aryl-5-(4-(methylsulfonyl)phenyl)-2-alkylthio derivatives (**37a–c**) and (**38a–c**). Method B involved the reaction of benzoin with excess ammonium thiocyanate in *n*-butanol to produce phenyl-substituted carbonothioate, followed by alkylation using alkyl iodide. The compound was further treated with oxone to obtain 2-alkylsulfonyl-1*H*-imidazole derivatives (**39a–c** and **40a–c**).

BRAF is a significant therapeutic target in melanoma. Literature studies have shown that several compounds having tri-substituted five membered aromatic heterocyclic compounds were found to be successful inhibitors of BRAF. Niculescu-Duvaz et al. [47] synthesized 2,4,5-tri-substituted imidazoles with different substituents in the fourth position, namely the naphthyl, benzothiophenyl, and benzofuranyl groups as BRAF inhibitors. Three assays were used to study the biological activity of synthesized compounds, viz. $_{V600E}$BRAF mutant enzyme inhibition assay (IC_{50}, BRAF), extracellular signal regulated kinase phosphorylation inhibition assay (IC_{50}, pERK), and BRAF-dependent WM266.4 cells growth inhibition assay (GI_{50}, SRB). After analyzing the results, it was

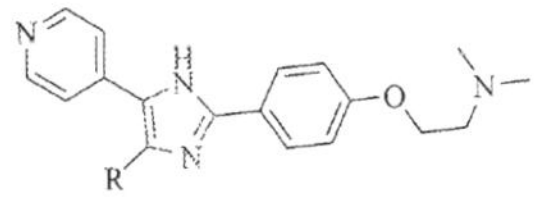

(41) Total 27 derivatives

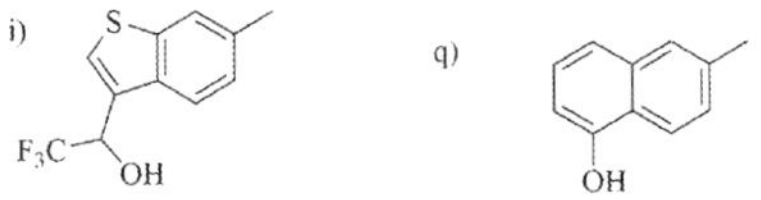

FIG. 2.22 Imidazole derivatives.

found that compounds (**41i**) with (BRAF IC_{50} = 190 nM and GI_{50} = 2100 nM) and (**41q**) with (IC_{50} = 9 nM and GI_{50} = 220 nM) were potent BRAF inhibitors (Fig. 2.22).

A series of 2-alkoxy-5-aryl-1*H*-imidazole scaffolds were synthesized and studied for their in vitro cancer cell growth inhibition [48]. The synthetic strategy involved nucleophilic aromatic substitution, Suzuki coupling, followed by ring closure metathesis reaction. The authors had three aims in order to study the anticancer activity of synthesized compounds: (a) to identify and optimize 5-aryl-1*H*-imidazoles as potent anticancer agents; (b) to characterize in vitro cytostatic versus cytotoxic anticancer study of synthesized compounds; and (c) to study their mechanism of action in melanoma cells. The compounds showed in vitro growth inhibitory concentration (IC_{50}) in the range of less than 100 μM to single-digit μM. Among all the synthesized compounds, compound (**42i**) was found to be the most potent in terms of showing cytostatic rather than cytotoxic anticancer effects in melanoma cells (Fig. 2.23). Furthermore, compound (**42i**) was tested by 60 cancer cell lines and compared with the > 763,000 NCI molecule database. NCI COMPARE algorithm did not show any correlation between its growth inhibition profiles and the NCI database compound profile.

In 2017, a new structural class of tri-substituted imidazole with 7-azaindole scaffold as a hinge binding was fabricated, and in silico study and biological testing were carried out to determine their inhibition activity against EGFR [49]. A target hopping pathway from p38α MAPK inhibitor templates was used to discover the potent molecule with different structure. The imidazole

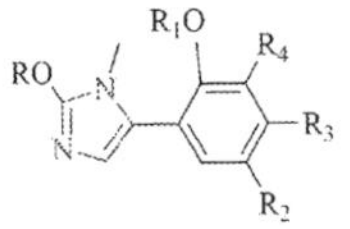

(42) Total 38 derivatives

42i) R_2=R_3=R_4=H; R_1=R=$(CH_2)_4CH$=CH_2

FIG. 2.23 Imidazole derivatives.

derivatives were synthesized as shown in Scheme 2.7. From the synthesized compounds, derivatives (**43d**) and (**43e**) were found to be the two reversible inhibitors of triple mutant L858R/T790M/C797S, with IC_{50} values in the low nanomolar range. Moreover, the authors also designed a kinome selective irreversible inhibitor (**44a**) having an IC_{50} value of 1 nM in opposition to the EGFR L858R/T790M double mutant.

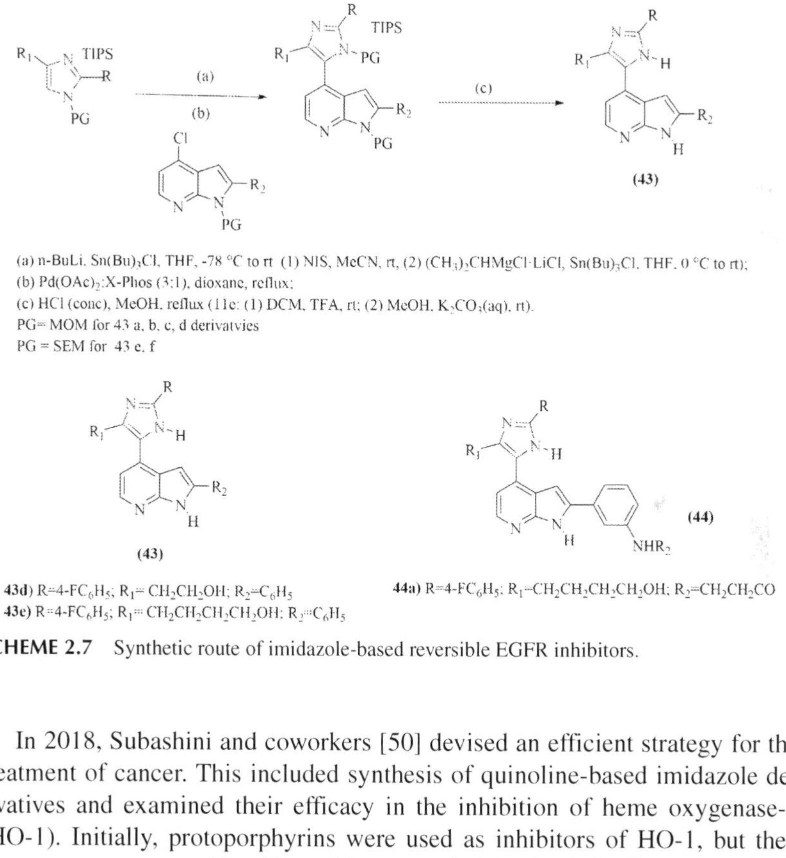

SCHEME 2.7 Synthetic route of imidazole-based reversible EGFR inhibitors.

In 2018, Subashini and coworkers [50] devised an efficient strategy for the treatment of cancer. This included synthesis of quinoline-based imidazole derivatives and examined their efficacy in the inhibition of heme oxygenase-1 (HO-1). Initially, protoporphyrins were used as inhibitors of HO-1, but they have some apparent side effects. Numerous imidazole derivatives were fabricated and applied as selective inhibitors of HO-1 with the least impact on the other heme-dependent enzymes. Azalanstat (**1**) is the first nonporphyrin inhibitor of HO. The synthetic pathway for quinoline-based imidazole derivatives is shown in Scheme 2.8. Among the synthesized compounds, compound (**46c**) was found to be the best inhibitor of HO-1, showing maximum inhibition of 87.06 (%) at minimum concentration of 9.2 μM. Compound (**46c**) exhibited viable inhibition like various other vital drugs. Among the synthesized compounds, compound (**46c**) was found to be the most potent and was also examined for its cytotoxicity against different cell lines, viz. A549, MG63, MCF7, and HEK293.

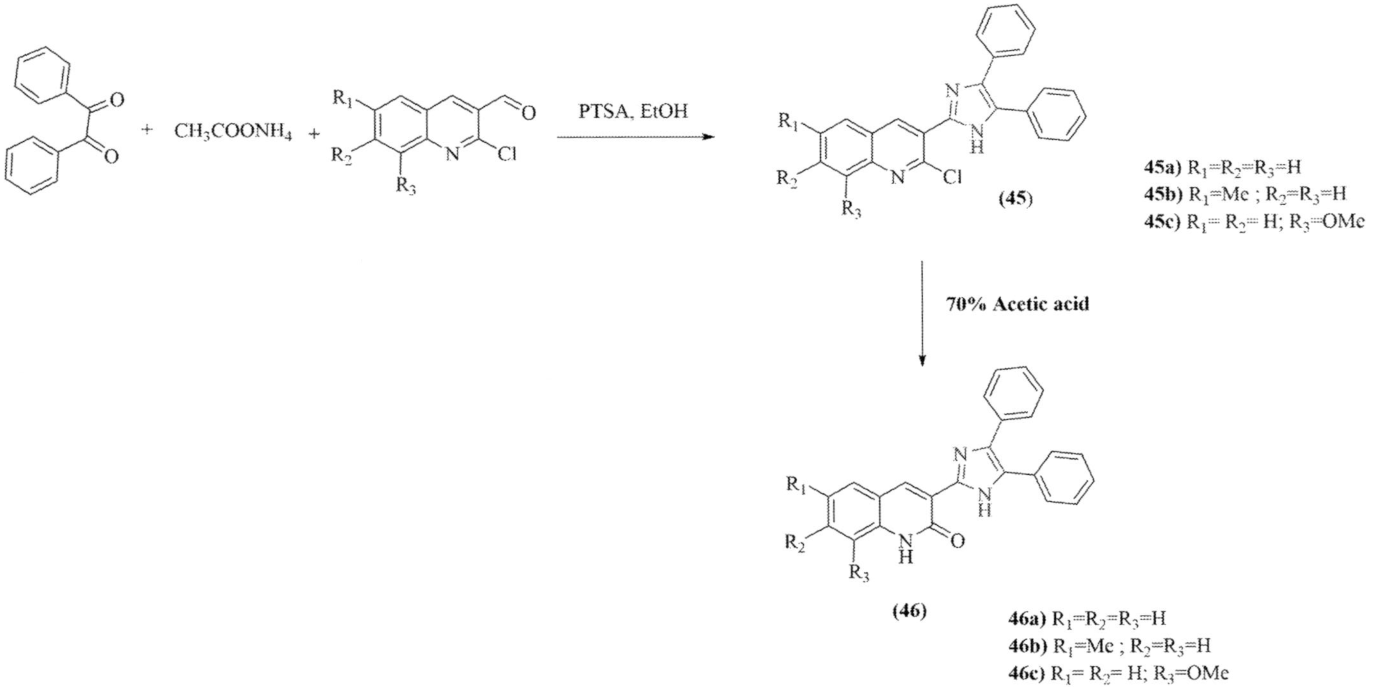

SCHEME 2.8 Synthesis of imidazole-substituted quinoline derivatives.

The compound (**46c**) was nonhazardous to the normal cell line HEK239. In the A549 cell line, compound (**46c**) showed an IC_{50} of 39.35 mM, while for MG63 and MCF7, IC_{50} was 62.03 and 50.10 mM, correspondingly.

Several enzymes are directly or indirectly involved in the modulation of gene expression that causes cancer. Histone deacetylase is one of these enzymes, involved in the remodeling of chromatin, and has a crucial role in managing the acetylation and deacetylation stages of chromatin. This eventually guarantees the epigenetic regulation of gene expression. Due to this, it became a probable target for cancer treatment. So histone deacetylase blockers materialized as a newfangled class of anticancer drugs. In March 2020, 14 imidazole-based derivatives (**47a–m**) were designed (Fig. 2.24) and the docking studies were performed to evaluate the epigenetic modulator application targeting histone-modifying enzymes [51]. These proteins—HDAC 2 (3MAX), HDAC 1 (4BKX), HDAC 4 (2VQM), HDAC 3 (4A69), HDAC 7 (3C10), HDAC 6 (3C5K), and HDAC 8 (3SFF)—were selected as drug targets. The compound (**47c**) (Fig. 2.25) firmly bound to the HDAC enzymes at their receptor regions with high binding score (− 11.039 kJ/mol). The product (**47c**) was further evaluated for in vitro anticancer cytotoxicity against A549 cell lines using MTT assay. It was found that compound (**47c**) diminished cell viability and blocked the anchorage-independent expansion of lung cancer cells. In addition to this, impact of this derivative on nuclear damage and stimulation of apoptosis was studied. After the treatment of A549 cells with this derivative, distinctive morphological changes occurred in apoptosis, with shrinking of the cytoplasm and nuclear disintegration with no change in the cell membrane and tapered nucleus.

Previously tri-substituted imidazole-based scaffolds (**48**) were examined as inhibitors that approached the antidrug EGFR (L858R/T790M/C797S) mutant with nanomolar cytotoxicity through a reversible binding mechanism. Its mechanism of action was later explained by Heppner [52]. Various authors studied the binding modes of numerous tri-substituted imidazole inhibitors in combination with the EGFR kinase domain using X-ray crystallography to know their targeting of EGFR drug-resistant mutant C797S. Crystal structure analysis showed that inhibitors bound to both active and inactive EGFR kinase. This binding demonstrated that the imidazole scaffold acted as an H-bond acceptor for the ε-NH_4^+ group of the lysine (K745) in the "αC-helix out" inactive state and thus blocked its interaction via formation of particular imidazole N-methylation. This H-bonding connection with EGFR kinase was not achieved with osimertinib or WZ4002, but it was found in other C797S inhibitors. This made it a potential way for facilitating strong, reversible binding, rendering the tri-substituted imidazole derivatives as effective inhibitors.

2.1.4 *Tetra-substituted imidazole derivatives*

In 1989, Anderson et al. [53] synthesized bis(carbamate) derivatives of 1,2-substituted 4,5-bis(hydroxymethyl)imidazoles (**49**) from a diester through a two-step process. The synthesized compounds were anticipated for their

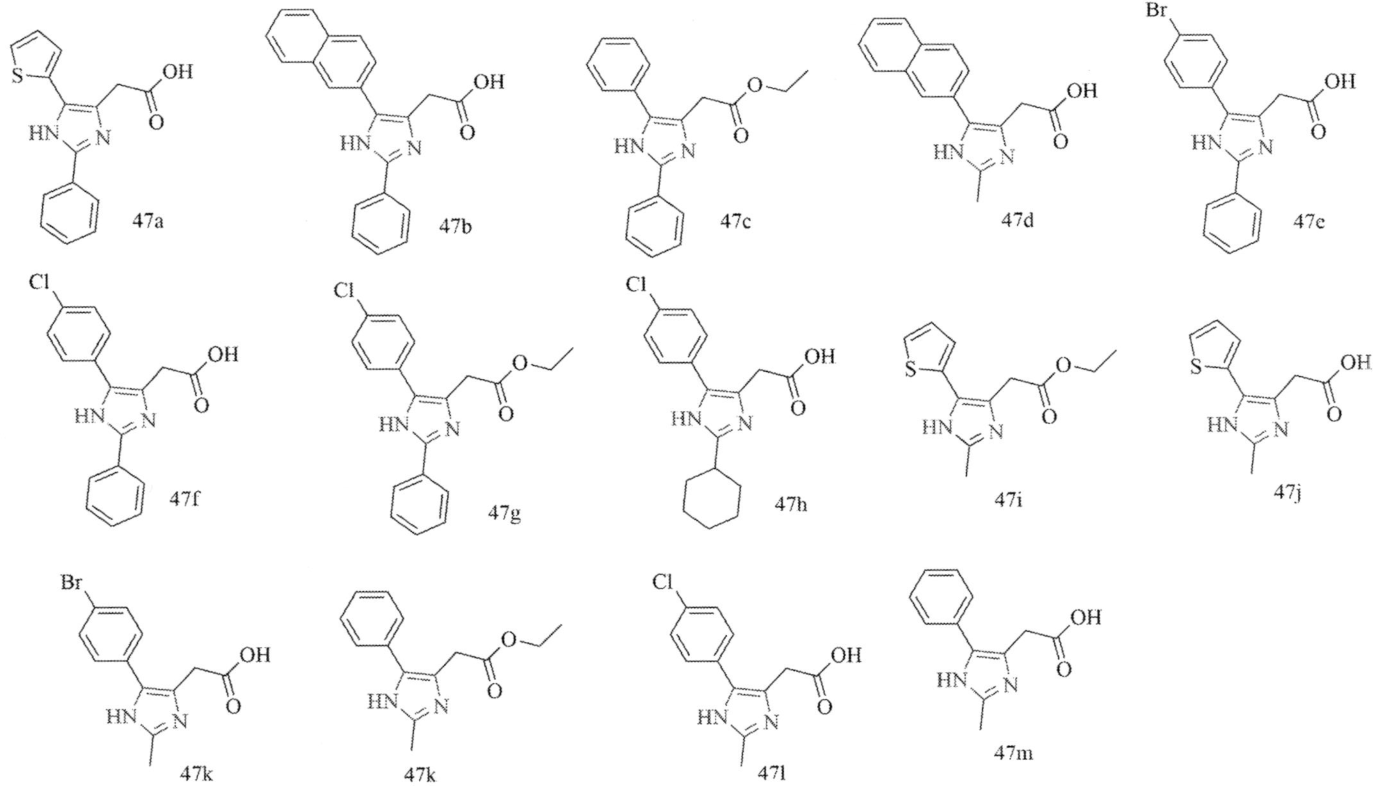

FIG. 2.24 Array of imidazole-based compounds.

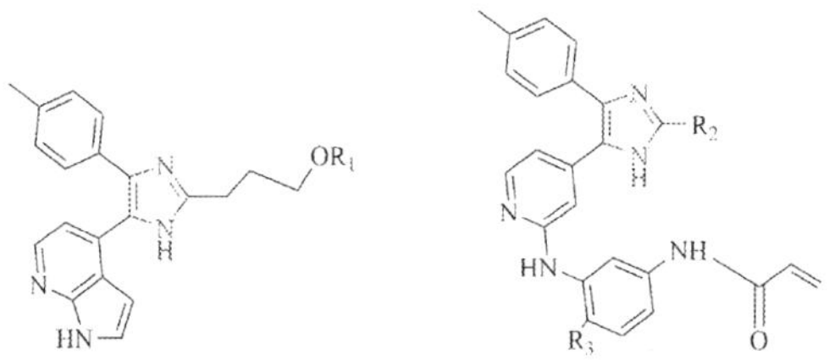

48a) R_1=H
48b) R_1=Me

48c) R_2=CH_2CH_2OH, R_3=OMe
48d) R_2=SCH_3, R_3=OMe
48e) R_2=$SCH_2CH_2OCH_3$, R_3=OMe

FIG. 2.25 Structures of tri-substituted imidazole inhibitors.

anticancer activity against P388 lymphocytic leukemia assay. According to the results, it can be concluded that electron-donating substituents gave active compounds whereas derivatives with electron withdrawing substituents were found to be inactive. The 2-(methylthio)-1-methyl derivative (carmethizole), (**49ai**), was found to be very active. The antitumor activity of carmethizole was also determined against four cell lines: MX-1 mammary xenograft, the human amelanotic melanoma cell line (LOX) xenograft, M5076 sarcoma, and L1210 lymphocytic leukemia (Fig. 2.26).

Later on, Atwell et al. [54] fabricated bis(hydroxymethyl)-substituted imidazoles and thioimidazoles, and estimated them for sequence-specific DNA alkylation and antitumor cytotoxicity. The synthesized derivatives were assessed for cytotoxicity against P388 murine leukemia cells in a culture, and the results, as IC_{50} values, are shown in Fig. 2.27. The bis(hydroxymethyl)imidazoles, bis(hydroxymethyl)thioimidazoles, and imidazole bis-(carbamates) are weak DNA alkylating agents.

In 2005, 1,2,4-triaryl-5-substiuted-1*H*-imidazoles were prepared and scrutinized for luciferase assay using ER positive MCF-7 breast cancer cells and anticancer activity against human breast cancer cell lines MCF-7 and MDA-MB 231. Compounds (**52b**) and (**52c**) effectively inhibited COX-1 and COX-2, and also reduced the growth of both cell lines. These compounds have equal $IC_{50}=8.4\,\mu M$ against COX-1 whereas (**52a**), (**52d**), and (**52e**) did not exert any effect on the cell lines. Compounds (**52b**) and (**52c**) abolished the growth of MCF-7 breast cancer cells (Fig. 2.28) [55].

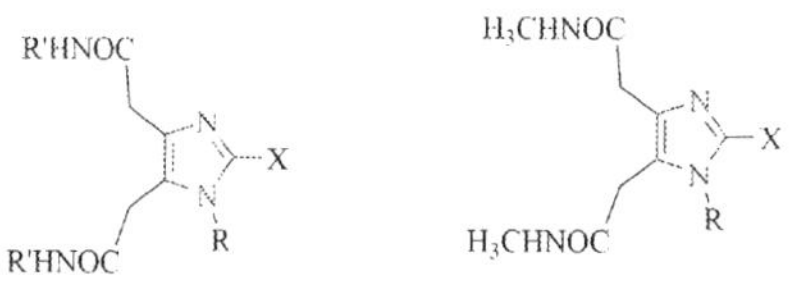

X=SMe
R=CH_3

49ai

49a) R=CH_3
49b) R=$CH(CH_3)_2$

FIG. 2.26 Biscarbamate analogs of imidazole.

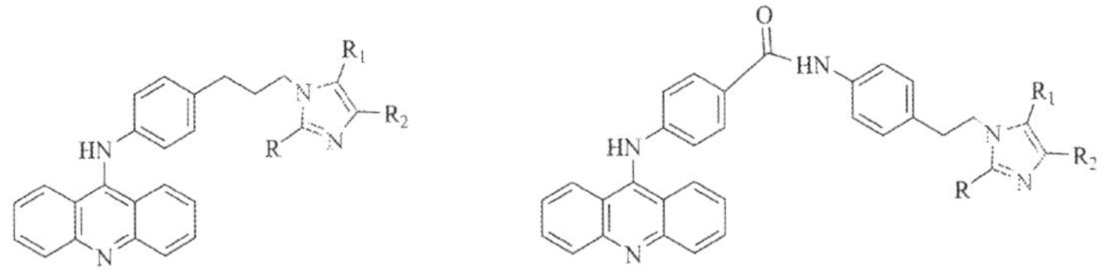

50a) R=H, R_1=CH_2OH IC_{50} (μM) =12
50b) R=H, R_1=$CH_2OCONHMe$ IC_{50} (μM) =16
50c) R=SMe R_1=R_2= COOEt Not determined
50d) R=SMe, R_1=R_2=CH_2OH IC_{50} (μM) =1.3
50e) R=SMe, R_1=R_2=$CH_2OCONHMe$ IC_{50} (μM) =0.8

51a) R=H, R_1=R_2=COOMe Not determined
51b) R=SMe, R_2=R_1=COOEt Not determined
51c) R=H, R_1=R_2= CH_2OH IC_{50} (μM) >20
51d) R=SMe, R_1=R_2=CH_2OH IC_{50} (μM) >20
51e) R=H, R_1=R_2=$CH_2OCONHMe$ IC_{50} (μM) > 20
51f) R=SMe, R_1=R_2=$CH_2OCONHMe$ IC_{50} (μM) = 11

FIG. 2.27 Biscarbamate imidazole derivatives and their IC_{50} values.

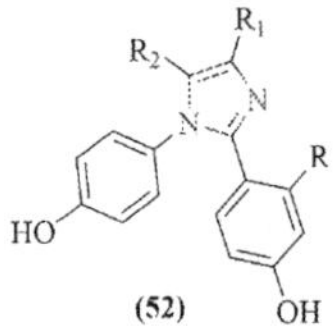

(52)

52a) R=R_2=H; R_1=4-OHPh
52b) R=R_2=H; R_1=2-Cl-4-OHPh
52c) R=Cl, R_2=R_1=2-Cl-4-OHPh
52d) R=H, R_2=Et, R_1=4-OHPh
52e) R=R_2=H, R_1=CF_3

FIG. 2.28 Tetra-substituted imidazole derivatives.

After a few years, the authors synthesized an array of C5-substituted 1,2,4-triaryl-1*H*-imidazoles and 1,2-substituted imidazole derivatives [56]. The gene stimulating characteristics were anticipated on estrogen receptor alpha positive MCF-7 breast cancer cells. Different synthetic routes were applied for the synthesis of these derivatives. The position of substituents greatly affected the activity, as negative results were obtained with 4,5-bis(4-hydroxyphenyl)-1*H*-imidazole-2,4,5-tris(4 hydroxyphenyl)imidazoles [57, 58]. Twenty-one derivatives were examined using calf uterine cytosol as a source of ERα, and for the luciferase assay, ERα positive MCF-7 cells were used. Derivatives without C5 alkyl substituents (**53a–d**) did not induce gene activation. The C5 position with an alkyl substituent improved the transcriptional activity. As per the SAR study, replacing the ethyl group with an aryl ring caused lower hormonal activity, i.e., compounds (**55**) were less active than compounds (**54**). The presence of OH groups on the N1/C4 aryl rings were more effective than on the C2/C4 rings whereas compounds with the OH group on the C2/N1 ring was inactive. The derivatives with chlorine substituent eradicated the activity. The presence of the OH group in the C5 phenyl ring decreased the activity (Fig. 2.29). This was due to the hindered hydrophobic interaction by the hydrophilic OH group demonstrating a definite orientation of the 1*H*-imidazoles in the ligand binding domain (LBD). The molecular docking of compound (**53j**) with 1ERE protein showed that crucial interactions were present. The C4 phenolic ring showed interaction with Arg394 and Glu35. The hydroxy group of the N1 phenolic ring

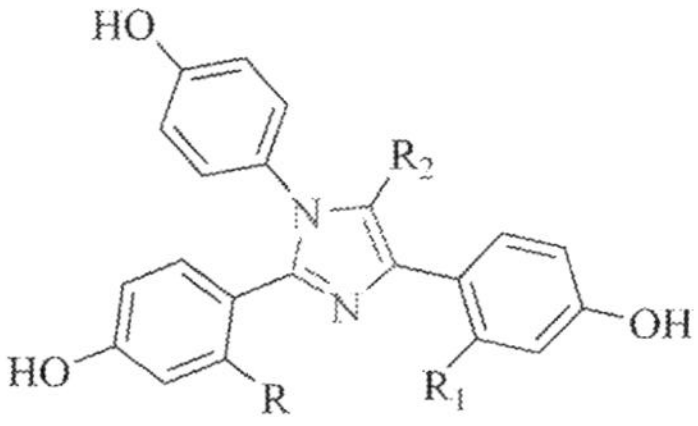

(53)

53a) $R=R_2=R_1=H$.
53b) $R=R_2=H$; $R_1=Cl$
53c) $R_2=R_1=H$; $R=Cl$
53d) $R=R_1=Cl$; $R_2=Cl$
53e) $R=R_1=H$; $R_2=Me$
53f) $R=Cl$, $R_1=H$; $R_2=Me$
53g) $R=H$, $R_1=Cl$; $R_2=Me$
53h) $R=R_1=Cl$; $R_2=Me$
53i) $R=R_1=H$; $R_2=Et$
53j) $R=Cl$, $R_2=H$, $R_1= Et$

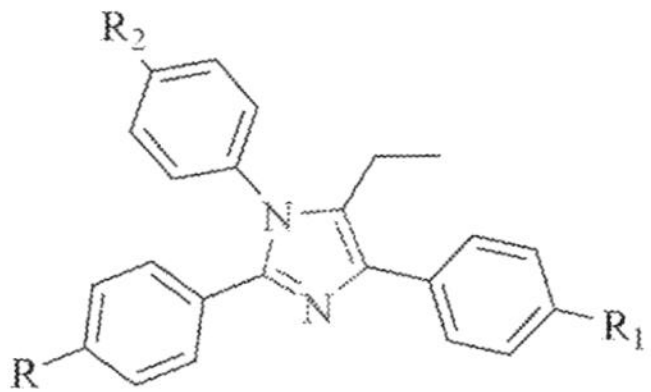

(54)

54a) $R_1=H$, $R=R_2=OH$
54b) $R_1=R_2=OH$, $R=H$
54c) $R_1=R=OH$, $R_2=H$
54d) $R_1=R_2=H$, $R=OH$

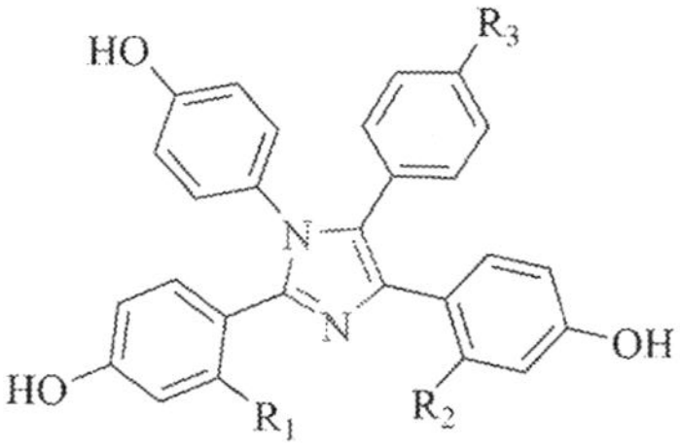

(55)

55a) $R_3=R_1=R_2=H$
55b) $R_1=Cl$, $R_3=R_2=H$
55c) $R_3=OH$, $R_1=R_2=H$
55d) $R_2=Cl$, $R_1=R_2=H$

FIG. 2.29 Tetra-substituted imidazole derivatives.

was oriented to His524. The third phenolic ring was linked to Met522. The C5 alkyl/aryl group was oriented in a hydrophobic binding pocket where Van der Waals interactions with Phe404, Leu391, Ile 424, and Leu346 were possible.

Later, in 2012, a series of 2-[4,5-dimethyl-1-(phenylamino)-1*H*-imidazol-2-ylthio]-*N*-(thiazol-2-yl)acetamide derivatives were synthesized by the reaction of equimolar ratio of 2-chloro-*N*-(thiazol-2-yl)acetamides and 4,5-dimethyl-1-(phenylamino)-1*H*-imidazole derivatives. All the synthesized derivatives were studied for anticancer activity using 60 different human tumor cell lines including nine cancer types. Three parameters—growth inhibitory activity (GI_{50}), cytostatic activity (IGI), and cytotoxic activity (LC_{50})—were studied for the compounds. Out of 32 compounds, seven were selected and the anticancer activity was done at a single conc. of 1×10^{-5}M. Afterward, three compounds (**56a**, **56b**, **56c**) were further studied for 10-fold dilutions of concentrations 10^{-5} to 10^{-9}M. Finally, compounds (**56a**) and (**56c**) showed the highest activity; this was probably due to the presence of the *p*-chlorophenyl group present at the fourth position of the thiazole ring (Fig. 2.30). Compound (**56b**) showed selective activity against melanoma cell lines, which was probably due to the –CH_3 group at the R_3 position [59].

In early 2020, a novel series of 1-(4-fluorobenzyl)-2,4,5-triphenyl-1*H*-imidazole derivatives were synthesized. The synthetic pathway is a multicomponent, one-pot strategy that involved 4-fluorobenzylamine, benzil, 4-substituted benzaldehydes, and CH_3COONH_4 in the presence of EtOH and a catalytic amount of acetic acid [60]. The synthesized compounds were also examined for anticancer activity by in silico and in vitro methods. The estrogen receptor 3ERT was selected as a protein and the anticancer activity was determined against MCF-7 cells by MTT assay, as illustrated in Table 2.10. The derivatives (Fig. 2.31) (**57a–d**) showed particular Van der Waals interactions with neighboring hydrophobic residues like MET 526, LEU 349, LEU 525, HIE 524, ALA 350, LEU 402, THIR 347, PHE 404, LEU 428, TRP 383, HIE 524, and LEU 384, in the Helix 12. The compounds stimulated cell apoptosis and reduced cell viability.

2.2 Imidazole hybrids as anticancer agents

The hybridization of heterocyclic scaffolds in one molecular entity has become an emerging approach for the production of novel and improved biologically

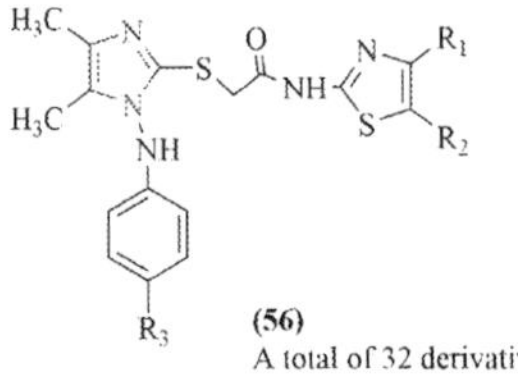

56a) R_3=R_2=H, R_1=4-ClPh
56b) R_2=H, R_3=Me, R_1=Ph
56c) R_2=H, R_3=OMe, R_1=4-ClPh

(56)
A total of 32 derivatives

FIG. 2.30 Tetra-substituted derivatives of imidazole.

TABLE 2.10 Anticancer effect of (57a–d) on MCF-7 cells on selective concentrations.

	Cell viability (%)			
Conc. (μM)	57a	57b	57c	57d
15.6	36	55	53	65
125	20	24	29	39
250	11	17	19	20
500	5	12	15	13

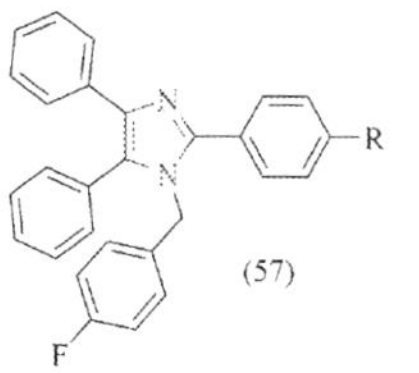

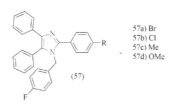

FIG. 2.31 Tetra-substituted derivatives.

active molecules. From the last 30 years, huge attention has been dedicated to the synthesis and biological evaluation of hybrid heterocyclic molecules. The imidazole moiety has been generally used as a privileged scaffold for the designing of therapeutic compounds of pharmaceutical or biological importance [61].

In 1999, imidazole-based benzodiazepine derivatives were designed and the inhibitory structure activity relationship was studied for farnesyltransferase; in vivo antitumor activity was also estimated using a Rat-1 tumor model [62]. The synthesized compounds were examined and the %T/C value of ≥ 125 was considered an active result. 2,3,4,5-Tetrahydro-1-(1*H*-imidazol-4-ylmethyl)-7-(4-pyridinyl)-4-[2-(trifluoromethoxy)benzoyl]-1*H*-1,4 benzodiazepine (**58a**) (Fig. 2.32) has an FT IC_{50} value of 24 nM, giving 85% phenotypic reversion of Ras transformed NIH 3T3 cells at 1.25 μM.

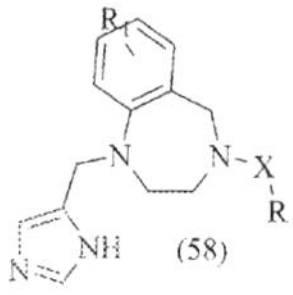

FIG. 2.32 Imidazole-based benzodiazepine.

The designing of a drug for the inhibition of double-stranded RNA-dependent protein kinase (PKR) characterizes an interesting approach for the protection of the brain. In general, the inhibition of kinase prompts the apoptotic process but could favor cell proliferation and tumorigenesis. It is therefore a huge challenge to fabricate a drug that causes inhibition of PKR without stimulating cell proliferation. The goal of the present study is to propose an animal model that displays high PKR activity in order to study further the effects of potential PKR's inhibitors in vivo. Due to this, in 2007, Ingrand and coworkers [63] investigated the brain levels of active and total PKR in rats at different ages by employing oxindole/imidazole derivative **C16** (**59**). Afterward, they estimated the effect of systemic injections of **C16** (Fig. 2.33) on both PKR/eIF2a and mTOR. The densitometry analysis of blots exposed an important dose-dependent decrease of the ratio p-PKRThr446/PKR that attained to 46%, 52%, and 58% of the control value after **C16** injections of 3.35, 33.5, and 167.5 μg/kg of body weight, correspondingly. Furthermore, these results suggested that an in vivo acute treatment with (**59**) was capable of obstructing the proapoptotic PKR/eIF2a signaling mechanism and highlighted that in vivo PKR inhibition was a viable pathway to modulate neurodegeneration.

In 2010, Niculescu-Duvaz and coworkers [64] devised an efficient strategy for the synthesis of novel tricyclic pyrazole-imidazole derivative. The synthesis of target compounds is a multistep synthesis that involved a Suzuki-coupling reaction. The fabricated pyrazole-imidazole derivatives are shown in Fig. 2.34. The synthesized compounds were examined for activity using three assays:

- inhibition of purified mutant BRAF activity in vitro;
- inhibition of oncogenic BRAF-driven extracellular regulated kinase (ERK) activation in BRAF mutant melanoma cell lines; and
- inhibition of propagation of cells.

Among the synthesized compounds, (**60a**) was a potent BRAF inhibitor. The compound triarylimidazole (**60j**) with a phenylpiperazine ring was the most proficient compound, with nanomolar activity in the mutant BRAF inhibition assay, the cellular pERK inhibition, and the mutant BRAF melanoma WM266.4 growth inhibition.

A series of 2-phenylbenzofuran-imidazole hybrid compounds were synthesized and studied for in vitro antitumor activity against a panel of different tumor cell lines [65]. The synthesis involved several steps. First, salicylaldehyde

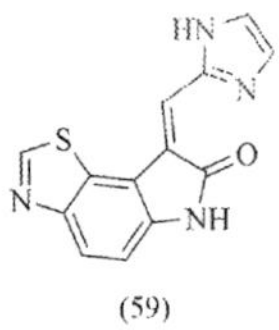

(59)

FIG. 2.33 Oxindole/imidazole derivative C16.

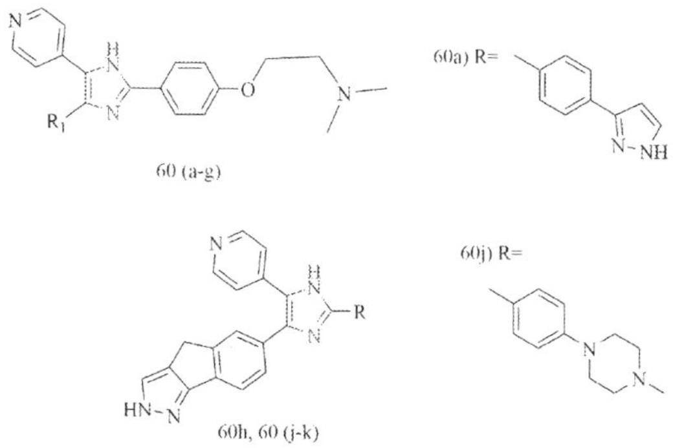

FIG. 2.34 Pyrazole-imidazole derivatives.

and 4-nitrobenzylbromide reacted to produce 2-(4-nitrophenyl)benzofuran. On its reduction with $SnCl_2$, 4-(benzofuran-2-yl)benzenamine was obtained. It further underwent reaction with ammonia, HCHO, and glyoxal to produce 1-(4-(benzofuran-2-yl)phenyl)-substituted imidazole. Finally, it reacted with phenacyl and alkyl halides in toluene to produce phenylbenzofuran-based imidazolium salts. The SAR studies showed that the groups, naphthylacyl or bromophenacyl at the third position of the imidazole ring were necessary for cytotoxic activity and therefore, compounds (**61j**) and (**61k**) proved to be the most potent antitumor compounds against all tested cell lines. Compound (**61k**) was found to be more active than cisplatin (reference drug) and showed cytotoxicity selectively against liver carcinoma (Fig. 2.35).

Further, a series of dibenzo[*b*,*d*]furan-imidazole hybrids were synthesized and evaluated for in vitro studies against a panel of human tumor cell lines, viz. leukemia, lung, colon, breast, and myeloid liver cell lines [66]. Cisplatin was used as a reference drug. Studies showed that hybrid compounds were more selective toward cancer cell lines. Molecular structure, water solubility, and charge distribution played important roles in determining the differences in cytotoxicity between neutral compounds and imidazolium salts. The results showed that the imidazolium salt hybrids (**62a**), (**62b**), (**62c**), (**62d**), (**62e**), and (**62f**), having benzimidazole ring/2-methyl imidazole ring and imidazolyl-3-position substituted with a 4-methoxyphenacyl group or naphthylacyl, showed high potency. Compounds (**62b**), (**62f**), and (**62g**) were the most potent with IC_{50} values of

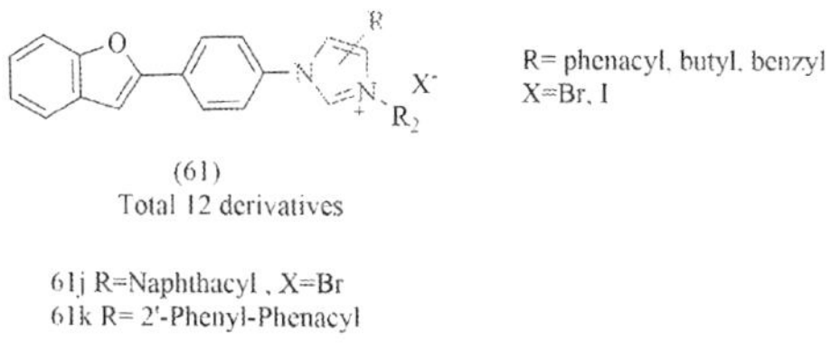

FIG. 2.35 Hybrids of imidazole and benzofuran.

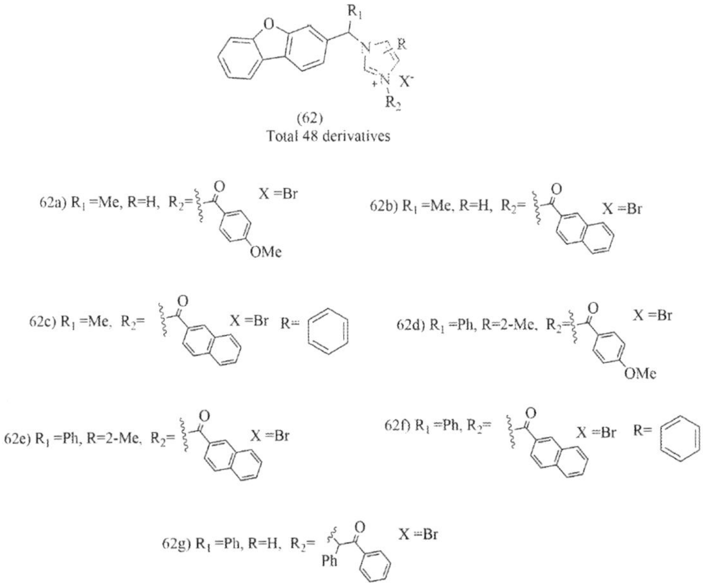

FIG. 2.36 Dibenzo[*b*,*d*]furane imidazole hybrid compounds.

0.52–3.86 μM against all human tumor cell lines. Compound (**62d**) was more selective against lung carcinoma and myeloid liver cell lines with IC_{50} values 7.2-fold and 5.9-fold lower than cisplatin (Fig. 2.36).

Sedic and colleagues designed nine novel 1,2,4-triazole and imidazole analogs of L-ascorbic acid and imino-ascorbic acid, and examined their antiproliferative activities against murine leukemia and human tumor cell lines, viz. cervical carcinoma, hepatocellular carcinoma, pancreatic carcinoma, human normal diploid fibroblasts, breast epithelial adenocarcinoma, colorectal carcinoma, and human T-cell acute lymphoblastic leukemia [67]. From the synthesized compounds, compounds (**63**) and (**64**) showed excellent cytostatic effects in all tumor cell lines (Fig. 2.37). These compounds showed high selectivity for human T-cell acute lymphoblastic leukemia cells. However, compound (**63**) was not found to be cytotoxic to human fibroblasts.

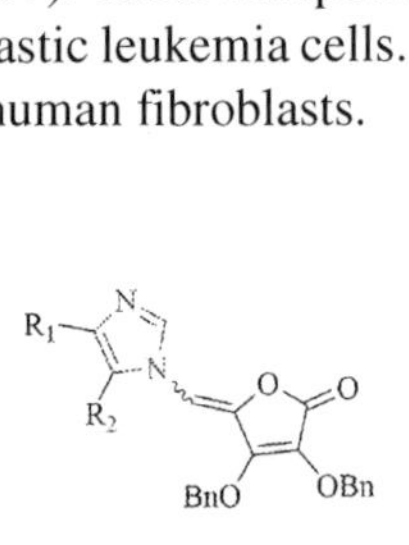

63) R_1=R_2=COOMe
64) R_1=R_2=CN

FIG. 2.37 Imidazole analogs of L-ascorbic acid.

In 2013, the antitumor activity of 2-benzylbenzofuran imidazole hybrids (38 derivatives) was studied against a panel of human tumor cell lines (leukemia, colon carcinoma, breast carcinoma, lung carcinoma, myeloid liver carcinoma) [68]. Cisplatin was used as a reference drug. After evaluating the biological studies and structure activity relationship, the presence of benzimidazole ring and substitution with naphthylacyl or 4-methoxyphenacyl group at the third position of the imidazole ring was found to be vital for the antitumor activity (Fig. 2.38). Among the synthesized compounds, (**65**) and (**66**), having 4-methoxyphenacyl or naphthylacyl groups at the third position of benzimidazole, were found to be potent antitumor agents with IC_{50} values of 1.02–3.57 μM against all tested human tumor cell lines and showed cytotoxic activities selectively against myeloid carcinoma and breast carcinoma, respectively.

Fused heterocycles of benzofuran and imidazole were synthesized and appraised against different cell lines HL-60, SW480, A549, MCF-7, and SMMC-7721 [69]. To fabricate the compounds, a schematic pathway was employed using benzofuran as a precursor, as explained in Scheme 2.9. The SAR studies showed that the presence of 2-methyl or 2-ethyl imidazole ring and substitution of the naphthylacyl or methoxyphenacyl group at the third position of imidazole was necessary for varying cytotoxic activity. Among the synthesized compounds, the derivatives with a naphthylacyl (**67a**) or 4-methoxyphenacyl (**67b**) at the third position of the imidazole ring showed comparable potency in vitro compared with standard cisplatin. Fascinatingly, the other hybrid compounds having naphthacyl substituent at the third position of 2-methyl-imidazole and 4-methoxyphenacyl substituent at the third position of 2-methyl-imidazole were found to be more potent derivatives, having IC_{50} values less than 5.0 μM against all human tumor cell lines under investigation, and more cytotoxic than cisplatin. The SAR studies showed that substituents significantly affected the activity as the substituent (R) on the imidazole ring followed the order of activity as 2-methylimidazole > 2-ethylimidazole > benzimidazole > imidazole, and for N-substituent R_2 the order is phenacyl > 2-bromobenzyl > benzyl > alkyl.

Pyrrole-imidazole polyamides (PIP) are tiny artificial molecules that distinguish and form noncovalent bonds to the minor groove of DNA, followed by blocking of DNA-protein interactions with high affinity and sequence

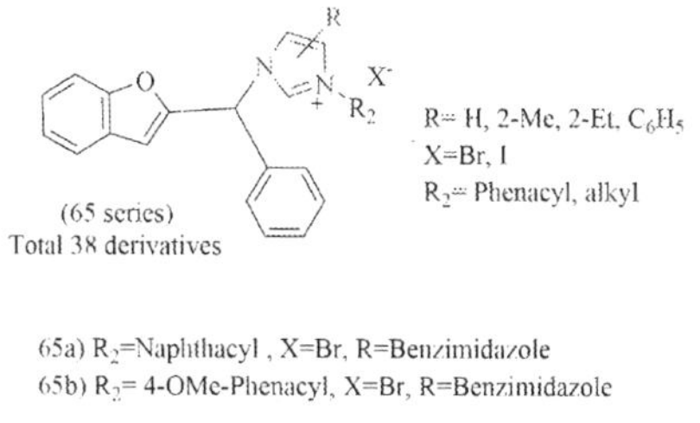

FIG. 2.38 2-Benzylbenzofuran imidazole hybrids.

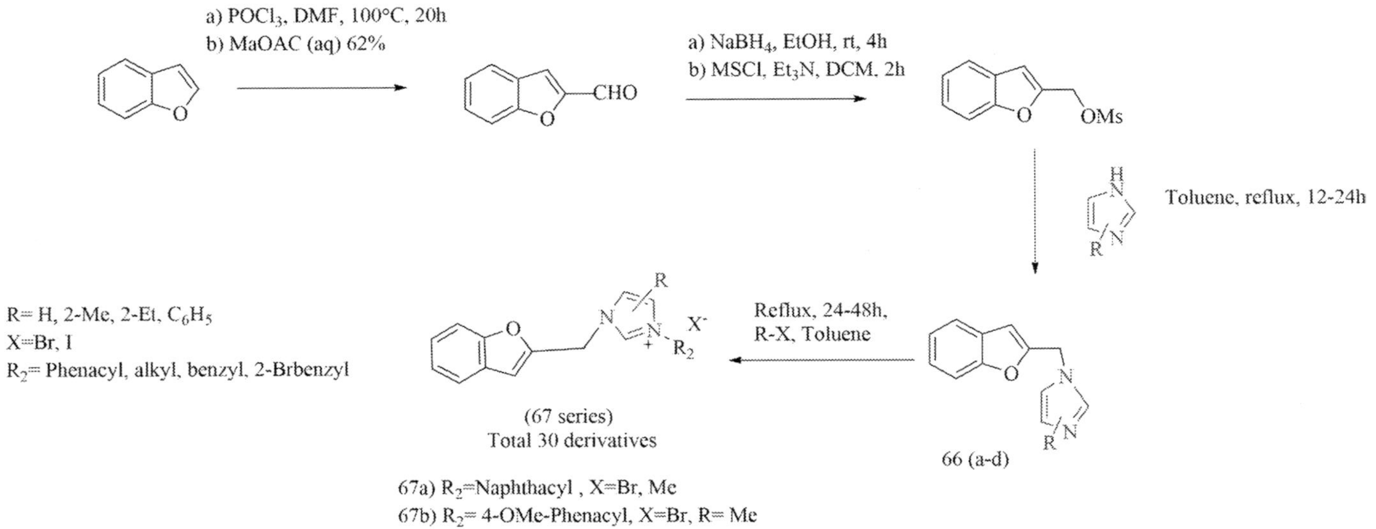

SCHEME 2.9 Synthesis of benzofuran imidazole-based heterocycles.

specificity. These outcomes suggested that targeted PIP might be a possible gene blocker for the treatment of diseases like cancer [70]. In 2014, Obinata et al. [71] fabricated pyrrole-imidazole polyamides that formed noncovalent bonds to the small groove of DNA with good affinity and specificity. These synthetic molecules were designed in such a way that they targeted the TMPRSS2-ERG translocation breakpoints and were also evaluated against human prostate cancer cells. It was observed that this synthetic molecule suppressed the cell proliferation and cancer growth of androgen-sensitive LNCaP prostate cancer cells. Additionally, it diminished the endogenous ERG expression and tumor growth in in vitro models, and also stimulated apoptosis in in vivo models.

Later on in the same year, pyrrole imidazole polyamides were also used to target the DNA that are specifically pointing the base pairs of the REL/ELK1 binding site in the EVI1 minimal promoter. EVI1 (human ectopic viral integration site 1) is an oncogenic transcription factor well-known for its vital role in several malignant cancers. Two sequences of pyrrole imidazole polyamide PIP1 or PIP2 were designed (Fig. 2.39) and their impacts on the expression level of ectopic EVI1 were studied [72]. PIP1 extensively inhibited MDA-MB-231 cells with increasing final concentrations to 2 and 10 mm. On the other hand, PIP2 did not decrease the EVI1 mRNA level appreciably within the given concentration range. This result established the significance of sequence specificity. PIP1 affected several EVI1-regulated genes like LTBP1, LOXL3, etc. that played a vital role in tumor development. In addition, PIP1 demonstrated strong antimetastatic characteristics in MDA-MB-231 breast cancer cells. These outcomes anticipated that PIP1 may act as a selective inhibitor of EVI1 and stimulated the new prospects to expand novel antitumor agents.

In 2016, hair pin polyamides MLH1–16 (Py-Im-β-Im-Im-Py-γ-Im-Py-β-Im-Py-Py) were designed to target inhibition of DNA methylation in the human MLH1 gene. MLH1 is a mismatch repair gene that gets hypermethylated in the late stages of cancer. Abnormal DNA methylation of gene is the main reason for major epigenetic changes that cause hypermethylation of promoter CpG islands, thereby inactivating the tumor suppressor genes and being responsible for cancerous tumor growth. The fabricated PPIs were assessed against DNA methylation and their efficacy was studied in colorectal cancer. MLH1–16 caused selective inhibition of DNA methylation in vitro. PPI acted as an antagonist of CpG methylation in living cells [73].

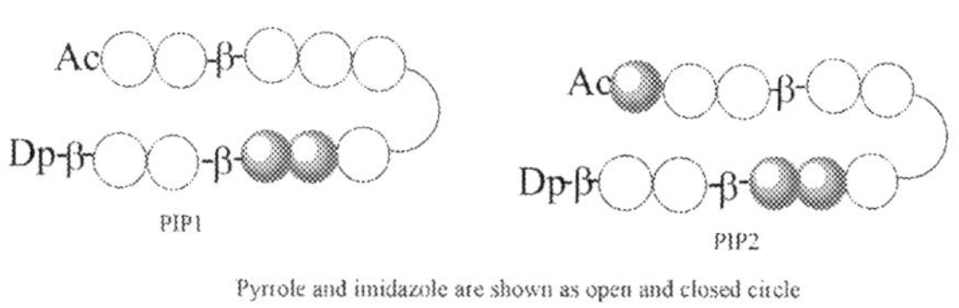

FIG. 2.39 Pyrrole and imidazole polyamides (PPI1 and PPI2).

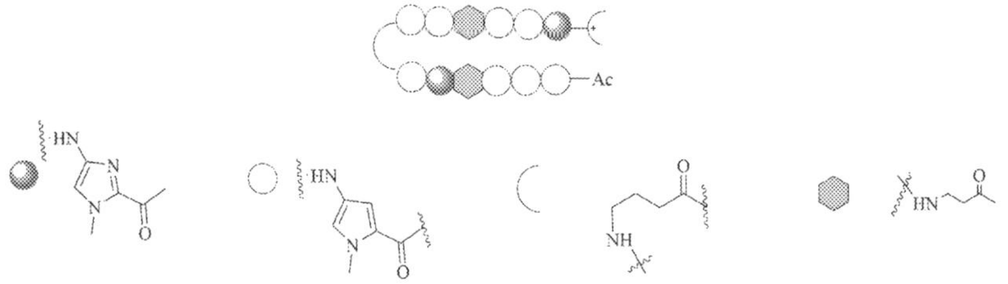

FIG. 2.40 Structural design of pyrrole-imidazole-polyamide.

Later on, in 2019, Malinee et al. designed pyrrole-imidazole polyamides (SOX2i) (Fig. 2.40) and assessed their anticancer activity; the authors found that SOX2i can significantly modify the expression of SOX2 at the protein and mRNA level in various human cancer cell lines. SOX2 is a central transcription factor that regulates different oncogenic activities like cell proliferation, metastasis, and avoiding of apoptosis. It also showed an antimetastatic property and depressed the expression of genes related with stemness and EMT. The study found that SOX2i as a DNA-based programmable small scaffold has a tendency to control the chief regulatory factors related with metastasis and tumorigenesis [74].

PIPs were also used to treat oncogenesis. Oncogenesis was caused by the well-known disease Epstein-Barr virus (EBV). It proficiently converts resting B cells to the everlastingly increasing lymphoblastoid cell line (LCL) in vitro, and robustly favors cancer pathogenesis. EBV causes latent infection and is related with numerous types of lymphomas and cancers. To inhibit these, Cheng and coworkers [75] considered and prepared two polyamides that targeted OriP sequences to inhibit EBNA1-OriP binding (Fig. 2.41).

The frame of PIPs is built up of linked *N*-methylpyrrole (Py) and *N*-methylimidazole (Im) in a hairpin like structure. A head-to-head Py-Py pair particularly targets the A/T or T/A base pairs, whereas Im/Py base pair directs

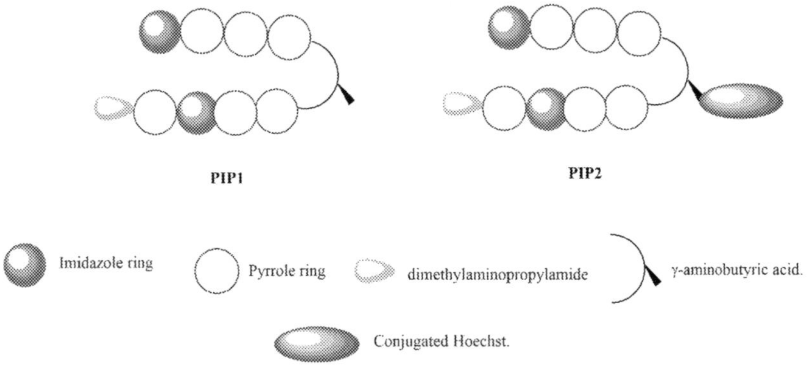

FIG. 2.41 Diagram of design and structure of polyamides.

toward G/C base pairs. Direction-oriented PIPs are permeable to biological membranes and can interfere with the binding of proteins to target DNA sequences.

The authors found that the Ht-conjugated polyamide PIP2 was proficient in displacing EBNA1 from OriP and blocked EBV-positive cells proliferation; thus it interrupted virus replication and viral protein expression. Due to this, EBV-positive cell proliferation and EBV-positive tumor growth in mice xenograft tumor models were significantly suppressed. This was confirmed by measuring the number of living cells by a trypan blue dye assay. In EBV-positive cells, the living cell treated with PIP2 was considerably reduced in comparison to PIP1 and the control treatment. MTT assay was also performed in order to reassure the results and GI_{50} values were determined in eight different cell lines. The results are summarized in Table 2.11.

In 2019, pyrrole imidazole polyamides were investigated to inhibit the E2F1. This is a recognized transcription factor that balances the tumor decreasing protein and programmed cell cycle progression. It is upregulated in chronic myeloid leukemia (CML) and leukemia stem cells (LSCs), and downstream of BCR-ABL1 signaling pathways. Pyrrole imidazole polyamides were fabricated in such a way that they could downregulate the E2F1-controlled gene expression. For this, they targeted a DNA sequence having 100 base pairs upstream of the E2F1 consensus sequence. The designed PIP was able to retard the growth of BC-CML cells [76].

IDO1 (indoleamine 2,3-dioxygenase) is a heme-containing enzyme that catalyzes the oxidative breaking of the C2–C3 indole double bond to form *N*-formylkynurenine. Inhibition of IDO1 is a striking immunotherapeutic pathway for the treatment of multiple types of cancers. The expression of IDO1 by diverse tumor cells caused depletion of tryptophan in the microenvironment and subsequent blocking of T-cell proliferation. Brant and group investigated the phenylimidazole scaffolds as inhibitors of IDO1. In this study, the authors

TABLE 2.11 GI_{50} (μM) of PIPs to different cell lines.

Cell line	PIP 1	PIP 2
Daudi (EBV +)	12.88 ± 1.58	4.02 ± 0.53
Raji (EBV +)	29.44 ± 3.56	8.00 ± 2.60
B95-8 (EBV +)	30.20 ± 5.42	9.42 ± 3.03
Nalm-6 (EBV −)	41.88 ± 4.77	22.18 ± 3.29
RPMI-8226(EBV −)	43.89 ± 3.24	24.50 ± 4.30
MOLT-4 (EBV −)	30.27 ± 4.01	25.00 ± 3.97

reported the structure-based fabrication of two associated arrays of compounds: *N*1-substituted 5-indolimidazoles and *N*1-substituted 5-phenylimidazoles. The synthesis of derivatives is shown in Scheme 2.10: the later derivatives were synthesized using Van Leusen imidazole synthesis. The synthesized compounds were examined for their efficacy against the IDO1 and IC_{50} values for selective potent compounds are summarized in Tables 2.12 and 2.13. An in silico study has also been done using 4PK5 protein and it was observed that compound **68a** showed interactions in pocket B and also have interactions with Arg 231. The most potent derivative from the **68**th series compounds (Fig. 2.42) was (**68k** with $IC_{50}=1.6\,\mu M$, $EC_{50}=9.97$) and from the **69th** series (Fig. 2.43) was found to be (**69m**) with ($IC_{50}=33.8\,\mu M$, $EC_{50}=260\,\mu M$). From the computational studies, it can be concluded that the synthesized compounds possessed crucial interaction, H-bonding, and hydrophobic interactions with the protein [77].

Negi et al. [78] demonstrated the synthesis of imine/amide-imidazole derivatives from 5-amino-4-cyano-*N*1-substituted benzyl imidazole and substituted aldehydes using microwave irradiation at 120°C in the presence of methanol as a solvent. The precursor, 5-amino-4-cyano-*N*1-substituted benzyl imidazole, was synthesized using 2,3-diaminomaleonitrile. The synthesis of imine/amide-imidazole derivatives (**70**) is shown in Scheme 2.11. The synthesized

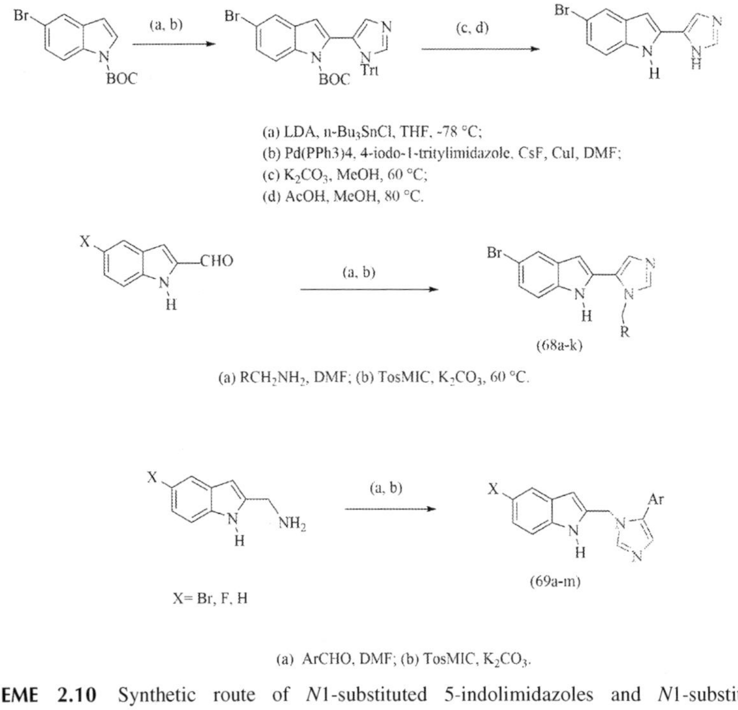

SCHEME 2.10 Synthetic route of *N*1-substituted 5-indolimidazoles and *N*1-substituted 5-phenylimidazoles.

TABLE 2.12 IC_{50} values for selective *N*1-substituted 5-indoleimidazoles derivatives.

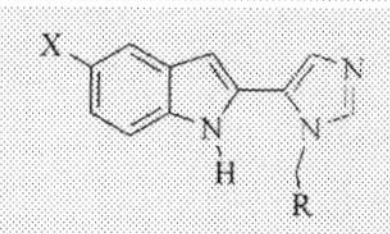

(68 series)

FIG. 2.42 *N*1-Substituted 5-indoleimidazoles derivatives.

Compound	X	R	IC_{50} (μM)	EC_{50} (μM)
68f	Br	4-CN	3.81	6.92
68j	Br	2-OH	6.40	5.01
68k	Br	3-OH	1.60	9.97

TABLE 2.13 IC_{50} values for selective *N*1-substituted 5-phenylimidazoles derivatives.

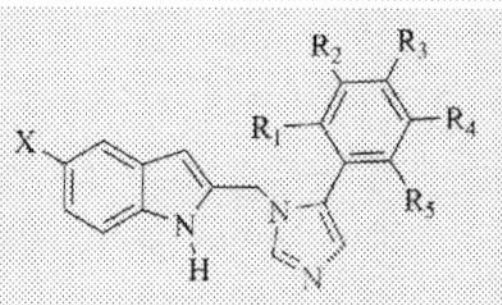

(69)

FIG. 2.43 *N*1-Substituted 5-phenylimidazoles derivatives.

Compound	X	R_1	R_2	R_3	R_4	R_5	IC_{50} (μM)	EC_{50} (μM)
69a	H	H	H	H	H	H	4.44	9.12
69b	Br	H	H	H	H	H	1.25	18.7
69c	Br	OH	H	H	H	H	0.180	1.45
69d	Br	OH	H	H	F	H	0.100	1.04
69e	Br	OH	H	H	Cl	H	0.0379	0.890
69f	F	OH	H	H	H	H	0.322	0.480
69l	F	OH	H	H	F	H	0.113	0.320
69m	F	OH	H	H	Cl	H	0.0338	0.260

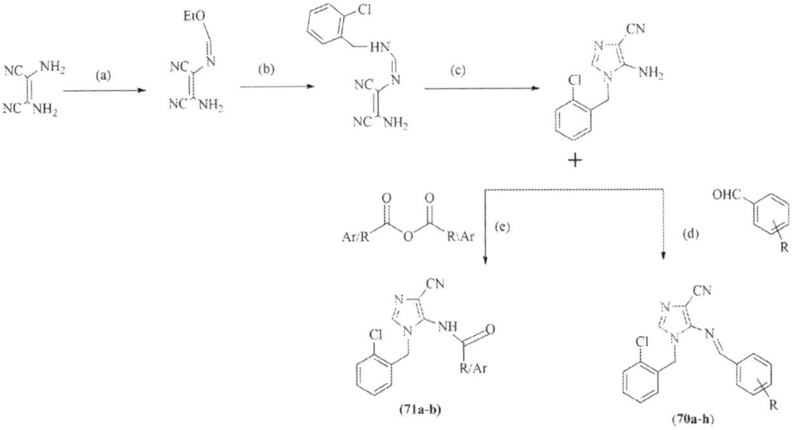

Reagents and conditions: (a) $CH(OEt)_3$, dry 1,4-dioxane, reflux, 8–10 h; (b) o-Cl-$PhCH_2NH_2$, EtOH, aniline hydrochloride (1 mol %), rt, 5–6 h; (c) 1 M KOH, rt, 7–8 h; (d) MWI, MeOH, 120°C; (e) MWI, 120°C

SCHEME 2.11 Synthesis of imine/amide-imidazole derivatives.

compounds were examined for their antiproliferative activity in vitro against a panel of seven cancer cell lines, comprising of A-459 (lung), HeLa (cervix), H-460 (liver), Hep-G2 (liver), PC3 (prostate), MCF-7 (breast), and IGROV-1 (ovary). The anticancer cytotoxicity was measured through changes in the ROS status in the cell. Increments in ROS levels are usually linked with variation in mitochondrial membrane potential that prompts its depolarization and leads to excretion of cytochrome-c into cytoplasm, thereby inducing apoptosis. The synthesized compounds were active against only three cell lines and their IC_{50} values are summarized in Table 2.14. The synthesized compounds exhibited good anticancer activity in vitro with less micromolar IC_{50} values in comparison to the standard drug Etoposide against the anticipated lung and liver cancer cell lines, and with no cytotoxicity to normal cells.

War et al. synthesized substituted imidazole-linked thiazolidinone derivatives by a multistep process and studied their DNA-binding ability. The desired product was prepared from the starting material 2-((3-(imidazol-1-yl)propyl)amine), which initially reacted with $ClCH_2COCl$ and NH_4SCN to form 2-((3-(imidazol-1-yl)propyl)amino)-thiazolidin-4-ones, which further condensed with aromatic aldehydes to furnish 2-((3-(imidazol-1-yl)propyl)amino)-5-(substituted benzylidene)thiazolidin-4-ones **(72a–k)** (Scheme 2.12). DNA-binding study results displayed that **(72e)** revealed strong bonding with a minor groove of DNA ($Kb = 0.18 \times 10^2$ L/mol). The protein 1G3X cocrystallized with an acridine-peptide drug was chosen for searching the intercalative mode of **(72e)**. An in silico molecular study also validated their pharmacological properties due to their strong binding (ΔG, − 8.5 kcal/mol) via H-bonding, hydrophobic interactions, and Van der Waals forces. Thus it

TABLE 2.14 Antiproliferative activity of the compounds (70a–h) and (71a–b).

Compound	R	A-549 (lung) (μM)	Hep-G2 (liver) (μM)	H-460 (liver) (μM)
70a	2-OH-3,5-$(Cl)_2$	7.5	12	11.2
70b	3,4-$(OMe)_2$	> 50	> 50	–
70c	2,5-$(OMe)_2$	18	20	28.3
70d	4-Cl	30.8	31	28.1
70e	4-OH	35.1	38.2	> 40
70f	2-NO_2	37	40	–
70g	3-Indole	10.9	15	14.6
70h	3-OMe-4-OH	39	> 40	–
71a	C_6H_5	16.5	17.1	13.4
71b	CH_3	20.3	28	33.9
Etoposide	–	18.2	20.9	30

can be concluded that the synthesized molecules have power to connect at the minor groove of DNA and could exert an anticancer effect by changing the structure of DNA [79].

Protein p53 is a tumor suppressor and the "protector of the genome." It is inactivated by mutations in the TP53 gene or possibly by the overexpression of its negative regulators, oncoproteins MDM2/MDMX in almost all types of cancer. The function of p53 can be regularized by disturbing the interaction of p53-MDM2/MDMX using a drug that acts as an antagonist. It can be considered as an efficient nongenotoxic anticancer therapy. This has become a popular approach to develop anticancer agents. With this in mind, in 2017, Twarda-Clapa et al. [80] synthesized 1,4,5-trisubstituted imidazoles (**73**) (Fig. 2.44) as inhibitors of p53-MDM2/MDMX, which worked as tumor suppressor oncoproteins. Most of the compounds showed high inhibition and compound (**73s**) was found to be highly effective as a tight binder of MDM2 and stimulated p53, and finally arrested cell cycle with inhibition of cell growth.

In 2017, the famous Deferasirox drug was modified with methoxy and imidazole scaffolds for the better delivery of drugs. It was further incorporated with PEG-*b*-CL polymeric micelles to improve its pharmacological characteristics [81]. It is an iron-chelating drug that reduces the iron overload in patients suffering from β-thalassemia and decreases tumor growth. According to the study, cancer cells caused disturbance in the iron metabolism, due to which

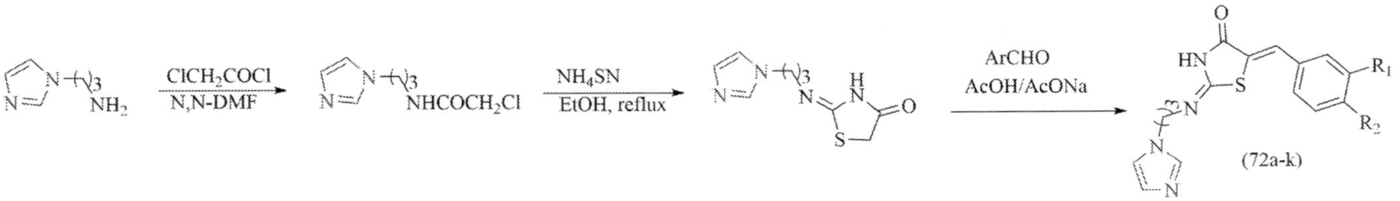

For most potent compound 72e $R_1 = NO_2$, $R_2 = H$

SCHEME 2.12 Synthesis of imidazole-linked thiazolidinone derivatives.

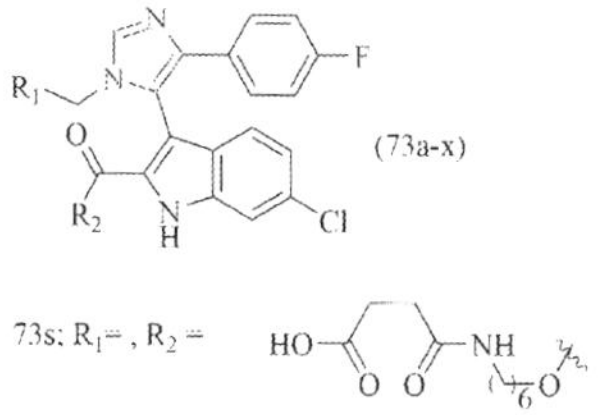

FIG. 2.44 Imidazole derivatives.

augmentation in transferrin receptor (influx iron) and diminishing of ferroportin (efflux iron) on the cancer cell membranes took place. Hence inhibition of iron metabolism is crucial for cancer treatment. The systemic layout for the synthesis of composites is shown in Scheme 2.13. All these composites demonstrated narrow size distribution, with PDI < 0.25. It was observed that the synthesized compounds possessed IC_{50} values in the range of micromolar to submicromolar against HepG2, L929, and PC-3 cell lines as shown in Table 2.15. Imidazole-based Deferasirox increased the pK_a value to 6.8 and micelles loaded with it caused faster release of drug at low pH in comparison to Deferasirox. For these micelles, 50% release was achieved in just 22 h at pH 4.5 in comparison to more than 7 days at pH 7.4.

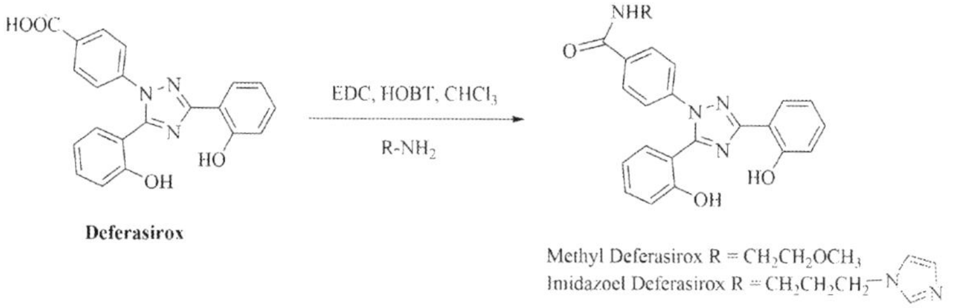

SCHEME 2.13 Schematic layout for Deferasirox derivatives.

Singh and coworkers [82] devised an efficient, MCR synthetic strategy for the synthesis of naphthalimide and phenanthro[9,10-*d*]imidazole conjugates. Initially, acenaphthene was used as a precursor and reacted with couple of reagents to furnish 6-bromo-1*H*,3*H*-benzo[*d*] isochromene-1,3-dione (A). On the other hand, 4-formylphenylboronic acid was further reacted with 9,10-phenanthrenequinone and CH_3COONH_4 in the presence of CH_3COOH to give boronates of 1*H*-phenanthro[9,10-*d*]imidazol-2-yls (B). Compounds (A) and (B) underwent a Suzuki-Miyaura cross-coupling reaction to form the desired products (**74**) (Scheme 2.14).

These compounds were estimated for in vitro antiproliferative activity against a panel of nine human cancer cell lines (leukemia, central nervous system [CNS], nonsmall cell lung, colon, melanoma, prostate, ovarian, renal, and breast). Among all the synthesized derivatives, compound (**74i**) was found to be

TABLE 2.15 IC_{50} of Deferasirox and its composites in the free form and micelle formulation.

IC_{50} values (μM)					
	PC-3		HepG2		L929
Compound	Free drug	Micelle	Free drug	Micelle	Micelle
Deferasirox	8.44 ± 0.48	10.23 ± 1.12	13.56 ± 2.80	12.67 ± 3.92	11.18 ± 1.11
Methyl Deferasirox	3.77 ± 1.08	5.50 ± 0.63	5.02 ± 1.59	6.46 ± 1.50	4.14 ± 1.07
Imidazole Deferasirox	0.27 ± 0.03	0.66 ± 0.12	2.32 ± 0.59	1.43 ± 0.69	3.13 ± 1.09

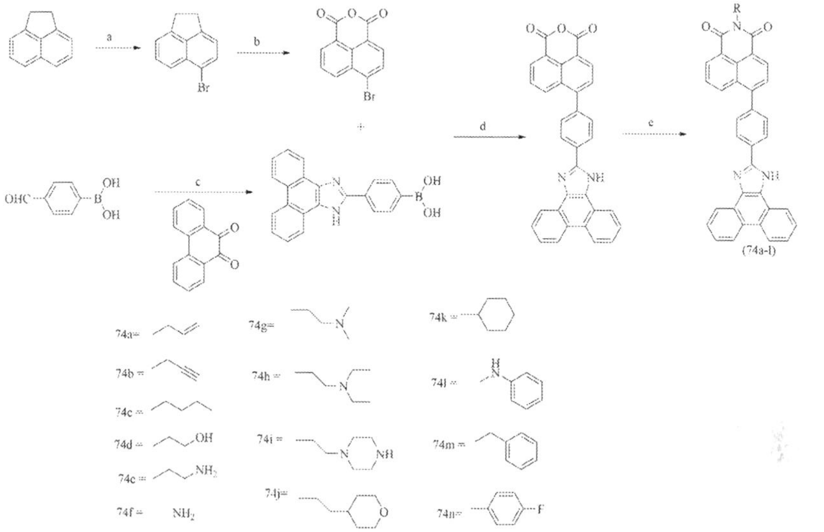

Reagents and conditions: (a) NBS, DMF, rt, 3 h, 94%; (b) Sodium dichromate, AcOH, reflux, 2.5 h, 72%; (c) Ammonium acetate, AcOH, reflux, 10 h, 58%; (d) $Pd(PPh_3)_4$, K_2CO_3, CH_3CN:water (9:1), N2, reflux, 12 h, 72%; (e) RNH_2, ethanol, reflux, 7–9 h, 78–91%.

SCHEME 2.14 Synthesis of naphthalimide and phenanthro[9,10-*d*]imidazole conjugates.

the most active, with a range of growth inhibition from − 55.78 to 94.53. The cytotoxicities of the most active compounds is illustrated in Table 2.16.

In order to study the binding interactions, compound (**74i**) was auto docked with DNA having PDB Id: 1BNA. The minimum binding energy of compound (**74i**) was found to be − 12.2 kcal/mol. Furthermore, the O atom of the naphthalimide ring of compound (**74i**) exhibited H-bonding interaction with sugar spine of DNA connected with cytosine (DC-9, A chain). Correspondingly, the N atom of piperazine formed H-bonding interaction with the phosphate backbone connected with thymine (DT20 of B chain). Furthermore, the naphthalimide and phenanthro[9,10-*d*]imidazole rings also demonstrated hydrophobic interactions with DNA.

TABLE 2.16 Cytotoxicities of selected compounds against A549, HeLa, and MCF7 cancer cell lines.

	IC_{50} values (μM)		
Compound	A549	MCF-7	HeLa
74h	81.34 ± 0.81	49.49 ± 0.46	72.84 ± 0.62
74i	33.53 ± 0.13	25.33 ± 0.15	28.34 ± 0.22
74n	90.07 ± 1.25	79.96 ± 0.95	84.29 ± 0.88

In April 2020, 1,2,3-triazole-imidazole (**75a–f**) and tetrazole derivatives (**76**) were synthesized by azide-alkyne cycloaddition reaction catalyzed by a novel nano copper catalyst and Ugi-azide four-component method (Scheme 2.15) [83]. Some of the prepared derivatives were also examined for anticancer activity against MCF-7 cell line using MTT assay. It was interesting to note that the tetrazole derivatives (**76a–i**) were inactive against this cell line. The IC_{50} values for the triazole derivatives are summarized in Table 2.17.

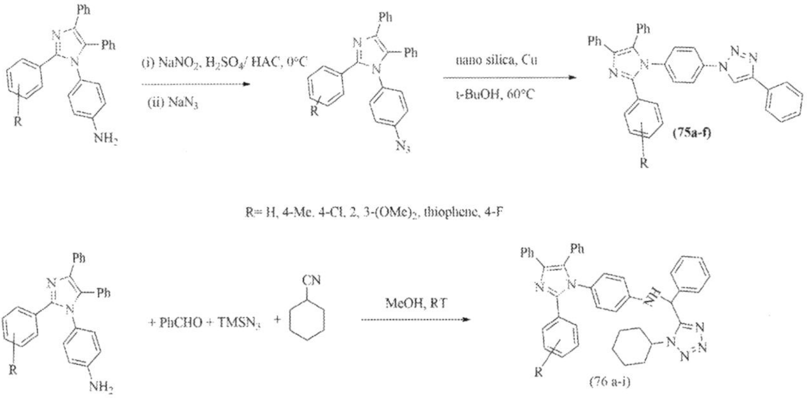

SCHEME 2.15 Synthesis of 1,2,3-triazole- and tetrazole imidazole composite.

A new series of bisimidazole and imidazo[1,2-a]pyridine has been synthesized by Schiff base dimers. SAR of all the compounds was studied against HeLa, MDA-MB-231, and ACHN cancer cell lines and the bis-imidazoles, imidazole-Schiff base hybrid derivatives (**77a–f**) and imidazopyridines (**78a–k**) presented significant anticancer activities against all cell lines [84]. The structures of all the derivatives and their synthesis are shown in Figs. 2.45 and 2.46. In particular, imidazopyridine derivatives demonstrated potential anticancer activity. Moreover, the unsubstituted derivatives, (**78a**) and (**78f**), exhibited the best anticancer activities among all the synthesized compounds. They were also found to be effective in treatment of cancer through in vivo study.

In 2016, quinoline-based substituted imidazole derivatives were synthesized and examined against different cell lines MCF-7, HeLa, MDA-MB-231, HuH-7, and EMT6/AR1, and it was interesting to note that the synthesized derivatives

TABLE 2.17 IC_{50} of triazole series treated on MCF-7 cells.

Sample	75a	75b	75c	75d	75e	75f	Methotrexate
IC_{50} (μg/mL)	376.5	424.4	261.8	1427	111.3	114.8	22

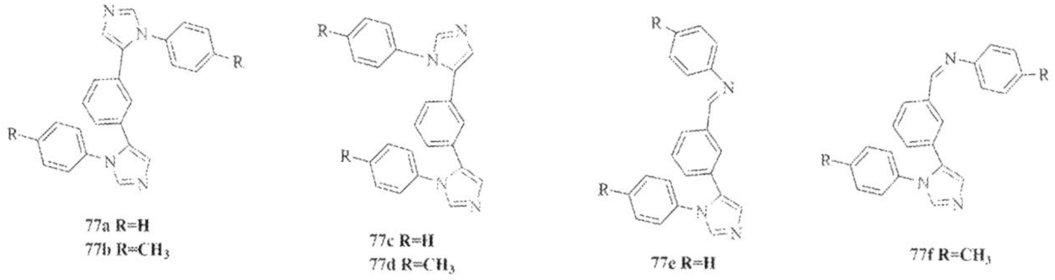

FIG. 2.45 Imidazole derivatives.

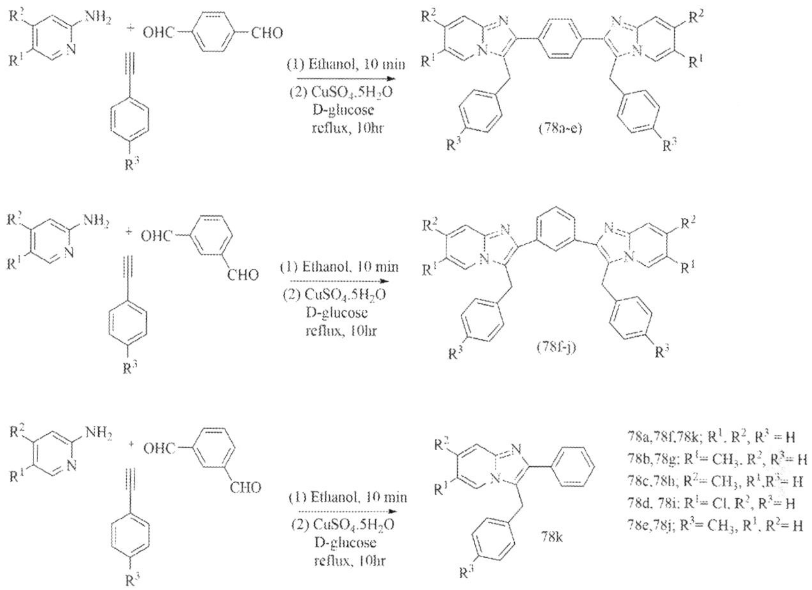

FIG. 2.46 Imidazopyridine dimmers.

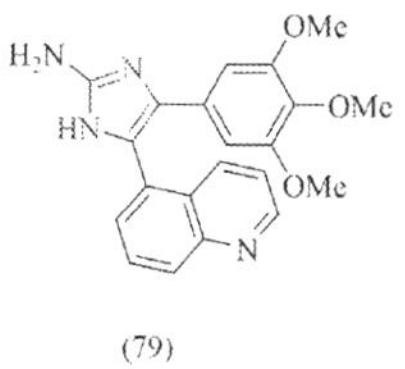

(79)

FIG. 2.47 Quinoline derivative of imidazole.

could inhibit the cell proliferation slightly more than the combretastatin A-4. The compound (**79**) (Fig. 2.47) also has a strong inhibitory effect on the movement of highly metastatic MDA-MB-231 cells, thereby causing a significant depolymerization of microtubules and mitotic block in MCF-7 cells, and blocking the tubulin polymerization in vitro [85].

2.3 Imidazole fused heterocycles

Heterocyclic scaffolds are the core structural unit of biologically active and pharmaceutically potent compounds. They play a vital role in many biological processes. Several drugs have heterocyclic moieties and as a result, immense efforts have been made to design green and efficient synthetic methods to fabricate these potent compounds. Since small and simple molecules often exhibit amazing and comprehensive pharmacological characteristics, scientists have been trying to compose the most imperative module of new active molecules with enhanced biological activity. To develop such an ideal structure, fused heterocycles have been demonstrated in the last couple of years. Imidazole fused heterocycles are one version of such entities that have a wide substrate scope and a plethora of biological activities [86].

This section is subdivided into two different parts:

a. imidazole fused heterocycles
b. benzimidazole fused heterocycles.

2.3.1 Imidazole fused heterocycles

The imidazole ring has a good substitution capacity and can be easily embedded with other rings. In 2013, Chen et al. [87] synthesized 34 derivatives of 2-thio-substituted anthra[1,2-*d*]imidazole-6,11-diones (Fig. 2.48) and studied them for cell proliferation, hTERT repressing action, and NCI 60-cell panel assay. From the synthesized derivatives, six compounds (**80a**, **80b**, **80c**, **80k**, **80o**, **81**) were selected by the National Cancer Institute (NCI). The anticancer activity and toxicity were studied using an NCI drug screen program of 60 human cancer cell lines including nine cancer cell types: colon, breast, ovarian, nonsmall cell lung, prostrate, leukemia, renal, melanoma, and CNS cancer. The leukemia, NSCLC, colon, and prostrate cell lines were sensitive to compounds (**80k**) and (**80o**) while CNS, melanoma, renal, ovarian, and breast cell lines were found to be less sensitive. Overall dose response curves showed that the leukemia subpanel was most sensitive for (**80k**) and (**80o**). Colon cancer was ranked as second most sensitive. Almost all compounds exhibited different cytostatic and cytotoxic activities for their potential applications as anticancer agents.

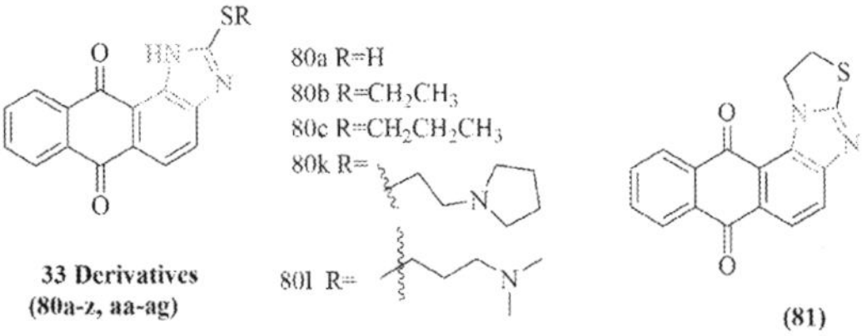

FIG. 2.48 2-Thio-substituted anthra[1,2-*d*]imidazole-6,11-diones derivatives.

Telomerase is a ribo-nucleoprotein and is made up of a catalytic subunit (human telomerase reverse transcriptase, hTERT) and a template RNA (human telomerase RNA component, hTERC). Expansion of telomeres by telomerase is necessary for unlimited propagation of most of the immortal and cancer cells. Due to the imperative function of telomerase in cellular immortalization, it has become an important molecular target for the drugs that are used to cure cancers that might have minimal side effects. For this reason in 2013, a library (40 derivatives) of anthra[1,2-*d*]imidazole-6,11-dione derivatives were prepared (Fig. 2.49) and estimated for hTERT expression and suppression of cancer cell growth, and telomerase inhibition in vitro [88]. All the compounds were examined against H1299 cells, PC-3 cells by MTT assay or sulforhodamine B assay (SRB assay), as well as telomerase inhibition using TRAP assay and SEAP assay. The compounds (**82a**), (**82f**), (**82j**), (**82n**), and (**82t**) showed cytotoxicity toward H1299 cell lines. It was interesting to note that with increase in the side chain R, the cytotoxicity toward the PC-3 cell increased. Six compounds—(**82d**), (**82g**), (**82p**), (**82x**), (**82aa**), and (**82ab**)—were selected by NIC and it was found that they inhibited the leukemia cells. Compound (**82t**) is the most potent inhibitor of telomerase activity among these compounds. IC_{50} values of some compounds are illustrated in Table 2.18. Compound (**82an**) appeared to show the best selectivity toward repressing hTERT expression.

In the designing of anticancer drugs, fused heterocycles played a crucial role due to enhancement of their biological and pharmacological profile. For this reason, Kayarmar and coworkers devised efficient azabicyclo[4.2.0]octa-1,3,5-trien-8-one analogs of 1*H*-imidazo[4,5-*c*]quinoline derivatives (**84a–f** and **85a–f**) (Fig. 2.50) [89]. The compounds were scrutinized against HeLa cells using a trypan blue exclusion assay method. An in silico study was also carried out on a protein with PDB Id: 1OJ0 in order to identify the interactions, and it was found that compounds were good inhibitors of β-tubulin. The IC_{50} values of compounds are also summarized in Table 2.19.

Imidazoacridinone derivative C-1311 (**86**) is an antitumor agent (Fig. 2.51). It causes inhibition of reactive DNA-topoisomerase II and is a selective inhibitor of recombinant FLT3. Skwarska and coworkers [90] further studied the impact of imidazolacridinone on leukemia cells with wild-type FLT3, FLT3-ITD mutant, and no FLT3 receptor. C-1311 inhibited FLT3 downstream pathways MAPK and AKT irrespective of FLT3 status. The treated FLT3-ITD included G1

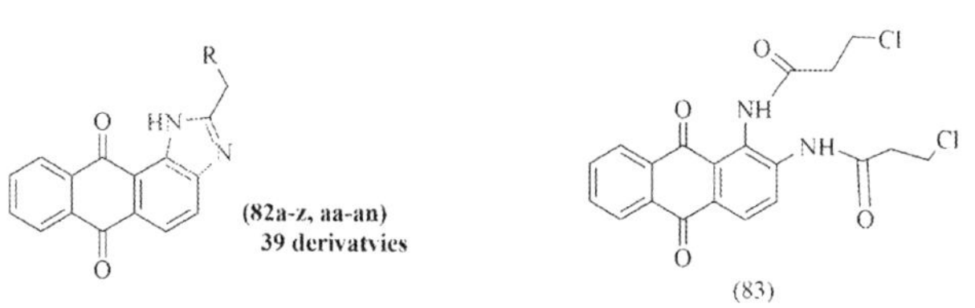

FIG. 2.49 Structure of anthra[1,2-*d*]imidazole-6,11-dione derivatives.

TABLE 2.18 Effects of anthra[1,2-*d*]imidazole-6,11-dione derivatives on cytotoxicity by MTT assay and repressing hTERT expression activity by SEAP assay and IC_{50} values against PC-3 cell for selected compounds.

		Cell type/H1299				PC-3 cell line
		MTT assay viability		SEAP activity		IC_{50} values
Compound	R	10 μM	100 μM	10 μM	100 μM	μM
82a	-Cl	0 ± 2	0 ± 1	0 ± 1	3 ± 1	5.1
82b	$-N(CH_3)_2$	64 ± 3	0 ± 0	45 ± 1	0 ± 1	8.8
82c	$-N(CH_3)CH_2CH_3$	65 ± 3	0 ± 0	39 ± 0	0 ± 1	7.5
82f	$-N(CH_3)(CH(CH_3)_2)$	23 ± 3	0 ± 1	1 ± 1	0 ± 2	6.7
82j	$-N(CH_2\text{-}2\text{-Pyridyl})_2$	13 ± 2	0 ± 2	0 ± 1	0 ± 2	6.4
82n	$-NHCH_2\text{-}4\text{-}CF_3C_6H_5$	11 ± 2	0 ± 1	8 ± 7	2 ± 1	7.6
82t	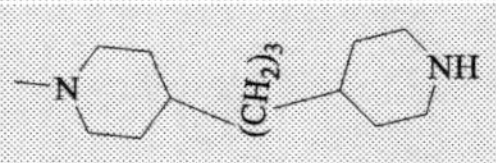	0 ± 0	0 ± 1	8 ± 0	3 ± 0	13.8
82v	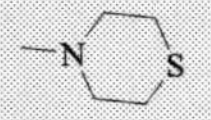	77 ± 4	1 ± 1	77 ± 2	5 ± 1	6.7
82x	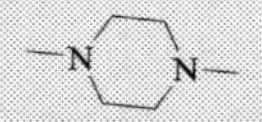	35 ± 4	0 ± 1	23 ± 1	6 ± 1	6.5
82y	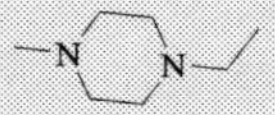	35 ± 4	0 ± 1	14 ± 1	0 ± 1	6.5
82ai	CH_2Cl	86 ± 3	40 ± 1	83 ± 5	8 ± 4	5.2
82an	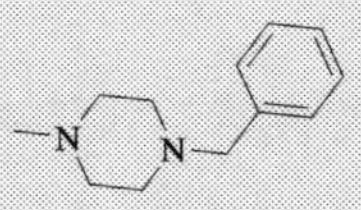	54 ± 9	6 ± 3	5 ± 5	0 ± 0	16

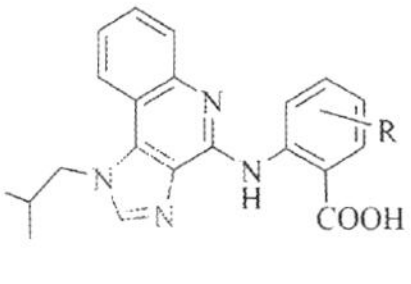

(84a-f)

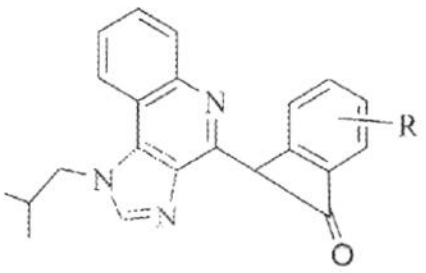

(85a-f)

FIG. 2.50 1*H*-Imidazo[4,5-*c*]quinoline derivatives.

TABLE 2.19 IC_{50} values of compounds against HeLa cell.

Compound	R	IC_{50} (μM)
84a	H	31 ± 1.63
84b	3-Br	19 ± 1.6
84c	5-Br	16.4 ± 1.4
84d	4-Cl	17.33 ± 1.25
84e	5-Cl	23.65 ± 2.03
84f	5-I	30.33 ± 2.5
85a	H	46 ± 2.45
85b	3-Br	16.3 ± 0.45
85c	5-Br	17.6 ± 0.37
85d	4-Cl	18.1 ± 0.98
85e	5-Cl	18.33 ± 1.27
85f	5-I	35.7 ± 1.79
Cyclophosphamide		4.2 ± 0.21

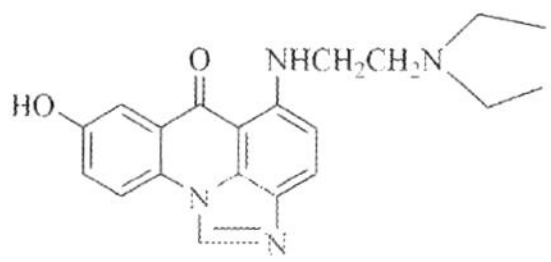

(86)

FIG. 2.51 Structure of C-1311.

and G2/M phase mutants, modest inhibition of Bcl-2, PARP cleavage, caspase-3 activation, and depolarization of mitochondria. C-1311 strongly decreased anti-apoptotic survivin mRNA, and also reduced protein expression, correlating well with improved apoptosis of FLT3-ITD cells. The mutated FLT3-ITD kinase is a good target for imidazoacridinone C-1311.

Cyclin-dependent kinases (CDKs) enzymes activate the cell cycle transition and regulate various factors of normal cell division. Therefore, 3*H*-imidazole[4,5-*c*]pyridine derivatives were designed for the treatment of cancer via inhibition of CDK2 [91]. The reaction proceeded via multistep synthesis; initially 4-hydroxypyridine underwent nitration and reacted with $SOCl_2$ and Et_3N to produce pyridine derivatives, which further reacted with HCl/$SnCl_2$ and $HC(OC_2H_5)_3$, ACN to form 4,6-dichloro-1-ethyl-1*H*-imidazo[4,5-*c*]pyridin-7-amine, and further reaction with H_2O_2, and H_2SO_4 yielded nitro derivative. This was then condensed with substituted amine/*N*,*N*-diisopropylethylamine and $SnCl_2$ in MeOH to produce the desired products (**87a–j** and **88a–f**) in good yields (Scheme 2.16). The results of an in vitro antiproliferation study revealed that most of the compounds showed high potency and (**87b**) displayed excellent **CDK2** inhibition (IC_{50}=21 nM) and cytotoxicity against A549, HL60, and HCT116 cell lines (Table 2.20).

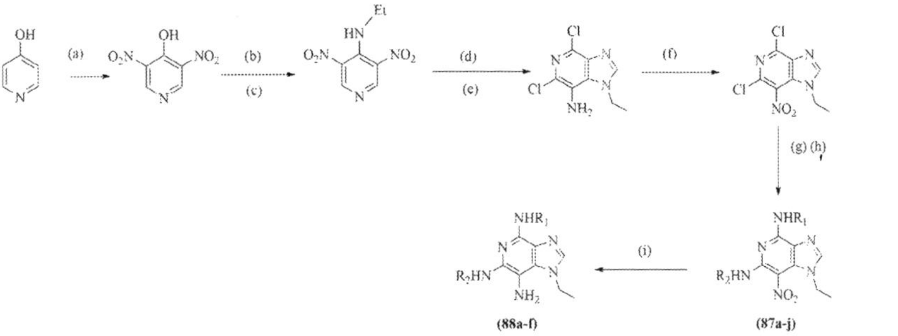

(a) HNO_3, H_2SO_4, reflux; (b) $SOCl_2$, reflux;
(c) $EtNH_2$, THF, rt; (d) HCl, $SnCl_2$, 70°C; (e) $HC(OC_2H_5)_3$, ACN; (f) H_2O_2, H_2SO_4; (g) R_1NH_2, DIEA, NMP, 80°C; (h) R_2NH_2, DIEA, NMP, 100°C;
(i) MeOH, $SnCl_2$, rt

SCHEME 2.16 Synthesis of 3*H*-imidazole[4,5-*c*]pyridine derivatives.

TABLE 2.20 IC_{50} values of selective compounds.

	IC_{50} values (μM)			
Compound	A549	HL60	HCT116	CDK2
87b	3.71	3.85	1.56	0.021
87c	12.60	> 50	13.85	8.55
87h	2.01	4.92	7.56	0.366
87i	2.38	21.57	20.20	0.415
87j	4.34	21.17	19.73	1.98
88c	2.67	15.41	16.18	14.46
CYC202	17.90	13.54	6.65	ND

Note: *ND*, not determined.

WDR5 is a component of a large family of WD40-repeat proteins. It is a chromatin-regulatory protein that gets overexpressed in several cancers like leukemia, breast, bladder, colorectal, pancreatic, and neuroblastoma. Due to this, it has become a probable epigenetic drug target to cure mixed-lineage leukemia. Wang et al. [92] introduced novel 2-aryl-6,7-dihydro-5-*H*-pyrrolo [1,2-*a*] imidazole derivatives (**89**) for the inhibition of WDR5-WIN-site, which was related with cell-cycle regulation by fragment-based methods and structure-based design (Fig. 2.52). The molecules displayed dose-dependent inhibition of the H3K4-methylation process and good growth inhibition against MLL-r-harboring cell lines.

Xiao et al. [93] synthesized a series of quinoline-imidazole composites (**90a–e**, **91a–e**, **92a–e**, **93a–e**, **94a–e**) (Fig. 2.53) and screened them for antitumor activity against four cancer cell lines—HepG2, A549, PC-3, and MCF-7—compared with a reference drug, sorafenib, and NVP-BEZ235. Most of the compounds displayed higher inhibition for the cell lines HepG2 and A549 than for MCF-7 and PC-3. The compounds (**93a**), (**93b**), (**94a**), and (**94b**) showed good inhibition of cancer cell proliferation. The activity against normal cell lines (WI-38) showed that derivatives (**93a**), (**93c**), and (**94a–c**) did not express any toxicity on WI-38. These results indicate that the presence of EWG on the quinoline ring increased their cytotoxicity, as depicted in Table 2.21.

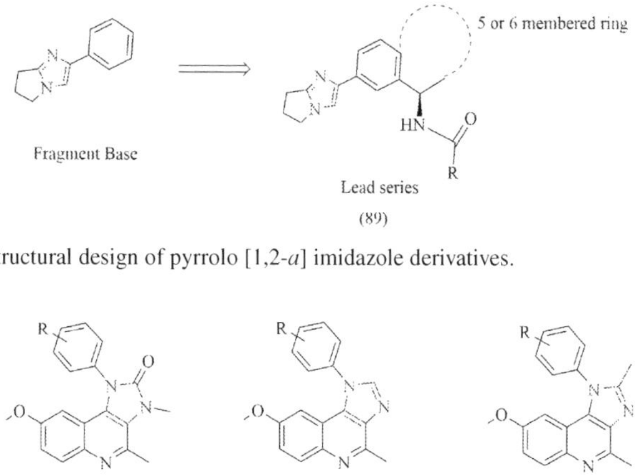

FIG. 2.52 Structural design of pyrrolo [1,2-*a*] imidazole derivatives.

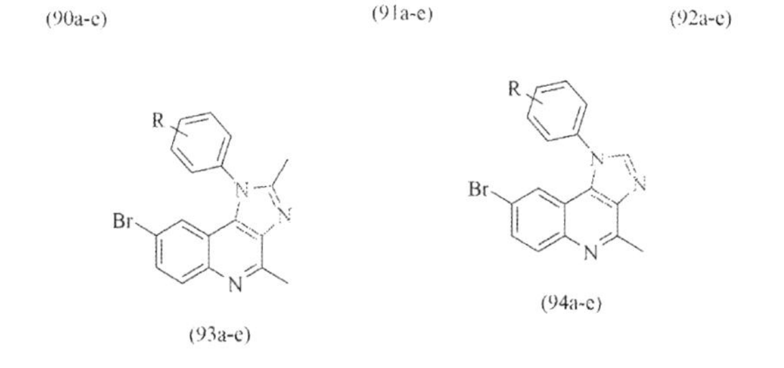

FIG. 2.53 Structures of quinoline-imidazole composites of series (**90–94**).

TABLE 2.21 The cytotoxicity of selective compounds of series (**90–94**).

		IC_{50} (μM)			
Compound	R	HepG2	A549	PC-3	MCF-7
90a	Isobutyronitrile	11.31 ± 1.30	39.66 ± 0.89	43.07 ± 1.43	32.52 ± 1.51
91b	OCH_3	68.54 ± 1.05	58.74 ± 0.94	73.73 ± 1.16	> 50
91e	4-Br-2-F	11.82 ± 1.12	28.03 ± 2.03	21.62 ± 3.89	43.29 ± 1.63
92b	OCH_3	15.54 ± 2.12	14.04 ± 1.18	16.54 ± 1.10	31.03 ± 1.49
92e	4-Br-2-F	6.08 ± 1.04	11.23 ± 1.02	15.36 ± 1.54	11.86 ± 1.07
93a	Isobutyronitrile	1.43 ± 1.02	13.43 ± 0.92	6.67 ± 0.99	> 50
93b	OCH_3	5.81 ± 1.04	4.67 ± 1.44	10.81 ± 1.21	6.21 ± 0.79
93c	Br	33.39 ± 1.43	> 50	31.68 ± 1.00	46.67 ± 1.66
93d	3-Cl-4-F	4.83 ± 1.07	3.93 ± 0.93	12.89 ± 0.98	16.17 ± 1.20
93e	4-Br-2-F	15.66 ± 1.05	7.68 ± 0.88	31.79 ± 2.14	18.51 ± 1.26
94a	Isobutyronitrile	2.42 ± 1.02	6.29 ± 0.99	5.11 ± 1.00	> 50
94b	OCH_3	16.37 ± 1.11	5.36 ± 1.02	13.20 ± 2.12	4.15 ± 0.62
Sorafenib	–	3.97 ± 0.13	6.53 ± 0.23	3.03 ± 0.11	4.21 ± 0.15
BEZ235		0.54 ± 0.13	0.36 ± 0.06	0.20 ± 0.01	0.14 ± 0.01

Zhang et al. [94] reported synthesis of pyrrole-imidazole derivatives by post-Ugi cascade reaction (Scheme 2.17) and evaluated them for anticancer activity against the human pancreatic cancer cell lines PANC and ASPC-1. The resultant anticancer activity of derivatives is depicted in Table 2.22. The library of synthesized compounds is depicted in Fig. 2.54. Among the series, compound **95e** exhibited the most promising and selective anticancer activity toward pancreatic cancer cell lines.

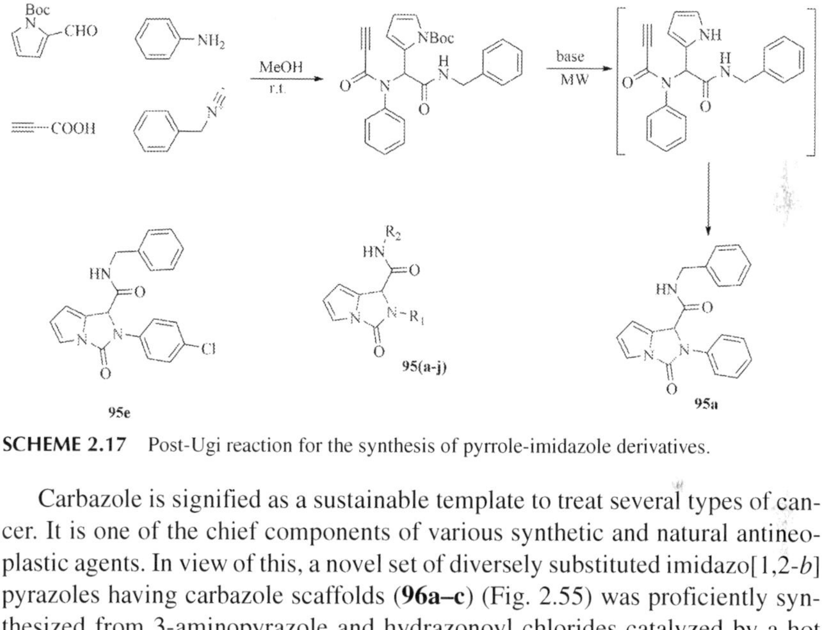

SCHEME 2.17 Post-Ugi reaction for the synthesis of pyrrole-imidazole derivatives.

Carbazole is signified as a sustainable template to treat several types of cancer. It is one of the chief components of various synthetic and natural antineoplastic agents. In view of this, a novel set of diversely substituted imidazo[1,2-*b*] pyrazoles having carbazole scaffolds (**96a–c**) (Fig. 2.55) was proficiently synthesized from 3-aminopyrazole and hydrazonoyl chlorides catalyzed by a hot alcoholic solution of triethylamine [95]. The compounds were also examined for their in vitro anticancer activity against three cell lines, MCF-7, HCT-116, and HepG-2, by MTT assay and also assessed toward normal cell RPE1 to examine their toxicity to normal cells (Table 2.23).

TABLE 2.22 Anticancer activities of compounds (95e–g) and (95i).

	IC_{50} (μM)			
Compound	A549	PC3	PANC	ASPC-1
95e	0.93 ± 0.34	0.83 ± 0.27	0.063 ± 0.09	0.062 ± 0.013
95f	10.53 ± 2.2	8.17 ± 1.7	31.87 ± 4.1	9.2 ± 0.12
95g	8.63 ± 1.6	8.09 ± 0.19	5.61 ± 0.14	6.21 ± 0.15
95i	12.31 ± 2.3	3.70 ± 0.22	7.85 ± 0.8	11.4 ± 1.1

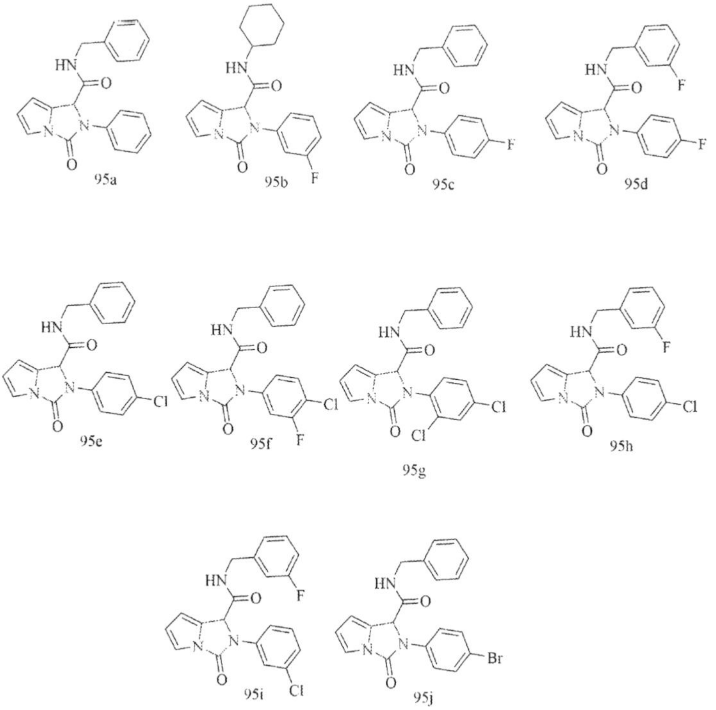

FIG. 2.54 Library of pyrrole-imidazole derivatives.

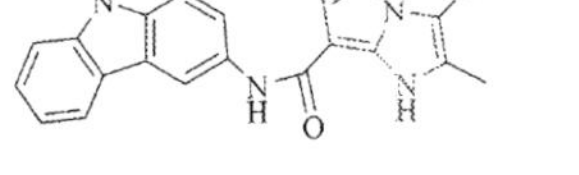

(96a-c)

FIG. 2.55 Imidazo[1,2-*b*]pyrazoles having carbazole scaffolds.

TABLE 2.23 IC_{50} values of synthesized compounds.

		IC_{50} (μM)			
Compound	Ar	HCT-116	Hep-G2	MCF-7	RPE1
96a	4-$CF_3C_6H_4$	12.31 ± 0.60	10.31 ± 0.20	14.01 ± 1.12	> 100
96b	4-ClC_6H_4	19.40 ± 1.60	12.97 ± 1.40	18.17 ± 1.40	> 100
96c	4-BrC_6H_4	> 100	> 100	> 100	> 100
Doxorubicin		0.90 ± 0.02	0.66 ± 0.02	0.64 ± 0.02	0.58 ± 1.12

2.3.2 Benzimidazole fused heterocycles

Heterocycles are chief structural composites of anticancer drugs in the market. In fact, approximately two-thirds of the available drugs that were approved by the Food and Drug Administration (FDA) between 2010 and 2017 contained heterocyclic moieties. One of these heterocycles, benzimidazole, is an important scaffold and has achieved extensive consideration in the area of modern medicinal chemistry. This structural unity is of considerable significance due to its widespread biological activities [96].

In 2009, pyridine-based benzimidazole derivatives were designed and evaluated for antiproliferative activity against K562 and CEM cells [97]. The cells were harvested for 48 h and 72 h of treatment, and treated with different concentration of derivatives (10, 100, 250 μM). IC_{50} values were estimated after 72 h. The compounds (**97a–j**) (Fig. 2.56) demonstrated moderate inhibition ($IC_{50} < 200$ μM) in the K562 cell line and very poor activity against the CEM cell line. Among the synthesized derivatives, (**97j**) showed the highest cytotoxicity with IC_{50} 140 ± 15.7 μM and 210 ± 19.8 μM for the K562 cell and CEM cell, respectively.

A series of novel C-3, C-16, and C-17 analogs were synthesized to explore the effect of small structural modifications of the molecule 3β-(hydroxy)-17-(1*H*-benzimidazol-1-yl)androsta-5,16-diene (galeterone, 98) on the inflection of the androgen receptor (AR) (Scheme 2.18). The structure activity relationship of the synthesized compounds established that benzimidazole moiety at C-17 is essential and the presence of hydrophilic and heteroaromatic groups at C-3 increased both antiproliferative and AR degrading activities. Compounds (**98c**), (**98a**), and (**98b**) were found to be the most potent antiproliferative compounds, with GI_{50} values of 0.87, 1.91, and 2.57 μM, respectively. Compound (**98c**) was found to have fourfold and eightfold more potency than (**98**) for antiproliferative and ARD activities, respectively. It was also discovered that compounds (**98**), (**98a**), (**98b**), and (**98c**) degraded both full-length and truncated ARs in CWR22rv1 human prostate cancer cells [98].

A series of 5,6-dichloro-1-β-D-ribofuranosyl benzimidazole (DRB) and benzo[*d*]imidazole as potent anticancer agents was synthesized by Alkahtani et al. [99]. The antiproliferative activity was assessed against HCT-116 colorectal carcinoma and MCF-7 breast carcinoma cells using MTT cytotoxicity assay. The structure activity relationship (SAR) was also established in order to ascertain the potency of designed compounds (Fig. 2.57). Compound (**99o**) was

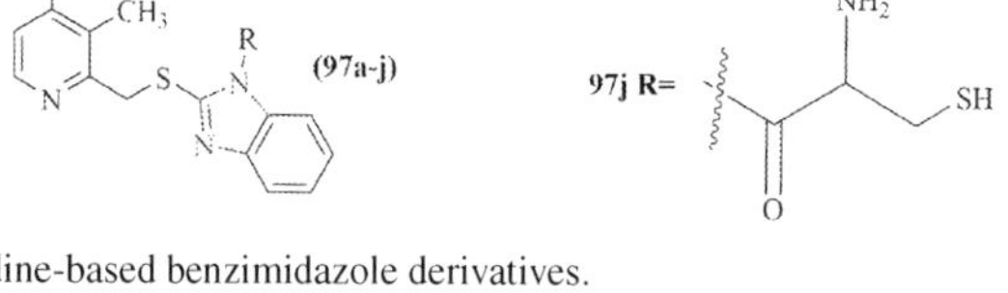

FIG. 2.56 Pyridine-based benzimidazole derivatives.

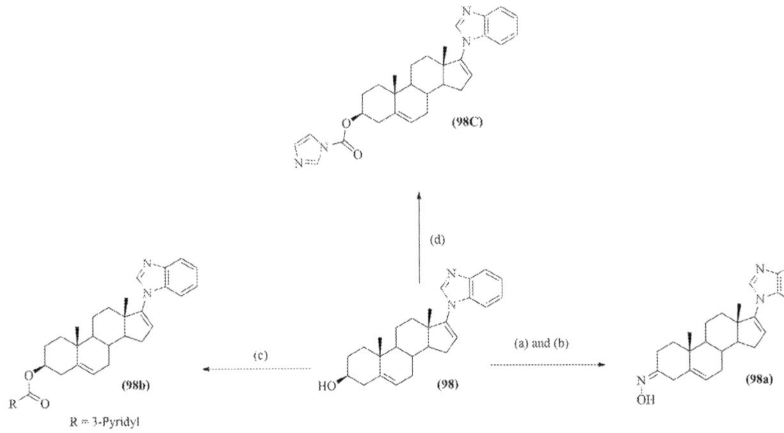

(a) Al(i-PrO)$_3$, 1-methyl-4-piperidone, toluene, reflux;
(b) Substituted hydroxylamine·HCl, sodium acetate, MeOH, EtOH, Ar, reflux (2-3 h);
(c) pyridinecarboxylic acid, 2-methyl-6-nitobenzoic anhydride, DMAP, TEA, THF, rt (1 h);
(d) 1,1'Œ-carbonylbisimidazole or 1,1'Œ-carbonylbis(2-methylimidazole) or 1,1'Œ-carbonyl-di-(1,2,4-triazole), CH_3CN, Ar, rt/reflux.

SCHEME 2.18 Synthesis of C-3 modified compound from (**98**).

(99a-r)
For 99o R_1= Me; R_2=Cyclopentyl

100a R_1= SOMe; R_2=Cyclopentyl
100b R_1= SO_2Me; R_2=Cyclopentyl

FIG. 2.57 Substituted benzimidazole derivatives.

found to induce cancer cell apoptosis. On replacing ribose moiety in DRB with cyclopentyl ring, the antiproliferative activity was slightly reduced compared to DRB. On further modification in R_1 with methyl/isopropyl group (**99o/99r**), the cellular potency in HCT-116 cells increased more than twofold (GI_{50}= 13 and 9 μM). MCF-7 cells were found to be resistant to both of these compounds. Compounds (**100a**) and (**100b**) with the R_1 group as methyl sulfinyl or methyl sulfonyl were found to be highly potent. Furthermore, the compounds were studied for caspase-3 activation assay in order to study their capability to induce apoptosis in cancer cells. However, compounds (**99o**), (**100a**), and (**100b**) showed significant caspase-3 activity.

Aurora A is associated with a family of mitotic serine/threonine kinases. It is implicated with important processes during mitosis and meiosis whose proper function is integral for healthy cell proliferation. It is overexpressed in several types of human cancers. Furthermore, overexpression of Aurora A kinase is connected with drug opposition and poor prognosis in numerous cancers. Hence, this has been an important anticancer target to cure human cancers. In 2016, Im and coworkers [100] reported the docking study from the surflex-dock module in SYBYL and 3D QSAR (quantitative structure activity relationship) studies of imidazole-based compounds (**101a–l**) (Fig. 2.58) against Aurora A kinase, which affected the apoptosis process in various cell lines. These molecules inhibited Aurora A kinase via controlling abnormal mitosis from arresting cell cycle at pseudo G1 state. The IC_{50} values are summarized in Table 2.24.

In one study, 6-(2-amino-1*H*-benzo[*d*]imidazol-6-yl)quinazolin-4(3*H*)-one derivatives (**102a–h**) were prepared in a multistep process. Firstly, substituted quinazolin-4(3*H*)-one derivatives were synthesized from 2-amino-5-bromobenzoic acid; these were then further reacted with borate ester of nitro aniline with a couple of reagents to furnish the desired product (Scheme 2.19) [101]. To anticipate the activity of the prepared compounds, MTT assay was used to estimate their antiproliferative activities against different cell lines, prostate cancer PC3, breast cancer MDA-MB-231, and neuroblastoma SH-SY5Y cells.

TABLE 2.24 Structure and biological activity of benzo[*d*]imidazoles.

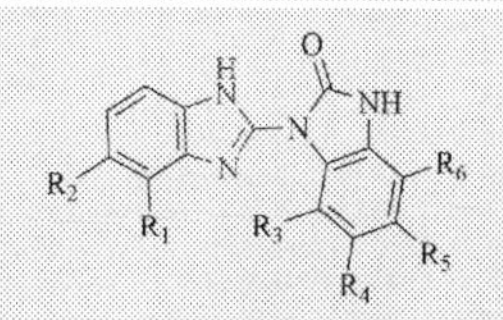

(101a-l)

FIG. 2.58 Benzimidazole derivatives.

Compound	R_1	R_2	R_3	R_4	R_5	R_6	IC_{50} (μM)
101a	H	H	H	H	H	H	7.0
101b	Me	H	H	H	H	H	14
101c	H	Me	H	H	H	H	7.8
101d	H	Br	H	H	H	H	9.0
101e	H	H	OH	H	H	H	15
101f	H	H	H	H	H	OH	21
101g	H	H	H	Cl	H	H	23
101h	H	H	H	OH	H	H	12
101i	H	H	H	COOH	H	H	3.5
101j	H	H	H	CONHMe	H	H	11
101k	H	H	H	H	OH	H	2.3
101l	H	H	H	H	COOH	H	3.5

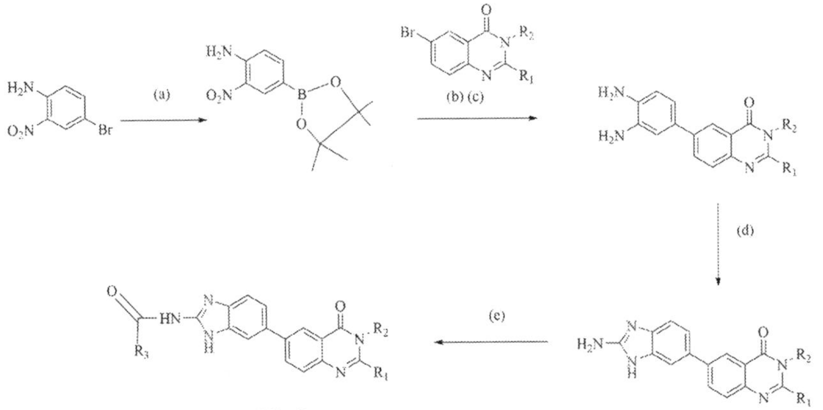

(102 a-h)

(a) bis(pinacolato)diborane, Pd(dppf)Cl_2, AcOK, dioxane, 100°C, 2 h;
(b) substituted quinazolin-4(3H)-one, Pd(dppf)Cl_2, K_2CO_3, dioxane/H_2O, 100 °C, 5 h;
(c) NH_2NH_2. H_2O, Pd/C, MeOH, 80°C, 2 h;
(d) CNBr, MeOH/H_2O. 60°C, 5 h;
(e) Acyl chloride, Et3N, THF, 0°C to ambient temperature, overnight

SCHEME 2.19 Synthesis of 6-(2-amino-1*H*-benzo[*d*]imidazole-6-yl)quinazolin-4(3*H*)-one derivatives.

The compound (**102h**) was found to be the most active against MDA-MB-231, with $IC_{50}=0.38\,\mu M$, and 1.09 μM for PC3 cells, 0.77 μM toward SH-SY5Y cells. Compound (**102h**) also inhibited colony formation slightly. The presence of electron withdrawing and bulky group decreased the cytotoxicity. Compound (**102h**) also underwent molecular docking studies using protein 3P9J. The N atom of the morpholine moiety, benzimidazoles, and the side chains of amide (**102h**) created conserved H-bonds with different amino acid residues like Ala213 in the kinase hinge region, the major sequence NH of Thr217 amino acid, and lysine of Lys162.

Hasanpourghadi and companions [102] introduced methyl 2-(5-fluoro-2-hydroxyphenyl)-1*H*-benzo[*d*]imidazole-5-carboxylate (MBIC) (**103**) (Fig. 2.59) as an effective drug candidate against inhibition of various cell lines HeLa, HCT-116, A549, HepG-2, and WRL-68 via arresting the cell cycle by MTT and tubulin polymerization assay. This MBIC molecule was prepared from the reaction of 5-fluoro-2-hydroxybenzaldehyde, *N*,*N*-dimethyl acetamide, and methyl

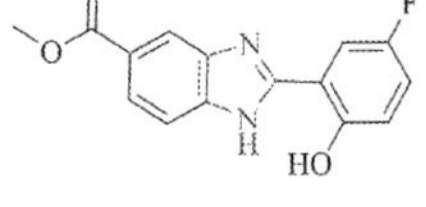

(**103**)

FIG. 2.59 Benzimidazole derivatives.

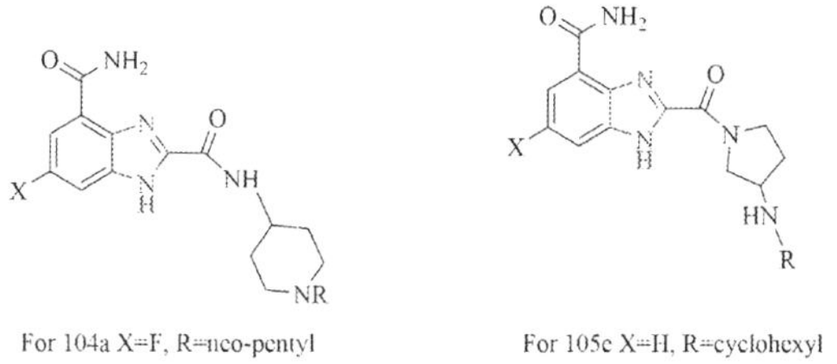

FIG. 2.60 2-Substituted 1*H*-benzo[*d*]imidazole-4-carboxamide derivatives.

3,4-diaminobenzoate (2 mmol) with $Na_2S_2O_5$ (2.4 mmol) at 100°C in good yields. In the presence of MBIC, HeLa cells underwent mitotic arrest comprising of multinucleation, displayed mitochondrial-dependent apoptotic proteins, interfered with tubulin polymerization, and showed synergistic effects against standard drugs, viz. doxorubicin, nocodazole, colchicine, and paclitaxel.

In 2017, Zhou et al. [103] synthesized 2-substituted 1*H*-benzo[*d*]imidazole-4-carboxamide derivatives and utilized them for antitumor activity via inhibition of poly(ADP-ribose)polymerase-1 (PARP-1). The in vitro study on breast cancer cells MX-1 results revealed that compounds (**104a**) and (**105e**) (Fig. 2.60) were highly active PARP-1 inhibitors with good potentiating effects on temozolomide (TMZ).

In the same year, benzimidazole-substituted-1,3-thiazolidin-4-one derivatives were designed and examined as anticancer drug candidates [104]. The reaction of 2-mercaptobenzimidazole with potassium hydroxide and ethyl chloroacetate produced imidazole derivatives, which on refluxing with hydrazine hydrate formed their hydrazide analogs. Then, it was refluxed with aldehydes (0.01 mol) in glacial acetic acid to prepare a Schiff base, and further condensed with thioglycolic acid using a $ZnCl_2$ catalyst, to form products (**106a–r**) in high yields (Scheme 2.20). The in vitro cytotoxic study was carried on human colorectal (HCT116) cell lines. All the synthesized compounds showed excellent activity against HCT116 cell lines compared to standard drugs. The compounds (**106l**) and (**106k**) exhibited high antiproliferative activity with IC_{50} = 0.00005 and 0.00012 μM/mL, respectively.

Furthermore, the authors designed a range of 2-(1-benzoyl-1*H*-benzo[*d*]imidazol-2-ylthio)-*N*-substituted acetamide analogs and studied their anticancer potency [105]. These hybrid compounds (**107a–t**) were prepared from the reaction of benzoyl chloride with 2-(1*H*-benzo[*d*] imidazol-2-ylthio)-*N*-substituted acetamide (Fig. 2.61). Among the synthesized molecules, compound (**107b**) possessed the highest cytotoxicity against MCF-7, with IC_{50} = 0.0047 μM/mL, and (**107j**) exhibited maximum cytotoxicity on HCT116 with IC_{50} = 0.0058 μM/mL.

In 2018, Wu and companions synthesized tanshinone-IIA (Tan-IIA) based imidazole derivatives (**108a–l**) and studied their inhibition against breast cancer (MDA-MB-231 cells) invasion and metastasis [106] (Fig. 2.62). The reductive

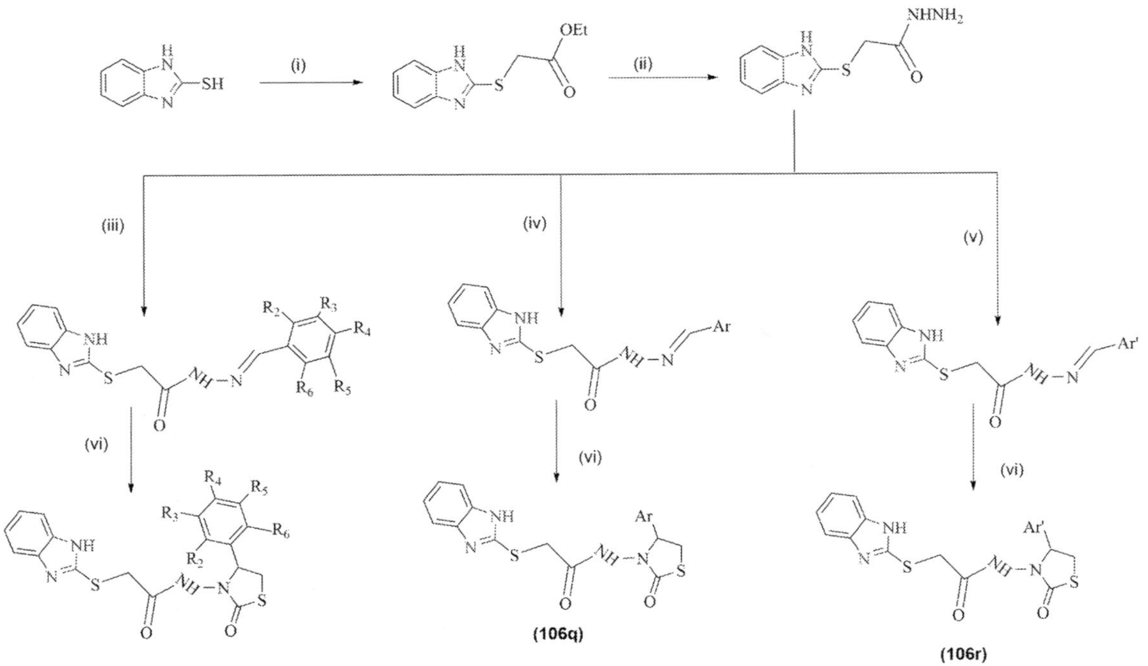

For 106k $R_2=R_5=R_6=H$; $R_3=OMe$, $R_4=OH$
For 106l $R_2=R_5=R_6=H$; $R_3=OEt$, $R_4=OH$

Reaction conditions: (i) Ethanol, ethyl chloroacetate, stirring for 24 h. (ii) Ethanol, hydrazine hydrate, reflux. (iii) Aryl aldehyde, ethanol, a few drops of glacial acetic acid. (iv) Cinnamaldehyde, ethanol, a few drops of glacial acetic acid. (v) 4-Hydroxy-naphthaldehyde, ethanol, a few drops of glacial acetic acid. (vi) Dioxane, thioglycolic acid, anhydrous zinc chloride, reflux

SCHEME 2.20 Synthesis of benzimidazole-substituted-1,3-thiazolidin-4-ones.

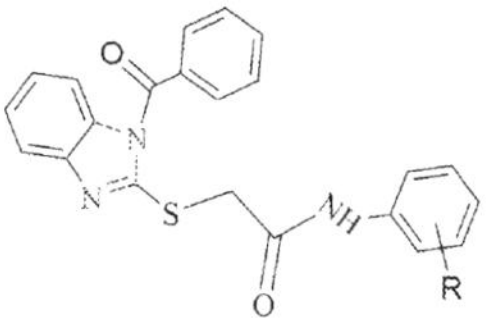

For 107b R=2-F
For 107j R=3-NO_2

2-(1-Benzoyl-1H-benzo[d]imidazol-2-ylthio)-N-substituted phenylacetamide derivatives **(107a-t)**

FIG. 2.61 2-(1-Benzoyl-1*H*-benzo[*d*] imidazol-2-ylthio)-*N*-substituted acetamide analogs.

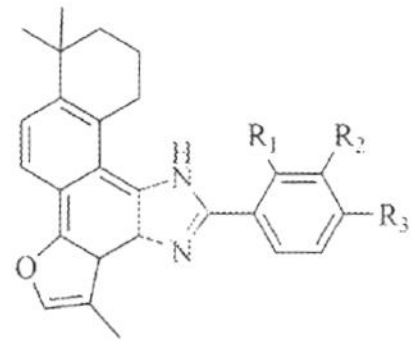

For 108l R_1=R_2=H; R_3=OH

Tanshinone-IIA Based Imidazole (108a-l)

FIG. 2.62 Tanshinone-IIA-based imidazole.

cyclization of aldehydes with Tan-IIA in basic medium under microwave irradiation yielded final, Tan-IIA based imidazole compounds. All the synthesized derivatives exhibited good activity, and compound (**108l**) displayed the highest blocking of proliferation, migration, and invasion of breast cancer cell lines with delaying of metastasis in a zebrafish xenograft model via arresting the S-phase pathway.

Also in 2018, a series of 3-(2-(1*H*-benzo[*d*]imidazol-2-ylthio)acetamido)-*N*-(substituted phenyl) benzamide derivatives was prepared and screened for antiproliferative activity against the human colorectal carcinoma cell lines (HCT116) from sulforhodamine B assay by Tahlan et al. [107] The desired product was prepared from the reaction of 3-(2-chloroacetamido) benzoic acid with 2-mercaptobenzimidazole, and after condensation with $SOCl_2$ produced 3-(2-(1*H*-benzo[*d*]imidazol-2-ylthio)acetamido)benzoyl chloride, which reacted with different amines and yielded final products (**109a–v**) (Fig. 2.63) in good yields. The molecule (**109q**) (IC_{50}=4.12 μM) showed higher inhibition activity compared to the reference drug, 5-FU (IC_{50}=7.69 μM).

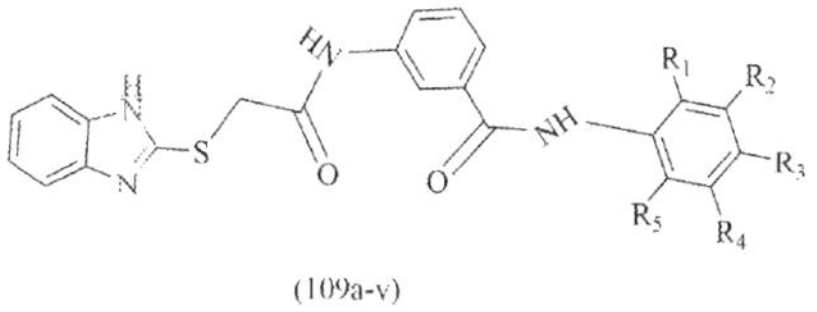

(109a-v)

(109q) R_1=R_2=R_3=R_5=H, R_4=OMe

FIG. 2.63 3-(2-(1*H*-Benzo[*d*]imidazol-2-ylthio)acetamido)-*N*-(substituted phenyl) benzamide derivatives.

Platelet-derived growth factor receptor β (PDGFRβ) is a trans-membrane receptor tyrosine kinase with greatly regulated cell expression. It exerts a vital role in angiogenesis and embryonic growth, and is responsible for the development of blood vessels, adipocytes, and kidneys. Cancer is related with abnormal expression and signaling of PDGFRβ, by the autocrine and paracrine stimulation of tumor cell growth. Consequently, PDGFRβ has gained substantial attention as an important anticancer drug target as well as being crucial for cancer imaging. The PDGFRβ imaging probe directs toward the intracellular domains, particularly adenosine triphosphate (ATP)-binding sites. In 2018, benzo[*d*]imidazole-quinoline composites (Fig. 2.64) were used to generate radioiodinated compounds as PDGFRβ-specific imaging probes [108]. Previously, the radioiodine and radiobromine derivatives of 1-{2-[5-(2-methoxyethoxy)-1*H*-benzo[*d*]imidazol-1-yl]quinolin-8-yl}piperidin-4-amine (CP-673451, IQP) were used for PDGFRβ imaging [109, 110].

The potency of synthesized derivatives was measured to reduce the viability of PDGFRβ-overexpressed (TR-PCT1) cell line and it was found that compounds (**110a**) and (**110d**) were the most potent for proliferation inhibition effects in PDGFRβ-positive cells. The iodination of (**110a**) and (**110d**) derivatives gave (**111**) and (**112**) derivatives. The derivative (**111**) exhibited superior inhibitory potency as a PDGFRβ inhibitor compared to (**112**).

Al-blewi and coauthors [111] devised an efficient, facile protocol for the synthesis of sulfonamide conjugated imidazole derivatives. The derivatives are regio-selective and synthesized in multiple steps. The initial step was regioselective propargylation catalyzed by Et_3N to give 2-substituted propargylated benzimidazole derivatives. The second step involved 1,3-dipolar cycloaddition

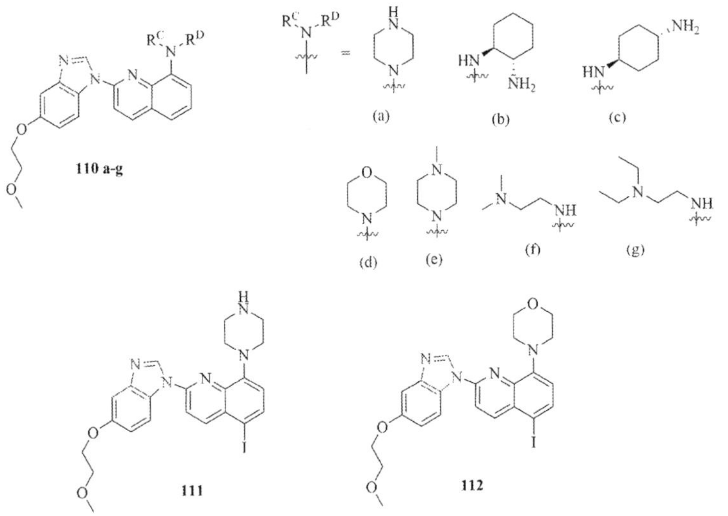

FIG. 2.64 Benzo[*d*] imidazole-quinoline composites.

reaction of azide building blocks of sulfa drugs with the propargylated benzimidazole derivative in the presence of Huisgen copper (I) to form the desired imidazole-based mono- and bis-1,4-disubstituted-1,2,3-triazole-sulfonamide conjugates (**113a–f** and **114a–f**) (Scheme 2.21). The fabricated compounds were examined for their anticancer activity against three different cell lines—PC-3, HepG2, and HEK293—and possessed remarkable activity with IC_{50} values in the range of 55–106 μM. The authors also examined the toxicity of compounds by in silico methods and it was found that the synthesized compounds were noncarcinogenic and possessed characters like a drug molecule.

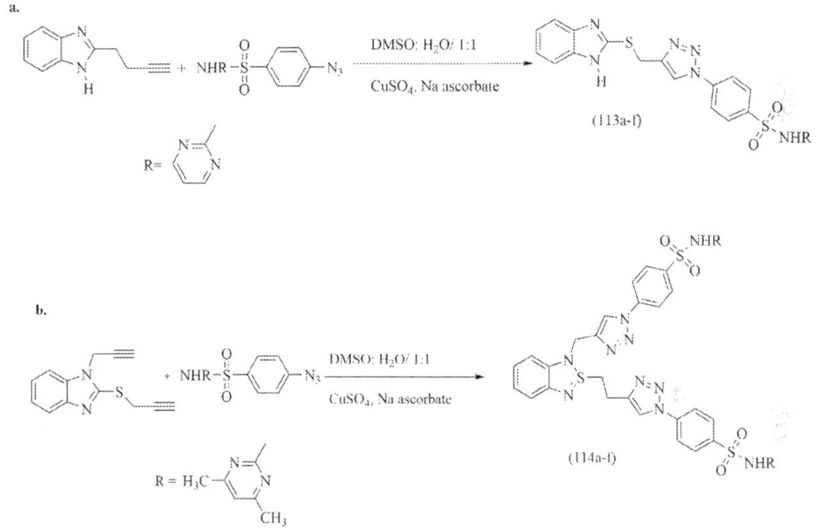

SCHEME 2.21 Synthesis of benzimidazole-sulfonamide conjugates.

Later on, in 2019, Tahlan et al. [112] described the synthesis of a library (30 derivatives) of benzimidazole derivatives through a multistep pathway (Scheme 2.22). At first, chloroacetyl chloride reacted with 2-mercaptobenzimidazole and formed 1*H*-benzo[*d*]imidazol-2-yl 2-chloroethanethioate. This further reacted with respective anilines and produced title products (**115a–o**). The intermediate formed reacted with hydrazine hydrate to form another intermediate that on reaction with various aldehydes gave products (**115p–z, aa–ad**). Moreover, these synthesized compounds were screened for their anticancer activity against HCT 116 cell lines. Among them, compound (**115w**) presented the most promising anticancer activity with an IC_{50} value of 0.46 μM, which was higher than that of standard drugs. The SAR study revealed that the presence of halogenated α, β-unsaturated aldehyde, i.e., 2-bromo-3-phenylacrylaldehyde, played an essential role in enhancing anticancer activity.

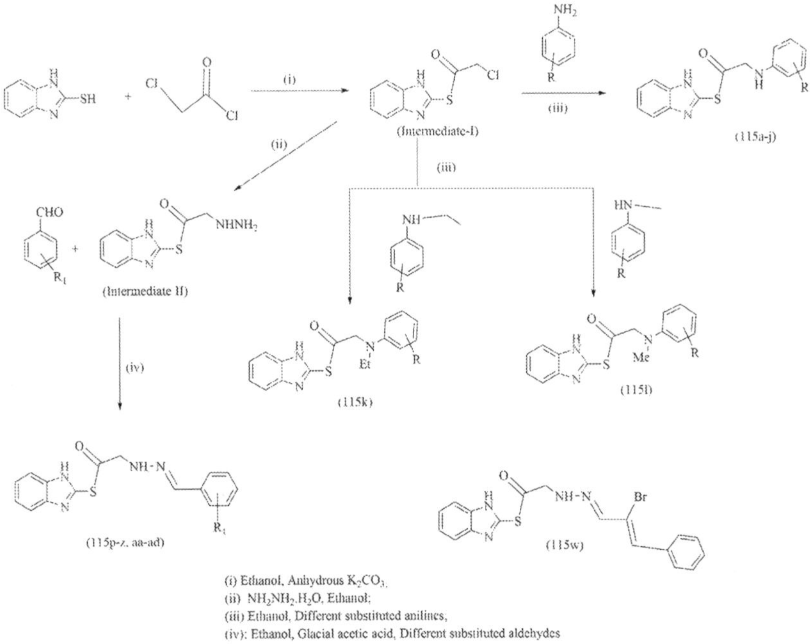

SCHEME 2.22 Benzimidazole derivatives.

Further in 2019, benzimidazole fused heterocycles were devised by Shi et al. [113]. The authors synthesized a series of 6-(2-(methylamino)ethyl)-5,6,7,8-tetrahydro-1,6-naphthyridin-2-amine derivatives and explored their CDK4/6 inhibitory activity against cancer (Scheme 2.23). Compound (**116c**) exhibited the most significant and selective activity toward CDK4/6 (IC_{50}=0.710/1.10 nM) compared to other kinases. It also exhibited good antiproliferative activity, desirable pharmacokinetic properties, and great metabolic properties.

Phosphatidylinositol-3-kinase (PI3K) controls several cellular procedures, viz. propagation of cells, apoptosis, growth, and autophagy. Class I PI3K is regularly mutated and overexpressed in a number of human cancers. Due to this, it has become a hit target for therapeutic treatment of cancer. In 2019, it was found that 1,6-disubstituted-1*H*-benzo[*d*]imidazole derivatives, synthesized from 4-bromo-2-fluoronitrobenzene through a multistep process, possessed good anticancer activity against three cell lines: MCF-7, T47D, and HCT116 (Fig. 2.65) [114]. SAR studies revealed that the activity of synthesized derivatives against HCT116 cell lines was greater for those derivatives that have a phenyl ring substituent at the third position than that of compounds containing pyrazolyl at the same position. It was also found that the activity of hydrophilic groups at third position was better than that of hydrophobic groups. Computational study was also performed with the selective (**117i**) molecule on protein with PDB ID: 3L08. The N atom of the benzimidazole ring created an H-bond with the side

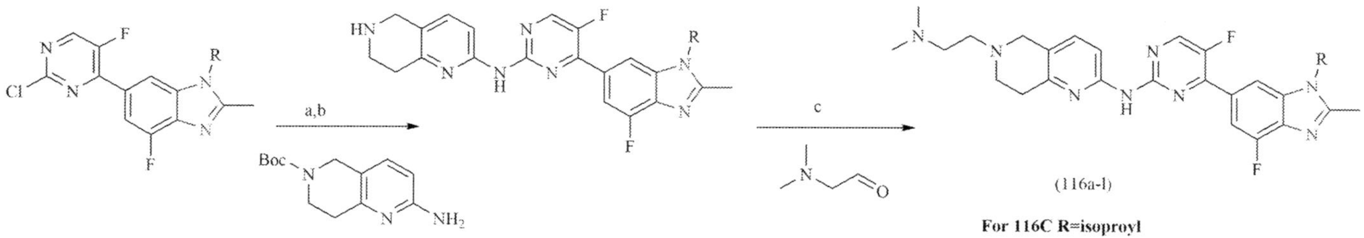

(116a-l)

For 116C R=isoproyl

(a) $Pd_2(dba)_3$. Xantphos. Cs_2CO_3, dioxane, 110°C;
(b) HCl, dioxane, rt;
(c) $NaBH(OAc)_3$, DCM.

SCHEME 2.23 Synthesis of 6-(2-(methylamino)ethyl)-5,6,7,8-tetrahydro-1,6-naphthyridin-2-amine derivatives.

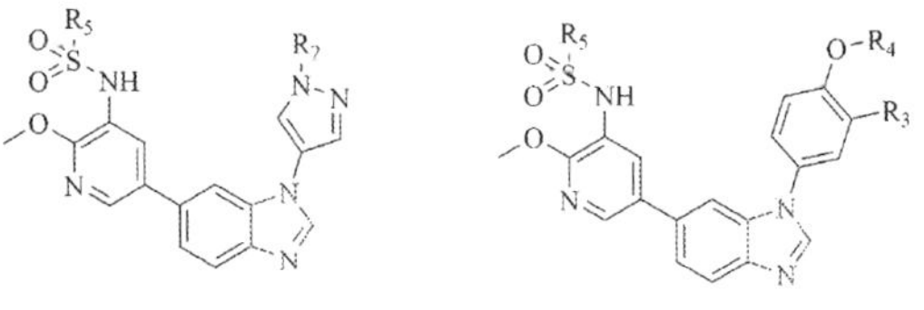

117a-h 117i-s

For 117i R_3=H, R_4= CH_2CH_2OH, R_5 =

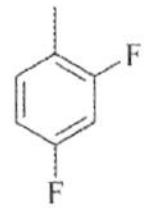

For 117l R_3=H, R_4= $CH_2CH_2CH_2OH$, R_5 =

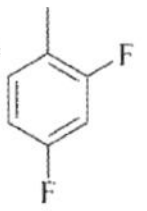

FIG. 2.65 1,6-Disubstituted-1*H*-benzo[*d*]imidazoles derivatives.

chain of amino acid Val882 in the hinge binder region of PI3K. Moreover, the oxygen atom of the OCH_3 group created an additional H-bond interaction with amino acid Lys833. The N atom of the pyridyl ring formed an H-bond with the conserved H_2O molecule. The H atom of the OH group formed a hydrogen bond interaction with Thr887. The in silico study results indicated that compound (**117i**) could fit into the binding site of PI3K kinase, which also indicated that this compound may be a potent PI3K inhibitor. The IC_{50} values for the potent compounds are summarized in Table 2.25.

The quest for a more potent, cost-effective anticancer drug led the researchers to investigate further in this direction. In view of this, several researchers are working on the design of a green, efficient, and sustainable pathway for the drug discovery.

Recently, a library of novel 1*H*-benzo[*d*]imidazole analogs (**118a–m** and **119a–m**) of dehydroabietic acid has been synthesized (Scheme 2.24) by Yang and colleagues [115], and these analogs were assessed for their in vitro

TABLE 2.25 Enzyme activities of (117i) and (117l) and HS-173 in Class I PI3K (IC_{50} values in nM).

	PI3Kα	PI3Kβ	PI3Kγ	PI3Kδ
		(IC50 values in nM)		
117i	0.50	1.9	1.8	0.74
117l	0.82	5.5	2.9	1.3
HS-173	1.1	53.2	104	97

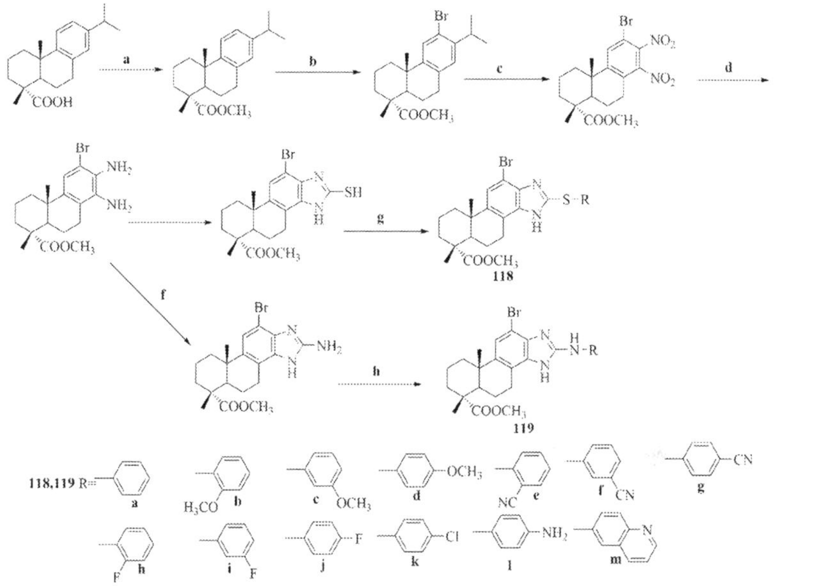

Reagents and conditions: (a) (1) $SOCl_2$, benzene, refluxe, 3h, (2) MeOH, reflux, 5h, 92%; (b) NBS, MeCN, rt,24h, 68%; (c) fuming HNO_3, H_2SO_4, 0°C, 40 min, 40%; (d) Fe, 2N HCl, HFIP, rt, 30 min, 52%; (e) Ethanol, H_2O, KOH, CS_2, reflux, 3h, 80%; (f) BrCN, Ethanol, H_2O, 0°C, 15 min, 80°C, 16h, 60%; (g) Corresponding aryl iodide, CuI, 1,10-phenanthroline, K_2CO_3, DMF, 140°C, 22h, 52-59%; (h) Corresponding aryl bromide, K_3PO_4, $Pd_2(dba)_3$, *t*-BuBrett-Phos, *t*-BuOH, 120°C, 5h, 58-70%

SCHEME 2.24 1*H*-Benzo[*d*]imidazole derivatives of dehydroabietic acid.

antiproliferative potencies on HepG2, MCF-7, HCT-116, and HeLa cancer cell lines. Most of the 2-arylamino-1*H*-benzo[*d*]imidazole derivatives showed significant antiproliferative activities, but in particular, (**119g**) possessed the most promising activity. In addition, derivative (**119g**) also exhibited potent and selective activity against PI3Kα (IC_{50} value of $0.012 \pm 0.002\,\mu M$) compared to other three isoforms (PI3Kβ, PI3Kγ, and PI3Kδ). Moreover, it has the ability to alleviate the potential of mitochondrial membrane, increase the level of intracellular ROS, enhance regulation of Bax, cleave the level of caspase-3 and caspase-9, and consequently cause apoptosis of HCT-116 cells.

In continuation of their previous work, the authors devised N-(1*H*-benzo[*d*]imidazole-2-yl)-benzamide/benzenesulfonamide derivatives of dehydroabietic acid through a multistep process, taking dehydroabietic acid as a precursor (Fig. 2.66) [116]. The derivatives were scrutinized for in vitro antiproliferative activity against three different cell lines (HeLa, HepG2, and MCF-7) and one human normal hepatocyte cell line (LO2) using MTT assay. The IC_{50} values for some of the potent derivatives are summarized in Table 2.26.

Among the synthesized derivatives, compound (**121h**) was found to be the most potent and the mechanism of action explained that derivative (**121h**) could inhibit the cell cycle of MCF-7 cells at the S-phase and provoked the apoptosis of MCF-7 cells in a ROS-mediated mitochondrial approach.

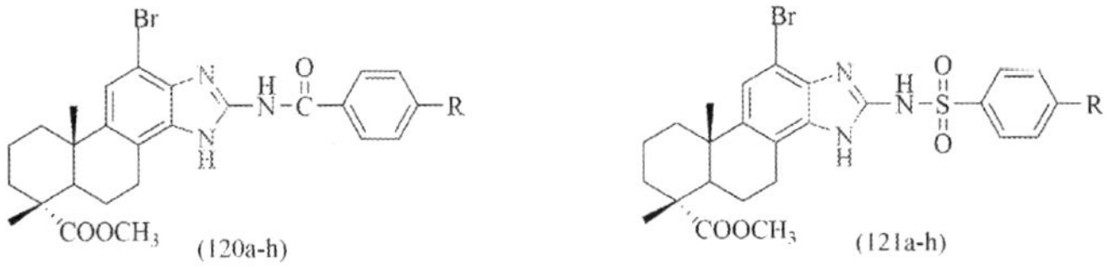

FIG. 2.66 1*H*-Benzo[*d*]imidazole derivatives of dehydroabietic acid.

TABLE 2.26 The IC_{50} values for some of the potent derivatives.

	IC_{50} (μM)			
Compound	MCF-7	HeLa	HepG2	LO2
120a	5.39 ± 0.79	> 50	12.75 ± 1.36	> 50
120g	7.32 ± 0.58	> 50	16.86 ± 2.09	> 50
120h	6.63 ± 0.32	> 50	19.61 ± 1.80	> 50
121a	18.37 ± 1.25	40.69 ± 2.38	> 50	> 50
121d	18.72 ± 2.05	22.55 ± 1.73	16.65 ± 1.82	> 50
121e	27.38 ± 2.71	38.39 ± 1.61	35.21 ± 2.17	> 50
121g	8.29 ± 0.25	21.17 ± 1.69	10.23 ± 1.07	45.87 ± 3.61
121h	0.87 ± 0.18	9.39 ± 0.72	8.31 ± 0.64	42.83 ± 3.18
Etoposide	0.77 ± 0.12	0.83 ± 0.22	0.62 ± 0.16	20.43 ± 0.82

To synthesize an anticancer drug, there are several receptors or enzymes that need to be blocked. DNA topoisomerase I is one of them, as it controls the topological structure of DNA in numerous cellular metabolic progressions and is an authorized target for the generation of antitumor agents. For the same, a library of new 2-[(5-(4-(5(6)-substituted-1*H*-benzimidazol-2-yl) phenyl)-1,3,4-oxadiazol-2-yl)thio]-1-(4-substitutedphenyl)ethan-1-one (**122a–p**) compounds has been prepared (Scheme 2.25) and appraised for the compounds' activity against DNA Topo I inhibition [117]. The cytotoxicity was also examined against two cell lines (A549, HepG2), and normal cell line NIH3T3. It was interesting to note that some compounds (**122a**, **122g**, **122i**) possessed greater cytotoxicity compared to the reference drug cisplatin. The agarose gel electrophoresis method was used to examine the inhibition efficiency of potent compounds with good IC_{50} values, against the topoisomerase I enzyme with the

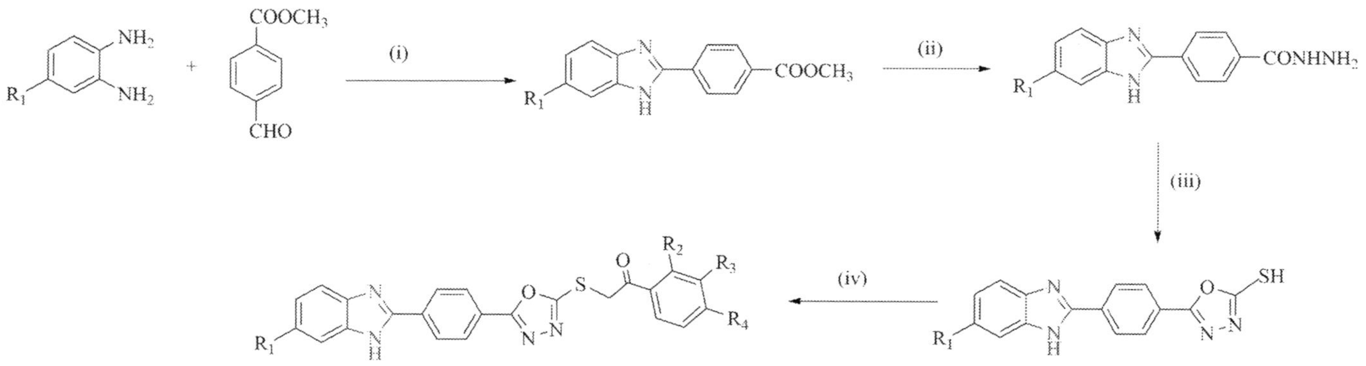

(122 a-s)

i: $Na_2S_2O_5$, DMF, MWI, 10 min,
ii: NH_2NH_2 x H_2O, EtOH, MWI, 10 min,
iii: CS_2/NaOH, EtOH, reflux, 8 h,
iv: appropriate phenacyl bromides, K_2CO_3, acetone, rt, 8 h

122a $R_1=R_2=R_3=R_4=H$
122g $R_1=H$, $R_2=F$, $R_3=H$, $R_4=F$
122i $R_1=Me$, $R_2=R_3=H$, $R_4=CN$

SCHEME 2.25 Benzimidazole-oxadiazole derivatives.

relaxation assay. It was surprising that compounds did not block the DNA Topo I inhibition. The Annexin VFITC assay confirmed that the potent derivatives induced cell death by apoptosis.

2.4 Imidazole-based metal complexes as anticancer agents

Coordination complexes play a crucial role in medicinal chemistry and enormous investigations have been carried out to design more cost-effective and potent complexes for the treatment of diseases.

In 1986, Keppler et al. [118] devised synthesis of imidazole-substituted ruthenium complex using $RuCl_3$, ethanol, imidazole, and 1 N HCl. Initially, $RuCl_3$ was mixed with 100 mL of ethanol and water and refluxed for 2 h. After that, imidazole dissolved in 1 N HCl was added dropwise and stirred for 3 min and evaporated at RT followed by addition of 1.8 mL of 8 N HCl, and the solution was heated to 85°C and then cooled by a cooling mixture of ice/$CaCl_2$ mixture. The synthesized compounds were examined for antitumor activity and the results showed that bis(imidazolium) (imidazole) pentachlororuthenate(III) increased the life span of the tumor-bearing animals up to T/C values of 150%–162%. In comparison to the synthesized ruthenium compounds, 5-fluorouracil was found to be less effective while cisplatin was found to be more effective in this experimental tumor model.

Groessl et al. [119] compared the antiproliferative activity of imidazolium-based Ru complex (123a) with other ruthenium complexes having indazole (**123b**), 1,2,4-triazole (**123c**), 4-amino-1,2,4-triazole (**123d**), and 1-methyl-1,2,4-triazole (**123e**) as ligands (Fig. 2.67). The anticancer activity was measured against two cell lines, human tumor cell lines HT-29 (colon carcinoma) and SK-BR-3 (mammary carcinoma), using MTT assay. The complex (**123b**) showed the most cytotoxicity due to greater protein binding interactions. The presence of a methyl group (**123e**) increased the lipophilic character thereby possessing similar cytotoxicity to the imidazole and triazole derivatives shown in Table 2.27.

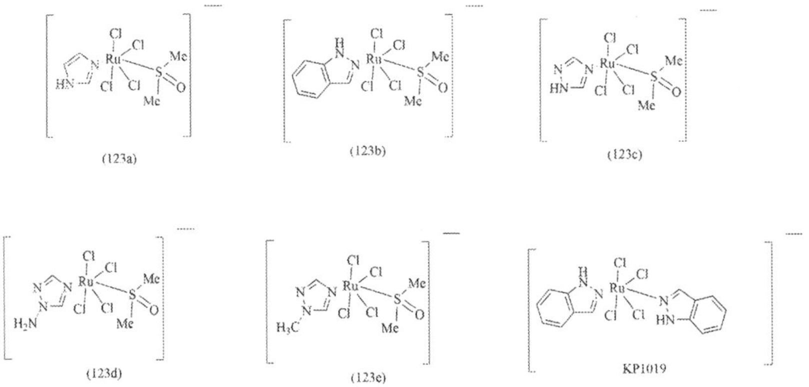

FIG. 2.67 Imidazole-based ruthenium complexes.

TABLE 2.27 IC_{50} values of synthesized complexes.

Compound	IC_{50} (μM)	(123a)	(123b)	(123c)	(123d)	(123e)
HT-29		339 ± 68	212 ± 22	322 ± 32	621 ± 5	315 ± 22
SK-BR-3 > 1000		472 ± 25	169 ± 10	415 ± 48	> 1000	517 ± 70

Skander and coworkers [120] fabricated N-heterocyclic carbine-based platinum complexes. For the synthesis of these complexes, Pt^0(NHC)(dvtms) complexes were used (Scheme 2.26). The synthesized compounds were examined against CEM T leukemia cells, lung NCI-H460 human cancer cell lines, A2780/DDP, CH1/DDP, and SKOV3, and some of the derivatives (**124b**, **124f**, **124h**, **124i**, **124l**, **124n**) were found to be more potent (IC_{50} < 2 μM) than cisplatin (3 μM) against CEM and H460 cells. Imidazole-derived NHC complexes demonstrated that they also worked with a similar mechanism of action as additional active trans-coordinated complexes. The compounds (**124a**), (**124b**), (**124d**), (**124f**), (**124h**), and (**124n**) also possessed cytotoxicity against cisplatin (> 10 μM)-resistant cell lines like A2780/DDP, CH1/DDP, and SKOV3 with IC_{50} values < 8 μM.

Later on, in 2014, Blunden and coworkers [121] developed a biocompatible amphiphilic block copolymer capable of self-assembling into polymeric micelles as drug delivery for NAMI-A drug (**125**) (Fig. 2.68). NAMI-A is an imidazole-based ruthenium complex tested under a phase II clinical trial. The fused polymeric micelles were examined for their cytotoxicity against ovarian cancer ovcar-3, pancreatic AsPC-1, and A2780 cancer cell lines. It was observed that NAMI-A copolymeric micelles were 1.5 times more cytotoxic than the NAMI-A molecule. The IC_{50} values for polymeric micelle and complexes are summarized in Table 2.28.

Moreover, the antimetastatic possibility was evaluated by estimating the inhibitory effects on the migration and invasion of three different cell lines characterized by differing degrees of malignancy, and it was found that antiangiogenic and antiinvasive properties were increased in the case of NAMI-A copolymeric micelles and followed the order MDA-MB-231 > MCF-7 > CHO.

Trifluoromethyl-based imidazole complexes (**126a–f**) were synthesized (Fig. 2.69) and studied for their pharmacological properties and cytotoxicity [122]. Various experimental studies revealed that lipophilicity controlled the solubility and affected protein interactions and oligomer formation. An in vitro study showed that most of the lipophilic molecules displayed good cytotoxicity and the presence of CF_3 group increased the cytotoxicity against A549 nonsmall cell lung carcinoma cells. The authors also concluded that an increase of lipophilicity enhanced the passive diffusion through cell membrane and boosted the concentration and activity of complexes in the cell.

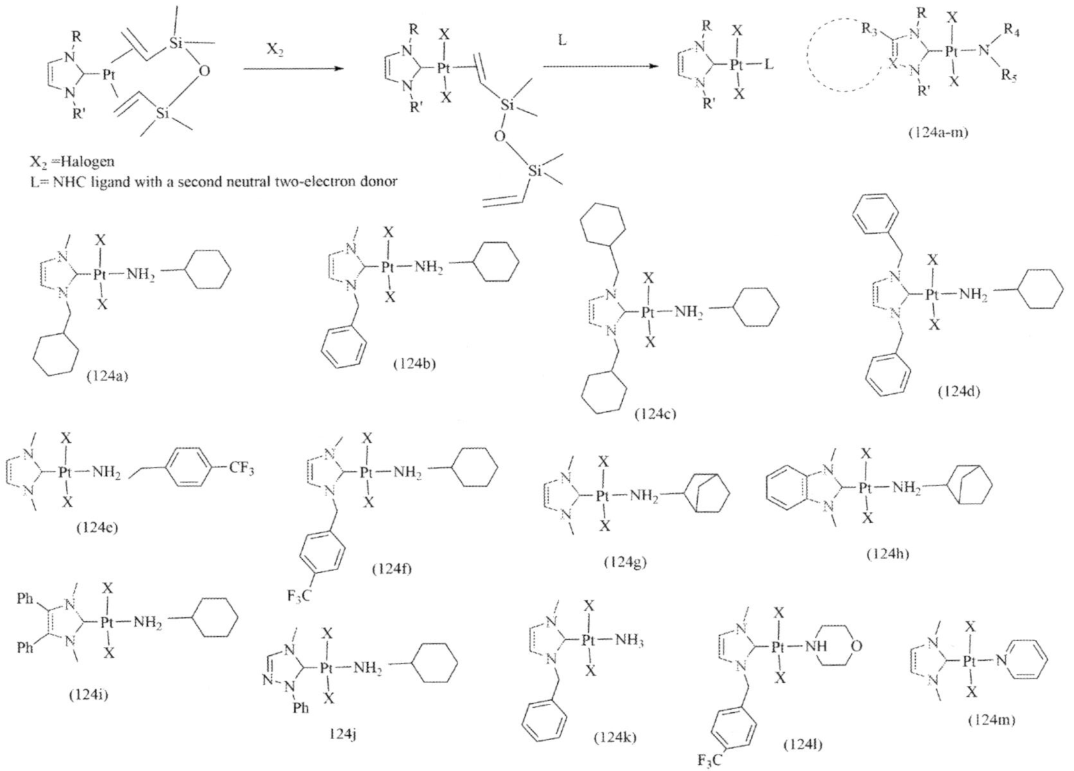

SCHEME 2.26 Synthesis of mixed NHC-amine complexes.

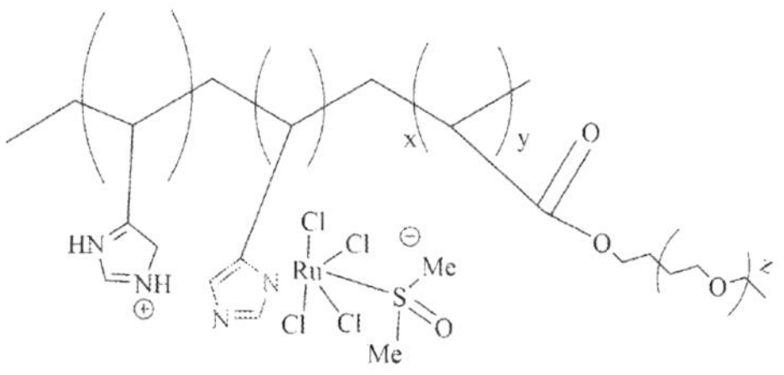

(125)

FIG. 2.68 Structure of NAMI-A.

TABLE 2.28 IC_{50} values with respect to the ruthenium and polymer concentration, against ovarian A2780 and OVCAR-3 and pancreatic AsPC-1 cancer cell lines.

	IC_{50} (μM) with respect to concentration of [Ru]		
	A2780	AsPC-1	OVCAR-3
NAMI-A	595.6 ± 28.5	601.7 ± 13.5	737.8 ± 38.0
P(NAMI-A)-PPEGMEA	438.7 ± 14.8	397.4 ± 13.5	433.4 ± 26.8
	IC_{50} (μM) with respect to concentration of [polymer]		
	A2780	AsPC-1	OVCAR-3
PVIm	0.37	0.51	0.70
PVIm-PPEGMEA	> 15.37	> 15.37	> 15.37
P(NAMI-A)-PPEGMEAa	9.77	8.86	9.65

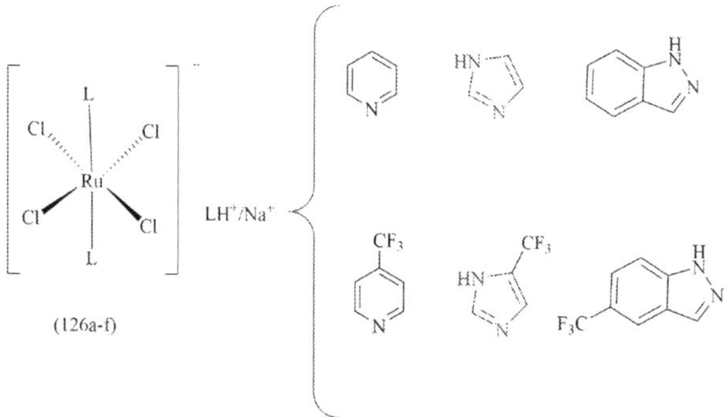

(126a-f)

FIG. 2.69 Keppler-type Ru(III) anticancer complexes and new trifluoromethyl derivatives.

In 2017, Pellei et al. [123] synthesized novel water-soluble sulfonated imidazole- and benzimidazole-based Cu(I) complex N-heterocyclic carbenes and studied their anticancer potency against various cell lines like ovarian, lung, cervical, carcinoma, melanoma, and colon. The reaction of various ligands like $HImBnPrSO_3$ $Na(4\text{-}Me)HImPrSO_3$ and $NaHBzImPrSO_3$ (**129**) with CuCl in H_2O/NaOH solution produced final complexes (**127–129**) in satisfactory yields (Fig. 2.70). NHC-copper complexes provoked damage of tumor cells, with IC_{50} values in the micromolar range. However, all the molecules could overcame cisplatin resistance.

Also in 2017, Rimoldi and companions [124] reported the synthesis of novel imidazole-based cationic platinum(II)complexes and studied their anticancer activity against breast cancer cell lines via MTT assay. The process of the synthesis of cationic platinum(II)complexes (Fig. 2.71) was based on associative ligand substitution using *N*-alkyl imidazole in DMF at 55°C with satisfactory yield. Among all the products, compound **129c** showed the highest cytotoxicity against MDA-MB-231 (IC_{50}=61.9mM) and was more potent than cisplatin on other MCF-7 (IC_{50}=79.9mM) and DLD-1 (IC_{50}=57.4mM) cell lines.

In 2019, Abdel-Rahmann and coworkers [125] fabricated Cr(II), Fe(II), and Cu(II) complexes using an imidazole-based ligand. For this, the imidazole

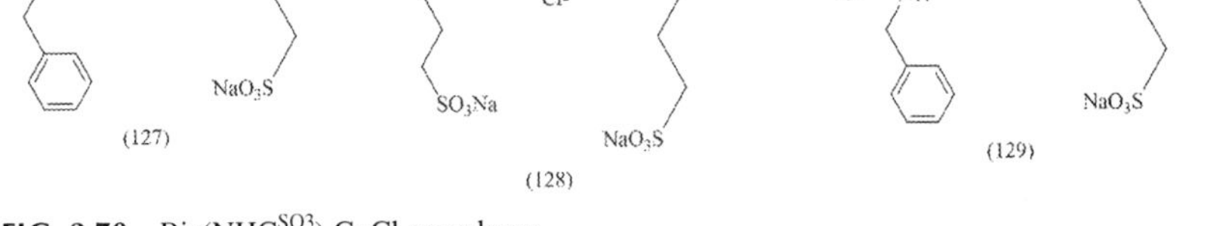

FIG. 2.70 Bis(NHC^{SO3}) CuCl complexes.

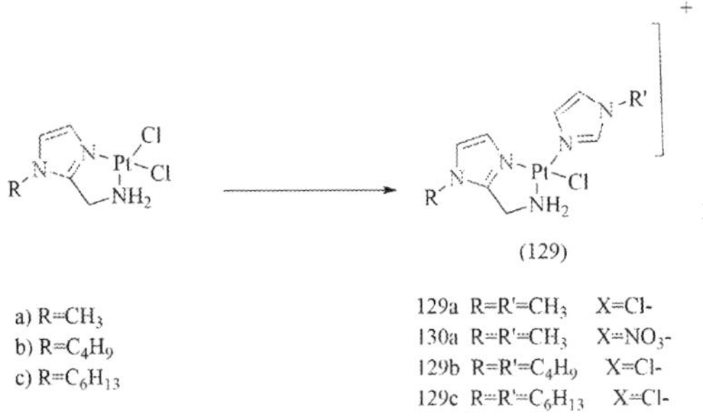

FIG. 2.71 Synthesis of cationic platinum(II) complexes.

ligand was prepared by the cyclocondensation of benzil, butan-1-amine, and 5-bromo-2-hydroxybenzaldehyde (Scheme 2.27). To prepare complexes, the imidazole ligand was dissolved in ethanol and then added to metal salt in EtOH. This reaction mixture was continuously stirred at 70°C for 1 h. The synthesized complexes $[Cr(Imd)Cl_2(H_2O)_2]$ (**131**), $[Fe(Imd)(NO_3)_2(H_2O)_2]$ (**132**), and $[Cu(Imd)Cl(H_2O)_3]$ (**133**) possessed distorted-octahedral geometry. The synthesized compounds were examined for their antiproliferative activity against different cell lines Hep-G2, MCF-7, and HCT-116.

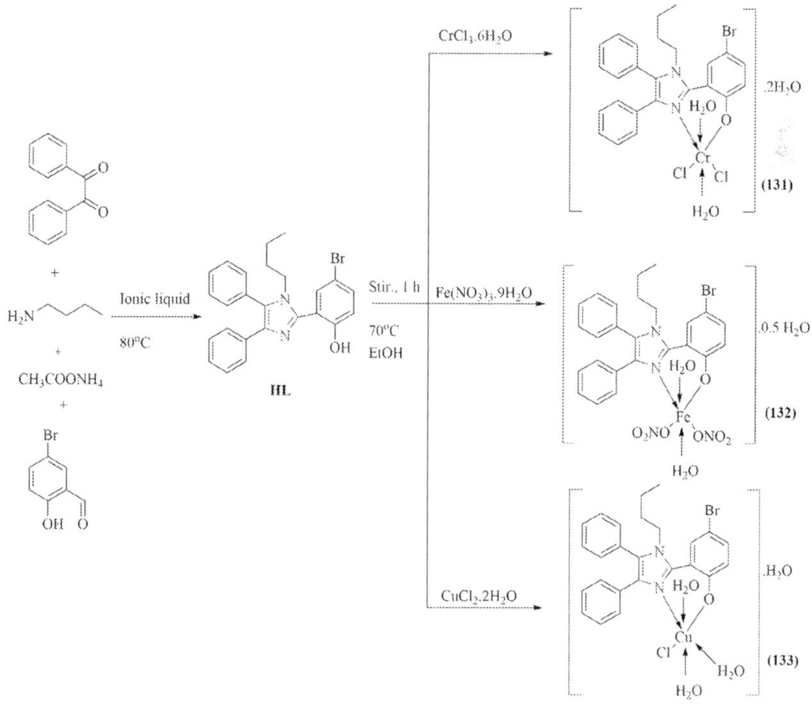

SCHEME 2.27 Synthesis of the aryl imidazole ligand HL and its CrL, FeL, and CuL compounds.

The IC_{50} values against the MCF-7 cell line were in the range of 10.5–40.24 mg/mL with a minimum IC_{50} value for $[Cu(Imd)Cl(H_2O)_3]$. The anticipated compounds showed good antiproliferative activity against the Hep-G2 cell line with IC_{50} values of 20.24–40.47 mg/mL. The derivatives showed least activity against colon carcinoma HCT-116 cell lines with IC_{50} values of 28.1–65.47 mg/mL. The nature of the ligand and the type of metal ions have significant impacts on the activity of the compounds. The cytotoxicity can be elucidated on the basis of a mechanism that the positive charge of the metal increased the acidity of coordinated pro-ligands that gave protons, causing more potent H-bonds and in turn augmenting the biological activity.

The IC_{50} order of the complexes for the MCF-7, HepG-2, and HCT-116 was as follows: (**133**) < (**131**) < (**132**) < (ligand).

In 2020, Salen/imidazole-based ligands (S) and their binary (MS) and ternary (MSI) complexes with various metals like Cd, Cu, Ni, Co, etc. were evaluated for their cytotoxicity [126]. The reaction of salicylaldehyde and ethylenediamine in ethanol (50 mL) yielded 2,20-{1,2-ethanediylbis[nitrilo(*E*)methylylidene]}diphenol (salen) and their binary compounds, which coordinated with metal salts [Cd(II), Cu(II), Ni(II) and Co(II), Al(III)] in aqueous ethanol at 70°C (Fig. 2.72). The in vitro cytotoxic study of synthesized compounds against human liver (Hep-G2) and breast (MDA-MB231) carcinoma cell lines also displayed good activity [126]. The IC_{50} values followed the order: cisplatin < CdSI (**137**) < CuSI (**136**) < salen < LaSI (**139**) < AlSI (**138**) < NiSI (**135**) < CoSI (**134**). Computational study was also performed using protein PDB ID: 3HB5 and crucial hydrophobic interactions with amino acid residues were found.

Also in 2020, Sharhan and coworkers [127] synthesized a library of benzimidazolium-acridine compounds (**140a–d**) (Fig. 2.73) and their silver complexes, and screened their anticancer activity against breast cancer cell line MCF-7 and nontumorigenic breast cell line MCF-10a. The whole range of compounds exhibited noncytotoxicity against the nontumorigenic cell line while silver coordination enhanced the cytotoxicity against cancer cell lines. The IC_{50} values are summarized in Table 2.29.

From a medicinal viewpoint, a sustainable approach to treat cancer includes the linking of a cytotoxic drug to a peptide scaffold to improve its bioavailability and biological profile. This designed anticancer pathway results in curative agents with augmented selectivity against tumors and reduced toxicity in ordinary cells. The neuro-endocrine hormone somatostatin and its analogs are

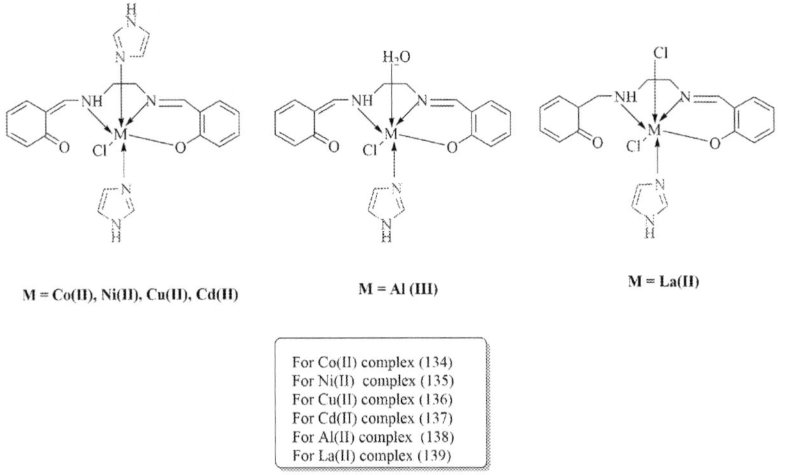

FIG. 2.72 Synthesis pathway for Co (II), Ni (II), Cu (II), Cd (II), Al (III), and La (III) mixed ligands of Salen and imidazole complexes.

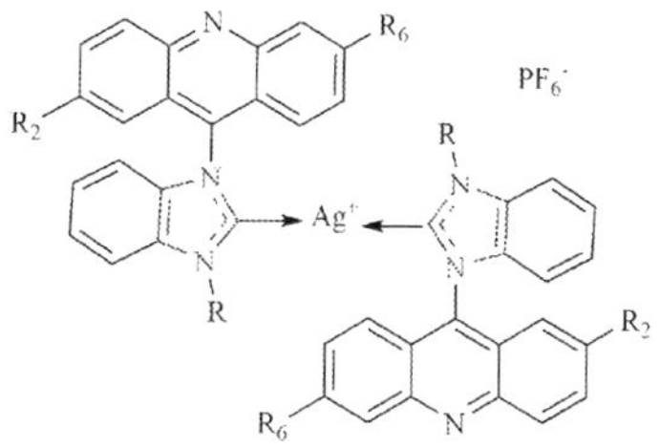

(140a-d)

FIG. 2.73 Benzimidazolium-acridine-based Ag-carbene complexes.

TABLE 2.29 Anticancer activities of synthesized complexes.

		Normal	Cancer MCF-7	
IC_{50} (mg/mL)	R	MCF-10a	(mg/mL)	μM
140a	R=Me, $R_2=R_6=H$	69±3	20±3	22
140b	R=Et, $R_2=R_6=H$	74±3	24±2	26
140c	R=Bn, $R_2=R_6=H$	80±3	22±3	21
140d	R=Me, R_2=OMe, R_6=Cl	78±4	23±3	22
Tamoxifen	–	85±6	11±1	30
Paclitaxel	–	75±2	6±1	7

one of the prominent natural receptor-binding peptides and have gained considerable attention since they have high affinity against five human receptors (sst_1–sst_5). Among the analogs, octreotide was combined with several cytotoxic drugs and new compounds were also synthesized. One such example is synthesis of derivatives of dicarba analogs of octreotide through a solid phase approach (Fig. 2.74) [128].

These conjugates were also examined for anticancer activity against MCF-7 and DU-145 cell lines. Only one conjugate (**141**) was found to be potent against these cell lines, with IC_{50} values of 63±2 μM in MCF-7 cells and 26±3 μM in DU-145 cells. Lastly, the mechanism of action of ruthenium-octreotide conjugate (**141**) with DNA was examined. After 48 h, ruthenation of 5′dCATGGCT was not observed and it was agreed that hydrolysis of Ru–Cl bond did not occur. Activation of the M–Cl bond is very important for the binding of

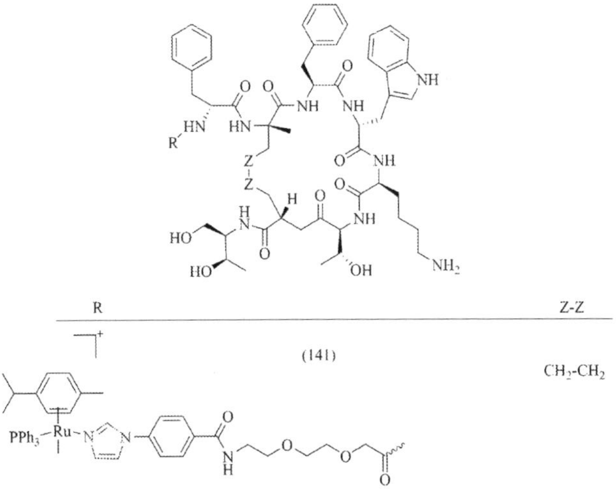

FIG. 2.74 Structure of the ruthenium complex conjugate molecules.

DNA. It showed greater priority for the guanine nucleobases, particularly those situated at the 5′ end in GG sequences. The sole metal-complexes were also examined for the anticancer activity in order to compare their efficacy with metal-complexes clubbed conjugates. It was astonishing that the sole metal complexes showed higher cytotoxicity and their activity decreased by the attachment of receptor-binding peptides.

Imidazole-based metal organic frameworks were used as a drug delivery system for several drugs to improve their pharmacokinetic profile and target-based pH-dependent drug release [129]. A zeolite-based imidazole framework (ZIF) for the drug delivery of doxorubicin, an anticancer drug, was devised by Adhikari et al. [130]. The authors devised two frames, ZIF-7 and ZIF-8, and studied their control release under different conditions and biomimetic system. The synthesis of ZIF-7 and ZIF-8 was done as per the previously reported papers [131, 132]. The encapsulation of drug to ZIF cannot be achieved by an in situ method because both are positively charged. For this, the drug solution was prepared by 2mL of 0.5mM drug, and this was mixed to 100mg of ZIFs and stirred for 48h. Afterward, the drug-incorporated ZIF was removed by centrifugation, washed with methanol, and dried. It was found that ZIF-7 did not excrete the drug and remain intact whereas ZIF-8 released the drug under acidic conditions. To overcome these issues, ZIF-7 and ZIF-8 were embedded with liposomes or micelles (Scheme 2.28), which facilitated the drug release, and ZIF-7 delivered the drug for 10h whereas ZIF-8 delivered the drug for a span of 3h. In acidic conditions, the ZIF framework got dissociated due to disconnection of metal ions and ligand. This gave a chance to release the drug at acidic pH. At pH 7.4, the release was less and the drug remained the same inside the framework of ZIF, and thus it did not affect the healthy cells.

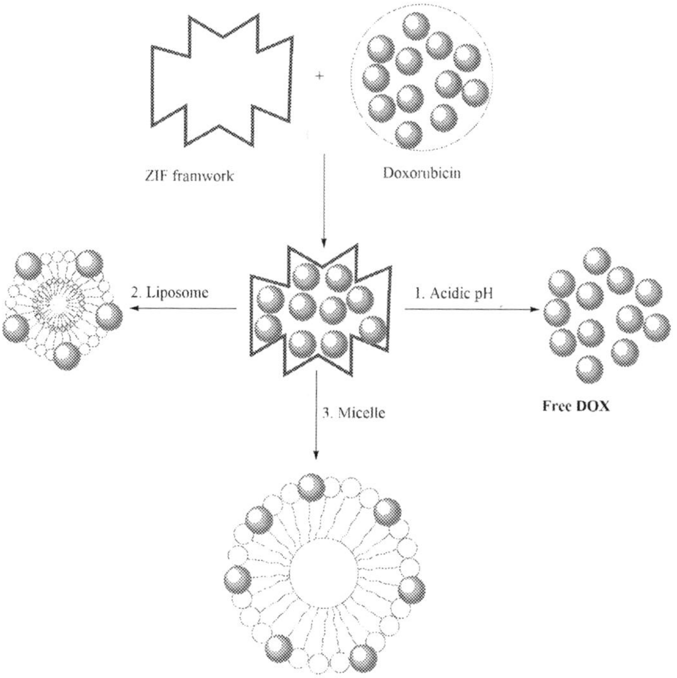

Encapsulation of DOX inside ZIFs Framework and (1) Its Release at Acidic pH Due to Dissociation of Coordination between Metal Ion and Ligands: (2, 3) Release of Drug by Liposome/Micelle Due to the Binding of Dox with Liposome/ Micelle

SCHEME 2.28 Encapsulation of DOX inside ZIFs framework.

5-Fluorouracil is a very famous anticancer drug used to cure several types of solid tumors including colon, breast, rectal, liver, pancreatic, gastric, bladder, and ovarian cancer. In 2019, a new nontoxic porous In^{III}-based metal-organic framework $[In(Hpbic)(pbic)](DMA)_2$ (1) (DMA= *N,N*-dimethylacetamide) was designed for drug delivery of 5-fluorouracil [133]. The organic framework was synthesized using solvothermal process from 2-(pyridin-4-yl)-1*H*-benzo[*d*] imidazole-5-carboxylic acid (H_2pbic). The drug loading capacity was found to be 32.6% and the drug 5-fluorouracil was mainly incorporated into the holes of the nanostructure.

The controlled excretion of drug in a virtual human body with liquid phosphate-buffered saline solution under different pH conditions was examined. CCK-8 assay was used to determine the inhibition of cells' viability and proliferation by 5-Fu@nanostructure on SW60 colon cancer. The composites reduced the cell proliferation with IC_{50} values of $2.6 \pm 0.14\,\mu M$. To expose the mechanism of action, Annexin V-FITC/PI assay and ROS level estimation was done after the treatment with 5-Fu@nanostructure. Later on, an in vivo xenograft model was used to study the volume of tumor and mice body weight. After 24 days of treatment, there was a significant decrease in the volume of tumors, and the weight of mice increased with decrease in tumor volume.

3. Conclusion and future challenges

Cancer is a curse for the present era. It is likely never to be ended and will remain a part of life, although a lot of research has been carried out. It is still one of the foremost causes of fatality around the globe. The upcoming trouble of cancer will possibly be even more in future, since people experience poor lifestyles with unhealthy diets, and less physical activity and more mental stress. Therefore, designing of efficient anticancer drugs having good selectivity and cost-efficiency with minimal side effects is one of the big challenges for the scientific community. Heterocyclic scaffolds are a core structural part of several anticancer drugs available on the market today. Among these, imidazole-based compounds possess enormous biological applications as well as unique structural characteristics. A huge number of review articles have been published to justify the medicinal importance of imidazole-based compounds as anticancer agents. They have proven to be capable candidates for target therapy and combination treatment of cancer. However, a great deal of investigation is still required to explore this moiety pharmacologically.

In this chapter, we have carried out exhaustive and in-depth study for the synthesis and characterization of imidazole analogs as anticancer agents to enrich the knowledge of researchers and readers about the widespread and intense activities going on in this field. The chapter was divided into various subsections, such as synthesis and biological profile of substituted imidazoles to imidazole fused heterocycles, benzimidazole-based compounds, metal complexes of imidazole, etc., for the easier understanding of the readers, and it adds to the systematic literature for the designing of imidazole-based drug templates. This chapter gives ample information and delivers updated literature on imidazole-based compounds using various cancer cell lines reported during the time period 2000–20, along with their mechanisms of action and structure activity relationships. However, there are still various limitations in the scope and number of clinical trials required to authenticate their beneficial effects on cancer. Moreover, there are numerous prospects that must be explored and scrutinized further:

- Future experimental study should emphasize human studies and in vivo analysis to fully assess and validate the biological potency of imidazole-based scaffolds.
- Most of the existing research carried out up to now has involved in vitro analysis, thus further clinical trials should be carried out to promote its beneficial uses practically.
- There is less information about the side effects of these derivatives and their toxicity.

Conflict of interest

The authors declare no conflict of interest, financial or otherwise.

Acknowledgment

The authors are grateful to the Department of Chemistry, Mohan Lal Sukhadia University, Udaipur, India for providing the necessary library facilities for carrying out the work. A. Sethiya is also thankful to UGC-MANF (201819-MANF-2018-19-RAJ-91971) for providing a Senior Research Fellowship to carry out this work. N. Sahiba and P. Teli are thankful to CSIR for providing a fellowship to carry out this work.

Funding source

This work was supported by UGC-MANF (201819MANF-2018-19-RAJ-91971) andCSIR[fileno.09/172(0088)2018-EMR-I],[fileno.09/172(0099)2019-EMR-I].

References

[1] Hill RA. Marine natural products. Annu Rep Prog Chem Sect B 2009;105:150–66.

[2] Jin Z. Muscarine, imidazole, oxazole, and thiazole alkaloids. Nat Prod Rep 2011;28:1143–91.

[3] Forte B, Malgesini B, Piutti C, Quartieri F, Scolaro A, Papeo G. A submarine journey: the pyrrole imidazole alkaloids. Mar Drugs 2009;7:705–53.

[4] Narasimhan B, Sharma D, Kumar P. Biological importance of imidazole nucleus in the new millennium. Med Chem Res 2011;20:1119–40.

[5] Shalini K, Sharma PK, Kumar N. Imidazole and its biological activities: a review. Chem Sin 2010;1:36–47.

[6] Bhatnagar A, Sharma PK, Kumar N. A review on "imidazoles": their chemistry and pharmacological potentials. Int J PharmTech Res 2011;3:268–82.

[7] Zhou CH, Gan LL, Zhang YY, Zhang FF, Wang GZ, Jin L, Geng RX. Review on super molecules as chemical drugs. Sci China Ser B 2009;52:415–58.

[8] Zhou CH, Zhang YY, Yan CY, Wan K, Gan LL, Shi Y. Recent researches in metal supra molecular complexes as anticancer agents. Anticancer Agents Med Chem 2010;10:371–95.

[9] Ali I, Lonea MN, Aboul-Enein HY. Imidazoles as potential anticancer agents. Med Chem Commun 2017;8:1742–73.

[10] Ali TFS, Iwamaru K, Ciftci HI, Koga R, Matsumoto M, Oba Y, Kurosaki H, Fujita M, Okamoto Y, Umezawa K, Nakao M, Hide T, Makino K, Kuratsu J, Abdel-Aziz M, Abuo-Rahma GEAA, Beshr EAM, Otsuka M. Novel metal chelating molecules with anticancer activity. Striking effect of the imidazole substitution of the histidine-pyridine-histidine system. Bioorg Med Chem 2015;23(17):5476–82.

[11] Boiani M, Gonzalez M. Imidazole and benzimidazole derivatives as chemotherapeutic agents. Mini Rev Med Chem 2005;5:409–24.

[12] Taniguchi M, Fujiwara K, Nakai Y, Ozaki T, Koshikawa N, Toshio K, Kataba M, Oguni M, Matsuda H, Yoshida Y, Tokuhashi Y, Fukuda N, Ueno T, Soma M, Nagase H. Inhibition of malignant phenotypes of human osteosarcoma cells by a gene silencer, a pyrrole–imidazole polyamide, which targets an E-box motif. FEBS Open Bio 2014;4:328–34.

[13] Gaba M, Mohan C. Development of drugs based on imidazole and benzimidazole bioactive heterocycles: recent advances and future directions. Med Chem Res 2016;25:173–210.

[14] Sethiya A, Agarwal DK, Agarwal S. Current trends in drug delivery system of curcumin and its therapeutic applications. Mini-Rev Med Chem 2020;20(13):1190–232.

[15] Cragg GM, Grothaus PG, Newman DJ. Impact of natural products on developing new anticancer agents. Chem Rev 2009;109:3012–43.

[16] Zhu XC, Lu WT, Zhang YZ, Reed A, Newton B, Fan Z, Yu HT, Ray PC, Gao R. Imidazole modified porphyrin as a pH-responsive sensitizer for cancer photodynamic therapy. Chem Commun 2011;47:10311–3.

[17] Kapuriya N, Kakadiya R, Dong HJ, Kumar A, Lee PC, Zhang XG, Chou TC, Lee TC, Chen CH, Lam K, Marvania B, Shah A, Su TL. Design, synthesis, and biological evaluation of novel water-soluble N-mustards as potential anticancer agents. Bioorg Med Chem 2011;19:471–85.

[18] Tang YD, Zhang JQ, Zhang SL, Geng RX, Zhou CH. Synthesis and characterization of thiophene derived amido bis-nitrogen mustard and its antimicrobial and anticancer activities. Chin J Chem 2012;30:1831–40.

[19] Ghosh S. Cisplatin: the first metal based anticancer drug. Bioorg Chem 2019;88:102925.

[20] Zhang L, Peng X, Damu GLV, Geng R, Zhou C. Comprehensive review in current developments of imidazole-based medicinal chemistry. Med Res Rev 2014;34(2):340–437.

[21] Jeong Y, Li H, Kim SY, Yun H, Baek KJ, Kwon NS, Kim D. Imidazole inhibits B16 melanoma cell migration *via* degradation of β-catenin. J Pharm Pharmacol 2010;62:491–6.

[22] Vlahakis JZ, Kinobe RT, Bowers RJ, Brien JF, Nakatsub K, Szareka WA. Synthesis and evaluation of azalanstat analogues as heme oxygenase inhibitors. Bioorg Med Chem Lett 2005;15:1457–61.

[23] Vlahakis JZ, Kinobe RT, Bowers RJ, Brien JF, Nakatsu K, Szarek WA. Imidazole-dioxolane compounds as isozyme-selective heme oxygenase inhibitors. J Med Chem 2006;49:4437–41.

[24] Roman G, Vlahakis JZ, Vukomanovic D, Nakatsu K, Szarek WA. Heme oxygenase inhibition by 1-Aryl-2-(1H-imidazol-1-yl / 1H-1,2,4-triazol-1yl) ethanones and their derivatives. Chem Med Chem 2010;5:1541–55.

[25] Sorrenti V, Guccione S, Giacomo CD, Modica MN, Acquaviva VPR, Basile L, Pappalardo M, Salerno L. Evaluation of imidazole-based compounds as heme oxygenase-1 inhibitors. Chem Biol Drug Des 2012;80:876–86.

[26] Sun B, Liu K, Han J, Zhao L, Su X, Lin B, Zhao D, Cheng M. Design, synthesis, and biological evaluation of amide imidazole derivatives as novel metabolic enzyme CYP26A1 inhibitors. Bioorg Med Chem 2015;23(20):6763–73.

[27] Stolinsky DC, Bogdon D, Solomon J, Bateman JR. Hexamethylmelamine (NSC-13875) alone and in combination with 5-(3,3-dimethyl-triazeno) imidazole-4-carboxamide (NSC-45388) in the treatment of advanced cancer. Cancer 1972;30:654–9.

[28] Cangir A, Morgan SK, Land VJ, Pullen J, Starling SA, Nitschke R. Combination chemotherapy with adramycin (NSC-123127) and dimethyl triazeno imidazole carboxamide (DTIC) (NSC-45388) in children with metastatic solid tumors. Med Pediatr Oncol 1976;2(2):183–90.

[29] Finklestein JZ, Albo V, Ertel I, Hammond D. 5-(3,3-Dimethyl-1-triazeno)imidazole-4-carboxamide (NSC-45388) in the treatment of solid tumors in children. Cancer Chemother Rep 1975;59(2):351–7.

[30] Perchellet EM, Perchellet J, Baures PW. Imidazole-4,5-dicarboxamide derivatives with antiproliferative activity against HL-60 cells. J Med Chem 2005;48:5955–65.

[31] Chen J, Wang Z, Li C, Lu Y, Vaddady PK, Meibohm B, Dalton JT, Miller DD, Li W. Discovery of novel 2-aryl-4-benzoyl-imidazoles targeting the colchicines binding site in tubulin as potential anticancer agents. J Med Chem 2010;53:7414–27.

[32] Chen J, Li C, Wang J, Ahn S, Wanga Z, Lu Y, Dalton JT, Miller DD, Li W. Synthesis and antiproliferative activity of novel 2-aryl-4-benzoyl-imidazole derivatives targeting tubulin polymerization. Bioorg Med Chem 2011;19:4782–95.

[33] Chen J, Ahn S, Wang J, Lu Y, Dalton JT, Miller DD, Li W. Discovery of novel 2-aryl-4-benzoyl-imidazole (abi-iii) analogues targeting tubulin polymerization as antiproliferative agents. J Med Chem 2012;55(16):7285–9.

[34] Xiao M, Ahn S, Wang J, Chen J, Miller DD, Dalton DT, Li W. Discovery of 4-Aryl-2-benzoyl-imidazoles as tubulin polymerization inhibitor with potent antiproliferative properties. J Med Chem 2013;56:3318–29.

[35] Hou Q, He C, Lao K, Luo G, You Q, Xiang H. Design and synthesis of novel steroidal imidazoles as dual inhibitors of AR/CYP17 for the treatment of prostate cancer. Steroids 2019;150:108384. https://doi.org/10.1016/j.steroids.2019.03.003.

[36] Pettit GR, Singh SB, Hamel E, Lin CM, Alberts DS, Garcia-Kendall D. Isolation and structure of the strong cell growth tubulin inhibitor combretastatin A-4. Experientia 1989;45:209–11.

[37] Nam NH. Combretastatin A-4 analogues as antitumor agents. Curr Med Chem 2003;10:1697–722.

[38] Singh R, Kaur H. Advances in synthetic approaches for the preparation of combretastatin-based anti-cancer agents. Synthesis 2009;2471–91.

[39] Jin Y, Qi P, Wang Z, Shen Q, Wang J, Zhang W, Song H. 3D-QSAR study of combretastatin A-4 analogs based on molecular docking. Molecules 2011;16:6684–700.

[40] Wang J, Woods KW, Li Q, Barr KJ, McCroskey RW, Hannick SM, Gherke L, Credo RB, Hui Y, Marsh K, Warner R, Lee JY, Zielinski-Mozng N, Frost D, Rosenberg SH, Sham HL. Potent, orally active heterocycle-based combretastatin A-4 analogues: synthesis, structure-activity relationship, pharmacokinetics, and *in vivo* antitumor activity evaluation. J Med Chem 2002;45:1697–711.

[41] Kim DK, Jang Y, Lee SH, Park H, Yoo J. Synthesis and biological evaluation of 4(5)-(6-alkylpyridin-2-yl)imidazoles as transforming growth factor-β type 1 receptor kinase inhibitors. J Med Chem 2007;50:3143–7.

[42] Li W, Hwang D, Song J, Chen C, Chuu J, Hu C, Lin H, Huang C, Huang C, Tseng H, Lin C, Chen T, Lin C, Wang H, Shen C, Chang C, Chao Y, Chen C. Synthesis and biological activities of 2-amino-1-arylidenamino imidazoles as orally active anticancer agents. J Med Chem 2010;53:2409–17.

[43] Dietrich J, Gokhale V, Wanga X, Hurley LH. Application of a novel [3+2] cycloaddition reaction to prepare substituted imidazoles and their use in the design of potent DFG-out allosteric B-Raf inhibitors. Bioorg Med Chem 2010;18:292–304.

[44] Heravi RE, Hadizadeh F, Sankian M, Afshari JT, Behravan J. Cyclooxygenase-2 inhibition by novel bisaryl imidazolyl imidazole derivatives increases Bax/Bcl-2 ratio and upregulates Caspase-3 gene expression in Caco-2 colorectal cancer cell line. Genes Genomics 2012;34:199–204.

[45] Tseng C, Li C, Chiu C, Hu H, Han C, Chen Y, Tzeng C. Combretastatin A-4 derivatives: synthesis and evaluation of 2,4,5-triaryl-1H-imidazoles as potential agents against H1299 (non-small cell lung cancer cell). Mol Divers 2012;16:697–709.

[46] Assadieskandar A, Amirhamzeh A, Salehi M, Ozadali K, Ostad SN, Shafiee A, Amini M. Synthesis, cyclooxygenase inhibitory effects, and molecular modelling study of 4-aryl-5-(4-(methylsulfonyl)phenyl)-2-alkylthio and -2-alkylsulfonyl-1H-imidazole derivatives. Bioorg Med Chem 2013;21:2355–62.

[47] Niculescu-Duvaz D, Niculescu-Duvaz I, Suijkerbuijk BMJM, Ménard D, Zambon A, Davies L, Pons J, Whittaker S, Marais R, Springer CJ. Potent BRAF kinase inhibitors based on 2,4,5-trisubstituted imidazole with naphthyl and benzothiophene 4-substituents. Bioorg Med Chem 2013;21:1284–304.

[48] Mathieu V, Berge EVD, Ceusters J, Konopka T, Cops A, Bruyère C, Pirker C, Berger W, Trieu-Van T, Serteyn D, Kiss R, Robiette R. New 5-aryl-1H-imidazoles display *in vitro* antitumor activity against apoptosis-resistant cancer models, including melanomas, through mitochondrial targeting. J Med Chem 2013;56(17):6626–37.

[49] Juchum M, Günther M, Döring E, Sievers-Engler A, Lämmerhofer M, Laufer S. Trisubstituted imidazoles with a rigidized hinge binding motif act as single digit nm inhibitors of clinically relevant EGFR L858R/T790M and L858R/T790M/C797S mutants: an example of target hopping. J Med Chem 2017;60:4636–56.

[50] Subashini G, Vidhya K, Arasakumar T, Angayarkanni J, Murugesh E, Saravanan A, Shanmughavel P, Mohan PS. Quinoline-based imidazole derivative as heme oxygenase-1 inhibitor: a strategy for cancer treatment. ChemistrySelect 2018;3:3680–6.

[51] Kandasamy S, Subramani P, Subramani S, Jayaraj JM, Prasanth G, Srinivasan K, Muthusamy K, Rajakannan V, Vilwanathan R. Design and synthesis of imidazole based zinc binding groups as novel small molecule inhibitors targeting histone deacetylase enzymes in lung cancer. J Mol Struct 2020;1214:128177.

[52] Heppner DE, Günther M, Wittlinger F, Laufer SA, Eck MJ. Structural basis for EGFR mutant inhibition by trisubstituted imidazole inhibitors. J Med Chem 2020;63(8):4293–305.

[53] Anderson WK, Bhattacharjee D, Houston DM. Design, synthesis, antineoplastic activity, and chemical properties of bis(carbamate) derivatives of 4,5-bis(hydroxymethyl)imidazole. J Med Chem 1989;32:119–27.

[54] Atwell GJ, Fan J, Tan K, Denny WA. DNA-directed alkylating agents. 7. Synthesis, DNA interaction, and antitumor activity of bis(hydroxymethyl)- and bis(carbamate)-substituted pyrrolizines and imidazoles. J Med Chem 1998;41:4744–54.

[55] Wiglenda T, Ott I, Kircher B, Schumacher P, Schuster D, Langer T, Gust R. Synthesis and pharmacological evaluation of 1*H*-imidazoles as ligands for the estrogen receptor and cytotoxic inhibitors of the cyclooxygenase. J Med Chem 2005;48:6516–21.

[56] Wiglenda T, Gust R. Structure-activity relationship study to understand the estrogen receptor-dependent gene activation of aryl- and alkyl-substituted 1H-imidazoles. J Med Chem 2007;50:1475–84.

[57] Gust R, Keilitz R, Schmidt K, von Rauch M. Synthesis, structural evaluation and estrogen receptor interaction of 4,5-bis(4-hydroxyphenyl) imidazoles. Arch Pharm Pharm Med Chem 2002;335:463–71.

[58] Gust R, Busch S, Keilitz R, Schmidt K, von Rauch M. Investigation on the influence of halide substituents on the estrogen receptor interaction of 2,4,5-tris(4-hydroxyphenyl)imidazoles. Arch Pharm Pharm Med Chem 2003;336:456–65.

[59] Duran M, Demirayak S. Synthesis of 2-[4,5-dimethyl-1-(phenylamino)-1H-imidazol-2-ylthio]-N-(thiazole-2-yl)acetamide derivatives and their anti cancer activities. Med Chem Res 2013;22:4110–24.

[60] Ganga M, Sankaran KR. Synthesis, spectral characterization, DFT, docking studies and cytotoxic evaluation of 1-(4-fluorobenzyl)-2,4,5-triphenyl-1H-imidazole derivatives. Chem Data Coll 2020;28:100412.

[61] Salahuddin, Shaharyar M, Mazumder A. Benzimidazoles: a biologically active compounds. Arab J Chem 2017;10(1):S157–73.

[62] Ding CZ, Batorsky R, Bhide R, Chao HJ, Cho Y, Chong S, Gullo-Brown J, Guo P, Kim SH, Lee F, Leftheris K, Miller A, Mitt T, Patel M, Penhallow BA, Ricca C, Rose WC, Schmidt R, Slusarchyk WA, Vite G, Yan N, Manne V, Hunt JT. Discovery and structure-activity relationships of imidazole-containing tetrahydrobenzodiazepine inhibitors of farnesyltransferase. J Med Chem 1999;42(25):5241–53.

[63] Ingrand S, Barrier L, Lafay-Chebassier C, Fauconneau B, Page G, Hugon J. The oxindole/imidazole derivative C16 reduces in vivo brain PKR activation. FEBS Lett 2007;581(23):4473–8.
[64] Niculescu-Duvaz D, Niculescu-Duvaz I, Suijkerbuijk BMJM, Ménard D, Zambon A, Nourry A, Davies L, Manne HA, Friedlos F, Ogilvie L, Hedley D, Takle AK, Wilson DM, Pons JF, Coulter T, Kirk R, Cantarino N, Whittaker S, Marais R, Springer CJ. Novel tricyclic pyrazole BRAF inhibitors with imidazole or furan central scaffolds. Bioorg Med Chem 2010;18(18):6934–52.
[65] Yang XD, Wan WC, Deng XY, Li Y, Yang LJ, Li L, Zhang HB. Design, synthesis and cytotoxic activities of novel hybrid compounds between 2-phenylbenzofuran and imidazole. Bioorg Med Chem Lett 2012;22(8):2726–9.
[66] Liu LX, Wang XQ, Yan JM, Li Y, Sun CJ, Chen W, Zhou B, Zhang HB, Yang XD. Synthesis and antitumor activities of novel dibenzo[b,d]furane imidazole hybrid compounds. Eur J Med Chem 2013;66:423–37.
[67] Wittine K, Babic MS, Makuc D, Plavec J, Pavelic S, Sedic M, Pavelic K, Leyssen P, Neyts J, Balzarini J, Mintas M. Novel 1,2,4-triazole and imidazole derivatives of L-ascorbic and imino-ascorbic acid: synthesis, anti-HCV and antitumor activity evaluations. Bioorg Med Chem 2012;20(11):3675–85.
[68] Wang XQ, Liu LX, Li Y, Sun CJ, Chen W, Li L, Zhang HB, Yang XD. Design, synthesis and biological evaluation of novel hybrid compounds of imidazole scaffold-based 2-benzylbenzofuran as potent anticancer agents. Eur J Med Chem 2013;62:111–21.
[69] Wan WC, Chen W, Liu LX, Li Y, Yang LJ, Deng XY, Zhang B, Yang XD. Synthesis and cytotoxic activity of novel hybrid compounds between 2-alkylbenzofuran and imidazole. Med Chem Res 2014;23:1599–611.
[70] Wang X, Nagase H, Watanabe T, Nobusue H, Suzuki T, Asami Y, Shinojima Y, Kawashima H, Takagi K, Mishra R, Igarashi J, Kimura M, Takayama T, Fukuda N, Sugiyama H. Inhibition of MMP-9 transcription and suppression of tumor metastasis by pyrrole-imidazole polyamide. Cancer Sci 2010;101(3):759–66.
[71] Obinata D, Ito A, Fujiwara K, Takayama KI, Ashikari D, Murata Y, Yamaguchi K, Urano T, Fujimura T, Fukuda N, Soma M, Watanabe T, Nagase H, Inoue S, Takahashi S. Pyrrole-imidazole polyamide targeted to break fusion sites in TMPRSS2 and ERG gene fusion represses prostate tumor growth. Cancer Sci 2014;105(10):1272–8.
[72] Syed J, Pandian GN, Sato S, Taniguchi J, Chandran A, Hashiya K, Bando T, Sugiyama H. Targeted suppression of EVI1 oncogene expression by sequence-specific pyrrole-imidazole polyamide. Chem Biol 2014;21(10):1370–80.
[73] Shinohara KI, Yoda N, Takane K, Watanabe T, Fukuyo M, Fujiwara K, Kita K, Nagase H, Nemoto T, Kaneda A. Inhibition of DNA methylation at the MLH1 promoter region using pyrrole–imidazole polyamide. ACS Omega 2016;1(6):1164–72.
[74] Malinee M, Kumar A, Hidaka T, Horie M, Hasegawa K, Pandian GN, Sugiyama H. Targeted suppression of metastasis regulatory transcription factor SOX2 in various cancer cell lines using a sequence-specific designer pyrrole–imidazole polyamide. Bioorg Med Chem 2020;28(3):115248.
[75] Cheng Z, Wang W, Wu C, Zou X, Fang L, Su W, Wang P. Novel pyrrole–imidazole polyamide Hoechst conjugate suppresses Epstein-Barr virus replication and virus-positive tumor growth. J Med Chem 2018;61(15):6674–84.
[76] Hayatigolkhatmi K, Padroni G, Su W, Fang L, Gómez-Castañeda E, Hsieh YC, Jackson L, Pellicano F, Burley GA, Jørgensen HG. An investigation of targeted inhibition of transcription factor activity with pyrrole imidazole polyamide (PA) in chronic myeloid leukemia (CML) blast crisis cells. Bioorg Med Chem Lett 2019;29(18):2622–5.

[77] Brant MG, Goodwin-Tindall J, Stover KR, Stafford PM, Wu F, Meek AR, Schiavini P, Wohnig S, Weaver DF. Identification of potent indoleamine 2,3-dioxygenase 1 (IDO1) inhibitors based on a phenylimidazole scaffold. ACS Med Chem Lett 2018;9(2):131–6.

[78] Negi A, Alex JM, Amrutkar SM, Baviskar AT, Joshi G, Singh S, Banerjee UC, Kumar R. Imine/amide–imidazole conjugates derived from 5-amino-4-cyano-N_1-substituted benzyl imidazole: microwave-assisted synthesis and anticancer activity *via* selective topoisomerase-II-a inhibition. Bioorg Med Chem 2015;23(17):5654–61.

[79] War JA, Srivastava SK, Srivastava SD. Synthesis and DNA-binding study of imidazole linked thiazolidinone derivatives. Luminescence 2017;32(1):104–13.

[80] Twarda-Clapa A, Krzanik S, Kubica K, Guzik K, Labuzek B, Neochoritis CG, Khoury K, Kowalska K, Czub M, Dubin G, Domling A, Skalniak L, Holak TA. 1,4,5-Trisubstituted imidazole-based p53 – MDM2/MDMX antagonists with aliphatic linkers for conjugation with biological carriers. J Med Chem 2017;60(10):4234–44.

[81] Theerasilp M, Chalermpanapun P, Ponlamuangdee K, Sukvanitvichai D, Nasongkla N. Imidazole-modified deferasirox encapsulated polymeric micelles as pH-responsive iron-chelating nanocarrier for cancer chemotherapy. RSC Adv 2017;7(18):11158–69.

[82] Singh I, Luxami V, Paul K. Synthesis of naphthalimide-phenanthro[9,10-*d*]imidazole derivatives: *in vitro* evaluation, binding interaction with DNA and topoisomerase inhibition. Bioorg Chem 2020;96:103631.

[83] Fakhree AA, Ghasemi Z, Rahimi M, Shahrisa A. Enhanced catalytic performance of copper iodide in 1,2,3-triazole-imidazole hybrid synthesis, and evaluation of their anti-cancer activities along with optical properties besides 1H-tetrazole-imidazole hybrids. Appl Organomet Chem 2020;35(9), e5773.

[84] Meenakshisundaram S, Manickam M, Pillaiyar T. Exploration of imidazole and imidazopyridine dimers as anticancer agents: design, synthesis, and structure–activity relationship study. Arch Pharm (Weinheim) 2019;352(12), e1900011.

[85] Chaudhary V, Venghateri JB, Dhaked HPS, Bhoyar AS, Guchhait SK, Panda D. Novel combretastatin-2-aminoimidazole analogues as potent tubulin assembly inhibitors: exploration of unique pharmacophoric impact of bridging skeleton and aryl moiety. J Med Chem 2016;59(7):3439–51.

[86] Vessally E, Soleimani-Amiri S, Hosseinian A, Edjlalid L, Bekhradnia A. New protocols to access imidazoles and their ring fused analogues: synthesis from *N*-propargylamines. RSC Adv 2017;7(12):7079–91.

[87] Chen TC, Yu DS, Huang KF, Fu YC, Lee CC, Chen CL, Huang FC, Hsieh HH, Lin JJ, Huang HS. Structure-based design, synthesis and biological evaluation of novel anthra[1,2-d]imidazole-6,11-dione homologues as potential antitumor agents. Eur J Med Chem 2013;69:278–93.

[88] Chen CL, Chang DM, Chen TC, Lee CC, Hsieh HH, Huang FC, Huang KF, Guh JH, Lin JJ, Huang HS. Structure-based design, synthesis and evaluation of novel anthra[1,2-d]imidazole-6,11-dione derivatives as telomerase inhibitors and potential for cancer polypharmacology. Eur J Med Chem 2013;60:29–41.

[89] Kayarmar R, Nagaraja GK, Bhat M, Naik P, Rajesh KP, Shetty S, Arulmoli T. Synthesis of azabicyclo[4.2.0]octa-1,3,5-trien-8-one analogues of 1H-imidazo[4,5-c]quinoline and evaluation of their antimicrobial and anticancer activities. Med Chem Res 2014;23:2964–75.

[90] Skwarska A, Augustin E, Beffinger M, Wojtczyk A, Konicz S, Laskowska K, Polewska J. Targeting of FLT3-ITD kinase contributes to high selectivity of imidazoacridinone C-1311 against FLT3-activated leukemia cells. Biochem Pharmacol 2015;95(4):238–52.

[91] Wu YZ, Ying HZ, Xu L, Cheng G, Chen J, Hu YZ, Liu T, Dong XW. Design, synthesis, and molecular docking study of 3H-imidazole[4,5-c]pyridine derivatives as CDK2 inhibitors. Arch Pharm (Weinheim) 2018;351(6), e1700381.

[92] Wang F, Jeon KO, Salovich JM, Macdonald JD, Alvarado J, Gogliotti RD, Phan J, Olejniczak ET, Sun Q, Wang S, Camper D, Yuh JP, Shaw JG, Sai J, Rossanese OW, Tansey WP, Stauffer SR, Fesik SW. Discovery of potent 2-aryl-6,7-dihydro-5-h-pyrrolo[1,2-a] imidazoles as WDR5-WIN-site inhibitors using fragment-based methods and structure-based design. J Med Chem 2018;61(13):5623–42.

[93] Xiao Z, Lei F, Chen X, Wang X, Cao L, Ye K, Zhu W, Xu S. Design, synthesis, and antitumor evaluation of quinoline-imidazole derivatives. Arch Pharm (Weinheim) 2018;351(6), e1700407.

[94] Zhang M, Ding Y, Qin HX, Xu ZG, Lan HT, Yang DL, Yi C. One-pot synthesis of substituted pyrrole–imidazole derivatives with anticancer activity. Mol Divers 2019. https://doi.org/10.1007/s11030-019-09982-z.

[95] Bondock S, Alqahtanic S, Foudaa AM. Synthesis and anticancer evaluation of some new pyrazolo[3,4-*d*][1,2,3]triazin-4-ones, pyrazolo[1,5-*a*]pyrimidines and imidazo[1,2-*b*]pyrazoles clubbed with carbazole. J Heterocycl Chem 2020. https://doi.org/10.1002/jhet.4148.

[96] Akhtar W, Khan MF, Verma G, Shaquiquzzaman M, Rizvi MA, Mehdi SH, Akhter M, Alam MM. Therapeutic evolution of benzimidazole derivatives in the last quinquennial period. Eur J Med Chem 2017;126:705–53.

[97] Ranganatha SR, Kavitha CV, Vinaya K, Prasanna DS, Chandrappa S, Raghavan SC, Rangappa KS. Synthesis and cytotoxic evaluation of novel 2-(4-(2,2,2-trifluoroethoxy)-3-methylpyridin-2-ylthio)-*1h*-benzo[d]imidazole derivatives. Arch Pharm Res 2009;32(10):1335–43.

[98] Purushottamachar P, Godbole AM, Gediya LK, Martin MS, Vasaitis TS, Kwegyir-Afful AK, Ramalingam S, Ates-Alagoz Z, Njar VCO. Systematic structure modifications of multitarget prostate cancer drug candidate galeterone to produce novel androgen receptor down-regulating agents as an approach to treatment of advanced prostate cancer. J Med Chem 2013;56(12):4880–98.

[99] Alkahtani HM, Abbas AY, Wang S. Synthesis and biological evaluation of benzo[d]imidazole derivatives as potential anti-cancer agents. Bioorg Med Chem Lett 2012;22(3):1317–21.

[100] Im C. Docking and three-dimensional quantitative structure–activity relationship analyses of imidazole and thiazolidine derivatives as Aurora A kinase inhibitors. Arch Pharm Res 2016;39(12):1635–43.

[101] Fan C, Zhong T, Yang H, Yang Y, Wang D, Yang X, Xu Y, Fan Y. Design, synthesis, biological evaluation of 6-(2-amino-1H-benzo[d]imidazole-6-yl)quinazolin-4(3H)-one derivatives as novel anticancer agents with Aurora kinase inhibition. Eur J Med Chem 2020;190:112108.

[102] Hasanpourghadi M, Karthikeyan C, Pandurangan AK, Looi CY, Trivedi P, Kobayashi K, Tanaka K, Wong WF, Mustafa MR. Targeting of tubulin polymerization and induction of mitotic blockage by methyl 2-(5-fluoro-2-hydroxyphenyl)-1 Hbenzo[d]imidazole-5-carboxylate (MBIC) in human cervical cancer HeLa cell. J Exp Clin Cancer Res 2016;35:58.

[103] Zhou J, Ji M, Zhu Z, Cao R, Chen X, Xu B. Discovery of 2-substituted 1 H-benzo[d]immidazole-4-carboxamide derivatives as novel poly(ADP-ribose)polymerase-1 inhibitors with *in vivo* antitumor activity. Eur J Med Chem 2017;132:26–41.

[104] Yadav S, Narasimhan B, Lim SM, Ramasamy K, Vasudevan M, Shah SAA, Selvaraj M. Synthesis, characterization, biological evaluation and molecular docking studies of 2-(1H-benzo[d] imidazol-2-ylthio)-*N*-(substituted 4-oxothiazolidin-3-yl) acetamides. Chem Cent J 2017;11(1):137.

[105] Yadav S, Lim SM, Ramasamy K, Vasudevan M, Shah SAA, Mathur A, Narasimhan B. Synthesis and evaluation of antimicrobial, antitubercular and anticancer activities of 2-(1-benzoyl-1 H-benzo[d] imidazol-2-ylthio)-N-substituted acetamides. Chem Cent J 2018;12(1):66. https://doi.org/10.1186/s13065-018-0432-3.

[106] Wu Q, Zheng K, Huang X, Li L, Mei W. Tanshinone-IIA-based analogs of imidazole alkaloid act as potentinhibitors to block breast cancer invasion and metastasis *in vivo*. J Med Chem 2018;61(23):10488–501.

[107] Tahlan S, Ramasamy K, Lim SM, Shah SAA, Mani V, Narasimhan B. Design, synthesis and therapeutic potential of 3-(2-(1 *H*-benzo[*d*]imidazol-2-ylthio)acetamido)-*N*-(substituted phenyl)benzamide analogues. Chem Cent J 2018;12(1):139. https://doi.org/10.1186/s13065-018-0513-3.

[108] Effendi N, Mishiro K, Takarada T, Yamada D, Nishii R, Shiba K, Kinuya S, Odani A, Ogawa K. Design, synthesis, and biological evaluation of radioiodinated benzo[*d*] imidazole-quinoline derivatives for platelet-derived growth factor receptor β (PDGFRβ) imaging. Bioorg Med Chem 2019;27(2):383–93.

[109] Effendi N, Ogawa K, Mishiro K, Takarada T, Yamada D, Kitamura Y, Shiba K, Maeda T, Odani A. Synthesis and evaluation of radioiodinated 1-{2-[5-(2-methoxyethoxy)-1*H*-benzo[*d*]imidazol-1-yl]quinolin-8-yl}piperidin-4-amine derivatives for platelet-derived growth factor receptor β (PDGFRβ) imaging. Bioorg Med Chem 2017;25(20):5576–85.

[110] Effendi N, Mishiro K, Takarada T, Makino A, Yamada D, Kitamura Y, Shiba K, Kiyono Y, Odani A, Ogawa K. Radiobrominated benzimidazole-quinoline derivatives as platelet-derived growth factor receptor beta (PDGFRβ) imaging probes. Sci Rep 2018;8:10369.

[111] Al-blewi FF, Almehmadi MA, Aouad MR, Bardaweel SK, Sahu PK, Messali M, Rezki N, El Ashry ESH. Design, synthesis, ADME prediction and pharmacological evaluation of novel benzimidazole-1,2,3-triazole-sulfonamide hybrids as antimicrobial and antiproliferative agents. Chem Cent J 2018;12(1):110. https://doi.org/10.1186/s13065-018-0479-1.

[112] Tahlan S, Kumar S, Ramasamy K, Lim SM, Shah SAA, Mani V, Pathania R, Narasimhan B. Design, synthesis and biological profile of heterocyclic benzimidazole analogues as prospective antimicrobial and antiproliferative agents. BMC Chem 2019;13(1):50. https://doi.org/10.1186/s13065-019-0567-x.

[113] Shi C, Wang Q, Liao X, Ge H, Huo G, Zhang L, Chen N, Zhai X, Hong Y, Wang L, Han Y, Xiao W, Wang Z, Shi W, Mao Y, Yu J, Xia G, Liu Y. Discovery of 6-(2-(dimethylamino)ethyl)-N-(5-fluoro-4-(4-fluoro-1- isopropyl-2-methyl-1H-benzo[d]imidazole-6-yl)pyrimidin-2-yl)-5,6,7,8-tetrahydro-1,6-naphthyridin-2-amine as a highly potent cyclin-dependent kinase 4/6 inhibitor for treatment of cancer. Eur J Med Chem 2019;178:352–64.

[114] Ding HW, Yu L, Bai MX, Qin XC, Song MT, Zhao QC. Design, synthesis and evaluation of some 1,6-disubstituted-*1H*-benzo[*d*] imidazoles derivatives targeted PI3K as anticancer agents. Bioorg Chem 2019;93:103283. https://doi.org/10.1016/j.bioorg.2019.103283.

[115] Yang YQ, Chen H, Liu QS, Sun Y, Gu W. Synthesis and anticancer evaluation of novel 1*H*-benzo[*d*]imidazole derivatives of dehydroabietic acid as PI3Kα inhibitors. Bioorg Chem 2020;100:102845. https://doi.org/10.1016/j.bioorg.2020.103845.

[116] Li AL, Yang YQ, Wang WY, Liu QS, Sun Y, Gu W. Synthesis, cytotoxicity and apoptosis-inducing activity of novel 1H-benzo[d]imidazole derivatives of dehydroabietic acid. J Chin Chem Soc 2020;67(9):1668–78.

[117] Çevik U, Osmaniye D, Çavuşoğlu BK, Sağlik BN, Levent S, Ilgin S, Can NO, Özkay Y, Kaplancikli ZA. Synthesis of novel benzimidazole–oxadiazole derivatives as potent anticancer activity. Med Chem Res 2019;28:2252–61.

[118] Keppler BK, Wehe D, Endres H, Rupp W. Synthesis, antitumor activity, and x-ray structure of bis(imidazolium) (imidazole) pentachlororuthenate(III), $(ImH)_2(RuImCl_5)$. Inorg Chem 1987;26(6):844–6.

[119] Groessl M, Reisner E, Hartinger CG, Eichinger R, Semenova O, Timerbaev AR, Jakupec MA, Arion VB, Keppler BK. Structure-activity relationships for NAMI-A-type complexes

(HL)[*trans*-RuCl4L(*S*-dmso)ruthenate(III)] (L = imidazole, indazole, 1,2,4-triazole, 4-amino-1,2,4-triazole, and 1-methyl-1,2,4-triazole): aquation, redox properties, protein binding, and antiproliferative activity. J Med Chem 2007;50(9):2185–93.

[120] Skander M, Retailleau P, Bourrie B, Schio L, Mailliet P, Marinetti A. N-heterocyclic carbene-amine Pt(II) complexes, a new chemical space for the development of platinum-based anticancer drugs. J Med Chem 2010;53(5):2146–54.

[121] Blunden BM, Rawal A, Lu H, Stenzel MH. Superior chemotherapeutic benefits from the ruthenium-based anti-metastatic drug NAMI-A through conjugation to polymeric micelles. Macromolecules 2014;47(5):1646–55.

[122] Chang SW, Lewis AR, Prosser KE, Thompson JR, Gladkikh M, Bally MB, Warren JJ, Walsby CJ. CF_3 derivatives of the anticancer Ru(III) complexes KP1019, NKP-1339, and their imidazole and pyridine analogues show enhanced lipophilicity, albumin interactions, and cytotoxicity. Inorg Chem 2016;55(10):4850–63.

[123] Pellei M, Gandin V, Marzano C, Marinelli M, Bello FD, Santini C. The first water-soluble copper(I) complexes bearing sulfonated imidazole- and benzimidazole-derived N-heterocyclic carbenes: synthesis and anticancer studies. Appl Organometal Chem 2017;32(3), e4185.

[124] Rimoldi I, Facchetti G, Lucchini G, Castiglioni E, Marchianò S, Ferri N. In vitro anticancer activity evaluation of new cationic platinum(II) complexes based on imidazole moiety. Bioorg Med Chem 2017;25(6):1907–13.

[125] Abdel-Rahman LH, Abdelhamid AA, Abu-Dief AM, Shehata MR, Bakheet MA. Facile synthesis, X-ray structure of new multi-substituted aryl imidazole ligand, biological screening and DNA binding of its Cr(III), Fe(III) and Cu (II) coordination compounds as potential antibiotic and anticancer drugs. J Mol Struct 2020;1200:127034. https://doi.org/10.1016/j.molstruc.2019.127034.

[126] Abdalla EM, Rahman LMA, Abdelhamid AA, Shehata MR, Alothman AA, Nafady A. Synthesis, characterization, theoretical studies, and antimicrobial/antitumor potencies of salen and salen/imidazole complexes of Co (II), Ni (II), Cu (II), Cd (II), Al (III) and La (III). Appl Organomet Chem 2020. https://doi.org/10.1002/aoc.5912, e5912.

[127] Sharhan O, Heidelberg T, Hashim NM, Al-Madhagi WM, Ali HM. Benzimidazolium-acridine-based silver N-heterocyclic carbene complexes as potential anti-bacterial and anti-cancer drug. Inorg Chim Acta 2020;504:119462. https://doi.org/10.1016/j.ica.2020.119462.

[128] Barragán F, Carrion-Salip D, Gómez-Pinto I, González-Cantó A, Sadler PJ, Llorens RD, Moreno V, González C, Massaguer A, Marchán V. Somatostatin Subtype-2 receptor-targeted metal-based anticancer complexes. Bioconjug Chem 2012;23(9):1838–55.

[129] Feng S, Zhang X, Shi D, Wang Z. Zeolitic imidazolate framework-8 (ZIF-8) for drug delivery: a critical review. Front Chem Sci Eng 2020. https://doi.org/10.1007/s11705-020-1927-8.

[130] Adhikari C, Das A, Chakraborty A. Zeolitic imidazole framework (ZIF) nanospheres for easy encapsulation and controlled release of an anticancer drug doxorubicin under different external stimuli: a way toward smart drug delivery system. Mol Pharm 2015;12:3158–66.

[131] Duan LN, Dang QQ, Han CY, Zhang XM. An interpenetrated bioactive nonlinear optical MOF containing a coordinated quinolone-like drug and Zn(II) for pH-responsive release. Dalton Trans 2015;44:1800–4.

[132] Burrows AD, Jurcic M, Keenan LL, Lane RA, Mahon MF, Warren MR, Nowell H, Paradowski M, Spencer J. Incorporation by coordination and release of the iron chelator drug deferiprone from zinc-based metal–organic frameworks. Chem Commun 2013;49:11260–2.

[133] Li HT, Song SJ, Pei XR, Lu DB. A In^{III}-MOF with imidazole decorated pores as 5-FU delivery system to inhibit colon cancer cells proliferation and induce cell apoptosis *in vitro* and *in vivo*. Z Anorg Allg Chem 2019;645(11):801–9.

Chapter 3

Recent advancements on imidazole containing heterocycles as antitubercular agents

Dinesh K. Agarwal[a], Jay Soni[b], Ayushi Sethiya[b], Nusrat Sahiba[b], Pankaj Teli[b], and Shikha Agarwal[b]
[a]*Department of Pharmacy, PAHER University, Udaipur, Rajasthan, India,* [b]*Synthetic Organic Chemistry Laboratory, Department of Chemistry, Mohanlal Sukhadia University, Udaipur, Rajasthan, India*

Abbreviations

DMF	dimethyl formamide
HIV	human immuno deficiency virus
IC	inhibitory concentration
LD	lethal dose
LJ	Löwenstein-Jensen
LORA	low oxygen recovery assay
MabA	combretastatin A-4
MABA	microplate alamar blue assay
MDR	multidrug resistant
MIC	minimum inhibitory concentration
MLC	minimum lethal concentration
Mtb	Mycobacterium tuberculosis
MtGS	*Mycobacterium tuberculosis* glutamine synthetase
REMA	Resazurin Microtiter Assay
SAR	structure-activity relationship
TAACF	Tuberculosis Antimicrobial Acquisition and Coordinating Facility
TB	tuberculosis
WHO	World Health Organization
XDR	extensively drug resistant

Imidazole-Based Drug Discovery. https://doi.org/10.1016/B978-0-323-85479-5.00002-2
Copyright © 2022 Elsevier Inc. All rights reserved.

1. Introduction

Tuberculosis, a highly communicable and infectious disease, is the ninth highest cause of death worldwide. It is caused by the bacillus *M. tuberculosis*. Globally, it affects about 10 million people every year and approximately 1.7 billion people are infected or at risk of developing TB [1, 2]. The "Global Tuberculosis Report 2020," published by the WHO, states that many high TB burden countries and many WHO regions are not on track to achieve the 2020 milestones of the End TB strategy. Globally, an overall reduction from 2015 to 2019 was 9% including a reduction of 2.3% between 2018 and 2019. The WHO European region and African region have almost reached the 2020 milestone, with reductions of 19% and 16% in the TB incidence rate and 31% and 19% reduction in TB deaths from 2015 to 2019, respectively [1]. A total of 78 countries are exactly on track to reach the 2020 milestone, including seven high TB burden countries. The high-level emergence of drug resistance toward anti-TB drugs is due to underreporting, underdiagnosis, and the variance in estimated incidence and exact enrollment cases [3, 4]. Therefore, there is an urgent need to expedite the pace in TB research via searching for highly effective, affordable, and nontoxic novel drugs with shorter regimens to reach the targets ascertained by the WHO. Moreover, the third pillar of the End TB strategy—"intensified research and innovation"—must be brought into action [5].

Heterocyclic compounds are the most privileged structures in organic chemistry and drug discovery. Among the biologically active heterocycles, imidazoles serve as important structural units in several natural products and drugs [6]. Heterocycles bearing imidazole units exhibit an array of pharmacological activities, namely antibacterial [7], anti-HIV [8], antimalarial [9], anticancer [10, 11], antitubercular [12], etc. Moreover, 4-nitroimidazoles, pretomanid (PA-824), delamanid have also shown excellent anti-TB activity (Fig. 3.1). Pretomanid (PA-824) is in Phase III clinical trials and delamanid has received approval for the treatment of MDR-TB infected patients [12–16]. Imidazopyridine, a telacebac (Q203) drug, is in Phase II clinical trials and benzimidazole, SPR720, is in Phase I clinical trials [5]. These drugs suggest imidazole as a useful pharmacophore for the treatment of TB-infected patients. Over the years, a large number of imidazole heterocycles have been synthesized and studied for their in vitro and in vivo antimycobacterial activities against both drug-sensitive and drug-resistant Mtb strains [17, 18].

The present chapter involves the recent advancements in the development of imidazole-containing derivatives, viz. substituted imidazoles, fused, hybrids, linker compounds, benzimidazoles, etc. as potential anti-TB agents over the last 10 years (2011–2021) (Fig. 3.2). Furthermore, the detailed study may direct future researchers for further rational development of imidazole derivatives as antitubercular agents.

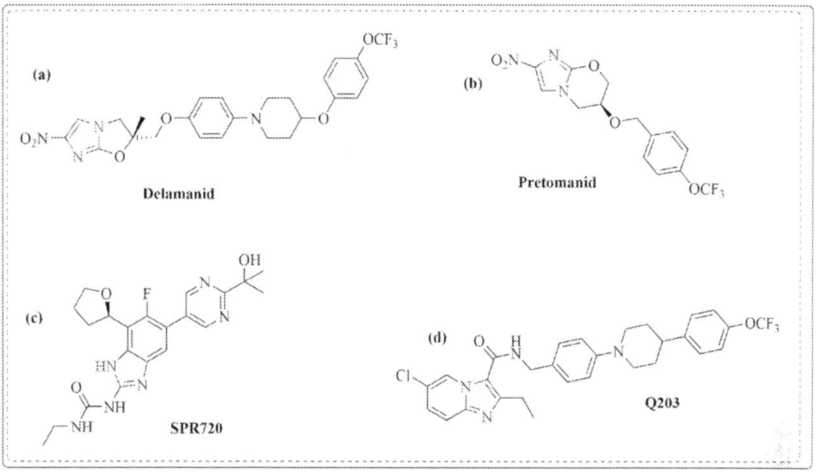

FIG. 3.1 Imidazole derivatives as antitubercular agents: (A) delamanid (approved); (B) pretomanid (PA-824) (Phase III clinical trials); (C) SPR720 (Phase I clinical trials); (D) Q203 (Phase II clinical trials).

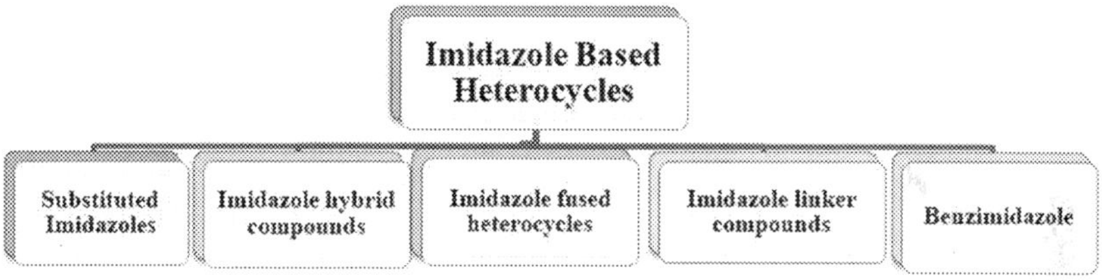

FIG. 3.2 Classification of imidazole-based heterocycles on the basis of their structural unit.

2. Imidazole-based heterocycles

2.1 Substituted imidazoles

Desai et al. [19] synthesized a series of 1,4-dihydropyridine-3,5-dicarbamoyl derivatives having an imidazole nucleus at the C-4 position in excellent yields using Hantzsch reaction and further screened them for antitubercular activity at 6.25 μg/mL against *M. tuberculosis* H37Rv by the Tuberculosis Antimicrobial Acquisition and Coordinating Facility (TAACF) in BACTEC 12B medium using the Microplate Alamar Blue Assay [20]. The synthesized derivatives included the advantage of two pharmacophores, imidazole and 1,4-dihydropyridine in a single molecular framework. The multicomponent Hantzsch reaction of substituted *N*-(aryl)-3-oxobutanamide(**1a–p**)with2-butyl-4-chloro-1*H*-imidazole-5-carbaldehyde and NH_3 in methanol solvent produced 4-(2-butyl-4-chloro-1*H*-imidazol-5-yl)-*N*3,*N*5-bis(aryl)-2,6-dimethyl-1,4 dihydropyridine-3,5 dicarboxamides (**3a–p**) in

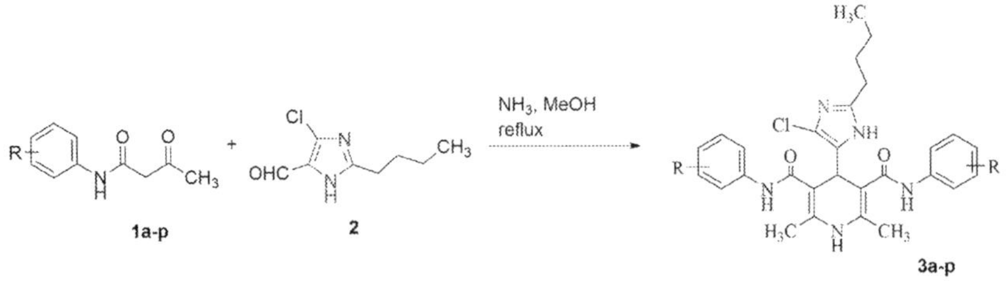

SCHEME 3.1 Synthesis of substituted imidazole derivatives.

high yields (Scheme 3.1). Among the synthesized compounds, derivatives with electron-withdrawing group (F, NO_2 at para position of phenyl ring), (**3j** and **3m**), were found to be the most active antitubercular agents. The presence of fluoro/nitro at the para position of the phenyl ring increased the lipophilicity and facilitated the compounds to diffuse through the biological membranes and to reach the site of action, and thus influenced the antitubercular activity positively. Furthermore, the potent antitubercular compounds were also accompanied by the low level of cytotoxicity and higher selective index (Table 3.1).

Gising et al. [21] synthesized 2-tert-butyl-4,5-diarylimidazole derivatives and assessed their antitubercular potential against *M. tuberculosis*. The MIC values of synthesized derivatives were calculated and compound **4** was found to be the best inhibitor with $IC_{50}=0.049\,\mu M$ on MtGS (*Mycobacterium tuberculosis* glutamine synthetase) and MIC = 2 μg/mL against *M. tuberculosis*. It was also found that the X-ray structure of compound **4** was bound to MtGS; furthermore, the 2-amino group formed an additional interaction with the hydroxyl oxygen of Ser280 in its primary binding mode instead of forming a H bond to

TABLE 3.1 In vitro antitubercular data of 1,4-dihydropyridine-3,5-dicarbamoyl derivatives bearing imidazole.

Compound	R	MIC (μg/mL)	Compound	R	MIC (μg/mL)
3a	H	–	**3i**	3-F	–
3b	2-CH_3	–	**3j**	4-F	0.2
3c	3-CH_3	–	**3k**	2-NO_2	3.13
3d	4-CH_3	–	**3L**	3-NO_2	–
3e	2-OCH_3	–	**3m**	4-NO_2	0.2
3f	3-OCH_3	–	**3n**	2-Cl	–
3g	4-OCH_3	–	**3o**	3-Cl	–
3h	2-F	6.25	**3p**	4-Cl	1.56

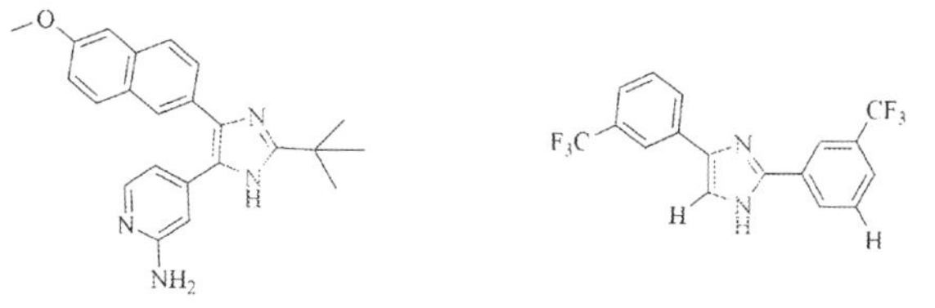

4 5

FIG. 3.3 2-Tert-butyl-4,5-diarylimidazole (**4**) and 2,4-diphenyl-1*H*-imidazole (**5**) derivatives.

the backbone carbonyl oxygen of Lys361 as calculated from docking studies (Fig. 3.3). A series of 2,4-diphenyl-1*H*-imidazoles were synthesized as effective antitubercular agents in a whole cell phenotypic assay against Mtb [22]. The MICs were established using Mtb H37Rv ATCC 27294 in MABA and LORA assays [20, 23]. Compound **5**, having m-CF_3 substitution at both the aromatic rings attached to the imidazole, was found to be 18 times more active than the substituted ones and was discovered to be highly efficient to inhibit the growth of Mtb at low micromolar concentrations (MIC = 1.7 μM) (Fig. 3.3).

A highly efficient synthesis of 2,4,5-trisubstituted imidazole derivatives **6** from 1-(4-nitrophenyl)-2-(4-(trifluoromethyl)phenyl)ethane-1,2-dione, substituted aldehydes and ammonium acetate using MoS_2-supported-calix[4]arene (MoS_2-CA4) nanocatalyst under solvent-free conditions was developed [24] (Scheme 3.2). The noteworthy features are economical, safe reaction profiles, solvent free, broad substrate scope, simple workup conditions, and excellent yields. The synthesized compounds were examined for their in vitro antitubercular activity against *M. tuberculosis* H37Rv. Compounds **6r** (four times more than pyrazinamide), **6c**, **6d**, **6f**, and **6m** were found to be more potent than the reference drugs, ciprofloxacin, pyrazinamide, and ethambutol. Further, these compounds were tested for inhibition of MabA (β-ketoacyl-ACP reductase) enzyme and cytotoxic activity against mammalian Vero cell line. The in silico studies were conducted on MabA (PDB ID: 1UZN) enzyme to predict protein interaction of synthesized compounds. Finally, the results of in vitro anti-TB activity and the docking study showed that the synthesized compounds have a strong anti-TB activity and can be adapted as lead compounds for future research (Table 3.2).

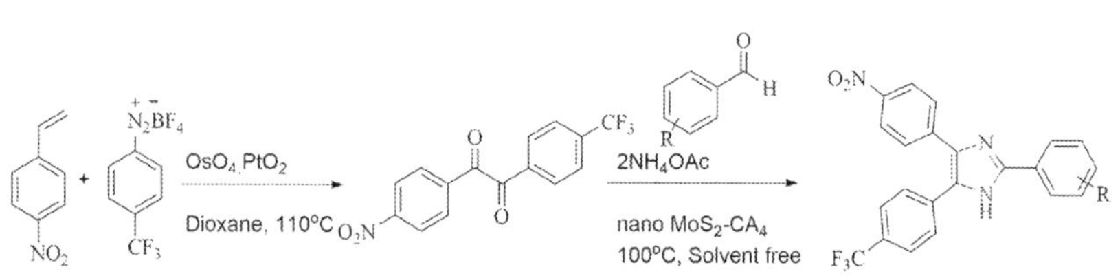

6

SCHEME 3.2 Synthesis of 2,4,5-trisubstituted imidazole derivatives (MoS_2-CA4) nanocatalyst.

TABLE 3.2 MabA enzyme inhibition activity of potent compounds (**6**).

Compound	IC_{50} (μM)
6c	46.4 ± 0.83
6d	63.8 ± 0.17
6f	95.1 ± 0.39
6m	28.3 ± 0.12
6r	21.7 ± 0.05
INH-NADP	20.2 ± 0.41

Note: All data are expressed in mean ± SEM (n = 3)

Lima et al. [25] synthesized imidazole modified pyrazinamide derivatives **7** and evaluated them against *M. tuberculosis* ATCC 27294 using the Microplate Alamar Blue Assay. The synthesized derivatives showed stronger antitubercular effect (IC_{50} = 50 μg/mL) than pyrazinamide derivatives (IC_{50} > 100 μg/mL). Furthermore, the synthesized derivatives were studied for their cell viabilities in infected and noninfected macrophages with Mycobacterium bovis Bacillus Calmette-Guerin, and compound **7** was found not to be cytotoxic to host cells (Fig. 3.4). Econazole, an antifungal drug, is one of the active drugs against replicating and nonreplicating Mtb. Econazole-derived nitroimidazoles were designed and investigated for antitubercular activity. Derivatives **8** and **9** were found to be active against Mtb. Compounds **9a** and **9f** (MIC = 0.5 μg/mL) were found to be more active than econazole (MIC = 16 μg/mL) [26] (Fig. 3.4).

Amini et al. [27] synthesized N3-(substituted phenyl)-N5-(substituted phenyl)-4-(4,5-dichloro-1H-imidazol-2-yl)-2-methyl-1,4-dihydropyridine-3,5-dicarboxamides **10** and evaluated them for antitubercular activity against *M. tuberculosis* strain using rifampicin as a reference drug (Scheme 3.3). The synthesized compounds exhibited low to moderate antitubercular activity. The percentage inhibitions of the synthesized compounds are shown in Table 3.3.

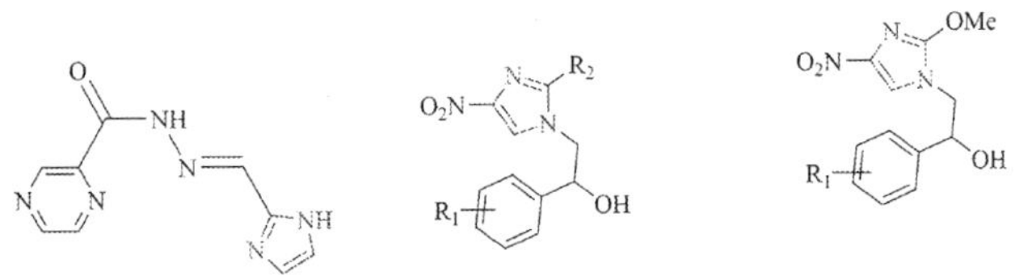

7 **8** **9**

R_1 9a= 2,4-diCl,
9f= 4-Ph

FIG. 3.4 Imidazole modified pyrazinamide derivatives (**7**) and Econazole-derived nitroimidazoles (**8** and **9**).

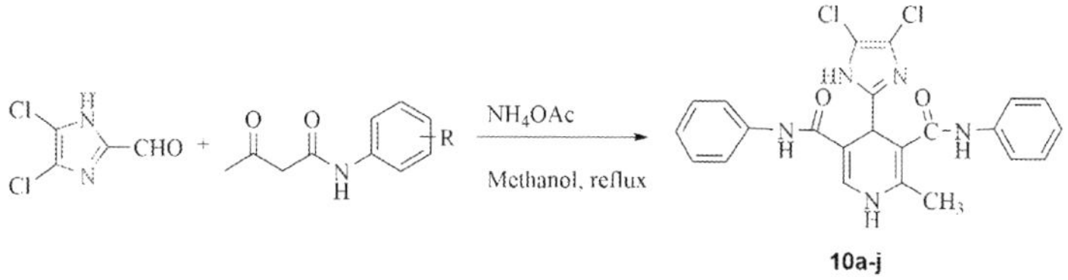

10a-j

SCHEME 3.3 Synthesis of substituted imidazole derivatives.

TABLE 3.3 Percentage inhibition of the synthesized imidazole derivatives.

Compound	R	Inhibition (%)
10a	H	9
10b	3-F	0
10c	4-F	13
10d	3-Cl	50
10e	4-Cl	12
10f	3,4-diCl	34
10g	3-Br	1
10h	4-Br	0
10i	3-NO_2	43
10j	4-NO_2	43
Rifampicin		> 98

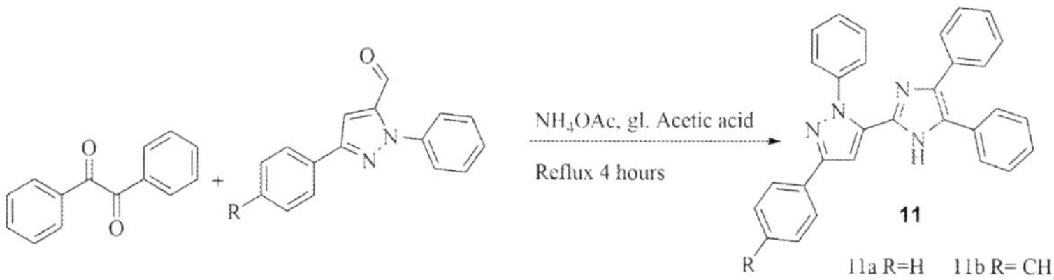

11

11a R=H 11b R= CH_3
11d R= NO_2 11e R=F
11f R= Br 11g R= Cl
11h R= OCH_3

SCHEME 3.4 Synthesis of tri-substituted imidazole hybrid derivatives.

A series of tri-substituted imidazole derivatives were synthesized by Radziszewski reaction using benzil, ammonium acetate, and 1-phenyl-3-(p-substituted phenyl)-1*H*-pyrazole-4-carbaldehydes (Scheme 3.4). The synthesized compounds were screened for their antimycobacterial activities against H37Rv strain of *M. tuberculosis* using isoniazid as a standard drug. Compounds

11d, **11e**, and **11g** showed significant antimycobacterial activity [28]. Other compounds showed moderate to poor antimycobacterial activity.

2.2 Hybrid Imidazoles

Imidazole-pyrazole hybrids bearing pyridine moiety were studied for in vitro antitubercular activity against MTB H37Rv strain Löwenstein-Jensen medium by Kalaria and colleagues [29]. All the synthesized hybrids showed moderate in vitro antitubercular activity. The growth inhibitory (GI) against MTB H37Rv was found to be 25%–96% at 250 μg/mL. Compounds **12**, **13**, **14a**, and **14b** exhibited higher inhibition and were further screened for MIC values. Compound **14a** (MIC = 25 mg/mL) was found to possess highest potency with 96% inhibition against *M. tuberculosis*. Compounds **13**, **12**, and **14b** showed 94%, 91%, and 92% inhibition, respectively. The SAR studies revealed that the substituents at imidazole and pyridine moieties influenced the anti-TB activity. The unsubstituted imidazole hybrids were found to be more potent with respect to methyl imidazole hybrids. The substituents at pyridine core affected the anti-TB activity in the following manner: 2-furyl > 4-OMePh > 4-(3′,4′-methylenedioxy) phenyl > 3-coumaryl > 2-thienyl (Fig. 3.5, Table 3.4).

A series of 5-imidazopyrazole hybrids incorporated polyhydroquinoline **15a–t** was evaluated for their anti-TB activity against MTB H37Rv at two concentrations, 250 and 100 μg/mL by Kalaria et al. [30]. All the synthesized hybrid compounds showed poor to excellent activity against MTB H37Rv with GI of 6%–94% and 7%–94% at 100 and 250 μg/mL, respectively. Hybrids **15a**, **15g**, and **15q** exhibited excellent activity (i.e., 90%, 94%, and 91% at 250 μg/mL) against *M. tuberculosis* H37Rv. Furthermore, compound **15g** showed the best activity at both the concentrations, i.e., 94% at 250 and 100 μg/mL. Other compounds were found to show moderate to poor inhibition. The SAR studies concluded that CN group helped in augmenting the antituberculosis activity against *M. tuberculosis* H37Rv (Fig. 3.6). Kalaria and colleagues [31] synthesized fused pyran derivatives bearing the 5-imidazopyrazole using microwave irradiations and studied them for *in vitro* antituberculosis activity against the *M. tuberculosis* H37Rv strain. The primary screening of synthesized compounds was performed

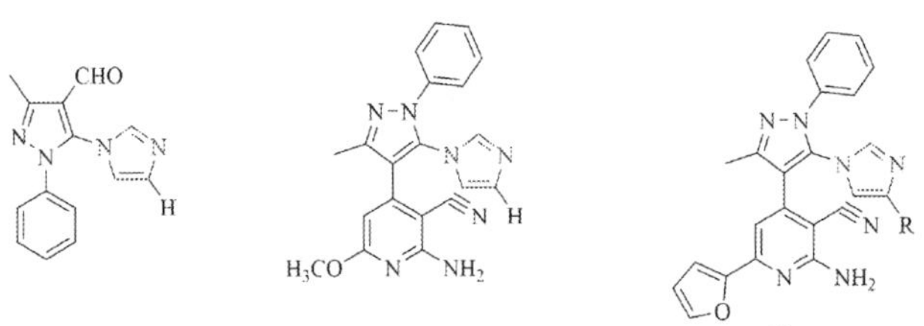

12 13 14

R= H (14a), R= CH_3,(14b)

FIG. 3.5 Imidazole-pyrazole hybrids bearing pyridine moiety.

TABLE 3.4 In vitro anti-TB activity of synthesized derivatives against *M. tuberculosis* H37Rv.

Entry	Inhibition (%)	MIC value (μg/mL)
12	91	100
13	94	62.5
14a	96	25
14b	92	100
Rifampicin	98	0.15
Isoniazid	99	0.20

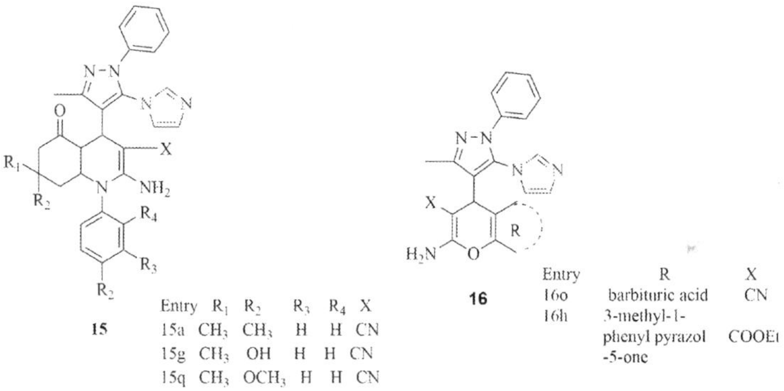

15

16

FIG. 3.6 5-Imidazopyrazole hybrids incorporated polyhydroquinoline (**15**) and fused pyran derivatives bearing 5-imidazopyrazole (**16**).

at 250 and 100 mg/mL using Löwenstein-Jensen medium with rifampicin and isoniazid as standard drugs. Compound **16o** showed 95% inhibition and 92% at 250 and 100 mg/mL concentration, respectively. Compound **16h** showed excellent activity, i.e., 92% at 250 and 100 mg/mL, respectively (Fig. 3.6).

Kishk and his group synthesized three series of imidazole derivatives [32] and screened them against *M. tuberculosis* H37Rv by Resazurin Microtiter Assay. In Series 1, compounds **17a** and **17b** (imidazole pyridyl derivatives) exhibited antitubercular activity with MIC = 12.5 μg/mL. In Series 2, compound **18** (5-((1*H*-imidazol-1-yl)methyl)-1-(4-methoxyphenyl)-3-methyl-1*H*-pyrazole) displayed antimycobacterial activity with an MIC of 25 μg/mL. In Series 3, most of the compounds displayed antimycobacterial activity and alkyl substituted aryl derivatives (**17c–g**) exhibited the most promising inhibitory activity in the range MIC = 0.781–1.562 μg/mL. By the Hansch analysis, a direct correlation between the lipophilicity and MIC of compounds was found and the increased lipophilicity might facilitate enhanced drug uptake across the *Mtb* lipid-rich cell wall which resulted in increased antimycobacterial activity. Further, the molecular modeling studies were performed with *M. tuberculosis* cytochrome P450 CYP121A1, a promising drug target for the treatment of tuberculosis. The highest binding affinity was found for branched alkyl derivatives, **17f**, **17g**, and **17h** (biphenyl) (Fig. 3.7).

2.3 Fused imidazoles

Ten compounds of imidazo[2,1-b][1,3,4]thiadiazoles **19** were designed and studied for in vitro antitubercular activity against *Mycobacterium tuberculosis* H37Rv strain with rifampicin as a reference, using the Microplate Alamar Blue Assay. Among the 10 compounds, derivatives bearing electron withdrawing groups at the sixth position of imidazo(2,1-b)-1,3,4-thiadiazole, (**19a**, **19b**, **19c**, **19d**, **19f**, **19i**), exhibited high antitubercular activity [33] (Fig. 3.8). Compound **19f**, 2-(1-methyl-1H-imidazol-2-yl)-6-(4-nitrophenyl)imidazo[2,1-b][1,3,4]thiadiazole, showed the maximum inhibitory activity, 98% with MIC of 3.14 μg/mL. The derivatives with electron donating groups, **19g** and **19h** and the unsubstituted **19e**, showed poor antitubercular activity. Compounds with high activity (**19a**, **19b**, **19c**, **19d**, **19f**, **19i**) were further studied for toxicity (IC_{50}) in a mammalian Vero cell line using MTT assay, and they showed IC_{50} values in the range of 8.5–12.3 μg/mL and exhibited antitubercular activity at noncytotoxic concentrations. Compound **19f** showed the lowest toxicity, with IC_{50} values of 9.8 μg/mL (SI = 3.12) (Table 3.5).

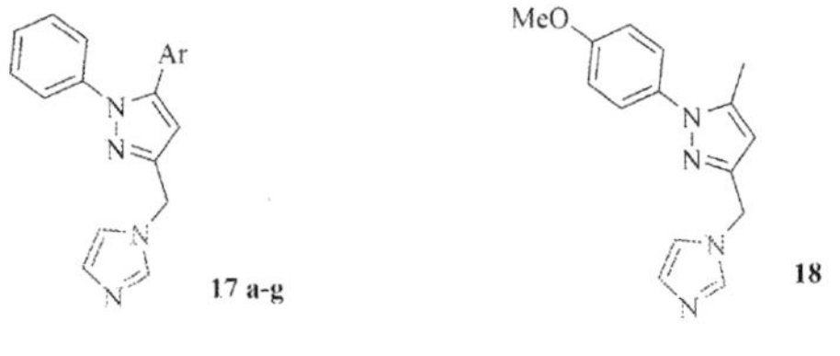

17 a-g = Ar = Ph, 4-Et Ph, 4-propyl Ph, 4-isopropyl Ph, 4-isobuPh, 4-tert.butyl Ph, biphenyl

FIG. 3.7 Imidazole hybrid derivatives.

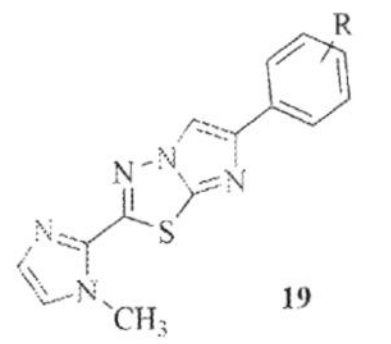

R= (a) 3-NO_2, (b) 4-Br, (c) 4-Cl, (d) 4-F, (e) H, (f) 4-NO_2, (g) 4-Me (h) 3-Me (i) 2,4-diCl

FIG. 3.8 Imidazo[2,1-b][1,3,4]thiadiazoles.

TABLE 3.5 Inhibition (%), MIC values, IC_{50} and SI values of synthesized fused compounds (19).

Compound	Inhibition (%)	Activity	MIC (μg/mL)	IC_{50}	SI (IC_{50}/MIC value)
19a	91	+	4.34	10.56	2.43
19b	94	+	5.78	11.4	1.97
19c	95	+	5.48	12.3	2.24
19d	90	+	4.86	8.5	1.74
19e	16	–	> 6.25	–	–
19f	98	+	3.14	9.8	3.12
19g	18	–	> 6.25	–	–
19h	30	–	> 6.25	–	–
19i	92	+	5.66	10.3	1.81
Rifampicin			0.125–0.25		> 10

Pandey and his coauthors [34] synthesized pyrido[1,2-*a*]imidazo-chalcones **20a–q** and evaluated their antitubercular activity against *M. tuberculosis* H37Rv strain taking ethambutol and pyrazinamide as standard drugs. The cytotoxicity of potent compounds was further studied in Vero cells (C1008) and mouse bone marrow derived macrophages (MBMDMϕ). Among the synthesized ones, compounds **20l**, **20p**, and **20q** were found to be the most active with MIC values of 7.89, 6.42, and 6.59 μM, respectively, in comparison to the standard drugs, ethambutol and pyrazinamide with MIC values of 9.78 and 101.53 μM without any toxicity (Fig. 3.9).

Tukulula and his colleagues [35] synthesized nitroimidazole and nitroimidazooxazine derivatives and examined them for antimycobacterial activity against the drug-sensitive H37Rv Mtb strain. The MCR series intermediates and target tetrazoles (**21** and **22**) exhibited MIC_{99} values in the range of 0.25–1.25 μM, excluding **22e** and **22h**. Compound **23** from the second series also showed potent antimycobacterial activity with an MIC_{99} value of 1.25 μM. Furthermore, these

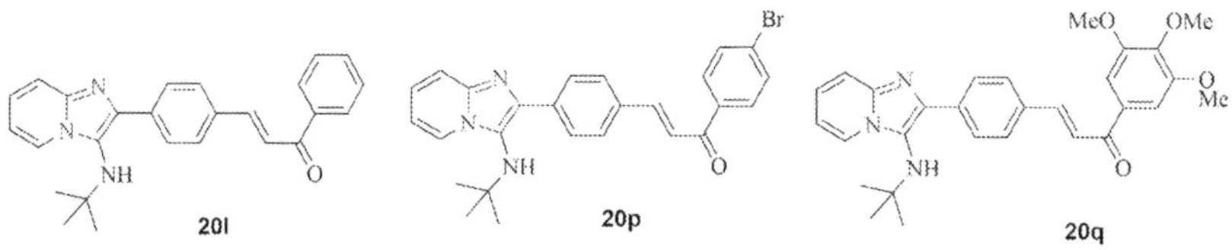

FIG. 3.9 Pyrido[1,2-*a*]imidazo-chalcones.

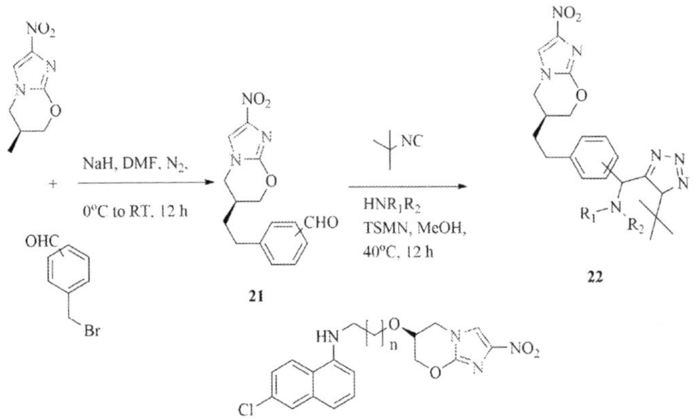

23 n= 2

SCHEME 3.5 Synthesis of nitroimidazole and nitroimidazooxazine derivatives.

compounds were found to be more active than the standard TB drug, kanamycin (MIC_{99} = 5.40 μM) (Scheme 3.5).

A novel series of triazolyl- and isoxazolyl-based NHIO (6-nitro-2,3-dihydroimidazooxazole) compounds were designed, synthesized, and evaluated against *M. tuberculosis* H37Rv [36]. Among the synthesized compounds, five compounds were studied for in vivo oral pharmacokinetics. Two compounds, **24g** and **25e**, showed a good PK profile. Compound **24g** exhibited 1.8 and 1 log CFU reduction with respect to the untreated and early control, respectively, in in vivo studies (Fig. 3.10).

A new class of drug to treat tuberculosis is nitroimidazooxazine PA-824. A series of 23 compounds of 2-nitroimidazooxazine was synthesized with

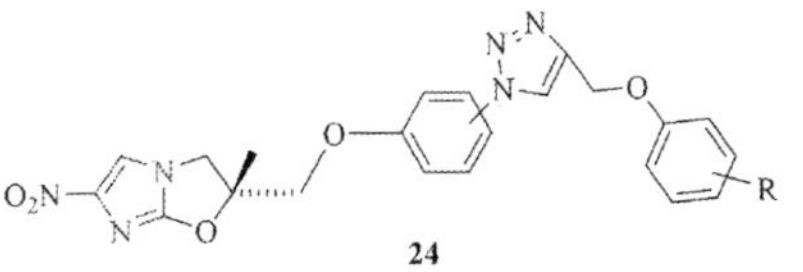

24

a-4-Me; b- 4-Et; c- 4-iso-pr; d-4-sec-but; e-4-F; f-4-CH_3; g-4-OCF_3
h-2-F; i-3-Cl; j-4-OCF_3; 1k-4-F; l-4-$OCHF_3$; m-4-f

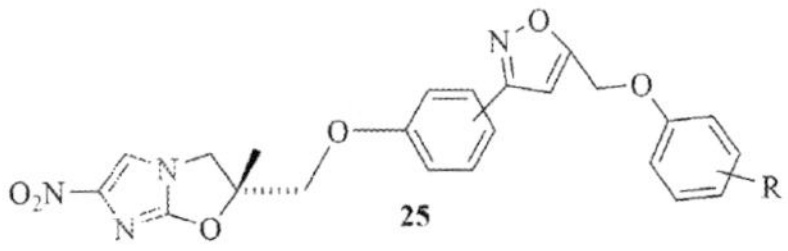

25

a-4-Me; b- 4-iso-pr; c-CF_3;d-4-OCH_3; e-4-OCH_3; f-4F; g-4-Br
h-3-F; i-2-F; j-2-Me; k-3-Cl; l- 4-iso-pr, 3-Me

FIG. 3.10 Triazolyl- and isoxazolyl-based NHIO (6-nitro-2,3-dihydroimidazooxazole) compounds.

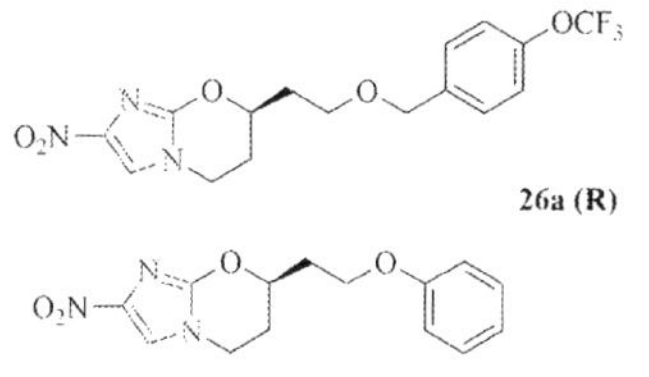

FIG. 3.11 2-Nitroimidazooxazine derivatives.

variations at C-7 position, i.e., benzyl ether, phenyl ether, benzyl carbonate, and phenyl carbamate as PA-824 analogs and further screened for antitubercular activity against Mtb (Fig. 3.11). Most of the synthesized compounds showed potency below micromolar concentrations with low toxicity. Compounds **26a (R)** with trifluoromethoxy benzyl group with MIC=0.078 μM was found to be five times more potent than PA-824 (MIC=0.390 μM) and the phenyl ether compounds, **27g (R)** showed the highest antimycobacterial activity against Mtb values with MIC=0.050 μM [37] (Table 3.6).

Pulipati et al. [38] synthesized a series of dibenzo[*b,d*]thiophene clubbed imidazo[1,2-*a*]pyridine carboxamides **28a–s (19 derivatives)** and were screened for in vitro antimycobacterial activity against *M. tuberculosis* H37Rv (ATCC27294) by an agar dilution method. The derivatives were synthesized through coupling of 2-dibenzo[*b,d*]thiophenyl imidazo [1,2-*a*]pyridine carboxylic acid with several benzyl amines. The designed moieties contain three segments: the first is imidazo[1,2-*a*]pyridine, an active pharmacophore from clinical antitubercular drug Q-203; the second is bioactive dibenzo[*b,d*]thiophene; and the third is appended substituted benzyl amines (Scheme 3.6). Out of the synthesized compounds, three compounds—**28k** (MIC: 0.78 μg/mL), **28e**, and **29n** (MIC: 1.56 μg/mL)—were found to be highly potent with lower cytotoxicity profile.

Triazole–imidazo[2,1-b][1,3,4]thiadiazole derivatives were designed by a molecular hybridization approach and synthesized via click chemistry reaction and screened for their antimycobacterial activity against *M. tuberculosis* H37Rv strain [39]. Out of the synthesized compounds, derivatives **29f** and **29n** exhibited significant growth inhibitory activity with a MIC of 3.125 μg/mL, and

TABLE 3.6 Antimycobacterial activity of 2-nitroimidazooxazine derivatives.

Compound	MIC (μM)	TC_{50} (μM)	Clog P
26a–R	0.078	> 100	3.21
27g–R	0.050	> 100	4.38
PA-824	0.390	> 100	2.79

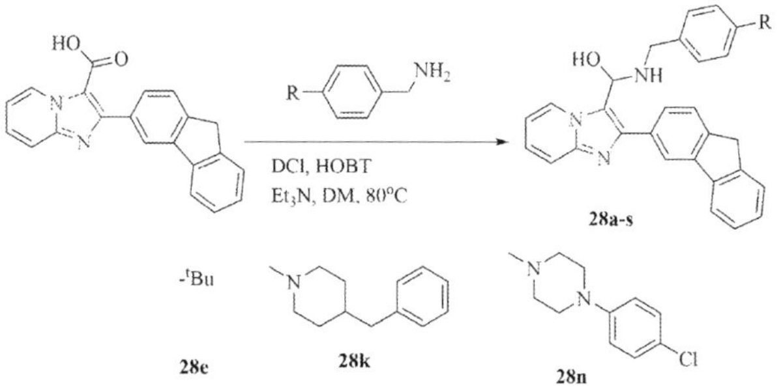

SCHEME 3.6 Synthesis of dibenzo[*b,d*]thiophene clubbed imidazo[1,2-*a*]pyridine carboxamides.

compound **29p** exhibited moderate activity with MIC of 6.25 μg/mL. The SAR study showed that the presence of ethyl, benzyl, or cyanomethylene groups on the 1,2,3-triazole ring and chloro substituent on the imidazo[2,1-b][1,3,4]thiadiazole ring increased the inhibition activity of the molecules. The active molecules also demonstrated positive drug-likeness score and their Clog*P* values are in the range of 2.2–2.9. Furthermore, the active molecules were not found to be toxic to normal cell lines (Fig. 3.12, Table 3.7).

Analogs of delamanid with high aqueous solubility and efficacy were synthesized involving replacement of proximal phenoxy linker of 2-nitroimidazole of delamanid by piperidine fused 5/6 membered ring heterocycles [40]. The MICs were determined against *M. bovis* ATCC 35737 and *M. tuberculosis* (M.tb H37Rv) under both aerobic and hypoxic conditions. The synthesized derivatives

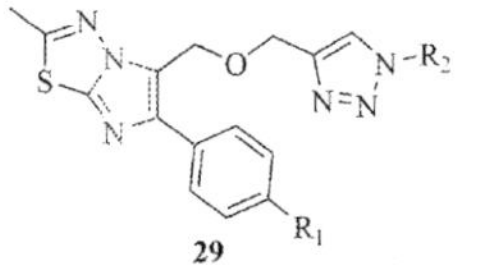

29

29f = OCH_3 (R_1), CH_2-C_6H_5 (R_2)
29n = Cl (R_1), CH_2-CH_3 (R_2)
29p = Cl (R_1), CH_2-CN (R_2)

FIG. 3.12 Triazole–imidazo[2,1-b][1,3,4]thiadiazole derivatives.

TABLE 3.7 Growth inhibitory activity of triazole–imidazo[2,1-b][1,3,4] thiadiazole derivatives.

Compound	Log P/Clog P	Drug-likeness score
29f	5.38/2.79	0.41
29n	4.67/2.25	0.85
29p	4.07/1.19	0.54

were found to be more hydrophilic than delamanid, as indicated by ClogP values. The compounds showed remarkable activity against *M. bovis*. Among the synthesized compounds, the tetrahydro-naphthyridine-linked nitroimidazoles exhibited excellent activity against both replicating (MABA) and nonreplicating (LORA) M.tb H37Rv with low cytotoxicity. Compound **30** showed better solubility and equivalent in vitro activity as compared to Delamanid. The high activity of fluoro substituted analogs, **30**, **31**, and **32**, revealed that the F atom is a determinant of both MABA and LORA activities in the tetrahydro-naphthyridine-linked nitroimidazoles (Fig. 3.13, Table 3.8).

Gawad et al. [41] synthesized 6-(4-nitrophenoxy)-2-substituted-1*H*-imidazo[4,5-*b*]pyridine derivatives and screened them for antitubercular activity against *M. tuberculosis* (H37Rv) ATCC 27294 using isoniazid as a reference. The in silico studies were conducted for synthesized derivatives with the enzyme (DprE1) to find the possible binding modes in comparison with the cocrystal reference molecules TCA1 and BTZ043. Among the synthesized compounds, derivatives **33c**, **33g**, **33i**, and **33v** showed good potency toward the *M. tuberculosis* strain. Gratifyingly, these compounds (**33c**, **33g**, **33i**, and **33v**) exhibited excellent docking scores (Fig. 3.14, Table 3.9).

Cherian et al. [42] synthesized derivatives at three positions of the 4-(trifluoromethoxy)-benzylamino tail and examined them for whole cell activity against replicating and nonreplicating *M. tuberculosis* (Mtb). The compound (*S*)-2-nitro-6-(4-(trifluoromethoxy)benzyloxy)-6,7-dihydro-5H-imidazo[2,1-b][1,3]oxazine named PA-824 (1) **34** demonstrated significant in vitro and in vivo

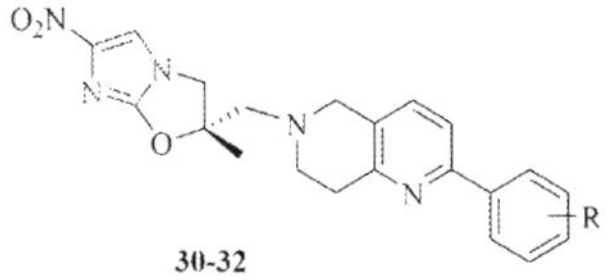

30-32

R = 4-F (30), 3,5-diF (31), 3,4,5-triF (32)

FIG. 3.13 Analogs of delamanid drug.

TABLE 3.8 MIC and cytotoxicity values of analogs of delamanid drug.

Compound	MIC (μg/mL)	MABA	LORA	CC_{50} (μg/mL)	Clog P
30	≤ 0.0625	< 0.031	1.23	> 32	5.25
31	≤ 0.0625	0.03	1.64	> 32	3.28
32	≤ 0.0625	0.08	0.76	> 32	3.28
Delamanid	≤ 0.0625	< 0.031	1.23	> 32	5.25
Rifampicin	–	0.10	0.72	> 32	–

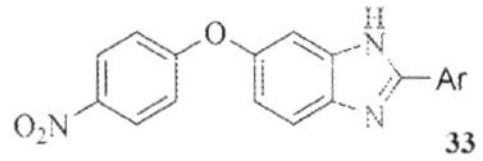

33

33c = 2,3 diOHPh
33g = 2,6-diOMePh
33i = 3-NO_2Ph
33v = 5-NO_2Ph

FIG. 3.14 Imidazo[4,5-*b*]pyridine derivatives.

TABLE 3.9 Antitubercular activity of synthesized imidazo[4,5-*b*]pyridine derivatives.

Compound	Antitubercular activity MIC (μmol/L)	Docking score
33c	0.6	− 7.500
33g	0.5	− 7.698
33i	0.8	− 8.825
33v	2.6	− 6.657

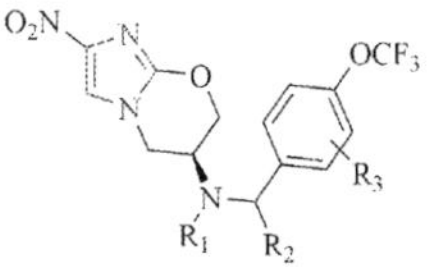

34

FIG. 3.15 Imidazo[2,1-b][1,3]oxazine derivatives.

antitubercular activity and is currently in clinical trials. The researchers also determined their kinetic parameters as substrates of the deazaflavin-dependent nitroreductase (Ddn) from Mtb that reductively activated these pro-drugs. These studies yielded several compounds with 40 nM aerobic and 1.6 μM anaerobic whole cell activity: 10-fold improvements over both characteristics from the parent compound (Fig. 3.15).

A series of novel imidazo[4,5-*c*]pyridine derivatives (IPD) containing amide/urea/sulfonamide was synthesized and studied for in vitro and in vivo antimycobacterial activity against *M. tuberculosis* by the agar dilution method. Some of the synthesized compounds showed significant activity against *M. tuberculosis*. Compounds **35**, **36**, and **37** were found to show excellent activity against *M. tuberculosis* [43]. The results showed that the antimycobacterial activity was in the order: amide > urea > sulfonamide (Fig. 3.16, Table 3.10). Furthermore, these compounds were further studied for toxicity (IC_{50}) in a mammalian Vero

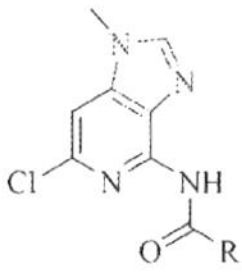

FIG. 3.16 Imidazo[4,5-*c*]pyridine derivatives.

TABLE 3.10 In vitro antimycobacterial activity against MTB H37Rv.

Compound	MTB (MIC) (μM)
35	0.25 ± 0.04
36	0.25 ± 0.16
37	0.25 ± 0.11
Rifampicin	0.12 ± 0.03
Isoniazid	0.36 ± 0.12
Ciprofloxacin	4.71 ± 0.07
Ethambutol	7.64 ± 0.15

TABLE 3.11 In vivo activity data against MIC ATCC35801 in mice.

Compound	Lungs (log CFU ± SEM)	Spleen (log CFU ± SEM)
Control	7.92 ± 0.13	8.97 ± 0.10
35 (50 mg/kg)	5.49 ± 0.01	5.41 ± 0.18
36 (50 mg/kg)	6.11 ± 0.17	6.35 ± 0.13
37 (50 mg/kg)	5.45 ± 0.09	5.29 ± 0.16
Isoniazid (25 mg/kg)	5.73 ± 0.03	4.79 ± 0.18

cell line and IC_{50} values of the active compounds **35**, **36**, and **37** were found to be 169.07, 183.35, and 162.07 μM, respectively. Additionally, in the in vivo animal studies, compounds **35**, **36**, and **37** decreased the bacterial load in lung and spleen tissues at the dose of 50 mg/kg body weight (Table 3.11).

Warekar and coworkers [44] synthesized 4-(4-nitro-phenyl)-2-phenyl-1,4-dihydro benzo[4,5]imidazo[1,2-*a*]pyrimidine-3-carboxylic acid ethyl ester derivatives by one-pot condensation of ethyl benzoylacetate, substituted aromatic aldehydes, and 2-amino benzimidazole using 260 mol% of citric acid

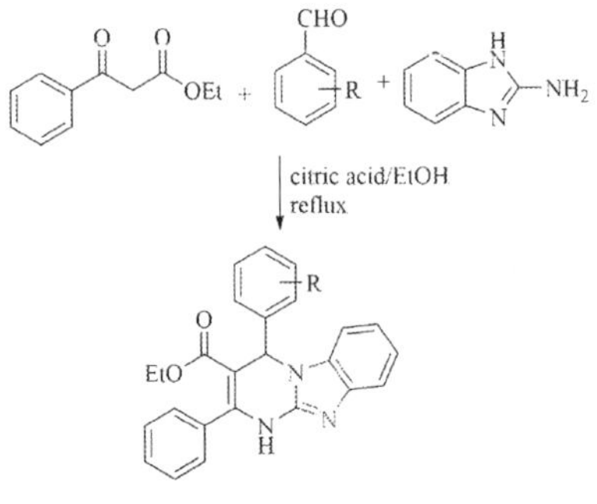

SCHEME 3.7 Synthesis of imidazopyrimidine derivatives.

(Scheme 3.7). The synthesized compounds were further examined for in vitro antimycobacterial activity against *M. tuberculosis* H37Rv (ATCC-27294) at concentrations of 0.8–100 µg/mL using standard drugs such as pyrazinamide, streptomycin, and ciprofloxacin. Among all the synthesized compounds, compounds **38** and **39** exhibited significant activity with a minimum inhibitory concentration (MIC) of 25 µg/mL. The results confirmed that all compounds illustrated activity at concentrations of 50 and 100 µg/mL. In this series, compounds **38** and **39** showed effective activity at 25 µg/mL against the *M. tuberculosis* H37Rv strain.

Palkar et al. [45] synthesized 18 analogs of 2-substituted-5,6-diarylsubstituted imidazo(2,1-b)-1,3,4-thiadiazoles **40a–r** by the reaction of 2-amino-5-substituted-1,3,4-thiadiazoles and substituted α-bromo-1,2-(p-substituted)diaryl-1-ethanones 4a–e. The synthesized compounds were tested for their in vitro antitubercular activity against *M. tuberculosis* H37Rv using an Alamar Blue susceptibility test and the activity was expressed as MIC in µg/mL. Out of the synthesized compounds, compound **40h** with MIC = 1.25 µg/mL exhibited excellent antitubercular activity compared to other synthesized compounds and reference drugs, isoniazid and gatifloxacin. Compounds **40c**, **40f**, and **40g** also displayed significant antitubercular activity. Furthermore, some of the active compounds were also assessed for their cytotoxic activity (IC_{50}) against a mammalian Vero cell line using MTT assay. The results showed that these derivatives exhibited antitubercular activity at noncytotoxic concentrations (Fig. 3.17, Table 3.12).

A series of 14 imidazo[1,2-*a*]pyridine-3-carboxamide derivatives was synthesized and assessed against *M. tuberculosis* H37Rv. The minimum inhibitory concentrations (MICs) of 12 compounds were found to be ≤ 1 µM against replicating bacteria [46]. Five compounds (**41**, **42**, **44**, **45**, and **46**) exhibited MIC values in the range of ≤ 0.006 µM. Furthermore, compounds **43** and **46** were screened against a panel of MDR and XDR drug-resistant clinical Mtb strains and compound **46** was found to be highly potent, surpassing clinical candidate PA-824 by nearly 10-fold (Fig. 3.18).

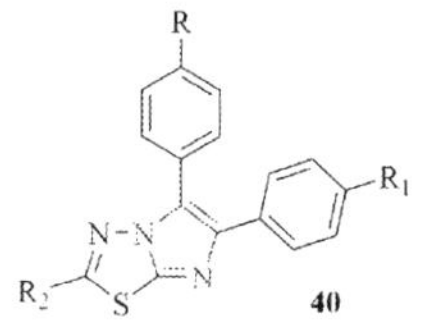

40c = R=OCH_3, R_1= OCH_3, R_2=CF_3
40f = R=H, R_1= SCH_3, R_2= SO_2NH_2
40g = R=OCH_3, R_1= SCH_3, R_2= SO_2NH_2
40h = R=OCH_3, R_1= OCH_3, R_2= SO_2NH_2

FIG. 3.17 2-Substituted-5,6-diarylsubstituted imidazo(2,1-b)-1,3,4-thiadiazoles.

TABLE 3.12 Antimycobacterial activity of synthesized derivatives (40a–r).

Compound	MIC	IC_{50} (μM)
40c	1.25	147.3
40f	2.5	192.2
40g	2.5	213.1
40h	1.25	246.6
Isoniazid	0.25	> 450
Gatifloxacin	1.0	> 550

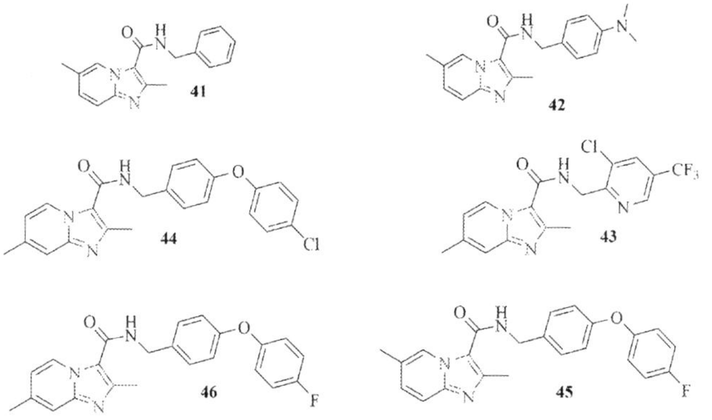

FIG. 3.18 Imidazo[1,2-*a*]pyridine-3-carboxamide derivatives.

Makwane et al. [47] synthesized 10-(2-(substituted phenyl)imidazo[2,1-b][1,3,4]thiadiazol-6-yl)-10H-phenothiazines **47** and evaluated them for antitubercular activity by the (LJ) agar method against the *M. tuberculosis* H37Rv strain using isoniazid as a reference drug (Fig. 3.19). The percentage inhibition and MIC values of synthesized derivatives **47a–j** are presented in Table 3.13.

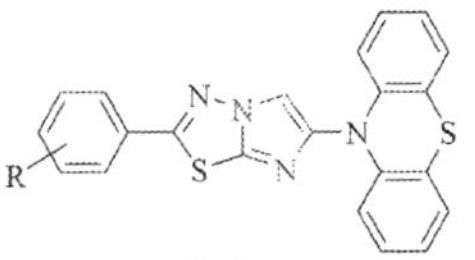

47a-j

FIG. 3.19 Imidazo[2,1-b][1,3,4]thiadiazol-6-yl)-10H-phenothiazines.

TABLE 3.13 Antitubercular activity of synthesized derivatives 47a–j.

Compound	R	Anti-TB activity inhibition (%) ppm 25	Anti-TB activity inhibition (%) ppm 50	Anti TB activity MIC (μg/mL)
47a	C_6H_5	22	45	12
47b	2-ClC_6H_4	32	79	75
47c	3-ClC_6H_4	36	80	65
47d	4-ClC_6H_4	32	78	7
47e	2-BrC_6H_4	29	73	10
47f	3-BrC_6H_4	30	76	85
47g	4-BrC_6H_4	30	75	9
47h	2-$NO_2C_6H_4$	28	82	55
47i	3-$NO_2C_6H_4$	27	84	4
47j	4-$NO_2C_6H_4$	32	83	5

Syed et al. [48] synthesized 6-(4-substituted phenyl)-2-(3,5-dimethyl-1*H*-pyrazol-1-yl)imidazo [2,1-b][1,3,4] thiadiazoles and evaluated them for antitubercular potential against *M. tuberculosis* strain. Compounds **48a**, **48b**, **49a**, **50a**, and **51a** showed the most potent antitubercular activity and the MIC values are presented in Table 3.14 (Scheme 3.8).

2.4 Linker imidazole

Pandey et al. [49] synthesized imidazole-based compounds **52–54** via reaction of simple imidazoles with alkyl halides or alkyl halocarboxylate using tetrabutylammonium bromide (TBAB) (Fig. 3.20). The derivatives were screened against the avirulent strain *M. tuberculosis* H37Ra and the virulent strain *M. tuberculosis* H37Rv at concentrations ranging from 3.25 to 50 mg/mL using

TABLE 3.14 Antitubercular activity of synthesized fused compounds.

Compound	MIC (μg/mL)
48a	10
48b	10
49a	10
49b	25
50a	10
50b	25
51a	10
51b	25
51c	25
51e	25
Streptomycin	7.5

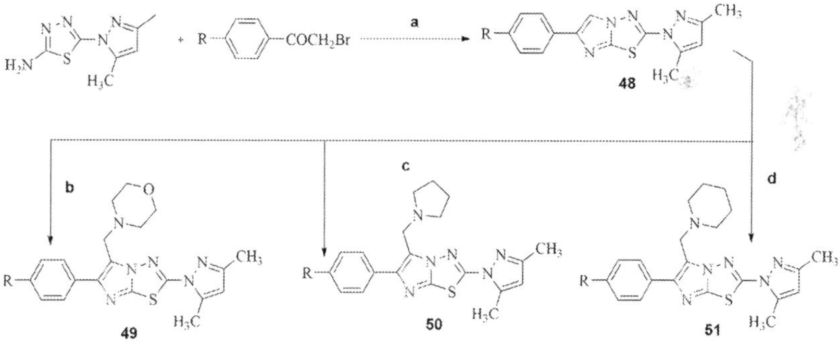

Reaction conditions: **a)** dry ethanol, reflux, 12h **b)** morpholine, HCHO, AcOH, MeOH, 8h **c)** pyrrolidine, HCHO, AcOH, MeOH, 8h **d)** piperidine, HCHO, AcOH, MeOH, 8h

SCHEME 3.8 Synthesis of 6-(4-substituted phenyl)-2-(3,5-dimethyl-1*H*-pyrazol-1-yl)imidazo [2,1-b][1,3,4] thiadiazoles.

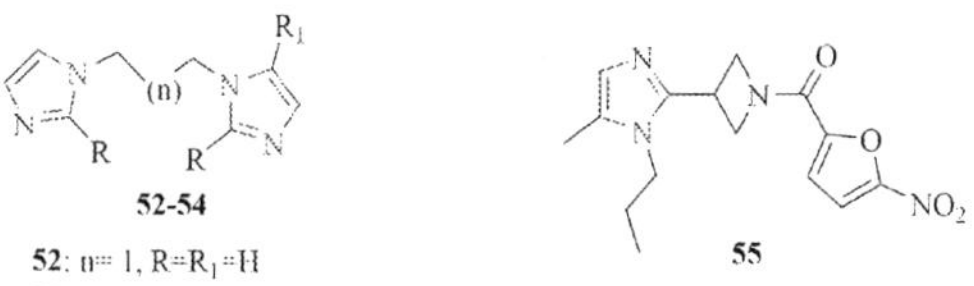

52: n= 1, R=R_1=H
53: n= 3, R=R_1 =H
54: n= 1, R= Propyl, R_1= H

FIG. 3.20 Imidazole-based compounds.

MABA, the agar dilution method and BACTEC assay. The data revealed that most of the compounds exhibited antitubercular activity with MIC, in the range of 12.5–25 mg/mL. Compound **54** showed excellent in vitro antitubercular activity with MIC, 6.25 mg/mL against the virulent strain *M. tuberculosis* H37Rv, and shall serve as lead for further optimization and studies.

A series of compounds involving linking of 5-nitrofuran-2-oyl and substituted imidazole moiety by a diverse set of aminoalkyl linkers were synthesized and studied for antimycobacterial activity against the drug-sensitive *M. tuberculosis* H37Rv strain [50]. Compound **55** demonstrated promising activity (MIC 0.8 mg/mL) against the drug-sensitive Mtb H37Rv strain in vitro. The findings of toxicological evaluation showed that compound **55** possessed low toxicity for mice ($LD_{50}=900.0 \pm 83.96$ mg/kg). The compound also showed the same activity to neurotoxic cycloserine when it was used as part of a four-drug combination therapy. The results concluded that compound **55** belonged to a low toxicity group (class 4) as per the classification of Hodge and Sterner [51] (Fig. 3.20).

In 2011, new analogs of antitubercular drug PA-824 **56** were synthesized bearing ether linkers of different size and flexibility with increased metabolic stability and high efficacy. Both α-methyl substitution and removal of the benzylic methylene were broadly tolerated and exhibited an eightfold better efficacy than the parent drug. Extended linkers (propenyloxy, propynyloxy, pentynyloxy) exhibited higher potencies against replicating M.tb (monoaryl analogs), and propynyl ethers were found to be the most effective under anaerobic (nonreplicating) conditions. For benzyloxybenzyl and biaryl derivatives, aerobic activity was found to be maximal with the original (OCH_2) linker. One of the derivatives, the propynyloxy-linked compound, was discovered to be 89-fold more efficient than the parent drug in the acute model, and it was found to be slightly superior to the antitubercular drug OPC-67683 in a chronic infection model [52] (Fig. 3.21).

A series of alkylated/aminated 2-methyl-5-nitroimidazoles and nitroimidazole-7-chloroquinoline derivatives were synthesized and evaluated for antitubercular activity against *M. tuberculosis* and their cytotoxicity was also studied regarding the J774 murine macrophage cell line. The synthesized compounds exhibited significant activities with minimal cytotoxicity. Compound **57**,

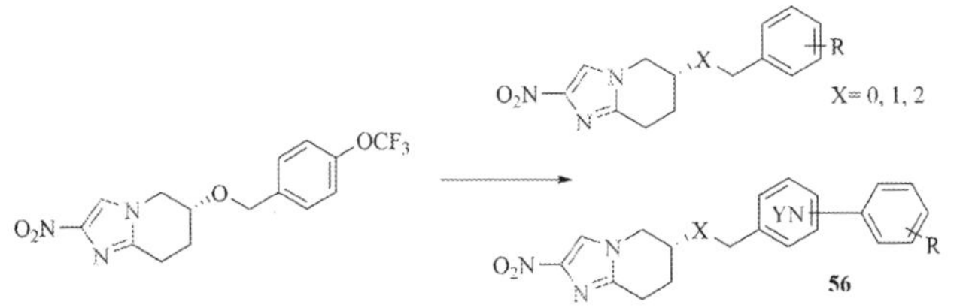

56

X= O, OCH_2, OCHMe, $O(CH_2)_3$, $O(CH_2)_2O$, $OCH_2CH{=}CH$, $OCH_2C{\equiv}C$, $O(CH_2)_3C{\equiv}C$

FIG. 3.21 New analogs of antitubercular drug PA-824.

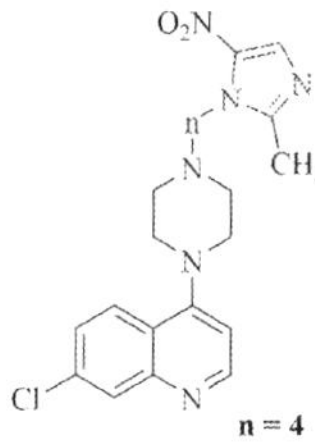
57

FIG. 3.22 Linker imidazole compound.

nitroimidazole-7-chloroquinoline conjugate, with butyl chain as a linker, was found to be the most potent with an MIC_{50} of 2.2 μg/mL [53] (Fig. 3.22).

2.5 Benzimidazole

A range of N-((2-(pyridin-4-yl)1*H*-benzo[*d*]imidazol-1-yl)methyl)-substituted-aminopyridine and N-((2-(3-(trifluoromethyl)styryl)-1*H*-benzo[*d*]imidazol-1-yl)methyl)-substituted-aminopyridine derivatives was evaluated for their in vitro antitubercular activity [54]. The change in the position of the nitrogen atom in the pyridine ring attached 2-(3-(trifluoromethyl)styryl)-1*H*-benzo[*d*] imidazole played an important role to attain better antimycobacterial activity as well as an in vitro leishmanicidal effect. Compound **58a** (MIC = 3.125 μM) and **58d** (MIC = 6.25 μM) showed potency against the *M. tuberculosis* H37Rv strain (Fig. 3.23).

In 2013, benzimidazole-based complexes **59** were synthesized using ethyl 3-amino-4-(4-(2-((2-amino-4-(ethoxycarbonyl)phenyl)amino)ethyl) piperazin-1-yl)benzoate and sodium bisulfate adducts in the presence of DMF under N_2 atmosphere for 1–2 days and examined for anti-TB activity against *M. tuberculosis* H37Rv (MTB-H37Rv) using broth dilution method as reported by Collins et al. [20]. The MIC values were determined using *M. tuberculosis* (MTB-H37Rv) and INH-resistant *M. tuberculosis* (INHR-MTB) strains [55] (Fig. 3.24, Table 3.15).

A range of benzyl-1*H*-imidazole-1-carbodithioate derivatives were synthesized and their anti-TB efficiency was evaluated [56]. Two compounds not only

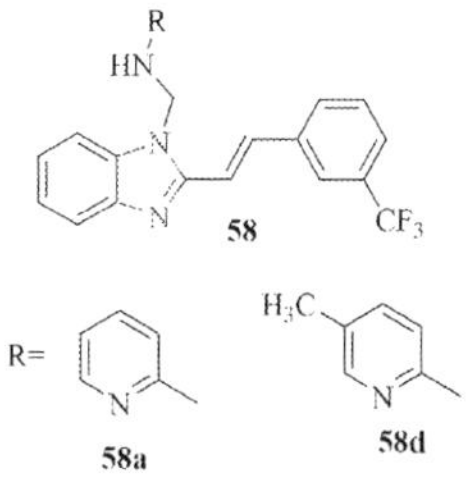

FIG. 3.23 Benzimidazole derivatives.

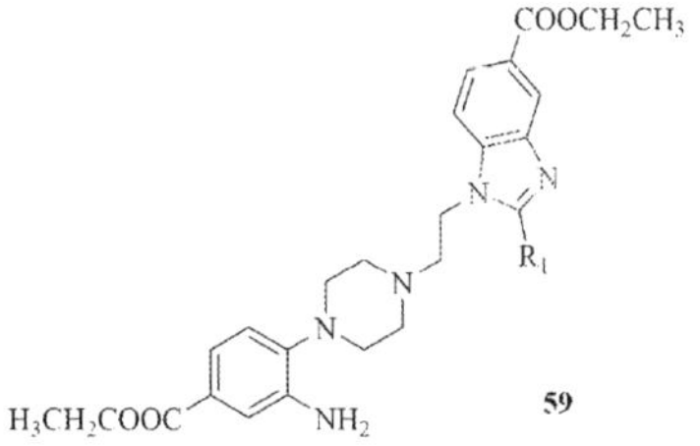

59

FIG. 3.24 Benzimidazole-based complexes.

TABLE 3.15 Antimycobacterial activity of benzimidazole-based compounds.

Compound	MIC (μM)
59a	6.219
59b	0.195
59c	4.320
59d	0.135
59e	6.195
59f	6.219
59g	0.112
Isoniazid	0.73

were active against *M. smegmatis* but also were found to be active against *M. tuberculosis*. All the synthesized compounds displayed high inhibition efficiency, and among them, a few compounds were found to have low MIC value ranges: 2–40 μg/mL. A lower MIC was found for B1HI1C (**61**) (2.0 μg/mL) than AP1C (**60**) (40.0 μg/mL) for *M. smegmatis*. B1HI1C was found to be more active (MIC value 5.0 μg/mL) than AP1C (MIC value 100.0 μg/mL) against *M. tuberculosis* H37Rv (Table 3.16).

Benzimidazole derivatives were synthesized from 2-[2-(pyridin-3-yl)-1*H*-benzimidazol-1-yl] acetohydrazide and aromatic acid in the presence of $POCl_3$ under reflux for 8–10 h [57]. Their anti-TB activity was studied against *M. tuberculosis* at concentrations ranging from 0.8 to 100 μg/mL (vaccine strain, H37Rv strain, ATCC-27294) using the Microplate Alamar Blue Assay process (MABA). Compound **62a**, an oxadiazole ring having a phenyl ring at the fifth position, was the most effective against *M. tuberculosis* with an MIC value of 1.6 μg/mL. Similarly, compounds **62b**, **62d**, and **62e** exhibited moderate anti-TB activity, having MICs of 12.5 μg/mL (Fig. 3.25, Table 3.17).

TABLE 3.16 Anti-TB efficiency of benzyl-1*H*-imidazole-l-carbodithioate derivatives.

	MIC (μg/mL)	
	M. smegmatis	*M. tuberculosis*
Compound	mc^2115	H37Rv
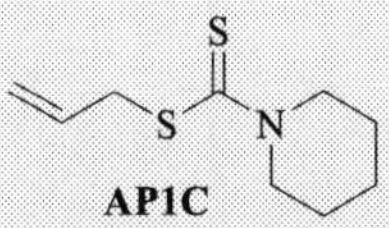 AP1C 60	40.0	100.0
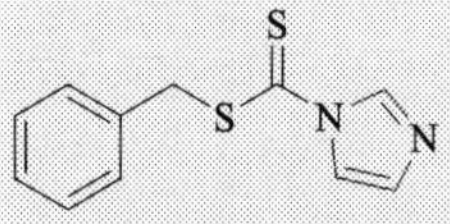 B1HI1C 61	2.0	5.0

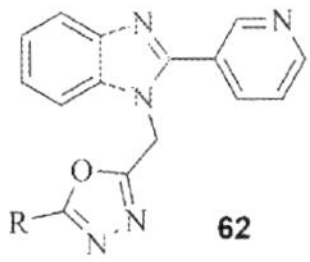

62

$R = C_6H_5$ (a)
4-ClC_6H_4 (b)
4-OHC_6H_4 (c)
4-$OCH_3C_6H_4$ (d)
3-$NO_2C_6H_4$ (e)

FIG. 3.25 Benzimidazole derivatives.

TABLE 3.17 Antitubercular activity of synthesized compounds (62a–e).

Compound	MIC value (μg/mL)
62a	1.60
62b	12.
62c	5.0
62d	12.5
62e	12.5

A novel series of 2-heterostyrylbenzimidazole derivatives **63** (14 derivatives) was synthesized and their anti-TB efficiency was evaluated *in vitro* using *M. tuberculosis* H37Rv by the micro-dilution method using a Löwenstein-Jensen medium (LJ). Several compounds were found to be slightly more active than streptomycin and ethambutol [58]. Compound **63f** displayed the greatest activity, with an MIC value of 16 μg/mL. Compounds with electron donating groups (Cl, O, S, etc.) on different aromatic aldehydes were found to be more active in inhibiting *M. tuberculosis*. Docking studies were also carried out and compound **63f** showed good affinity for the enzyme. Comparison was made with the binding energies of the standard drugs, amoxicillin (− 34.28 kcal/mol) and ciprofloxacin (− 28.20 kcal/mol). Among all the designed compounds, compound **63f** showed the highest binding energy with two amino acid interactions with protein Lys160 and Val187 (− 9.80 kcal/mol) (Fig. 3.26).

A new series of triheterocycles containing indole–benzimidazole-based 1,2,3-triazole hybrids were synthesized and screened in vitro for their inhibitory potency against *M. tuberculosis* H37Rv using the Resazurin Microtiter Assay (REMA) [59]. Compound **64h** came out as the best antitubercular drug candidate by inhibiting the growth of the MTB strain with MIC = 3.125 μg/mL) (control rifampicin MIC = 0.04 μg/mL and isoniazid MIC = 0.38 μg/mL) due to the presence of a nitro group on the phenyl ring at the second position and was highly favored for antitubercular activity. Compound **64b** (MIC = 6.25 μg/mL) with chloro substitution, compound **64i** (MIC = 6.25 μg/mL) with trifluoromethyl substitution, and compound **64j** (MIC = 12.5 μg/mL) with benzyl substitution exhibited moderate antitubercular activity (Fig. 3.27 and Table 3.18).

Nandha et al. [60] synthesized 2-((1H-imidazol-1-yl) methyl)-6-substituted-5-fluoro-1H-benzo[*d*]imidazoles **65** and evaluated them for antitubercular activity against *M. tuberculosis* strain by MABA using isoniazid as a reference

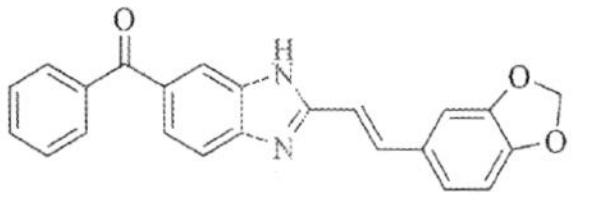

63f

FIG. 3.26 2-Heterostyrylbenzimidazole derivatives.

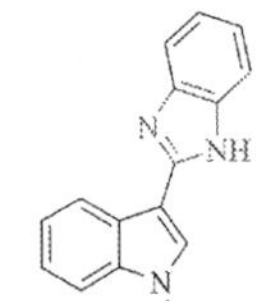

64

64a = Phenyl
64b = 2-chlorophenyl
64c = 4-chlorophenyl
64d = 3-methylphenyl
64e = 4-methylphenyl
64f = 2-methylphenyl
64g = 3-methoxyphenyl
64h = 2-nitrophenyl
64i = 3(trifluoromethyl)phenyl
64j = benzyl

FIG. 3.27 Indole–benzimidazole-based 1,2,3-triazole hybrids.

TABLE 3.18 Antimycobacterial activity of indole–benzimidazole-based 1,2,3-triazole hybrids (MIC).

Compound	*M. tuberculosis* H37Rv MIC (μg/mL)
64a	25
64b	6.25
64c	12.5
64d	25
64e	50
64f	50
64g	50
64h	3.125
64i	6.25
64j	12.5
Rifampicin	0.04
Isoniazid	0.38

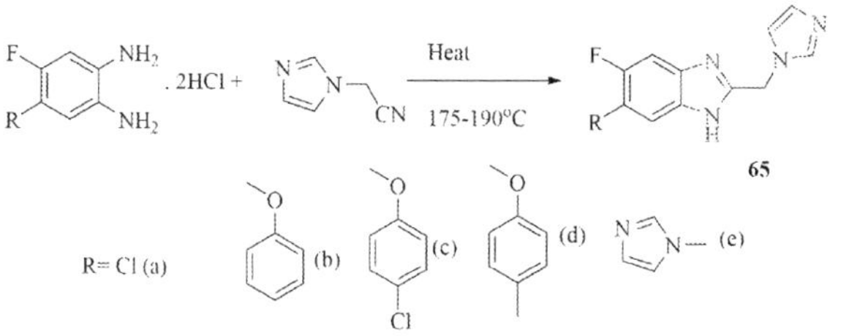

SCHEME 3.9 Synthesis of 2-((1H-imidazol-1-yl) methyl)-6-substituted-5-fluoro-1H-benzo[*d*] imidazoles.

drug (Scheme 3.9). The antitubercular activity of synthesized derivatives is presented in Table 3.19.

Nandha et al. [61] synthesized 6-(benzo[d][1,3]dioxol-5-yl-oxy)-2-substituted-5-fluoro-1H-benzo[d] imidazoles **66**, **67** and evaluated them for antitubercular activity against *M. tuberculosis* (ATCC27294) by MABA using streptomycin, ciprofoxacin, and pyrazinamide as reference drugs (Fig. 3.28). The MIC values are presented in Table 3.20.

2-(1-benzoyl-1*H*-benzo[*d*]imidazole-2-ylthio)-2-ylthio)-*N*-substituted acetamide derivatives (20) were synthesized and studied for their in vitro and in vivo

TABLE 3.19 Antitubercular activity of benzimidazole derivatives **65**.

Compound	MIC (μg/mL)
65a	100
65b	50
65c	25
65d	50
65e	125
Isoniazid	0.78

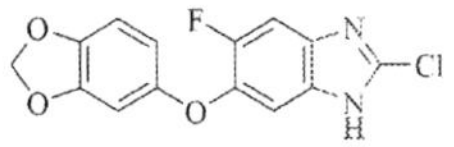

66

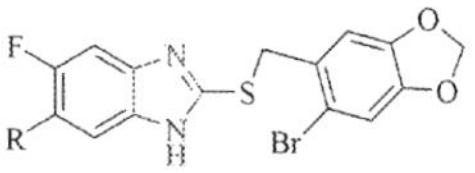

67

FIG. 3.28 Benzimidazole derivatives **66** and **67**.

TABLE 3.20 MIC values of benzimidazole derivatives **66** and **67**.

Compound	MIC (μg/mL) MABA	Compound	MIC (μg/mL) MABA
66a	50	**67a**	100
66b	50	**67b**	50
66c	50	**67c**	50
66d	50	**67d**	50
66e	50	**67e**	50
66f	50	**67f**	50
Streptomycin	6.25	**67g**	50
Pyrazinamide	3.12	**67h**	25
Ciprofloxacin	3.12	**67i**	50

antitubercular activity against *Mycobacterium tuberculosis* H37Rv (NCFT/TB/537) [62]. The synthesized derivatives were studied for zone of inhibition, MIC, and MLC values. The structure-activity relationship studies showed that compounds bearing electron donating groups (OCH_3, Me) at para position influenced the antitubercular activity significantly in comparison to ortho and meta

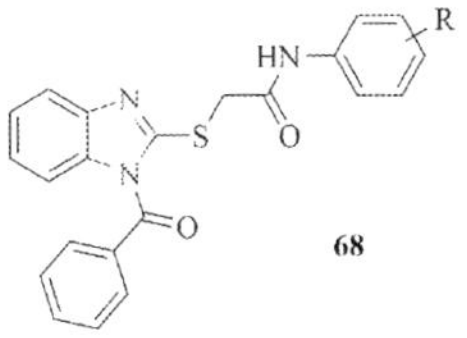

68

FIG. 3.29 2-(1-Benzoyl-1H-benzo[*d*]imidazole-2-ylthio)-2-ylthio)-N-substituted acetamide derivatives.

positions. Regarding in vivo studies, the dose 5.67 mg/Kg was found to be toxic and fatal in mice models infected with *M. tuberculosis*. The LD_{50} values varied from 1.81 to 3.17 mg/kg body weight of the mice and a dose of 1.34 mg/kg was found to be safer for each of the synthesized derivatives. The active derivatives were further studied for their capacity to inhibit the mycobacterial enzymes, namely pantothenate synthetase, isocitrate lyase, and chorismate mutase. Derivative **68** inhibited the mycobacterial isocitrate lyase, pantothenate synthetase, and chorismate mutase activity to 64.56%, 60.12%, and 58.23%, respectively, in comparison to inhibition by streptomycin sulfate of 75.12%, 77.06%, and 79.56%, respectively (Fig. 3.29).

3. Conclusion

In the current era, tuberculosis has become a worldwide burden as resistance has developed to the existing drugs. US-FDA has recently approved drugs, viz. pretomanid (2019), delamanid, and bedaquiline (2017), due to their better activity against MDR-TB in a limited manner. However, these drugs have caused significant adverse effects. The exploration of drug-resistant TB treatment, i.e., development of new drugs or repurposing existing ones, has expanded tremendously in recent years. Due to the problems and continuous efforts related to tuberculosis, it is anticipated that the treatment and survival rate of TB patients will progress in the future. Presently, 23 molecules are undergoing clinical trials and it is hoped that they will contribute to a new regimen that may be suitable for all types of tuberculosis patients.

The present chapter included recent progress in the discovery and development of imidazole-containing derivatives for the treatment of tuberculosis. Moreover, several potent molecules including substituted imidazoles, fused derivatives, benzimidazoles, linker compounds, hybrid compounds, etc. have been discussed at length to inspire researchers about future prospects in drug discovery for tuberculosis. The discovery of new drug targets and novel methods of targeted drug delivery for anti-TB drugs is anticipated shortly with the potential of improving drug efficacy, lessening drug dosing, and diminishing adverse effects.

Conflict of interest

The authors declare no conflict of interest, financial or otherwise.

Acknowledgments

The authors are grateful to the Department of Chemistry, Mohan Lal Sukhadia University, Udaipur, India for providing the necessary library facilities for carrying out the work. A. Sethiya is also thankful to UGC-MANF for providing a Senior Research Fellowship to carry out this work. N. Sahiba and P. Teli are thankful to CSIR for providing a fellowship to carry out this work.

Funding source

This work was supported by UGC-MANF (201819-MANF-2018-19-RAJ-91971). CSIR [file no. 09/172(0088)2018-EMR-I], [file no. 09/172(0099)2019-EMR-I].

References

[1] World Health Organization. Global tuberculosis report 2020. WHO; 2020. https://www.who.int/publications/i/item/9789240013131. 978-92-4-001313-1.

[2] Tetali SR, Kunapaeddi E, Mailavaram RP, Singh V, Borah P, Deb PK, Venugopala KN, Hourani W, Tekade RK. Current advances in the clinical development of anti-tubercular agents. Tuberculosis 2020;125:101989. https://doi.org/10.1016/j.tube.2020.101989.

[3] Daley CL. The global fight against tuberculosis. Thorac Surg Clin 2019;29:19–25.

[4] Caminero JA, Cayla JA, García-García J-M, García-P'erez FJ, Palacios JJ, Ruiz-Manzano J. Diagnosis and treatment of drug-resistant tuberculosis. Arch Bronconeumol Engl Ed 2017;50:1–9. https://doi.org/10.1016/j. arbr.2017.07.005.

[5] Umumararungu T, Mukazayire MJ, Mpenda M, Mukanyangezi MF, Nkuranga JB, Mukiza J, Olawode EO. A review of recent advances in anti-tubercular drug development. Indian J Tuberc 2020;67:539–59. https://doi.org/10.1016/j.ijtb.2020.07.0.

[6] Daraji DG, Prajapati NP, Patel HP. Synthesis and applications of 2-substituted imidazole and its derivatives: a review. J Heterocycl Chem 2019;56(9):2299–317. https://doi.org/10.1002/jhet.3641.

[7] Hu Y, Shen YF, Wu XH, Tu X, Wang GX. Synthesis and biological evaluation of coumarin derivatives containing imidazole skeleton as potential antibacterial agents. Eur J Med Chem 2018;143:958–69.

[8] Serrao E, Xu ZL, Debnath B, Christ F, Debyser Z, Long YQ, Neamati N. Discovery of a novel 5-carbonyl-1H-imidazole-4-carboxamide class of inhibitors of the HIV-1integrase-LEDGF/p75 interaction. Bioorg Med Chem 2013;21:5963–72.

[9] Vlahakis JZ, Kinobe RT, Nakatsu K, Szarek WA, Crandall IE. Anti-plasmodium activity of imidazole-dioxolane compounds. Bioorg Med Chem Lett 2006;16:2396–406.

[10] Akhtar J, Khan AA, Ali Z, Haider R, Yar MS. Structure-activity relationship (SAR) study and design strategies of nitrogen-containing heterocyclic moieties for their anticancer activities. Eur J Med Chem 2017;125:143–89.

[11] Bistrović A, Krstulović L, Harej A, Grbčić P, Sedić M, Koštrun S, Pavelić SK, Bajić M, Raić-Malić S. Design, synthesis and biological evaluation of novel benzimidazole amidines as potent multi-target inhibitors for the treatment of non-small cell lung cancer. Eur J Med Chem 2018;143:1616–34.

[12] Fan Y, Jin X, Huang Z, Yu H, Zeng Z, Gao T, Feng L. Recent advances of imidazole-containing derivatives as anti-tubercular agents. Eur J Med Chem 2018;25(150):347–65. https://doi.org/10.1016/j.ejmech.2018.03.016.

[13] Xu Z, Gao C, Ren QC, Song XF, Feng LS, Lv ZS. Recent advances of pyrazole-containing derivatives as anti-tubercular agents. Eur J Med Chem 2017;139:429–40.

[14] Xu Z, Zhang S, Gao C, Zhao F, Lv ZS, Feng LS. Isatin hybrids and their anti-tuberculosis activity. Chin Chem Lett 2017;28:159–67.

[15] Fan YL, Wu JB, Cheng XW, Zhang FZ, Feng LS. Fluoroquinolone derivatives and their anti-tubercular activities. Eur J Med Chem 2018;146:554–63.

[16] Chety S, Ramesh M, Snghillay A, Solian MES. Recent advancements in the developments of anti-tuberculosis drugs. Bioorg Med Chem Lett 2017;27:370–86.

[17] Siwach A, Verma PK. Synthesis and therapeutic potential of imidazole containing compounds. BMC Chem 2021;15(1):12. https://doi.org/10.1186/s13065-020-00730-1.

[18] Showalter DH. Recent progress in the discovery and development of 2-nitroimidazooxazines and 6-nitroimidazooxazoles to treat tuberculosis and neglected tropical diseases. Molecules 2020;25:4137. https://doi.org/10.3390/molecules25184137.

[19] Desai NC, Trivedi AR, Somani HC, Bhatt KA. Design, synthesis and biological evaluation of 1,4-dihydropyridines derivatives as potent antitubercular agents. Chem Biol Drug Des 2015;86:370–6.

[20] Collins LA, Franzblau SG. Microplate alamar blue assay versus BACTEC 460 system for high- throughput screening of compounds against *Mycobacterium tuberculosis* and *Mycobacterium avium*. Antimicrob Agents Chemoter 1997;41:1004–9.

[21] Gising J, Nilsson MT, Odell LR, Yahiaoui S, Lindh M, Iyer H, Sinha AH, Srinivasa BR, Larhed M, Mowbray SL, Karleń A. Trisubstituted imidazoles as *Mycobacterium tuberculosis* glutamine synthetase inhibitors. J Med Chem 2012;55(6):2894–8.

[22] Pieroni M, Wan B, Zuliani V, Franzblau SG, Costantino G, Rivara M. Discovery of antitubercular 2,4-diphenyl-1H-imidazoles from chemical library repositioning and rational design. Eur J Med Chem 2015;100:44–9.

[23] Cho SH, Warit S, Wan B, Hwang CH, Pauli GF, Franzblau SG. Low-oxygen recovery assay for high-throughput screening of compounds against non replicating *Mycobacterium tuberculosis*. Antimicrob Agents Chemother 2007;51:1380–5. https://doi.org/10.1128/AAC.00055-06.

[24] Raghu MS, Kumar CBP, Prasad KNN, Prashanth KM, Kumarswamy KY, Chandrasekhar S, Veeresh B. MoS2 – calix[4]arene catalyzed synthesis and molecular docking study of 2,4,5 trisubstituted imidazoles as potent inhibitors of *Mycobacterium tuberculosis*. ACS Comb Sci 2020;22:509–18.

[25] Lima CHS, Henriques MGMO, Candea ALP, Lourenco MCS, Bezerra FAFM, Ferreira ML, Kaiser CR, de Souza MVN. Synthesis and antimycobacterial evaluation of N-(E) heteroaromaticpyrazine-2-carbohydrazide derivatives. Med Chem 2011;7(3):245–9.

[26] Lee SH, Kim S, Yun MH, Lee YS, Cho SN, Oh T, Kim P. Synthesis and antitubercular activity of monocyclic nitroimidazoles: insights from econazole. Bioorg Med Chem Lett 2011;21:1515–8.

[27] Amini M, Navidpour L, Shafiee A. Synthesis and antitubercular activity of new N, N-diaryl-4-(4, 5-dichloroimidazole-2-yl)-1, 4-dihydro-2, 6-dimethyl-3, 5-pyridine dicarboxamides. DARU J Pharm Sci 2008;16(1):9–12.

[28] Bhatt BH, Sharma S. Synthesis, characterization, and biological evaluation of some tri-substituted imidazole/thiazole derivatives. J Heterocycl Chem 2015;52(4):1126–31. https://doi.org/10.1002/jhet.1992.

[29] Kalaria PN, Satasia SP, Avalani JR, Raval DK. Ultrasound-assisted one-pot four-component synthesis of novel 2-amino-3 cyanopyridine derivatives bearing 5-imidazopyrazole scaffold and their biological broadcast. Eur J Med Chem 2014;83:655–64.

[30] Kalaria PN, Satasia SP, Raval DK. Synthesis, characterization and pharmacological screening of some novel 5-imidazopyrazole incorporated polyhydroquinoline derivatives. Eur J Med Chem 2014;78:207–16.

[31] Kalaria PN, Satasia SP, Raval DK. Synthesis, characterization and biological screening of novel 5-imidazopyrazole incorporated fused pyran motifs under microwave irradiation. New J Chem 2014;38:1512–20.

[32] Kishk SM, McLean KJ, Sood S, Smith D, Evans JWD, Helal MD, Gomaa MS, Salama I, Mostafa SM, de Carvalho LPS, Levy CW, Munro AW, Simons C. Design and synthesis of imidazole and triazole pyrazoles as *Mycobacterium tuberculosis* CYP121A1 inhibitors. ChemistryOpen 2019;8:995–1011.

[33] Patel HM, Noolvi MN, Sethi NS, Gadad AK, Cameotra SS. Synthesis and antitubercular evaluation of imidazo[2,1-b][1,3,4]thiadiazole derivatives. Arab J Chem 2017;10:996–1002.

[34] Pandey AK, Sharma R, Purohit P, Dwivedi R, Chaturvedi V, Chauhan PMS. Synthesis of pyrido[1,2-a]imidazo-chalcone via 3-component groebkeblackburn-bienayme reaction and their bioevaluation as potent antituberculosis agents. Chem Bio Interface 2016;6(5):290–9.

[35] Tukulula M, Sharma RK, Meurillon M, Mahajan A, Naran K, Warner D, Huang J, Mekonnen B, Chibale K. Synthesis and antiplasmodial and antimycobacterial evaluation of new nitroimidazole and nitroimidazooxazine derivatives. ACS Med Chem Lett 2012;4(1):128–31.

[36] Munagala G, Yempalla KR, Singh S, Sharma S, Kalia NP, Rajput VS, Kumar S, Sawant SD, Khan IA, Vishwakarma RA, Singh PP. Synthesis of new generation triazolyl- and isoxazolyl-containing 6-nitro-2,3- dihydroimidazooxazoles as anti-TB agents: in vitro, structure–activity relationship, pharmacokinetics and in vivo evaluation. Org Biomol Chem 2015;13:3610–24.

[37] Kang Y, Park C, Shin H, Singh R, Arora G, Yu C, Lee Y. Synthesis and anti-tubercular activity of 2-nitroimidazooxazines with modification at the C-7 position as PA-824 analogs. Bioorg Med Chem Lett 2015;25(17):3650–3. https://doi.org/10.1016/j.bmcl.2015.06.060.

[38] Pulipati L, Sridevi JP, Yogeeswari P, Sriram D, Kantevari S. Synthesis and antitubercular evaluation of novel dibenzo[b,d]thiophene tethered imidazo[1,2-a]pyridine-3-carboxamides. Bioorg Med Chem Lett 2016;26(13):335–3140.

[39] Ramprasad J, Nayak N, Dalimba U, Yogeeswari P, Sriram D. One-pot synthesis of new triazole—imidazo[2,1-b][1,3,4]thiadiazole hybrids via click chemistry and evaluation of their antitubercular activity. Bioorg Med Chem Lett 2015;25(19):4169–73.

[40] Tao X, Gao C, Huang Z, Luo W, Liu K, Peng K, Ding CZ, Li J, Chen S, Yua LT. Discovery and evaluation of novel nitrodihydroimidazooxazoles as promising anti-tuberculosis agents. Bioorg Med Chem Lett 2019;29:2511–5.

[41] Gawad J, Bonde C. Synthesis, biological evaluation and molecular docking studies of 6-(4-ni trophenoxy)-1H-imidazo[4,5-b] pyridine derivatives as novel antitubercular agents: future DprE1 inhibitors. Chem Cent J 2018;12:138.

[42] Cherian J, Choi I, Nayyar A, Manjunatha UH, Mukherjee T, Lee YS, Boshoff HI, Singh R, Ha YH, Goodwin M, Lakshminarayana SB, Niyomrattanakit P, Jiricek J, Ravindran S, Dick T, Keller TH, Dartois V, Barry CE. Structure–activity relationships of antitubercular nitroimidazoles. 3. Exploration of the linker and lipophilic tail of ((S)-2-Nitro-6,7-dihydro-5H-imidazo[2,1-b][1,3] oxazin-6-yl)-(4 trifluoromethoxybenzyl)amine (6-amino PA-824). J Med Chem 2011;54:5639–59.

[43] Madaiah M, Prashanth MK, Revanasiddappa HD, Veeresh B. Synthesis and evaluation of novel imidazo[4,5-c]pyridine derivatives as antimycobacterial agents against *Mycobacterium tuberculosis*. New J Chem 2016;40:9194–204. https://doi.org/10.1039/C6NJ02069K.

[44] Warekara PP, Patila PT, Patila KT, Jamale DK, Kolekar GB, Anbhule PV. Ecofriendly synthesis and biological evaluation of 4-(4-nitro-phenyl)-2-phenyl-1,4-dihydro-benzo[4,5]imid-

azo [1,2-a]pyrimidine-3-carboxylic acid ethyl ester derivatives as an antitubercular agents. Synth Commun 2016;46(24):2022–30.

[45] Palkar MB, Noolvi MN, Maddi VS, Laxmivenkat MG, Nargund G. Synthesis, spectral studies and biological evaluation of a novel series of 2-substituted-5,6-diarylsubstituted imidazo(2,1-b)-1,3,4-thiadiazole derivatives as possible anti-tubercular agents. Med Chem Res 2012;21:1313–21.

[46] Moraski GC, Markley LD, Cramer J, Hipskind PA, Boshoff H, Bailey MA, Alling T, Ollinger J, Parish T, Miller MJ. Advancement of Imidazo[1,2-a]pyridines with Improved Pharmacokinetics and nM Activity vs. *Mycobacterium tuberculosis*. ACS Med Chem Lett 2013;4:675–9.

[47] Makwane S, Dua R. Synthesis and antitubercularacitivity of new imidazo [2,1-B][1, 3, 4]-thiadiazole-phenothiazine derivatives. Arch Org Inorg Chem Sci 2018;3(4):391–7.

[48] Syed MA, Ramappa AK, Alegaon S. Synthesis and evaluation of antitubercular and antifungal activity of some novel 6-(4-substituted aryl)-2-(3, 5-dimethyl-1H-pyrazol-1-yl) imidazo[2,1-b][1,3,4]-thiadiazole derivatives. Asian J Pharm Clin Res 2013;6(3):47–51.

[49] Pandey J, Tiwari VK, Verma SS, Chaturvedi V, Bhatnagar S, Sinha S, Gaikwad AN, Tripathi RP. Synthesis and antitubercular screening of imidazole derivatives. Eur J Med Chem 2009;44:3350–5.

[50] Krasavin M, Lukin A, Vedekhina T, Manicheva O, Dogonadze M, Vinogradova T, Zabolotnykh N, Rogacheva E, Kraeva L, Yablonsky P. Conjugation of a 5-nitrofuran-2-oyl moiety to aminoalkylimidazoles produces non-toxic nitrofurans that are efficacious in vitro and in vivo against multidrug-resistant *Mycobacterium tuberculosis*. Eur J Med Chem 2018;157:1115–26.

[51] Gosselin RE, Smith RP, Hodge HC, Braddock JE. Clinical toxicology of commercial products. Baltimore: Williams & Wilkins; 1984. p. 427.

[52] Thompson AM, Sutherland HS, Palmer BD, Kmentova I, Blaser A, Franzblau SG, Wan B, Wang Y, Ma Z, Denny WA. Synthesis and structure_activity relationships of varied ether linker analogues of the antitubercular drug (6s)-2-nitro-6-{[4-(trifluoromethoxy)benzyl]oxy}-6,7-dihydro-5H imidazo-[2,1-b][1,3]oxazine (PA-824). J Med Chem 2011;54:6563–85.

[53] Shalini VA, Kremer L, Kumar V. Alkylated/aminated nitroimidazoles and nitroimidazole-7-chloroquinoline conjugates: synthesis and anti-mycobacterial evaluation. Bioorg Med Chem Lett 2018;28:1309–12.

[54] Patel VM, Patel NB, Chan-Bacab MJ, Rivera G. N-Mannich bases of benzimidazole as a potent antitubercular and antiprotozoal agents: their synthesis and computational studies. Synth Commu 2020;50(6):858–78. https://doi.org/10.1080/00397911.2020.1725057.

[55] Yoon YK, Ali MA, Choon TS, Ismail R, Wei AC, Kumar RS, Osman H, Beevi F. Antituberculosis: synthesis and antimycobacterial activity of novel benzimidazole derivatives. Biomed Res Int 2013;2013:1–6. https://doi.org/10.1155/2013/926309.

[56] Goutam M, Koushik M, Rituparna D, Shubhra MR, Indranil R, Balaram M, Kumar SA. Allyl piperidine-1-carbodiothioate and benzyl 1H-imidazole 1 carbodithioate: two potential agents to combat against mycobacteria. J Appl Microbiol 2021;130(3):786–96.

[57] Bhati S, Kumar V, Singh S, Singh J. Synthesis, characterization, antimicrobial, anti-tubercular, antioxidant activities and docking simulations of derivatives of 2-(pyridin-3-yl)-1Hbenzo[d] imidazole and 1,3,4-oxadiazole analogy. Lett Drug Des Discovery 2020;17(8):1047–59.

[58] Anguru MR, Taduri AK, Bhoomireddy RD, Jojula M, Gunda SK. Novel drug targets for *Mycobacterium tuberculosis*: 2-heterostyrylbenzimidazoles as inhibitors of cell wall protein synthesis. Chem Cent J 2017;11(1):68.

[59] Ashok D, Gundu S, Aamate VK, Devulapally MG. Conventional and microwave-assisted synthesis of new indole-tethered benzimidazole-based 1,2,3-triazoles and evaluation of their antimycobacterial, antioxidant and antimicrobial activities. Mol Divers 2018;22:769–78.

[60] Nandha B, Nargund LG, Nargund SL, Bhat K. Design and synthesis of some novel fluorobenzimidazoles substituted with structural motifs present in physiologically active natural products for antitubercular activity. Iran J Pharm Res 2017;16(3):929–34.

[61] Nandha B, Nargund LVG, Nargund SL, Kuntal H. Design and synthesis of imidazolylmethyl substituted fluorobenzimidazoles for antitubercular and antifungal activity. J Chem Pharm Res 2014;6(1):530–9.

[62] Yadav S, Lim SM, Ramasamy K, Vasudevan M, Shah SAA, Mathur A, Narasimhan B. Synthesis and evaluation of antimicrobial, antitubercular and anticancer activities of 2-(1-benzoyl-1H-benzo[d] imidazol-2-ylthio)-N-substituted acetamides. Chem Cent J 2018;12:66. https://doi.org/10.1186/s13065-018-0432-3.

Chapter 4

Imidazole derivatives: Impact and prospects in antiviral drug discovery

Pankaj Teli, Nusrat Sahiba, Ayushi Sethiya, Jay Soni, and Shikha Agarwal
Synthetic Organic Chemistry Laboratory, Department of Chemistry, Mohanlal Sukhadia University, Udaipur, Rajasthan, India

Abbreviations

AIDS	acquired immune deficiency syndrome
BKPyV	BK human polyomavirus type 1
BMZ	benzimidazole
BVDV	bovine viral diarrhea virus
CC_{50}	50% cytotoxic concentration
CMV	cytomegalovirus
CV	coxsackie virus
CVB	coxsackievirus B
DENV	dengue virus
EC_{50}	half maximal effective concentration
FHV	flock house virus
FIPV	feline infectious peritonitis virus
HAART	highly active antiretroviral therapy
HBV	hepatitis B virus
HCV	hepatitis C virus
HIV	human immune deficiency virus
HPV	human papilloma virus
HSV	herpes simplex virus
IAV	influenza A virus
IC_{50}	half maximal inhibitory concentration
IMPDH	inosine-5′-monophosphate dehydrogenase
MDBK cells	Madin-Darby bovine kidney cells
MDCK cells	Madin-Darby canine kidney cells
MERS HCoV	Middle-East respiratory syndrome human coronavirus
MPA	mycophenolate acid
Mpro	main protease
NNRTI	nonnucleoside reverse transcriptase inhibitors
PI-3V	parainfluenza-3 virus

Imidazole-Based Drug Discovery. https://doi.org/10.1016/B978-0-323-85479-5.00001-0
Copyright © 2022 Elsevier Inc. All rights reserved.

PIV-3	parainfluenza virus type 3
PTV	Punta Toro virus
RSV	respiratory syncytial virus
RT	reverse transcriptase
RV	reovirus
SAR	structure activity relationship
SARS-CoV-2	severe acute respiratory syndrome human coronavirus-2
SI	selective index
SV	sindbis virus
SVCV	spring viremia of carp virus
TMV	tobacco mosaic virus
TMV-CP	tobacco mosaic virus coat protein
VSV	vesicular stomatitis virus
VV	vaccinia virus
WHO	World Health Organization
YFV	yellow fever virus
ZIKV	Zika virus

1. Introduction

Microbes, invisible to the human eye, tend to threaten humans, not only in medical terms but also in terms of disrupting the social and economic aspects of life as depicted by the present-day COVID-19 pandemic. This pandemic has proven how these small viruses can harm the very existence of humans. It has shaken the social, economic, and medical backbone of every nation all around the world [1]. Viruses cause several epidemic diseases and generate havoc for the entire world (Table 4.1). They enter human bodies via various paths, like oral paths, the nasal tract, through the skin, or via any external wound [2].

Data analysis has shown that viral infections alone cause mortality of around 2 million globally [3, 4]. The virulent behavior does not stop here; viruses tend to change their genomic structure and become resistant to the drugs used to stop their multiplication and infection rate [5, 6]. Various viruses are present in our surroundings, yet only a handful are recognized and characterized. Human immune deficiency virus (HIV), hepatitis virus, influenza virus, human papilloma virus (HPV), herpes simplex virus (HSV), and coronavirus are some of the pathogenic viruses that have caused large-scale mortality (Table 4.1). Thus, the identification and generation of antiviral drugs are essential to human well-being.

Antiviral drugs can be either natural or chemically synthesized. Curcumin extracted from turmeric is a trending natural antiviral drug that shows its potent antiviral property upon various viruses including parainfluenza virus type 3 (PIV-3), feline infectious peritonitis virus (FIPV), vesicular stomatitis virus (VSV), herpes simplex virus (HSV), flock house virus (FHV), and respiratory syncytial virus (RSV) [7]. Enfuvirtide, maraviroc, indinavir, acyclovir, foscarnet, abacavir, lamivudine, tenofovir, adefovir, entecavir, telbivudine, tenofovir,

TABLE 4.1 Major epidemics caused by virus strains.

Name	Time period	Death toll	Type/prehuman host
Japanese smallpox epidemic	735–737	1 million	Variola major virus
New World smallpox outbreak	1520 onwards	56 million	Variola major virus
Yellow fever	Late 1800s	100,000–150,000 (US)	Virus/mosquitoes
Russian flu	1889–90	1 million	Believed to be H2N2 (avian origin)
Spanish flu	1918–19	40–50 million	H1N1 virus/pigs
Asian flu	1957–58	1.1 million	H2N2 virus
Hong Kong flu	1968–70	1 million	H3N2 virus
HIV/AIDS	1981–present	25–35 million	Virus/chimpanzees
SARS	2002–03	770	Coronavirus/bats, civets
Swine flu	2009–10	200,000	H1N1 virus/pigs
Ebola	2014–16	11,000	Ebolavirus/wild animals
MERS	2015–present	850	Coronavirus/bats, camels
COVID-19	2019–present	2.7 million (Johns Hopkins University estimate as of March 16, 2021)	Coronavirus—unknown (possibly pangolins)

camptothecin, ribavirin, and interferons (siRNA) are examples of synthetic antiviral drugs [8]. For the management of morbidities and mortalities incurred by viruses, pharmaceutical departments are in a constant race to develop new bioactive moieties out of which heterocyclic compounds are in the limelight [9].

Nitrogen-based heterocycles are readily available in nature with diverse biological activities and similarities with various bioactive drugs [10]. Some nitrogen-based heterocycles, e.g., imidazoles, are often considered potent drugs in clinical practices. They have been in long run due to their amphoteric nature, i.e., they can act as an acid and base at the same time and further increase their potency [11–13]. Imidazoles and their fused derivatives are five-membered cyclic structures and their structure gives them a unique identity in the field of antiviral drugs [13–16]. Special structural features of imidazole and benzimidazole ring with their desirable electron-rich characters help them to bind with various targets and give them an advantage over other known moieties [17, 18].

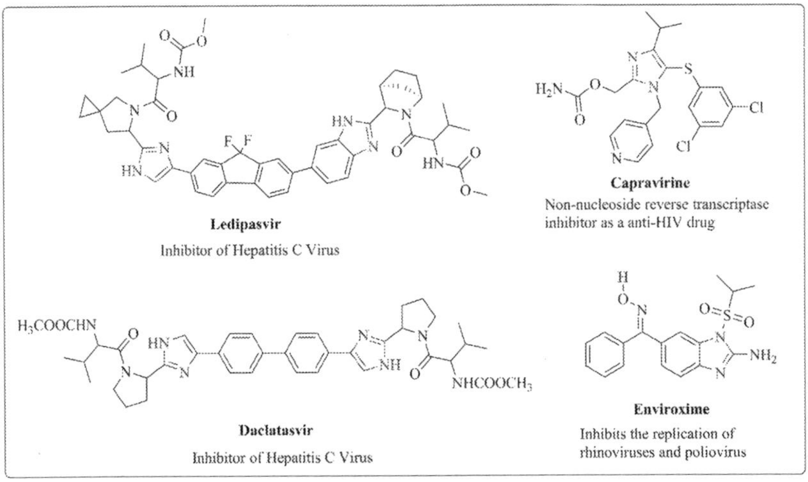

FIG. 4.1 Several imidazole-based antiviral drugs.

These moieties are not specifically antiviral but also possess therapeutic action regarding various other diseases. Extensive research has been carried out to find other imidazole derivatives or imidazole-containing moieties to aid the medical department [19–22]. Some imidazole-based antiviral drugs are depicted in Fig. 4.1. This chapter illustrates recent attempts regarding the antiviral activity of imidazole-based moieties. Furthermore, the structure–activity relationships of various imidazole derivatives against different virus strains are reported for the development of significant antivirals.

2. Imidazole derivatives and their action against different viruses

Different potent and drug-like imidazole derivatives have been depicted against several virus strains like ZIKV, HIV, HPC, SARS CoV-2, influenza, dengue, etc. (Fig. 4.2).

2.1 Zika virus

Zika virus (ZIKV) belongs to the *Flaviviridae* family and causes congenital abnormalities in fetuses and newborns and upregulated a number of microcephaly cases [23, 24]. Worldwide, more than 2 billion people are at risk of ZIKV and the WHO declared ZIKV a public health emergency in 2016. Moreover, ZIKV is responsible for ophthalmological complications in adults and neural-inflammatory diseases such as Guillain-Barré syndrome [25]. Current data show that it can be sexually transmitted without any signs in tests over a long period [26]. Currently there are no virus-specific drugs or medications available

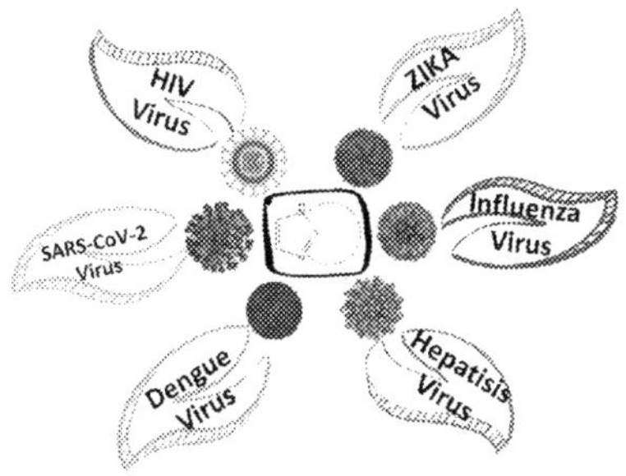

FIG. 4.2 Imidazole derivatives against various strains of viruses.

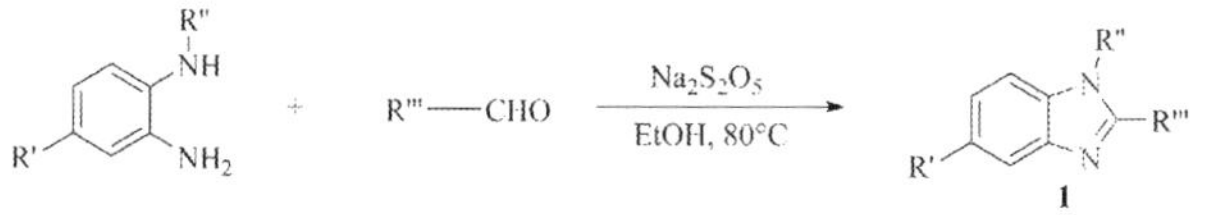

SCHEME 4.1 Synthesis of benzimidazole derivatives.

to medicate ZIKV-infected patients. Several of the imidazole derivatives have been examined for effective treatment against ZIKV.

A library of 50 structurally diverse benzimidazole derivatives were synthesized by a one-pot condensation of 1,2-phenylenediamines with several aromatic aldehydes using sodium metabisulfite as a catalyst under mild conditions (Scheme 4.1) and assessed for their inhibitory activity against Zika virus. Compound **1** was found to be the most promising (EC_{50} value = 1.9 ± 1.0 μM) against the African ZIKV strain in Huh-7 (SI > 37) and neural stem cells (SI = 12). The SAR studies demonstrated that the heteroaromatic ring at the C-2 position and 4-OCH_3-benzyl, 3-pyridinylmethyl or 2-Cl-benzyl at the N-1 position with the presence of CF_3 group at the C-5 position of the benzimidazole ring showed immense pharmacological profile against ZIKA virus *viz.* compound **1b** (EC_{50} value = 24.7 ± 2.0 μM), **1c** (EC_{50} value = 13.3 ± 1.1 μM), **1d** (EC_{50} value = 7.5 ± 1.1 μM), **1e** (EC_{50} value = 18.5 ± 1.1 μM), **1f** (EC_{50} value = 48.3 ± 1.3 μM), and **1g** (EC_{50} value = 6.1 ± 1.2 μM). Moreover, naphthalene conjugated to benzimidazole with Cl at N-1 and CF_3 at the C-5 position exhibited the highest antiviral activity toward ZIKA strains with SI values less than 37 in Huh-7 that are more comparable to the reference, mycophenolate acid (MPA) [27] (Scheme 4.1, Fig. 4.3).

A novel series of 34 compounds of 1*H*-benzo[*d*]imidazole-5-carboxamide derivatives was drafted, synthesized, and screened to investigate their anti-yellow fever virus (YFV) and anti-ZIKA virus activity. Compounds **2a–g** were found to be efficient against YFV in low micromolar range using human Vero cells and hepatoma Huh-7 cells. The SAR study was explored and it was found that alteration of the carboxylic acid groups and 5-carboxylate ester into amide group enhanced the inhibitory action against YFV. Among all

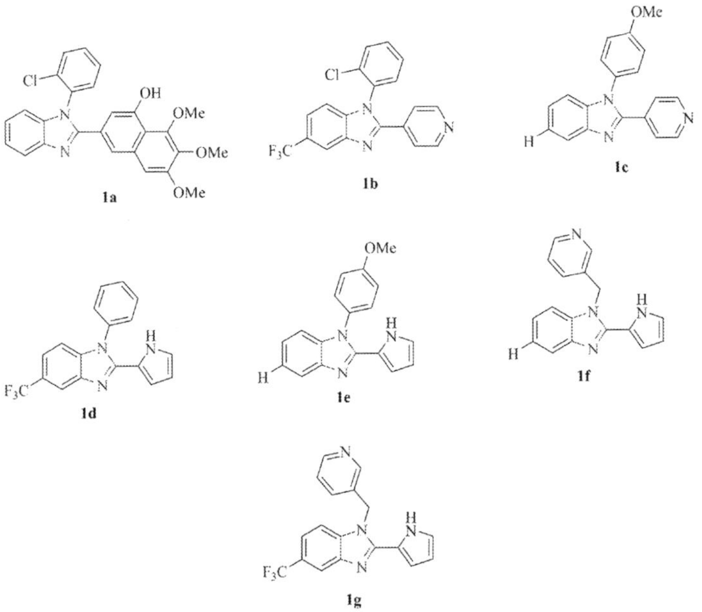

FIG. 4.3 Biological active compound **1** against ZIKA virus.

the synthesized compounds, compound **2a** proved to be effective against YFV ($EC_{50}=1.7\pm0.8\,\mu M$ on Huh-7 cells, $EC_{50}=1.2\pm0.02\,\mu M$ on Vero cells), as well as ZIKA virus ($EC_{50}=4.5\pm2.1\,\mu M$) [28] (Fig. 4.4).

2.2 Influenza

Influenza continues to be a highly contagious and barely inhibited human infection. Worldwide, 500 million people suffer every year from the flu with about 2 million fatalities [29, 30]. The influenza A virus is responsible for the two of the four reported rare pandemics and is the cause of recurring epidemic outbreaks, remaining a continuous risk to socioeconomic development and public health.

Available medications for the treatment of influenza mainly focus on some influenza protein targets such as antivirals for M2 ion channels (rimantadine and amantadine) and neuraminidase (peramivir, oseltamivir, laninamivir, and zanamivir) and the vaccines for hemagglutinin [31]. Currently most of the influenza A virus strains have shown strong resistance to these drugs [32–36]. Therefore, there is a crucial demand for novel antiinfluenza agents, particularly after the 2009 H1N1 (swine flu) and 2013 H7N9 outbreaks [37, 38].

A series of novel 2-substituted 7,8-dihydro-6*H*-imidazo[2,1-*b*][1,3]benzothiazol-5-ones (**3a–k**) were synthesized by cyclohexane-1,3-diones and assessed for their cytotoxicity and antiviral activity against influenza virus A/Puerto Rico/8/34 (H1N1) in MDCK cells. The three compounds **3i–k**,

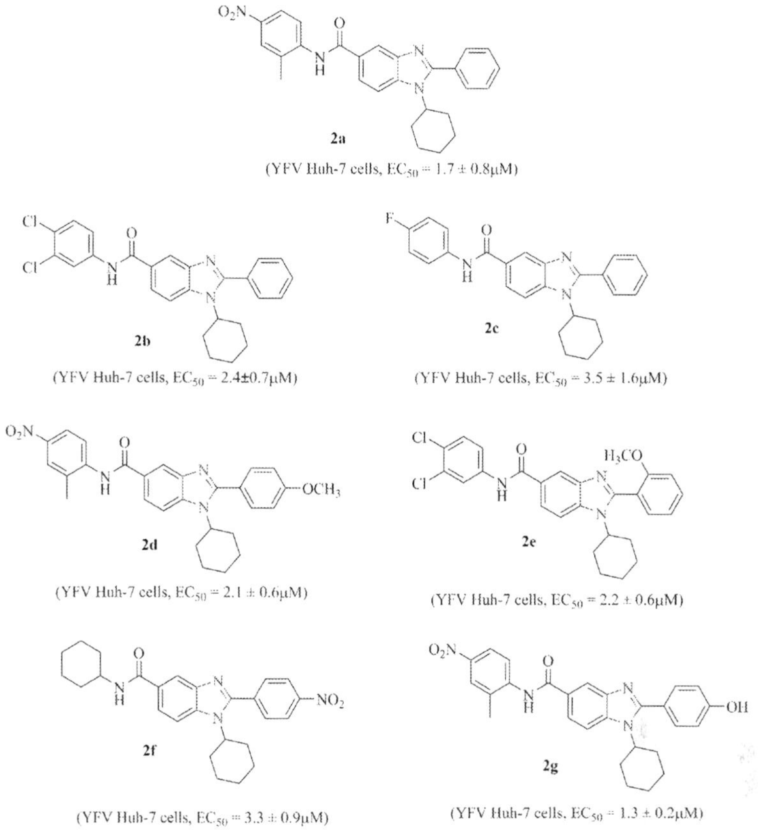

FIG. 4.4 Active anti-ZIKV benzimidazole derivatives.

containing a thiophene ring, presented the most promising virus-inhibiting activity and low toxicity profile. The analog **3j** demonstrated the highest antiviral activity against influenza virus with CC_{50}: > 1000 μM, SI = 77 [39] (Scheme 4.2, Fig. 4.5).

A series of compounds was developed by the conjugation of imidazole moiety with pinanamine derivatives and evaluated for their antiinfluenza activity against the amantadine-sensitive virus A/M2 wild-type virus A/HK/68 and amantadine-resistant strain A/WSN/33. Most of the compounds exhibited inhibitory activity against the amantadine-sensitive virus at a very low concentration by blocking the A/M2-WT ion channel. Compound **4** afforded the inhibition of A/M2 wild-type virus A/HK/68 as well as the amantadine-resistant strain A/WSN/33, with IC_{50} values of 2.5 mM and 3.4 mM, respectively [40] (Scheme 4.3).

A total of 250,000 pure chemicals and semipurified fractions from natural extracts were evaluated by throughput screening for antiviral activity against

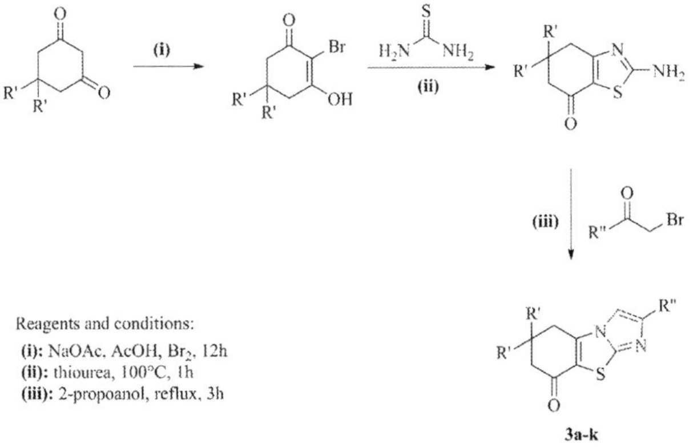

SCHEME 4.2 Synthesis of 2-substituted 7,8-dihydro-6*H*-imidazo[2,1-*b*][1,3]benzothiazol-5-ones.

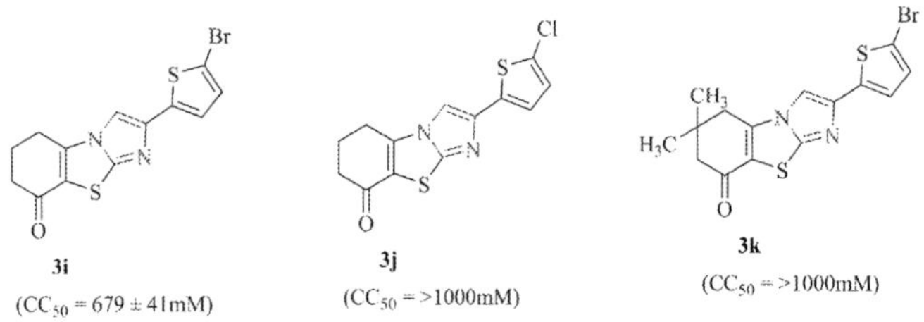

FIG. 4.5 Active antiinfluenza derivatives.

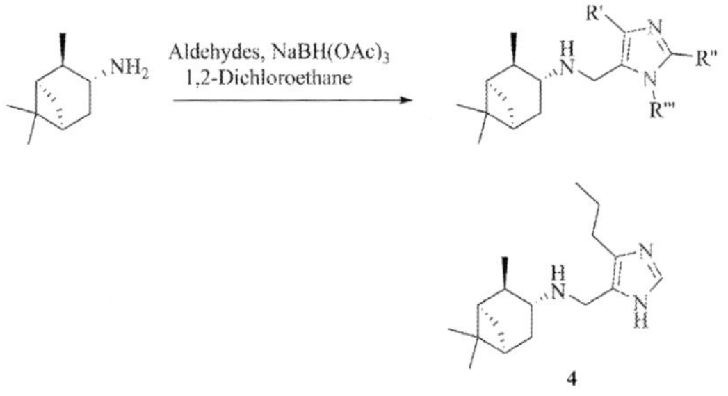

SCHEME 4.3 Synthesis of imidazole moiety with pinanamine derivatives and most active compound **4**.

M2 proton channel of the influenza A virus. Twenty-one compounds were found to be active, viz. amantadine, rimantadine, 13 related adamantanes, and six nonadamantanes. Two imidazole-based compounds, **5a** and **5b**, also exhibited antiviral activity against influenza A virus with EC_{50} values of 0.3 and 0.4 μM, respectively [41] (Fig. 4.6).

Five imidazole alkaloids were extracted from the marine sponge *Pericharax heteroraphis* and evaluated for their antiviral activity against H1N1 influenza

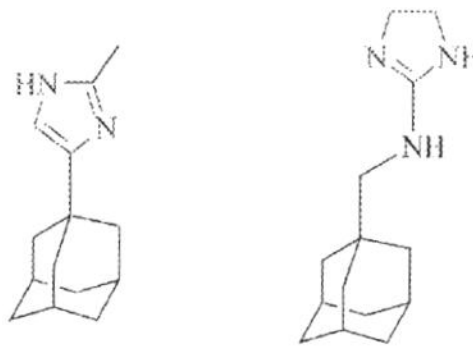

5a 5b

FIG. 4.6 Various imidazole containing antiviral agents against the influenza virus.

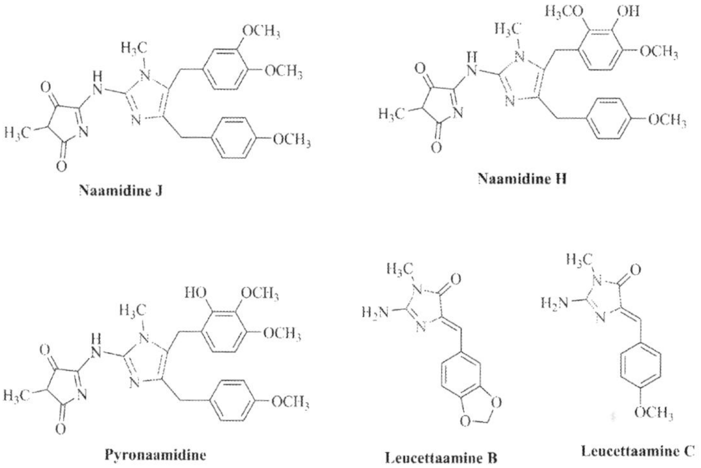

FIG. 4.7 Alkaloids extracted from the marine sponge *Pericharax heteroraphis*.

A virus (IAV) [42]. All the alkaloids have a central 2-aminoimidazole ring substituted at the C-4 and C-5 positions by one or two functionalized benzyl groups. Only alkaloid leucettamine C exhibited weak inhibitory activity against the H1N1 virus with an inhibition rate of 33% while the positive control drug, ribavirin, showed an inhibition rate of 65% (Fig. 4.7).

2.3 SARS-COVID

The first coronavirus (SARS-CoV-2) infection was reported in Wuhan (China) in December 2019, and spread all over the world [43–47]. SARS-CoV-2 belongs to the Betacoronaviruses family like Middle-East Respiratory Syndrome Human Coronavirus (MERS HCoV) and Severe Acute Respiratory Syndrome Human Coronavirus (SARS-CoV-1) [48]. Current studies on SARS-CoV-2 have demonstrated that the chymotrypsin-like protease, 3CL hydrolase, or main protease (Mpro) of SARS-CoV-2 play a crucial role in the life cycle of coronavirus and hence the inhibition of Mpro can provide significant therapeutic treatment against COVID-19 infection [49, 50].

A docking study was performed on 18 imidazole analogs attached with 7-chloro-4-aminoquinoline against coronavirus (SARS-CoV-2) via binding to the active site of SARS-CoV-2 main protease. The study showed that the compounds **6a**, **6b**, and **6c** have greater binding energy with SARS-CoV-2 main protease than other imidazole derivatives, and the two drugs, hydroxychloroquine and chloroquine, caused potent antiviral activity against COVID-19. Furthermore, the study indicated that the molecules with electronegative atoms and more than three cycles have high affinity toward binding site of the protease due to halogen interaction, formation of π-bonds, and hydrogen bonding [51] (Fig. 4.8).

Four new azo imidazole derivatives **7a–d** were synthesized by the condensation reaction of amino functionalized imidazole compounds with azo-coupled ortho-vaniline precursor and the molecular docking studies of these ligands were carried out against the main protease (6LU7) of novel coronavirus (COVID-19). The results displayed good binding energies for derivatives **7a–d** (−7.7 kcal/mol for **7a**, −7.0 kcal/mol for **7b**, −7.9 kcal/mol for **7c**, and −7.9 kcal/mol for **7d**) and promising inhibitory activity of all ligands against the main protease (M ^{pro}) of SARS-CoV- 2 [52] (Scheme 4.4).

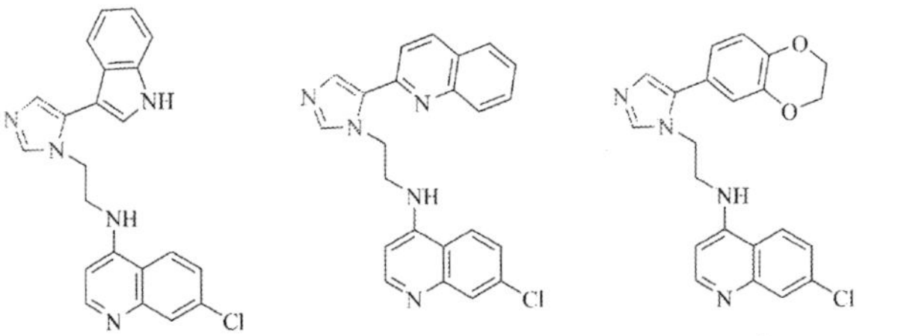

6a Affinity = -8.5 Kcal/mol

6b Affinity = -8.3 Kcal/mol

6c Affinity = -8.3 Kcal/mol

FIG. 4.8 Active imidazole analogs against coronavirus.

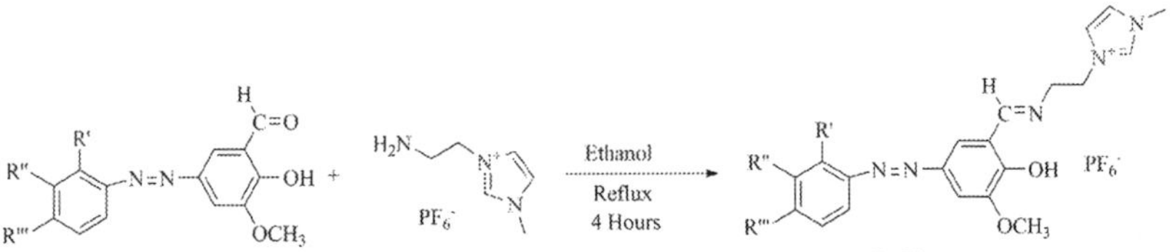

7a-7d

7a: R' = H, R'' = H, R''' = H
7b: R' = Cl, R'' = H, R''' = H
7c: R' = H, R'' = H, R''' = Cl
7d: R' = H, R'' = H, R''' = F

SCHEME 4.4 Synthesis of azo imidazole derivatives.

2.4 Dengue

Dengue is the most prevalent arthropod-borne viral infection in the world, especially in tropical and subtropical areas, and is estimated to infect 96 million people annually [53]. The genome of DENV is composed of 10.7 kb, positive, single-stranded RNA with four different stereotypes (DENV-1 to DENV-4) [54]. The RNA-dependent RNA polymerase plays a pivotal role in the synthesis of viral genome, and it is thus determined as an effective drug target [55]. Immense efforts have been made to find the potent antiviral chemotherapeutics or vaccine against DENV, but without success as yet, and disease treatment is therefore restricted to supportive care. Further, more biologically active molecules have been explored for the treatment of DENV. In this search, a series of imidazole 4,5-dicarboxamide derivatives has been synthesized (Scheme 4.5) and evaluated for their inhibitory action against dengue virus using high-throughput screening assay of dengue virus-2 replicon. Some of these derivatives have a tendency to inhibit dengue virus (DENV) and yellow fever virus (YFV) in the micromolar range. In particular, compound **8b** showed the highest antiviral activity against YFV ($EC_{50} \pm 1.85\,\mu M$) while compound **8c** showed the most potent inhibitory activity against DENV in Vero cells ($EC_{50} \pm 1.93\,\mu M$) [56] (Fig. 4.9).

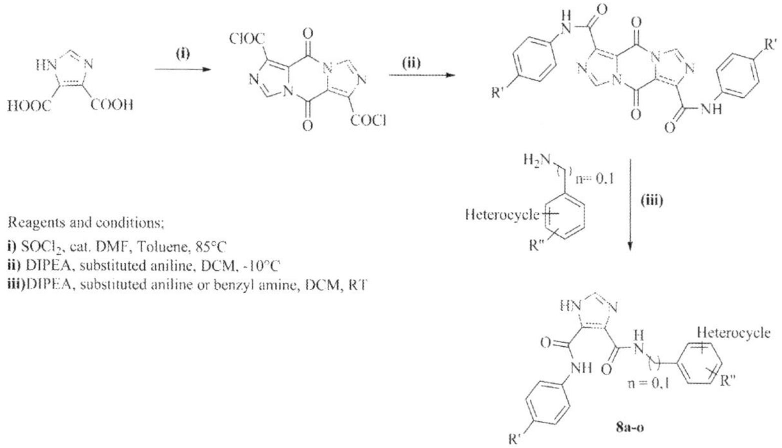

SCHEME 4.5 Synthesis of imidazole 4,5-dicarboxamide derivatives.

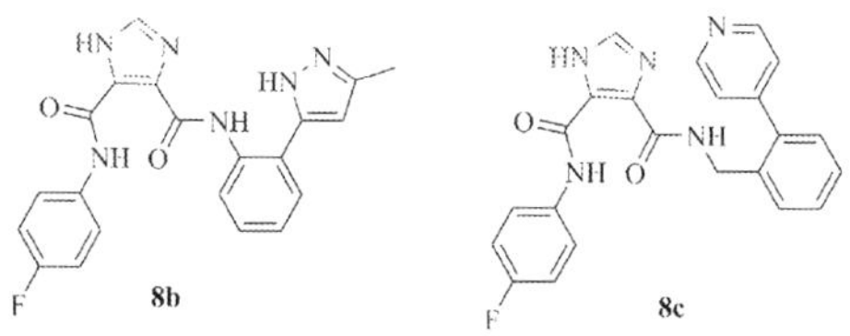

FIG. 4.9 Active imidazole 4,5-dicarboxamide derivatives against dengue virus.

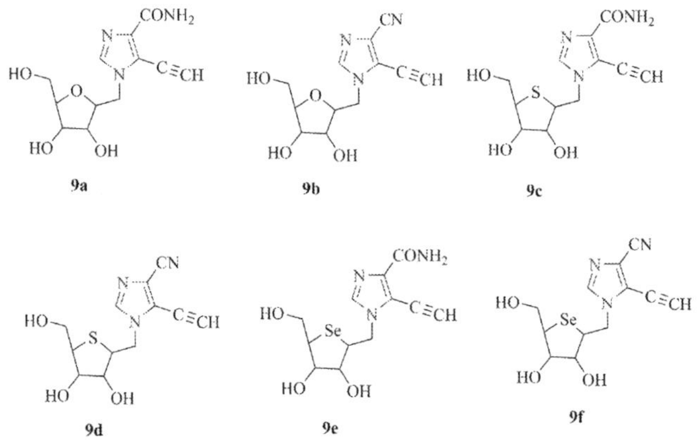

FIG. 4.10 Different imidazole-based anti-DENV agents.

A nucleoside series has been evaluated to detect the effective antiviral compounds that inhibited the replication of DENV. Compound 5-ethynyl-(1-β-D-ribofuranosyl)imidazole-4-carboxamide (**9a**) and its 4- carbonitrile derivative (**9b**) were found to be lead compounds against DENV, but due to cytotoxicity, more derivatives were developed. As a result, 4′-thio and 4′-seleno derivatives of **9a** and **9b**, i.e., **9c–9f**, were prepared. These derivatives presented a positive regulation on DENV replication inhibition without any sign of toxicity [57] (Fig. 4.10).

The antiviral activity of the metal complex [Cu(2,4,5-triphenyl-1*H*-imidazole)$_2$(H_2O)$_2$].Cl_2 was investigated through inhibition of replication of DENV-2 in Vero cells. The complex was found to be significant in inhibiting the growth of DENV-2 with an IC_{50} value of 98.62 μg/mL and exhibited a low cytotoxicity value (CC_{50}=300.36 μg/mL) against Vero cells [58].

2.5 Hepatitis

Hepatitis C (HCV) is associated with both mild and acute liver disease and may lead to chronic states like cirrhosis, hepatocellular carcinoma, and liver failure [59, 60]. Globally, approximately 150 million people have been infected by HCV with an estimated 3–4 million new cases occurring annually [61]. The conventional treatment of HCV infection by pegylated interferon and ribavirin also possesses toxicity [62, 63] and in 2011, boceprevir and telaprevir were accepted for the medication of chronic hepatitis C genotype 1 infection [64, 65]. Moreover, due to the drug-resistant nature of HCV, there is an urgent need to discover novel drugs for treatment.

A library of novel 2-aminoalkylsubstituted 6-chloro- or 5,6-dichloro-*1H*-imidazo[4,5-*b*]pyridines has been developed and assessed for their anti-HBV activity in an efficient infectious environment. Compounds **10d**, **11a**, **12a**, **12b**, and **12d** showed promising activity against HBV. The monochloro diethylaminoethyl-substituted derivative **12d** was found to be the most active to show anti-HBV effect to pegylated interferon α2b while compounds **13a–d** did not show significant activity. All the compounds were capable of reducing HBV rcDNA, cccDNA, and pgRNA levels, and possessed significant anti-HBV activity [66] (Scheme 4.6).

A series of imidazole analogs was investigated for their significant HCV NS5B polymerase inhibition activity through their binding at the active site of PDB ID: 2DXS. Some compounds exhibited good binding scores; in particular, compounds **14a** ($-84.12\,kJ\,mol^{-1}$) and **14b** ($-65.47\,kJ\,mol^{-1}$) showed high interaction with promising inhibitory activity toward HCV polymerase [67] (Fig. 4.11).

Twenty imidazole-coumarin conjugates were synthesized by linking imidazole with coumarin derivatives by $-SCH_2-$ moiety (Scheme 4.7) and screened for their antiviral activity against HCV. Among all the synthesized compounds,

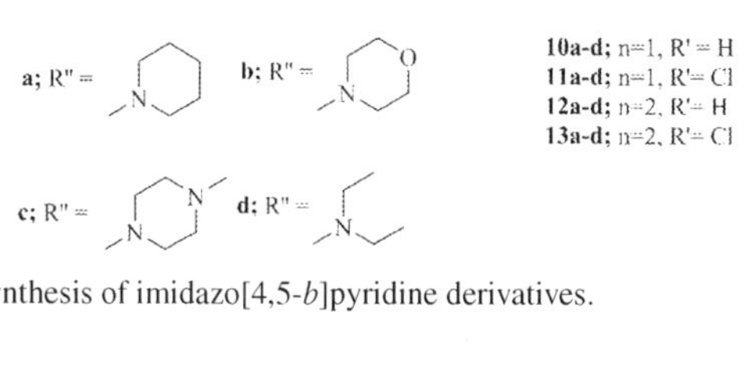

SCHEME 4.6 Synthesis of imidazo[4,5-*b*]pyridine derivatives.

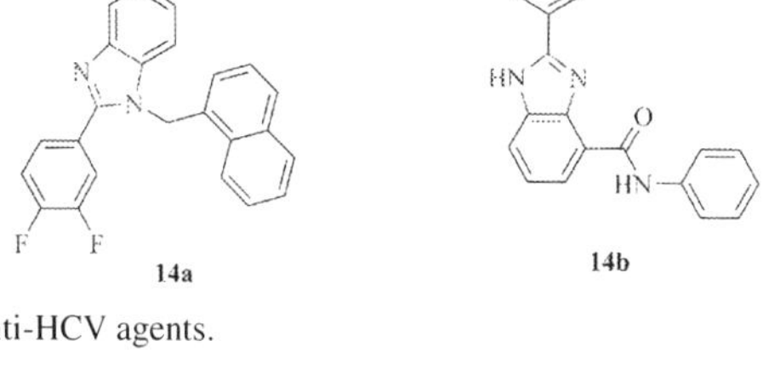

14a **14b**

FIG. 4.11 Potent anti-HCV agents.

	R''	R'''
a	H	H
b	H	F
c	H	Cl
d	H	Br
e	OMe	H

15	R'	R''	R'''
a	H	H	H
b	H	H	F
c	H	H	Cl
d	H	H	Br
e	H	OMe	H
f	Me	H	H
g	Me	H	Cl

SCHEME 4.7 Synthesis of imidazole-coumarin conjugates.

three derivatives (**15b**, **15d**, and **15e**) showed immense anti-HCV activity with EC_{50} values of 7.2, 5.1, and 8.4 μM combined with SI values of 12, 15, and 21, respectively. Furthermore, the SAR study was explored and it was established that the parent imidazole analog with an N–H proton afforded a larger SI value, and inclusion of different groups into the coumarin ring enhanced selectivity as well as potency of the conjugates [68].

Novel 1*H*-1,2,4-triazole and imidazole L-ascorbic acid and imino-ascorbic acid derivatives were synthesized (Scheme 4.8) and their antiviral activity was evaluated against HCV through their inhibitory activity on a Huh 5.2 replicon. Compound **16a** was established as the most promising agent (EC_{50} value = 36.6 μg/mL and CC_{50} value more than 100 μg/mL) against replication of the HCV virus by inhibiting IMPDH, a major target for antiviral activity [69].

Twenty-five new imidazole NH-substituted daclatasvir-modified conjugates were synthesized in order to enhance pharmacokinetic properties and potency against HCV and assessed in a HCV genotype 1b replicon. Among all the synthesized compounds, 2-oxoethyl acetate substituted compound **17** (Fig. 4.12) demonstrated comparable anti-HCV potency (EC_{50} = 0.08 nM) with respect to the lead drug daclatasvir. Prodrug **17** showed similar exposure to the lead compound in vivo and also behaved as an ideal candidate for a gradual and continuous release of daclatasvir [70]. The entire structure of HCV p7 protein was described and the allosteric site on the channel periphery for the specific drug–protein interactions was determined. Furthermore, the structure guided diverse inhibitory small molecules with high activity and selectivity against HCV was examined. Compound **18** (Fig. 4.12) was found to be effective against HCV by forming two H-bonding interactions with backbone carbonyl of Gly46 and the

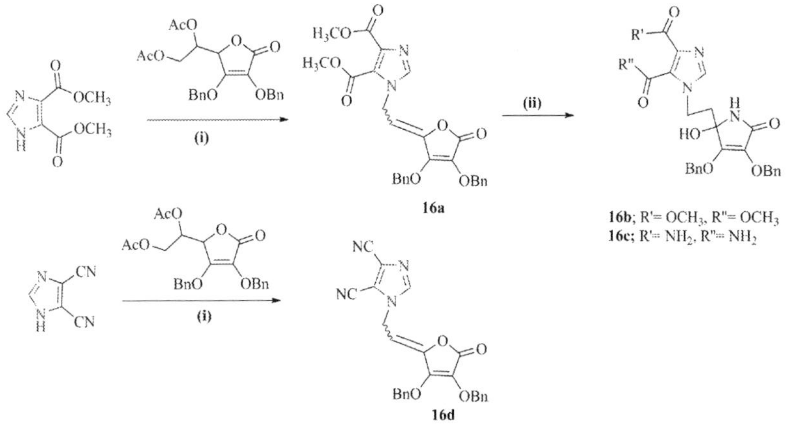

Reagents and conditions: **(i)** HMDS/$(NH_4)_2SO_4$/argon/reflux 12 h; TMS-triflate/dry acetonitrile/ 55-70 °C **(ii)** NH_3/MeOH-dioxane/rt/ 24 h; **(iii)** BCl_3/CH_2Cl_2/ -78°C/2 h.

SCHEME 4.8 Synthesis of imidazole L-ascorbic acid derivatives.

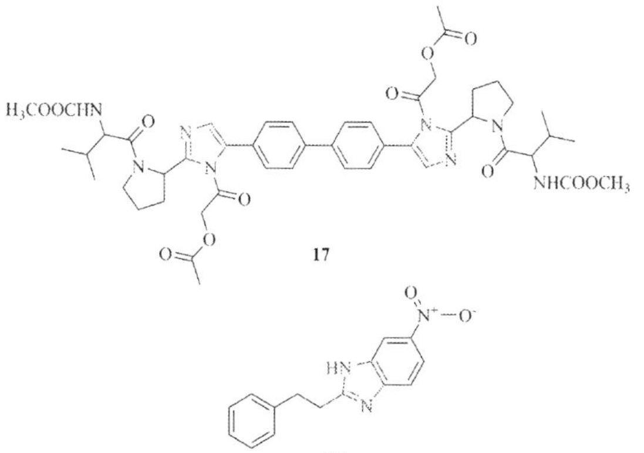

FIG. 4.12 Active imidazole derivatives **17** and **18** against HCV.

hydroxyl group of Tyr45, and depressed the rimantadine resistance polymorphism at submicromolar concentrations [71].

2.6 HIV

Human immunodeficiency virus (HIV) was discovered in the mid-1980s as a responsible agent for acquired immune deficiency syndrome (AIDS) [72]. Highly active antiretroviral therapy (HAART) was employed in 1995 as a treatment for HIV/AIDS. The main component of HAART is Efavirenz, which has a tendency to inhibit the HIV-1 reverse transcriptase (RT), the enzyme responsible for interchange of viral RNA into double-stranded DNA [73]. However, HAART could not eradicate the virus. Viral resistance emerged toward this mechanistic class of inhibitors [74–76], and urged the demand of new drugs with unique resistance profiles for complete care of patients with the virus.

A series of new 1-substituted-5-aryl-1*H*-imidazole was synthesized by cycloaddition of *para* toluenesulfonylmethyl isocyanide with imines and aldehydes using microwave irradiation and all the synthesized compounds were screened for their anti-HIV activity via Alpha Screen HIV-1 IN-LEDGF/p75 inhibition assay. Six imidazole-based derivatives (**21c**, **21f**, **22c**, **22f**, **25a**, and **25b**) showed promising inhibitory activity, i.e., more than 50% inhibition at 10 μM against the HIV strain. Furthermore, the SAR study indicated that the two aromatic rings and *N*-heterocyclic moiety played crucial roles in inhibition and directed the HIV-1 IN and LEDGF/p75 protein–protein interaction [77] (Scheme 4.9).

A library of novel nonnucleoside reverse transcriptase inhibitors related to imidazole-amide conjugates was described and evaluated for their antiviral activity toward HIV-1, along with the resilient Y188L-mutated virus. The ligand-protein interaction was optimized for key H-bonding motif using X-ray crystallography and compound **26** (Fig. 4.13) demonstrated enormous antiviral

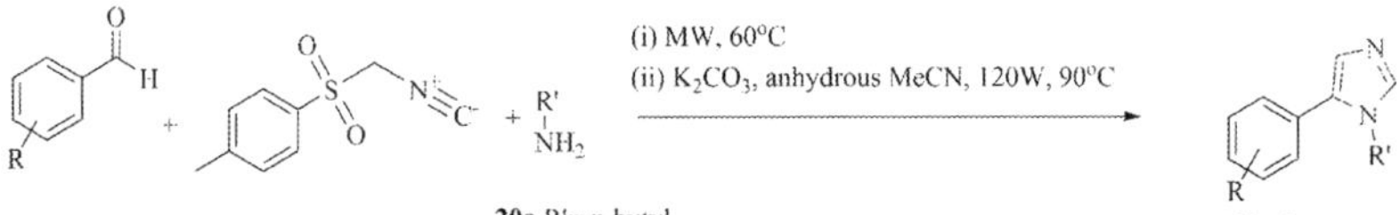

19a R = 4-F
19b R = 2-Cl
19c R = 3-OMe
19d R = 2,4-OMe
19e R = 3,4-OCH2O-
19f R = 4-*t*-butyl
19g R = 3,4-(OCH2)2O-
19h R = 2-NO2
19i R = 4-NO2
19j R = H

20a R' = *n*-butyl
20b R' = cyclohexyl
20c R' = 4-BrPh
20d R' = 2,4-ClPh
20e R' = 2,4-FPh
20f R' = 2,4-BrPh

21a-i
22a-h
23a-c
24a-d
25a-c

SCHEME 4.9 Synthesis of 1-subtituted-5-aryl-1*H*-imidazole derivatives.

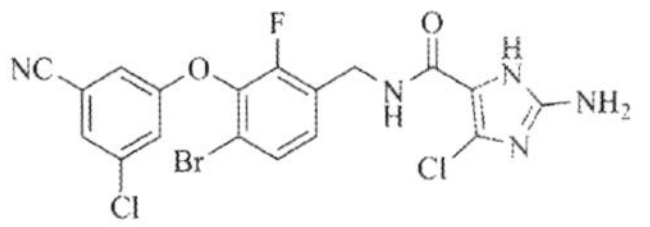

26

FIG. 4.13 Active imidazole-amide conjugate against HIV.

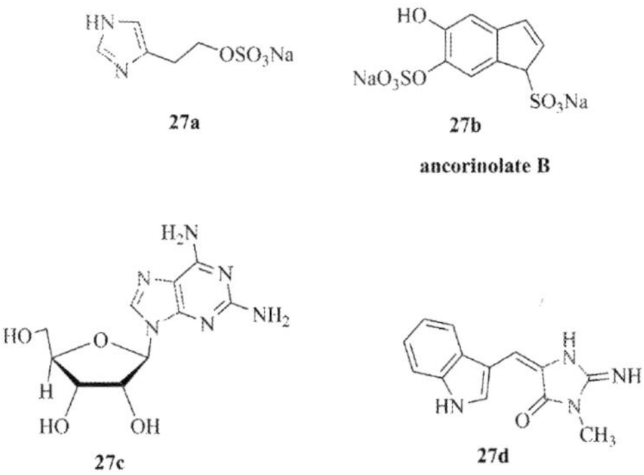

27a
27b **ancorinolate B**
27c **2-aminoadenosine**
27d **2'-de-*N*-methylaplysinopsin**

FIG. 4.14 Isolated imidazole sulfates from the sponge **Dercitus japonensis.**

activity ($EC_{50} < 1$ nM) against a huge series of NNRTI-resistant viruses with a good pharmacokinetic profile [78].

One novel imidazole sulfate (**27a**) and three known derivatives (**27b–d**) were isolated from the sponge *Dercitus (Halinastra) japonensis* and evaluated for their antiviral activity against HIV. Among all compounds, only **27b** was found to be active against HIV with an IC_{50} value of 109 μM and a CC_{50} value of more than 2.84×10^2 μM [79] (Fig. 4.14).

Through the high-throughput screening program, 4-(phenylcarbamoyl)-1*H*-imidazole-5-carboxylic acid (**28**) was chosen as a selective and significant inhibitor of the interaction between LEDGF/p75 and HIV-1 IN (IC_{50} value = 6 ± 4 μM). Furthermore, the SAR study explored active groups in the synthesized compounds and a library of nontoxic 5-carbonyl-1*H*-imidazole-4-carboxamide inhibitors of LEDGF/p75 and HIV-1 IN interaction was synthesized. Compound **28** showed good interactions with protein by forming two H-bonds and inhibited the replication of HIV-1 by depressing the interaction of HIV1 IN to LEDGF/p75 [80] (Fig. 4.15).

A series of imidazole thioacetanilide derivatives was synthesized and screened for their anti-HIV activity. Among all derivatives, compounds **29e** (EC_{50}=0.18 μM), and **29b** (EC_{50}=0.20 μM) presented the most potent inhibition of HIV-1 compared to the reference drugs, nevirapine and delavirdine. Moreover, the SAR study demonstrated that the aryl ring attached to imidazole moiety and the hydrophobicity of the aryl group played a crucial role for binding affinity between active binding site and the inhibitors, and thus modified the biological potency [81] (Scheme 4.10) (Fig. 4.16).

2.7 Miscellaneous

A library of various imidazo[1,2-*a*]pyrrolo[3,2-*c*]pyridines was synthesized and assessed for their anti-BVDV activities in MDBK cells. Furthermore, modification in structure at positions 2, 3, 7, and 8 were performed to enhance the

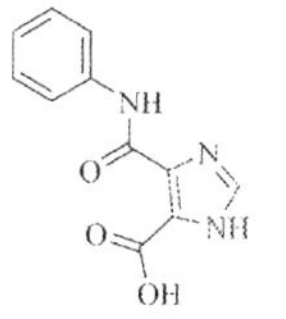

28

FIG. 4.15 Active 4-(phenylcarbamoyl)-1*H*-imidazole-5-carboxylic acid against HIV.

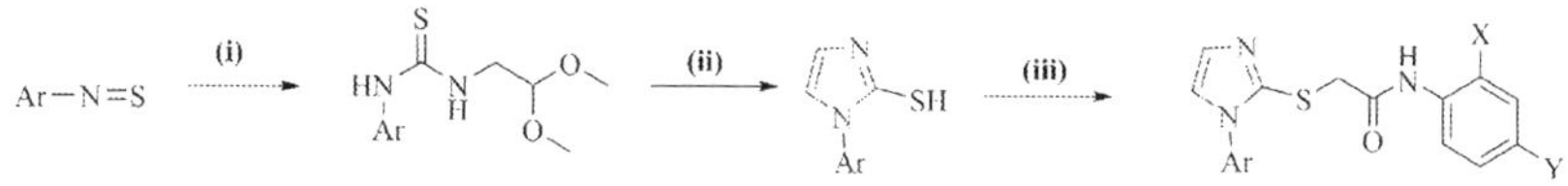

Ar= naphthalen-1-yl, 4-chlorophenyl
p-tolyl, 4-methoxyphenyl

29a-f

Reagents and conditions: **(i)** 2,2-dimethoxyethanamine, EtOH/petroleum ether; **(ii)** 5 M HCl; **(iii)** $ClCH_2CONHPh$, Na_2CO_3 or NaOH, EtOH

SCHEME 4.10 Synthesis of imidazole thioacetanilide derivatives.

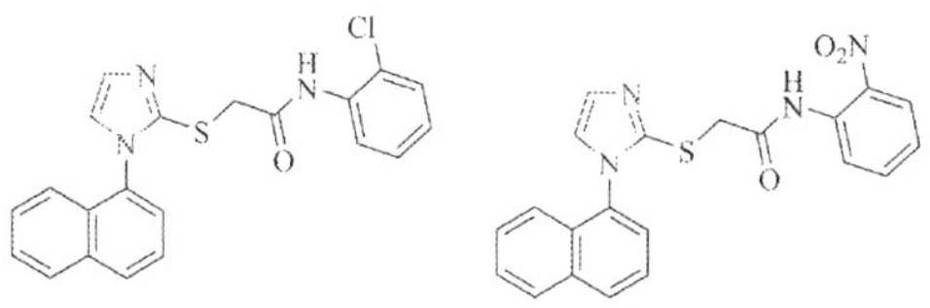

29b **29e**

FIG. 4.16 Potent imidazole thioacetanilide derivatives against HIV.

potency against BVDV. The SAR study concluded that substitution on the pyrrole ring did not affect the activity while modification at position C-3 reduced the antiviral activity. Among all the synthesized compounds, compounds **30a–f** demonstrated potent anti-BVDV activity [82] (Fig. 4.17).

A library of 2-oxoimidazolidines derivatives was synthesized and their antiviral activities were assessed against BK human polyomavirus type 1 (BKPyV) in vitro. The bioassay study identified that derivatives **31b** and **31a** demonstrated moderate antiviral potency against BKPyV with EC_{50} values of 5.4 and 5.5 μM, respectively, that were comparable to those of the standard drug cidofovir. Compound **31b** showed the same selective index and cytotoxicity to cidofovir while compound **31a** has high toxicity and a less selective index than the reference drug [83] (Fig. 4.18).

The antiviral activity of 7-(6-(2-methyl-imidazole))-coumarin (**32**) was examined against spring viremia of carp virus (SVCV) in zebra fish. The data indicated that compound **32** was able to inhibit the half-life in the early stage of viral infection (1–4 days) and reduced the viral titer in fish body by suppressing the SVCV glycoprotein gene expression. Moreover, compound **32** enhanced the expression of interferon genes (IFNγ, IFNφ1, IFNφ2, and RIG-1) in nonviral infected zebra fish and strengthened the immune response. Compound **32** also displayed antioxidant protection on fish by raising the levels of antioxidant-related enzyme activities and gene transcription in SVCV-infected zebra fish [84] (Scheme 4.11).

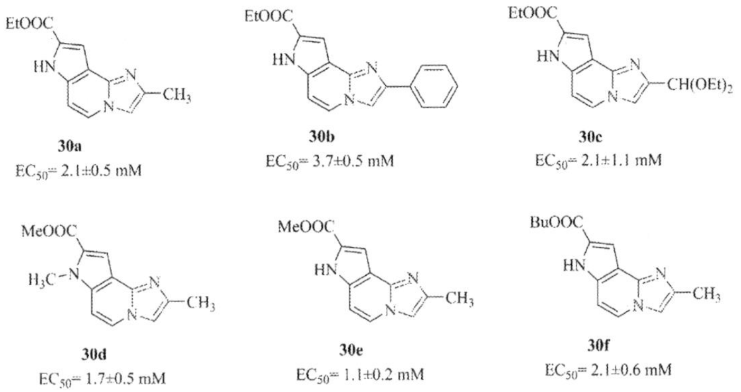

FIG. 4.17 Biologically active imidazo[1,2-*a*]pyrrolo[3,2-*c*]pyridines against BVDV.

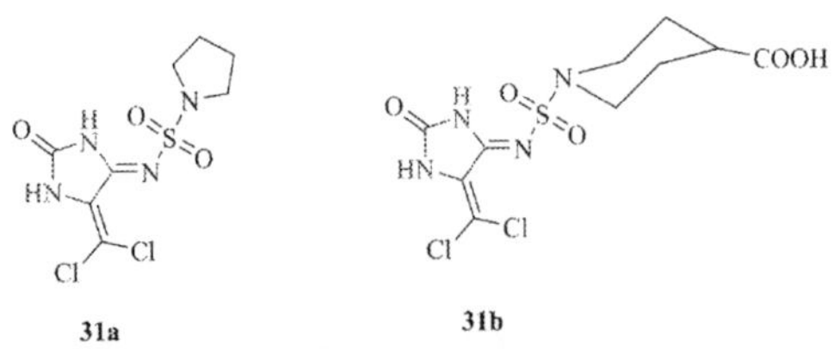

FIG. 4.18 2-oxoimidazolidines derivatives with anti-BK human polyomavirus.

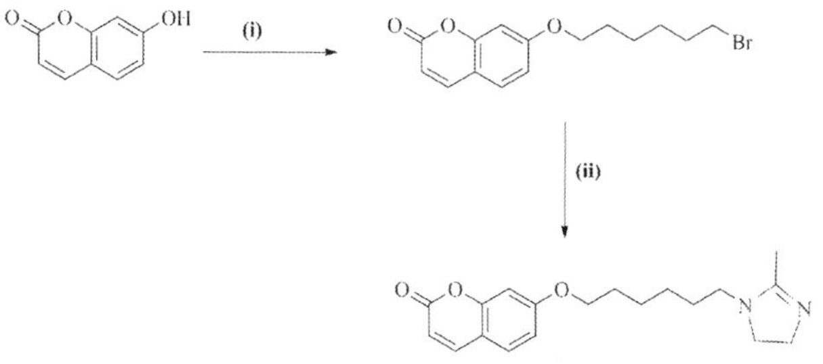

32

Conditions and reagents
(i) 1,6-dibromohexane,K_2CO_3, triethylamine, dry acetone, 60 °C, 20–24 h;
(ii) 2-methylimidazole, K_2CO_3, CH_3CN, r.t.,20–24 h.

SCHEME 4.11 Synthesis of 7-(6-(2-methyl-imidazole))-coumarin derivatives.

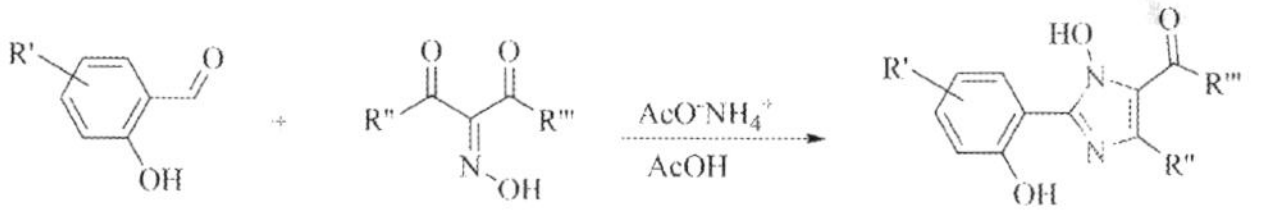

33a-h

	R'	R''	R'''
33a	H	Me	Me
33b	H	Me	OEt
33c	H	$CH_2C(Me)_2$	CH_2
33d	H	Ph	NHPh
33e	5-Cl	Me	Me
33f	4-OCH_2Ph	Me	OEt
33g	5-Cl	Ph	NHPh
33h	5-Br	Ph	NHPh

SCHEME 4.12 Synthesis of a series of 1-hydroxyimidazole derivatives.

A series of 1-hydroxyimidazole derivatives was synthesized by the condensation of oximes with salicylaldehyde derivatives and ammonium acetate in glacial acetic acid and evaluated for their antiviral activity against vaccinia virus in Vero cell culture. The synthesized compounds presented good inhibitory activity and compound **33c** showed the most promising activity ($IC_{50}=1.29\pm0.09$ μg/mL) against the vaccinia virus. Furthermore, the SAR study showed that modification at the 2-hydroxyphenyl moiety of 1-hydroxyimidazoles led to enhanced cytotoxicity while the *N*-phenylcarbamoyl substituent in position 5 caused cytotoxicity and loss of inhibitory activity. The *N*-hydroxy group proved crucial for antiviral activity against the vaccinia virus [85] (Scheme 4.12).

An effective method was introduced for the synthesis of different chalcone derivatives and their antiviral activity was screened against TMV and CMV. The assay study illustrated that various compounds showed potential anti-CMV and anti-TMV activities in vivo. Specifically, compound **34** presented the most promising inactivating activity against TMV (EC_{50} value of 51.65 μg/mL), which was more than the reference drug ribavirin; it also behaved as an excellent protective and curative agent against CMV. Moreover, the molecular docking study was performed and four hydrogen bonds were found between

TMV coat protein and compound **34**, which confirmed the strong binding capacity to TMV-CP. The SAR study demonstrated that the substitution of an electron-releasing group at the 2-position of benzenesulfonamide aromatic cycles with less steric hindrance enhanced the antiviral activity [86] (Fig. 4.19). A series of 2-(substituted phenyl)-1*H*-imidazole and (substituted phenyl)-[2-(substituted phenyl)-imidazol-1-yl]-methanone derivatives was synthesized and evaluated for their antiviral activity against various virus strains such as vaccinia virus (VV), herpes simplex virus-1 (KOS) (HSV-1 KOS), herpes simplex virus-2 (G) (HSV-2G), Coxsackie virus B4 (CV-B4), vesicular stomatitis virus (VSV), respiratory syncytial virus (RSV), reovirus-1 (RV-1), Sindbis virus (SV), parainfluenza-3 virus (PI-3V), and Punta Toro virus (PTV). Among all the compounds, compounds **35a** and **35b** were found to be the most prominent antiviral agents against VV with EC_{50} values of 2 and 4 mg/mL, respectively. Moreover, compound **35b** showed good antiviral activity against HSV-1 KOS (EC_{50} = 59 mg/mL) and HSV-2G (EC_{50} = 50 mg/mL) [87] (Fig. 4.19).

Seventy-six 2-phenylbenzimidazole analogs were synthesized and screened for the cytotoxicity and antiviral activity toward a group of 10 RNA and DNA viruses. The compounds showed good antiviral activity against CVB-2, BVDV, Sb-1, HSV-1, and YFV. Among these compounds, compound **36a** exhibited an immense antiviral profile against VV (EC_{50} = 0.1 μM) and compounds **36b**, **36c**, and **36d** showed promising inhibitory activity against BVDV with EC_{50} values of 1.5, 0.8, and 1.0 μM, respectively [88] (Fig. 4.20).

The repurposing-based design of drugs was performed for the evaluation of antiviral activity of the imidazole molecules at sublethal doses to reduce Newcastle disease virus replication in vivo, in ovo, and in vitro. Chickens treated with the repurposed drug of imidazole developed antiviral type I interferon and exhibited absence of the virus [89]. In addition, the *N*-methylpyrrole–imidazole polyamides exhibited significant antiviral activity against three different genotypes of HPV: HPV16 (in W12 cells), HPV18 (in Ker4–18 cells), and HPV31 (in HPV31 maintaining cells) [90].

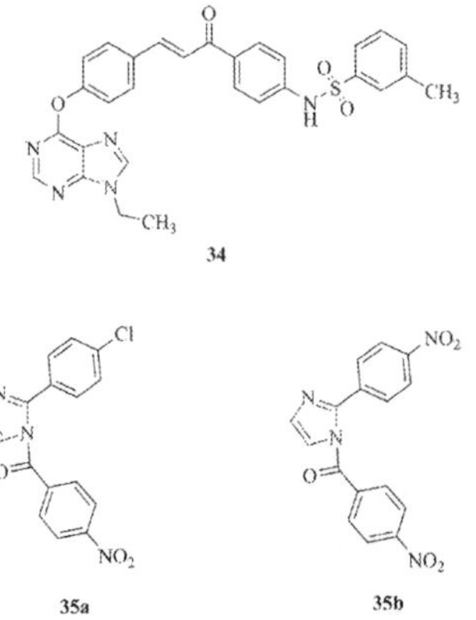

FIG. 4.19 Active antiviral agents containing imidazole moiety.

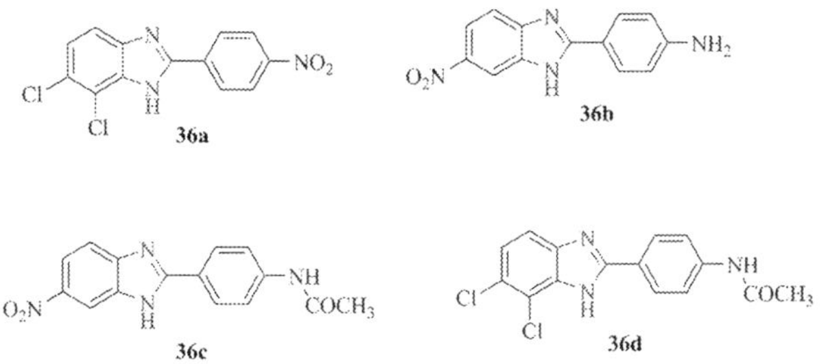

FIG. 4.20 2-Phenylbenzimidazole analogs as biologically potent antiviral compounds.

3. Conclusion

The present era is full of stressful life and poor eating habits and has decreased immunity, making our bodies optimum place for food and shelter for many pathogens, including viruses. Viruses not only disrupt daily life but also are contagious, and their genomic structure is constantly mutating. Many antiviral drugs are present to curb viral infections, but there are problems associated with available antiviral drugs including limited efficiency, toxicity, low bioavailability, and complex synthesis. Therefore, discovery of a new generation of active antiviral drugs with better drug activity and a good pharmacological profile seems to be challenging in pharmaceutical sciences and antiviral research.

Imidazoles have emerged as appealing scaffolds with exceptional structural features and noteworthy biological properties. The present chapter is focused on providing insights for the synthesis of novel imidazole-based antiviral agents. For some decades, numerous studies have been dedicated to the advancement of antiviral imidazoles. Several natural, semisynthetic, and synthetic imidazole derivatives have been described as potential antiviral agents against a wide range of viruses. This chapter systematically describes the mode of action of imidazoles on various viruses including novel coronavirus, Zika virus, HIV, hepatitis, dengue, etc. Subsections of this chapter also discussed synthesis, SAR, molecular docking, and biological profile of imidazoles and their derivatives, opening a new platform for easier understanding of readers, and motivating researchers to create new imidazole drug templates with ease. The chapter also provides abundant knowledge, ample information, and prospects about imidazoles with updated literature. Regardless of extensive work and promising outcomes on imidazole moiety as significant antiviral drugs, a few challenges and opportunities remain for researchers that need to be discussed:

- evaluation of the antiviral activity of numerous imidazole-based chemical derivatives and exploration of novel methodologies, biomarkers for determining the most relevant molecular targets, and appropriate mechanism of action of active analogs;

- promotion of more rational design of antivirals by determining X-ray crystallography of target-ligand complexes; and
- more efficient and effective methods such as computer-aided drug design, structure-based drug design, fragment-based drug design, etc. should be used for designing new antiviral molecules.

Acknowledgments

The authors are grateful to the Department of Chemistry, Mohan Lal Sukhadia University, Udaipur (Raj.), India, for providing necessary library facilities for carrying out the work. P.T. [file no. 09/172(0099)2019-EMR-I] and N.S. are deeply grateful to the Council for Scientific and Industrial Research (CSIR) (file no. 09/172(0088)2018-EMR-I), New Delhi and A.S. is thankful to UGC MANF (file no. 201819-MANF-2018-19-RAJ-9197) for providing a Senior Research Fellowship as financial support.

Conflict of interest

The authors declare no conflict of interest, financial or otherwise.

Funding source

This work was supported by CSIR [file no. 09/172(0099)2019-EMR-I] and [file no. 09/172(0088)2018-EMR-I] and UGC MANF (file no. 201819-MANF-2018-19-RAJ-9197).

References

[1] Ozili PK, Arun T. Spillover of COVID-19: impact on the global economy [Available at SSRN 3562570]; 2020.

[2] Breitbart M, Rohwer F. Here a virus, there a virus, everywhere the same virus? Trends Microbiol 2005;13(6):278–84.

[3] Colpitts CC, Verrier ER, Baumert TF. Targeting viral entry for treatment of hepatitis B and C virus infections. ACS Infect Dis 2015;1(9):420–7.

[4] Rivera A, Messaoudi I. Pathophysiology of Ebola virus infection: current challenges and future hopes. ACS Infect Dis 2015;1(5):186–97.

[5] Efrida E, Nasrul E, Parwati I, Jamsari J. New drug resistance mutations of reverse transcriptase Human immunodeficiency virus type-1 gene in first-line antiretroviral-infected patients in West Sumatra, Indonesia. Russ Open Med J 2018;7(2).

[6] Bangham J, Obbard DJ, Kim KW, Haddrill PR, Jiggins FM. The age and evolution of an antiviral resistance mutation in Drosophila melanogaster. Proc Biol Sci 2007;274(1621):2027–34.

[7] Zorofchian Moghadamtousi S, Abdul Kadir H, Hassandarvish P, Tajik H, Abubakar S, Zandi K. A review on antibacterial, antiviral, and antifungal activity of curcumin. Biomed Res Int 2014;2014.

[8] Kazmierski WM. Antiviral drugs: from basic discovery through clinical trials. John Wiley & Sons; 2011.

[9] Reyes-Arellano A, Gómez-García O, Torres-Jaramillo J. Synthesis of azolines and imidazoles and their use in drug design. Med Chem (Los Angeles) 2016;6:561–70.

[10] Gaba M, Singh S, Mohan C. Benzimidazole: an emerging scaffold for analgesic and anti-inflammatory agents. Eur J Med Chem 2014;76:494–505.
[11] DeSimone RW, Currie KS, Mitchell SA, Darrow JW, Pippin DA. Privileged structures: applications in drug discovery. Comb Chem High Throughput Screen 2004;7(5):473–93.
[12] Fei F, Zhou Z. New substituted benzimidazole derivatives: a patent review (2010–2012). Expert Opin Ther Pat 2013;23(9):1157–79.
[13] Ingle RG, Magar DD. Heterocyclic chemistry of benzimidazoles and potential activities of derivatives. Int J Drug Res Technol 2011;1:26–32.
[14] Narasimhan B, Sharma D, Kumar P. Biological importance of imidazole nucleus in the new millennium. Med Chem Res 2011;20(8):1119–40.
[15] Yadav G, Ganguly S. Structure activity relationship (SAR) study of benzimidazole scaffold for different biological activities: a mini-review. Eur J Med Chem 2015;97:419–43.
[16] Gaba M, Gaba P, Uppal D, Dhingra N, Bahia MS, Silakari O, Mohan C. Benzimidazole derivatives: search for GI-friendly anti-inflammatory analgesic agents. Acta Pharm Sin B 2015;5(4):337–42.
[17] Wright JB. The chemistry of the benzimidazoles. Chem Rev 1951;48(3):397–541.
[18] Bhatnagar A, Sharma PK, Kumar N. A review on "Imidazoles": their chemistry and pharmacological potentials. Int J PharmTech Res 2011;3(1):268–82.
[19] Steinman RA, Brufsky AM, Oesterreich S. Zoledronic acid effectiveness against breast cancer metastases—a role for estrogen in the microenvironment? Breast Cancer Res 2012;14(5):1–9.
[20] Ashley ES. Pharmacology of azole antifungal agents. Antifungal Ther 2010;199–218.
[21] Mishra R, Ganguly S. Imidazole as an anti-epileptic: an overview. Med Chem Res 2012;21(12):3929–39.
[22] Burnier M, Wuerzner G. Pharmacokinetic evaluation of losartan. Expert Opin Drug Metab Toxicol 2011;7(5):643–9.
[23] Rasmussen SA, Jamieson DJ, Honein MA, Petersen LR. Zika virus and birth defects—reviewing the evidence for causality. N Engl J Med 2016;374(20):1981–7.
[24] Lei J, Hansen G, Nitsche C, Klein CD, Zhang L, Hilgenfeld R. Crystal structure of Zika virus NS2B-NS3 protease in complex with a boronate inhibitor. Science 2016;353(6298):503–5.
[25] Desai SK, Hartman SD, Jayarajan S, Liu S, Gallicano GI. Zika virus (ZIKV): a review of proposed mechanisms of transmission and associated congenital abnormalities. Am J Stem Cells 2017;6(2):13.
[26] Rausch K, Hackett BA, Weinbren NL, Reeder SM, Sadovsky Y, Hunter CA, Schultz DC, Coyne CB, Cherry S. Screening bioactives reveals nanchangmycin as a broad spectrum antiviral active against Zika virus. Cell Rep 2017;18(3):804–15.
[27] Hue BT, Nguyen PH, De TQ, Van Hieu M, Jo E, Van Tuan N, Thoa TT, Anh LD, Son NH, La Duc TD, Dupont-Rouzeyrol M. Benzimidazole derivatives as novel zika virus inhibitors. ChemMedChem 2020;15(15):1453–63.
[28] Mitry M, El-Araby A, Neyts J, Kaptein S, Serya RA, Samir N. Molecular dynamic study and synthesis of 1H-benzo [d] imidazole-5-carboxamide derivatives as inhibitors for yellow fever and zika virus replication. Arch Pharm Sci ASU 2020;4(2):145–80.
[29] WHO Influenza update–310, 5 March 2018. http://www.who.int/.
[30] Halford B. Outsmarting influenza. C&EN 2018;96(11):42–7.
[31] Das K. Antivirals targeting influenza a virus. J Med Chem 2012;55(14):6263–77.
[32] Bright RA, Shay DK, Shu B, Cox NJ, Klimov AI. Adamantane resistance among influenza a viruses isolated early during the 2005-2006 influenza season in the United States. JAMA 2006;295(8):891–4.

[33] Wan XF, Carrel M, Long LP, Alker AP, Emch M. Perspective on emergence and re-emergence of amantadine resistant influenza A viruses in domestic animals in China. Infect Genet Evol 2013;20:298–303.
[34] Bashashati M, Marandi MV, Sabouri F. Genetic diversity of early (1998) and recent (2010) avian influenza H9N2 virus strains isolated from poultry in Iran. Arch Virol 2013;158(10):2089–100.
[35] Govorkova EA, Baranovich T, Seiler P, Armstrong J, Burnham A, Guan Y, Peiris M, Webby RJ, Webster RG. Antiviral resistance among highly pathogenic influenza A (H5N1) viruses isolated worldwide in 2002–2012 shows need for continued monitoring. Antiviral Res 2013;98(2):297–304.
[36] Salter A, Laoi BN, Crowley B. Emergence and phylogenetic analysis of amantadine-resistant influenza a subtype H3N2 viruses in Dublin, Ireland, over six seasons from 2003/2004 to 2008/2009. Intervirology 2011;54(6):305–15.
[37] WHO. Safety of Pandemic a (H1N1) Influenza Vaccines. vol. 85; 2010. p. 285–92.
[38] Mao C, Wu XY, Fu XH, Di MY, Yu YY, Yuan JQ, Yang ZY, Tang JL. An internet-based epidemiological investigation of the outbreak of H7N9 Avian influenza A in China since early 2013. J Med Internet Res 2014;16(9), e221.
[39] Galochkina AV, Bollikanda RK, Zarubaev VV, Tentler DG, Lavrenteva IN, Slita AV, Chirra N, Kantevari S. Synthesis of novel derivatives of 7, 8-dihydro-6H-imidazo [2, 1-b][1,3] benzothiazol-5-one and their virus-inhibiting activity against influenza A virus. Arch Pharm (Weinheim) 2019;352(2):1800225.
[40] Dong J, Chen S, Li R, Cui W, Jiang H, Ling Y, et al. Imidazole-based pinanamine derivatives: discovery of dual inhibitors of the wild-type and drug-resistant mutant of the influenza A virus. Eur J Med Chem 2016;108:605–15.
[41] Balgi AD, Wang J, Cheng DY, Ma C, Pfeifer TA, Shimizu Y, Anderson HJ, Pinto LH, Lamb RA, DeGrado WF, Roberge M. Inhibitors of the influenza A virus M2 proton channel discovered using a high-throughput yeast growth restoration assay. PLoS One 2013;8(2), e55271.
[42] Gong KK, Tang XL, Liu YS, Li PL, Li GQ. Imidazole alkaloids from the South China Sea sponge Pericharax heteroraphis and their cytotoxic and antiviral activities. Molecules 2016;21(2):150.
[43] Tosepu R, Gunawan J, Effendy DS, Lestari H, Bahar H, Asfian P. Correlation between weather and Covid-19 pandemic in Jakarta, Indonesia. Sci Total Environ 2020;725:138436.
[44] Wu C, Liu Y, Yang Y, Zhang P, Zhong W, Wang Y, Wang Q, Xu Y, Li M, Li X, Zheng M. Analysis of therapeutic targets for SARS-CoV-2 and discovery of potential drugs by computational methods. Acta Pharm Sin B 2020;10(5):766–88.
[45] Singh AK, Singh A, Shaikh A, Singh R, Misra A. Chloroquine and hydroxychloroquine in the treatment of COVID-19 with or without diabetes: a systematic search and a narrative review with a special reference to India and other developing countries. Diabetes Metab Syndr Clin Res Rev 2020;14(3):241–6.
[46] Kang D, Choi H, Kim JH, Choi J. Spatial epidemic dynamics of the COVID-19 outbreak in China. Int J Infect Dis 2020;94:96–102.
[47] Zhai P, Ding Y, Wu X, Long J, Zhong Y, Li Y. The epidemiology, diagnosis and treatment of COVID-19. Int J Antimicrob Agents 2020;55(5):105955.
[48] Elfiky AA. Anti-HCV, nucleotide inhibitors, repurposing against COVID-19. Life Sci 2020;248:117477.
[49] Ramajayam R, Tan KP, Liang PH. Recent development of 3C and 3CL protease inhibitors for anti-coronavirus and anti-picornavirus drug discovery. Biochem Soc Trans 2011;39(5):1371–5.

[50] Ren Z, Yan L, Zhang N, Guo Y, Yang C, Lou Z, Rao Z. The newly emerged SARS-like coronavirus HCoV-EMC also has an" Achilles' heel": current effective inhibitor targeting a 3C-like protease. Protein Cell 2013;4(4):248.

[51] Belhassan A, En-Nahli F, Zaki H, Lakhlifi T, Bouachrine M. Assessment of effective imidazole derivatives against SARS-CoV-2 main protease through computational approach. Life Sci 2020;262:118469.

[52] Chhetri A, Chettri S, Rai P, Sinha B, Brahman D. Exploration of inhibitory action of azo imidazole derivatives against COVID-19 main protease (Mpro): a computational study. J Mol Struct 2021;1224:129178.

[53] Bhatt S, Gething PW, Brady OJ, Messina JP, Farlow AW, Moyes CL, Drake JM, Brownstein JS, Hoen AG, Sankoh O, Myers MF. The global distribution and burden of dengue. Nature 2013;496(7446):504–7.

[54] Normile D. Surprising new dengue virus throws a spanner in disease control efforts. Science 2013;342(6157):415.

[55] Rawlinson SM, Pryor MJ, Wright PJ, Jans DA. Dengue virus RNA polymerase NS5: a potential therapeutic target? Curr Drug Targets 2006;7(12):1623–38.

[56] Saudi M, Zmurko J, Kaptein S, Rozenski J, Neyts J, Van Aerschot A. Synthesis and evaluation of imidazole-4, 5-and pyrazine-2, 3-dicarboxamides targeting dengue and yellow fever virus. Eur J Med Chem 2014;87:529–39.

[57] Okano Y, Saito-Tarashima N, Kurosawa M, Iwabu A, Ota M, Watanabe T, Kato F, Hishiki T, Fujimuro M, Minakawa N. Synthesis and biological evaluation of novel imidazole nucleosides as potential anti-dengue virus agents. Bioorg Med Chem 2019;27(11):2181–6.

[58] Sucipto TH, Martak F. Inhibition of dengue virus serotype 2 in Vero cells with [Cu (2,4,5-triphenyl-1H-imidazole)$_2$(H_2O)$_2$]. Cl_2. Infect Dis Rep 2020;12(S1):93–7.

[59] Alter MJ, Kruszon-Moran D, Nainan OV, McQuillan GM, Gao F, Moyer LA, Kaslow RA, Margolis HS. The prevalence of hepatitis C virus infection in the United States, 1988 through 1994. N Engl J Med 1999;341(8):556–62.

[60] Di Bisceglie AM. Natural history of hepatitis C: its impact on clinical management. Hepatology 2000;31(4):1014–8.

[61] Murphy DG, Sablon E, Chamberland J, Fournier E, Dandavino R, Tremblay CL. Hepatitis C virus genotype 7, a new genotype originating from Central Africa. J Clin Microbiol 2015;53(3):967–72.

[62] Oze T, Hiramatsu N, Mita E, Akuta N, Sakamoto N, Nagano H, Itoh Y, Kaneko S, Izumi N, Nomura H, Hayashi N. A multicenter survey of re-treatment with pegylated interferon plus ribavirin combination therapy for patients with chronic hepatitis C in Japan. Hepatol Res 2013;43(1):35–43.

[63] Kwong AD, Kauffman RS, Hurter P, Mueller P. Discovery and development of telaprevir: an NS3-4A protease inhibitor for treating genotype 1 chronic hepatitis C virus. Nat Biotechnol 2011;29(11):993–1003.

[64] Thibault PA, Wilson JA. Targeting miRNAs to treat hepatitis C virus infections and liver pathology: inhibiting the virus and altering the host. Pharmacol Res 2013;75:48–59.

[65] Hulskotte EG, Feng HP, Xuan F, Gupta S, van Zutven MG, O'Mara E, Wagner JA, Butterton JR. Pharmacokinetic evaluation of the interaction between hepatitis C virus protease inhibitor boceprevir and 3-hydroxy-3-methylglutaryl coenzyme A reductase inhibitors atorvastatin and pravastatin. Antimicrob Agents Chemother 2013;57(6):2582–8.

[66] Gerasi M, Frakolaki E, Papadakis G, Chalari A, Lougiakis N, Marakos P, Pouli N, Vassilaki N. Design, synthesis and anti-HBV activity evaluation of new substituted imidazo [4, 5-b] pyridines. Bioorg Chem 2020;98:103580.

[67] Patil VM, Gupta SP, Samanta S, Masand N. Virtual screening of imidazole analogs as potential hepatitis C virus NS5B polymerase inhibitors. Chem Pap 2013;67(2):236–44.

[68] Tsay SC, Lin SY, Huang WC, Hsu MH, Hwang KC, Lin CC, Horng JC, Chen I, Hwu JR, Shieh FK, Leyssen P. Synthesis and structure-activity relationships of imidazole-coumarin conjugates against hepatitis C virus. Molecules 2016;21(2):228.

[69] Wittine K, Babić MS, Makuc D, Plavec J, Pavelić SK, Sedić M, Pavelić K, Leyssen P, Neyts J, Balzarini J, Mintas M. Novel 1, 2, 4-triazole and imidazole derivatives of L-ascorbic and imino-ascorbic acid: synthesis, anti-HCV and antitumor activity evaluations. Bioorg Med Chem 2012;20(11):3675–85.

[70] Zong X, Cai J, Chen J, Wang P, Zhou G, Chen B, Li W, Ji M. Design and synthesis of imidazole N–H substituted amide prodrugs as inhibitors of hepatitis C virus replication. Bioorg Med Chem Lett 2015;25(16):3147–50.

[71] Foster TL, Thompson GS, Kalverda AP, Kankanala J, Bentham M, Wetherill LF, Thompson J, Barker AM, Clarke D, Noerenberg M, Pearson AR. Structure-guided design affirms inhibitors of hepatitis C virus p7 as a viable class of antivirals targeting virion release. Hepatology 2014;59(2):408–22.

[72] Le Douce V, Janossy A, Hallay H, Ali S, Riclet R, Rohr O, Schwartz C. Achieving a cure for HIV infection: do we have reasons to be optimistic? J Antimicrob Chemother 2012;67(5):1063–74.

[73] Vrouenraets SM, Wit FW, Tongeren JV, Lange JM. Efavirenz: a review. Expert Opin Pharmacother 2007;8(6):851–71.

[74] Delelis O, Thierry S, Subra F, Simon F, Malet I, Alloui C, Sayon S, Calvez V, Deprez E, Marcelin AG, Tchertanov L. Impact of Y143 HIV-1 integrase mutations on resistance to raltegravir in vitro and in vivo. Antimicrob Agents Chemother 2010;54(1):491–501.

[75] Anstett K, Brenner B, Mesplede T, Wainberg MA. HIV drug resistance against strand transfer integrase inhibitors. Retrovirology 2017;14(1):1–6.

[76] Clavel F, Hance AJ. HIV drug resistance. N Engl J Med 2004;350(10):1023–35.

[77] Rashamuse TJ, Harrison AT, Mosebi S, van Vuuren S, Coyanis EM, Bode ML. Design, synthesis and biological evaluation of imidazole and oxazole fragments as HIV-1 integrase-LEDGF/p75 disruptors and inhibitors of microbial pathogens. Bioorg Med Chem 2020;28(1):115210.

[78] Chong P, Sebahar P, Youngman M, Garrido D, Zhang H, Stewart EL, Nolte RT, Wang L, Ferris RG, Edelstein M, Weaver K. Rational design of potent non-nucleoside inhibitors of HIV-1 reverse transcriptase. J Med Chem 2012;55(23):10601–9.

[79] Hirade H, Haruyama T, Kobayashi N, de Voogd NJ, Tanaka J. A new imidazole from the sponge Dercitus (Halinastra) japonensis. Nat Prod Commun 2017;12(1):19–20.

[80] Serrao E, Xu ZL, Debnath B, Christ F, Debyser Z, Long YQ, Neamati N. Discovery of a novel 5-carbonyl-1H-imidazole-4-carboxamide class of inhibitors of the HIV-1 integrase–LEDGF/p75 interaction. Bioorg Med Chem 2013;21(19):5963–72.

[81] Zhan P, Liu X, Zhu J, Fang Z, Li Z, Pannecouque C, De Clercq E. Synthesis and biological evaluation of imidazole thioacetanilides as novel non-nucleoside HIV-1 reverse transcriptase inhibitors. Bioorg Med Chem 2009;17(16):5775–81.

[82] Chezal JM, Paeshuyse J, Gaumet V, Canitrot D, Maisonial A, Lartigue C, Gueiffier A, Moreau E, Teulade JC, Chavignon O, Neyts J. Synthesis and antiviral activity of an imidazo [1, 2-a] pyrrolo [2, 3-c] pyridine series against the bovine viral diarrhea virus. Eur J Med Chem 2010;45(5):2044–7.

[83] Kornii Y, Chumachenko S, Shablykin O, Prichard MN, James SH, Hartline C, Zhirnov V, Brovarets V. New 2-Oxoimidazolidine derivatives: design, synthesis and evaluation of anti-BK virus activities in vitro. Chem Biodivers 2019;16(10), e1900391.

[84] Liu L, Hu Y, Lu J, Wang G. An imidazole coumarin derivative enhances the antiviral response to spring viremia of carp virus infection in zebrafish. Virus Res 2019;263:112–8.
[85] Nikitina PA, Bormotov NI, Shishkina LN, Tikhonov AY, Perevalov VP. Synthesis and antiviral activity of 1-hydroxy-2-(2-hydroxyphenyl) imidazoles against vaccinia virus. Russ Chem Bull 2019;68(3):634–7.
[86] Zhou D, Xie D, He F, Song B, Hu D. Antiviral properties and interaction of novel chalcone derivatives containing a purine and benzenesulfonamide moiety. Bioorg Med Chem Lett 2018;28(11):2091–7.
[87] Sharma D, Narasimhan B, Kumar P, Judge V, Narang R, De Clercq E, Balzarini J. Synthesis, antimicrobial and antiviral evaluation of substituted imidazole derivatives. Eur J Med Chem 2009;44(6):2347–53.
[88] Tonelli M, Simone M, Tasso B, Novelli F, Boido V, Sparatore F, Paglietti G, Pricl S, Giliberti G, Blois S, Ibba C. Antiviral activity of benzimidazole derivatives. II. Antiviral activity of 2-phenylbenzimidazole derivatives. Bioorg Med Chem 2010;18(8):2937–53.
[89] Das M, Baro S, Kumar S. Evaluation of imidazole and its derivative against Newcastle disease virus infection in chicken: a drug repurposing approach. Virus Res 2019;260:114–22.
[90] Edwards TG, Koeller KJ, Slomczynska U, Fok K, Helmus M, Bashkin JK, Fisher C. HPV episome levels are potently decreased by pyrrole–imidazole polyamides. Antiviral Res 2011;91(2):177–86.

Chapter 5

Imidazole heterocycles: Therapeutically potent lead compounds as antimicrobials

Nusrat Sahiba, Ayushi Sethiya, and Shikha Agarwal
Synthetic Organic Chemistry Laboratory, Department of Chemistry, Mohanlal Sukhadia University, Udaipur, Rajasthan, India

Abbreviations

ADME properties	absorption distribution metabolism and excretion (pharmcokintetic properties)
Am B	amphotericin B
Cfn	ciprofloxacin
ERG	electron releasing group
EWG	electron withdrawing group
FCZ	fluconazole
IC50 value	half maximal inhibitory concentration
ITZ	itraconazole
MBC	minimum bactericidal concentration
MIC	minimum inhibitory concentration
MW	microwave
QSAR	quantitative SAR
RMSE	root mean square error
ROS	reactive oxygen species
SAR	structure activity relationship
TB bacteria	mycobacterium tuberculosis
ZOI	zone of inhibition

1. Introduction

The pioneering discovery of penicillin in 1940s started a new age of antimicrobial drug development which is valuable for human health and has greatly benefited human survival [1]. Since then, a large number of natural, synthetic, and semi-synthetic antimicrobial drugs and chemotherapeutics have been introduced for the treatment of various life-threatening diseases [2–5]. Alkaloids, coumarins,

Imidazole-Based Drug Discovery. https://doi.org/10.1016/B978-0-323-85479-5.00006-X
Copyright © 2022 Elsevier Inc. All rights reserved.

flavonoids, lipoids, macrolides, macrolactams, quinones, and terpenoids are classes of some naturally available antibiotics [6–9]. The majority of antimicrobial drugs belong to arsphenamines, β-lactams, sulfonamides, polypeptides, aminoglycosides, tetracyclines, amphenicols, lipopeptides, macrolides, oxazolidinones, glycopeptides, streptogramins, ansamycins, quinolones, triazoles, and lincosamides classes, which have changed the medical background for future innovations [10–16]. Antimicrobial agents destroy and inhibit the growth of microorganisms and are categorized based on their active target sites and their tasks. Most of the antibiotics work via inhibiting the synthesis of nucleic acids, protein, sterol, cell wall, and cytoplasmic membrane, and reducing the energy metabolism of microbes [17–19]. However, antimicrobial solutions cannot be achieved once and for all, as more and more antimicrobial resistance strains have emerged over time. Even though antimicrobial resistance is a natural phenomenon due to modification in genetic, metabolic level, and rapid progression in environmental conditions. Still injudicious use of antibiotics by physicians is the main cause of antimicrobial-resistance. Illiteracy, low-economics, overpopulation, poor sanitation, and hygienic condition are other considering factors [20–23]. Thus, there is significant concern among researchers and medical professionals regarding the design and generation of novel hybrid antimicrobial drug candidates.

N-containing heterocyclic compounds are widely dispersed in nature and are invaluable sources of various therapeutic agents [24–29]. Imidazole has two nitrogen atoms, planar structure, and aromaticity with high polarity and water solubility, which offers stereo-complexity and diverse functionality in a five-membered core structure. In 1872, Hobrecker synthesized 2,5 and 2,6-dimethylbenzimidazole, but he never imagined that these compounds would form such a prime structure [30, 31]. After a long time, in 1950, 5,6-dimethyl-1-(a-D-ribofuranosyl) benzimidazole was found as a structural unit of vitamin B12 and has received much attention from medicinal chemists [32]. This electronically rich heterocycle displays an amphoteric nature and thereby works as an acid as well as a base due to resonance interaction and also forms various weak interactions with therapeutic targets [33]. This ubiquitous behavior of imidazole makes it indispensable in the area of antimicrobial drug designing with broad pharmacological activity [34–39]. This pervasive pharmacophore is present in several natural compounds, viz. alkaloids, nucleic acids, histamine, biotin, histidine, and many more. A large number of antibiotics possess imidazole core structure viz. cimetidine, etomidate, ketoconazole, metronidazole, ornidazole, azomycin, oxiconazole, and clonidine and so on (Fig. 5.1). The aforementioned properties make this advantageous scaffold an epicenter of attraction for future researchers to design and discover novel imidazole-based antimicrobial drug-candidates. Thus, a wide range of antimicrobial imidazole derivatives has been synthesized in recent years along and their various modes of action studied [40–44].

In this chapter, the state-of-the-art innovation with a special emphasis on the antimicrobial features of imidazole pharmacophore has been addressed. An endeavor

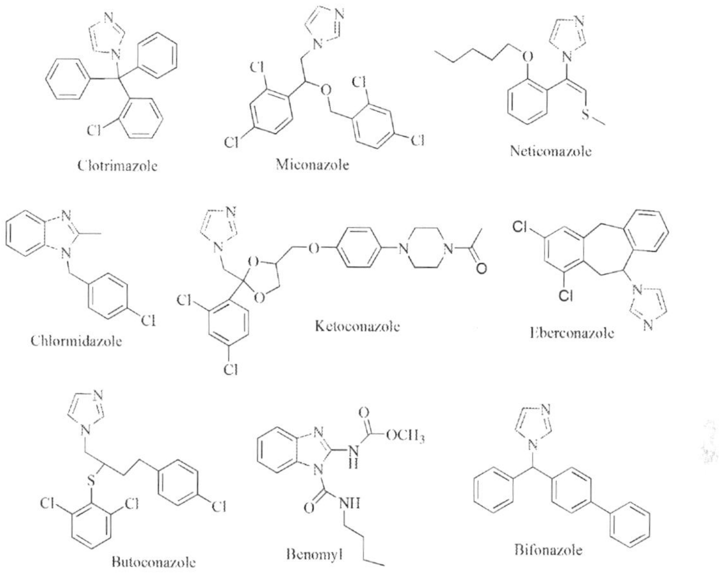

FIG. 5.1 Some clinically available imidazole-based antimicrobial drugs.

has been made to outline the superior and equally potent therapeutic effects including resistance strains against standard antibiotics with careful observation of structure–activity relationships with intriguing profiles. Finally, challenges, future viewpoints, and critical assessment have been presented. It is envisioned that this systematic study will help open new avenues in the quest for the rational design of safer, economical, and more potent antimicrobial drugs in future.

2. Imidazole-based antimicrobial compounds

For ease of understanding, antimicrobial active imidazole derivatives are classified here on the basis of their structural modifications (Fig. 5.2):

1. imidazole ring generation;
2. imidazole linked moieties;
3. imidazole metal complexes;
4. imidazolium ionic liquids and salts; and
5. imidazole-based polymers.

2.1 Based on imidazole ring generation

2.1.1 Imidazole

2.1.1.1 From natural product

Rokon Ul Karim and coworkers detected and isolated three novel imidazole compounds from the marine Gammaproteobacterium Microbulbifer

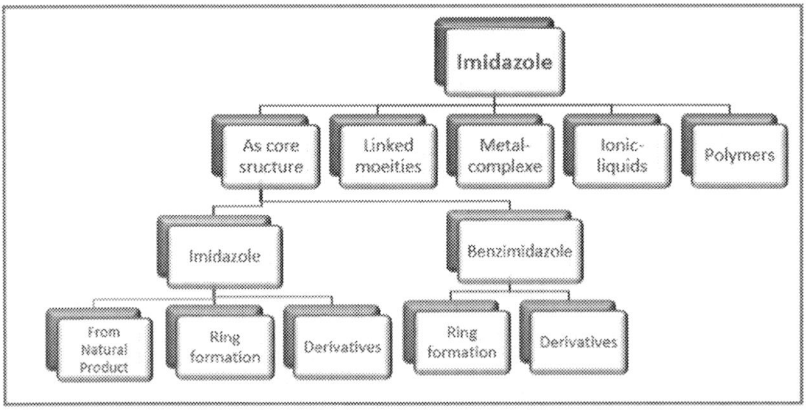

FIG. 5.2 Classification of biologically active imidazole derivatives as potent antimicrobials.

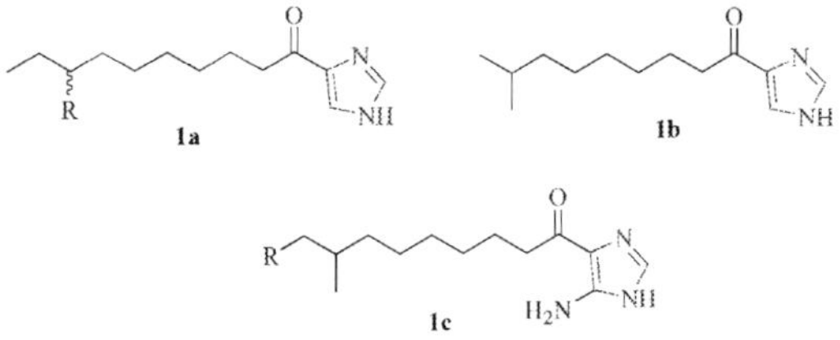

FIG. 5.3 Structure of biologically active bulbimidazoles.

species (sp.) and studied their antimicrobial efficiency against two gram + ve (*K. rhizophila* and *S. aureus*), three gram − ve (*C. albicans*, *G. cingulate*, and *T. rubrum*), and three fungal strains (*C. albicans*, *G. cingulate*, and *T. rubrum*) via a liquid microculture method [45]. The antimicrobial study revealed that these molecules showed good pathogen inhibition against most of the strains (0.78–12.5 μg/mL) and compounds **1a** and **1b** demonstrated relatively good activity compared to **1c** (Fig. 5.3).

2.1.1.2 Ring formation

A range of novel imidazole derivatives (**2a–2d**) was synthesized at ambient temperature from the reaction of substituted aldehyde, 1,2-diketone, and NH_4OAc using molecular I_2 catalyst via a grindstone technique (Scheme 5.1) and their antimicrobial activity was studied against *S. aureus*, *E. coli*, *K. pneumoniae*, *A. niger*, *A. flavus*, and *Candida*-6 species [46]. All derivatives displayed good antimicrobial activity at 25–100 μg/mL against standard antibiotics, amphotericin-B, and ciprofloxacin (Table 5.1).

A range of imidazo-[1,2-*a*]pyridine derivatives was synthesized and their antimicrobial efficiency was evaluated for bacterial species (*S. aureus*, *S. pyogenes*, *E. coli*, *P. aeruginosa* and fungal strains (*C. albicans*, *A. niger*, *A. clavatus*

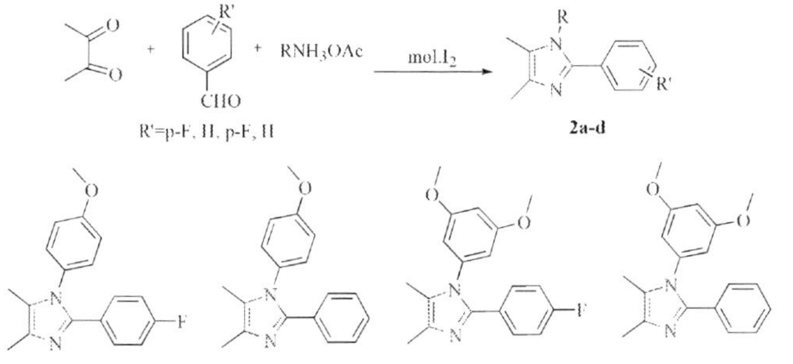

SCHEME 5.1 Synthesis of 2-aryl imidazole derivatives.

TABLE 5.1 In vitro study of 2-aryl imidazole derivatives MIC value (μg/mL).

Compounds	*S. aureus*	*E. coli*	*K. pneumoniae*	*Candida-6*	*A. niger*	*A. flavus*
2a	50	50	100	50	50	50
2b	50	25	100	25	100	25
2c	50	25	50	25	25	25
2d	25	50	100	25	25	25
Ciprofloxacin	25	50	100	–	–	–
Amphotericin-B	–	–	–	12.5	50	25

via a serial broth dilution method [47]. The desired compounds were prepared in multistep. Initially, the reaction of 4-methylpyridin-2-amine and 2-bromo-1-(aryl)ethanone derivatives in EtOH yielded imidazo-pyridine compounds, which were further derivatized using *N,N*-dimethyl amine and paraformaldehyde to furnish final products in good yields with high efficiency (Scheme 5.2). All the synthesized compounds displayed high inhibition efficiency, and among them, compounds **3f**, **3h**, **3j**, **3k**, and **3l** were found to be highly potent against

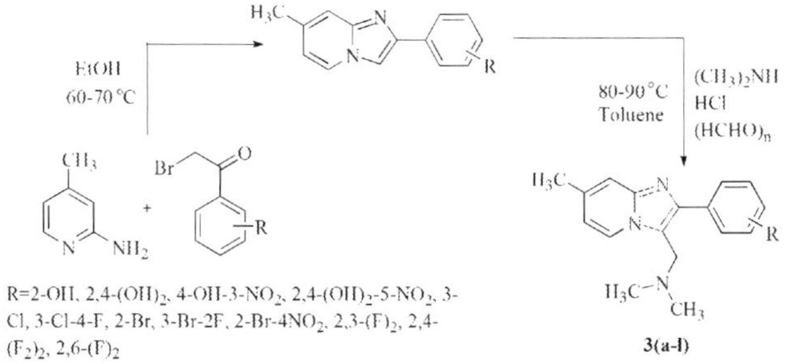

SCHEME 5.2 Synthesis of imidazo-[1,2-*a*]pyridine derivatives.

all pathogens with an MIC value range of 25–500 μg/mL, while others showed selective antimicrobial activity against particular strains in comparison with ampicillin and griseofulvin. The SAR study demonstrated that the electronic nature of substituents affected the antimicrobial potency, i.e., EWG (such as chloro and fluoro) and ERG (such as hydroxy) displayed more and less inhibition activity, respectively.

A mild and efficient synthesis of 1,4-diaryl-2-mercaptoimidazoles **4a–h** was demonstrated and their antimicrobial activity was studied against *S. aureus*, *E. coli*, *K. pneumoniae*, *P. aeruginosa*, *P. mirabilis*, *C. albicans*, and *A. niger* using an agar well diffusion process (Scheme 5.3) [48]. The desired products were prepared in high yields from the reaction of phenacylbromide with substituted anilines and were further treated with KSCN and *p*-TSA. Most of the synthesized compounds displayed good efficiency against ceftriaxone and fluconazole (FCZ). Most of the compounds presented higher potency against *P. aeruginosa* and *C. albicans* with > 2 mm ZOI compared to other strains, while **4h** (*p*-Br) exhibited broad-spectrum microbe inhibition. SAR results clarified that the imidazole center and -SH group are key components for antimicrobial activity, which was amplified in the presence of halogen and o-methyl substituents.

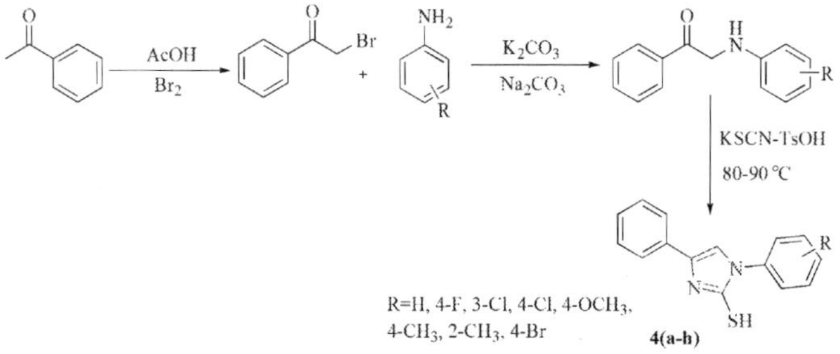

SCHEME 5.3 Synthesis of 1,4-diaryl-2-mercaptoimidazoles.

In 2014, Desai and coworkers introduced hybrid pyrimidine-based imidazole pharmacophores as antimicrobial agents. Here, tetrahydropyrimidine and hydrazine hydrate were refluxed in 1,4-dioxane, followed by reaction with KSCN in acidic solvent; they were then further refluxed with $ClCH_2COOH$ and CH_3COONa in glacial acetic acid to produce an imidazole ring structure (Scheme 5.4) [49]. The mechanism involved Knoevenagel condensation reaction with substituted aldehydes in EtOH to afford desired imidazole derivatives with high efficiency. An in vitro antimicrobial study was performed against bacteria (*S. aureus*, *S. pyogenes*, *E. coli*, and *P. aeruginosa*) and fungus (*C. albicans*, *A. niger*, and *A. clavatus*) using ciprofloxacin and griseofulvin as standard antibiotics. The data revealed that most of the compounds showed moderate to high activity, although some compounds displayed selectivity in inhibition pattern; for example, **5b**, **5d**, **5f**, and **5j** showed high inhibition for *E. coli* and

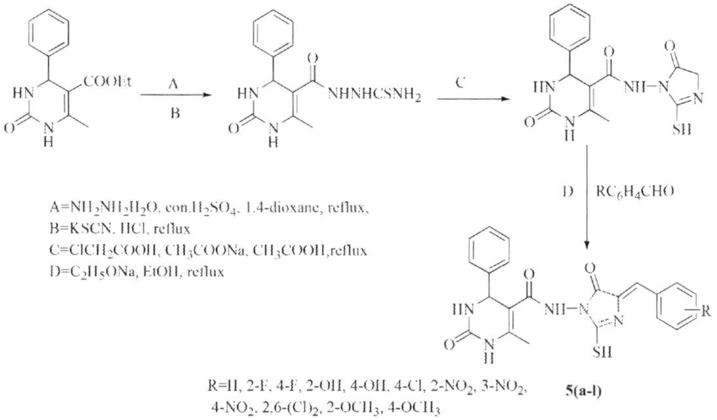

SCHEME 5.4 Pyrimidine-based imidazole synthesis.

TABLE 5.2 *In vitro* study of selective pyrimidine-based imidazole derivatives MIC value (μg/mL).

		Bacterial strain				Fungal strain		
S. no.	-R	*E.c.*	*P.a.*	*S.a.*	*S.p.*	*C.a.*	*A.n.*	*A.c.*
5b	2-F	50	500	50	100	100	200	12.5
5d	2-OH	50	100	250	500	200	500	500
5f	4-Cl	25	250	50	100	1000	100	200
5j	2,6-$(Cl)_2$	25	100	12.5	500	100	50	100
Ciprofloxacin		25	25	50	50	–	–	–
Griseofulvin		–				500	100	100

compounds **5b** and **5f** for *S. aureus*. Compounds **5b** and **5j** exhibited the highest antifungal activity against tested strains (Table 5.2).

Imidazol-2-thiones **6(a–d)** were prepared and screened for their potency against various bacterial strains, *B. cereus*, *E. coli*, *P. aeruginosa*, *S. aureus*, and *E. faecalis*, using an agar plate diffusion technique with a reference drug, gentamicin. Imidazole-2-thiones were synthesized from intramolecular cyclization of *N,N′*-di-substituted thiourea ketal derivatives in an acidic medium (Scheme 5.5) [50]. All compounds displayed good inhibition against *S. aureus* (128–256 μg/mL) bacterium and showed lower activity for other pathogens.

ZrO_2-β-cyclodextrin assisted solvent-free synthesis of imidazole derivatives and their applications in antibacterial study was demonstrated by Girish and

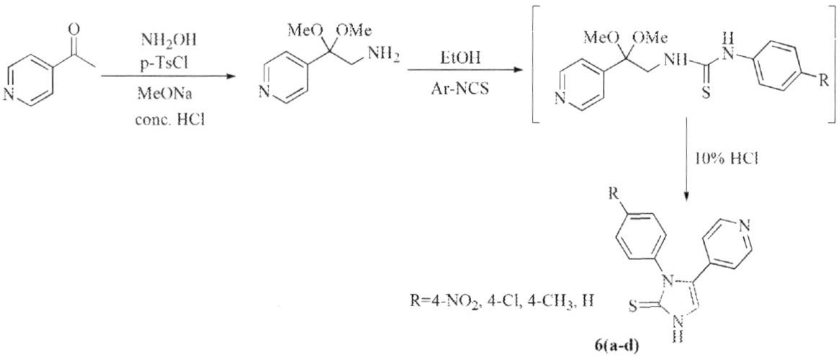

SCHEME 5.5 Synthesis of *N,N′*-di-substituted imidazole-2-thione derivatives.

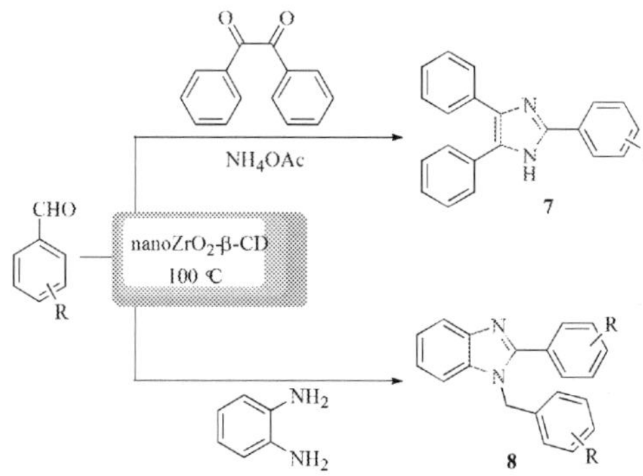

SCHEME 5.6 ZrO_2-β-cyclodextrin-assisted imidazole synthesis.

companions (Scheme 5.6) [51]. The reaction of benzil, aromatic aldehydes, and NH_4OAc in solvent-free conditions produced 2,4,5-trisubstituted imidazoles and *o*-phenylenediamine was reacted with aldehydes to furnish 1,2-di-substituted benzimidazoles with high atom economy with good recyclability of catalyst. The in vitro study was done against six bacterial strains: *K. pneumoniae*, *B. subtilis*, *P. aeruginosa*, *E. aeruginosa*, *S. flexneri*, and *S. typhi*. Most of the compounds exhibited moderate antibacterial activity against *B. subtilis* (ZOI; 8–15) and low inhibition against other pathogens.

The synthesis of methyl 2-(4-chlorophenyl)-7*a*-((4-chlorophenyl) carbamothioyl)-1-oxo-5,5-diphenyl-3-thioxo-hexahydro-1*H*-pyrrolo[1,2-*e*] imidazole-6-carboxylates (**9**) via cyclization of pyrrolidine derivatives with 4-chlorophenyl isothiocyanate was reported and screened as antimicrobial agents (Scheme 5.7) [52]. In vitro antimicrobial activity was accomplished against bacterial (*B. subtilis*, *E. coli*, *S. aureus*, *A. baumannii*, and *A. hydrophila*) and fungal strains (*M. tuberculosis* H37Rv) using ampicillin and ethambutol as standard drugs, respectively. These molecules showed high antibacterial potency in the range of 31.25–62.5 μg/mL and good fungal inhibition with an MIC value of 40 μg/mL.

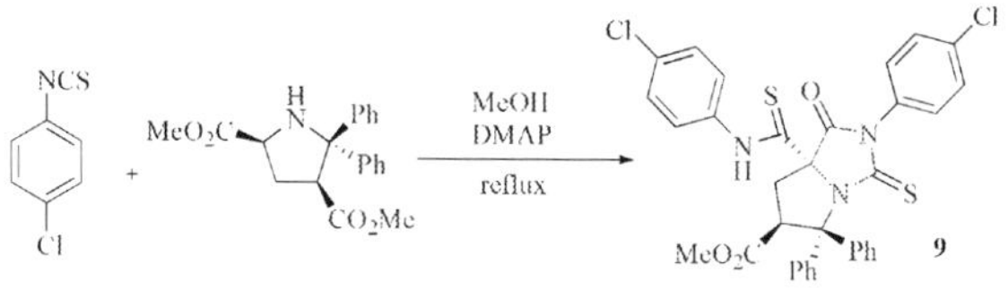

SCHEME 5.7 Synthesis of imidazole derivative.

The novel regioselective and one-pot, microwave-assisted synthesis of 4-aryl-2-methyl-*N*-phenacylimidazoles were synthesized and tested for their antifungal potency [53]. The reaction of α-bromoacetophenones and acetamidine hydrochloride yielded corresponding imidazoles and further reduction produced *N*-(2-hydroxyethyl)imidazoles in high yields (Scheme 5.8). In vitro antifungal study was performed against *C. albicans* and *C. neoformans* through a broth microdilution method using Amphotericin B as a reference drug. All of the compounds showed high inhibition activity against *C. neoformans* compared to *C. albicans*, whereas ketone group **10** (IC50 = 15.6–62.5 mg/mL) displayed higher activity than alcoholic substitution **11** (IC50 = 31.2–250 mg/mL). Dihalogenated derivatives of **10** exhibited the highest fungal inhibition with IC_{50} values of 15.6 mg/mL against *C. neoformans*.

The synthesis of imidazo[1,2-*a*]pyridine derivatives from the reaction of maleimide (base) and 2,3 diaminopyridine (nucleophile) was reported, and their antimicrobial activity was investigated different pathogens like methicillin-sensitive *S. aureus* (MSSA), methicillin-resistance *S. aureus* (MRSA), *P. aeruginosa*, *E. faecalis*, *E. coli*, *M. luteus*, and *C. albicans* IPA via the well diffusion method using gentamycin and nystatin as standards (Scheme 5.9) [54]. The study revealed that most of the compounds displayed promising antimicrobial efficiency and displayed 12–15 mm ZOI against *M. luteus* microbes. Compounds **12a** and **12b** showed higher antifungal potency than the reference drug and compounds **12b** and **12d** possessed the highest bactericidal inhibition against MRSA.

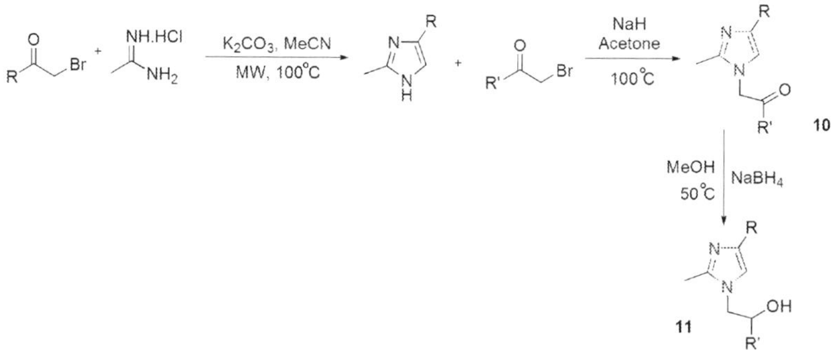

SCHEME 5.8 Synthesis of *N*-substituted 4-aryl-2-methylimidazole derivatives.

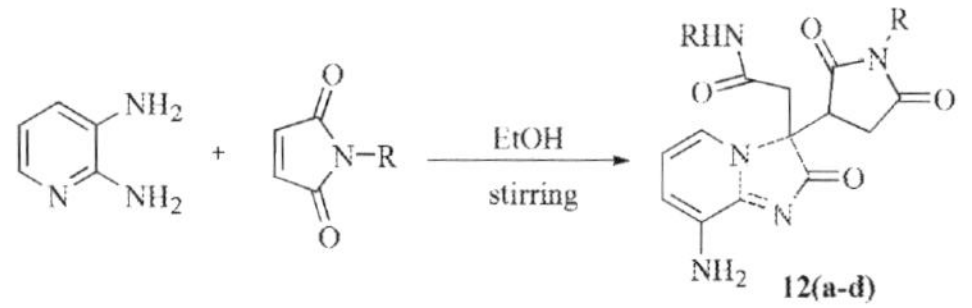

12(a-d)

R=H, CH_3, C_2H_5, C_6H_5

SCHEME 5.9 Synthesis of dihydroimidazo[1,2-*A*]pyridine derivatives via simple methodology.

The catalyst-free synthesis of novel imidazole derivatives from the reaction of amines with alkyl/aryl isothiocyanate in isopropyl alcohol to obtain good yields was demonstrated by Basha and companions and further screened as antibacterial agents against *B. subtilis*, *S. aureus*, *K. pneumoniae*, and *E. coli* via the agar well diffusion method (Scheme 5.10). Most of the compounds displayed good antimicrobial activity compared to ciprofloxacin, the reference drug [55]. Compounds **13a**, **13c**, **13d**, **13e**, **13f**, and **13i** revealed the highest pathogenic inhibition with maximum ZOI and low MIC range value, 6.62–21.42 μg/mL (Table 5.3).

Substituted 1,4-diaryl-2-mercaptoimidazoles were synthesized and studied for their broad-spectrum antimicrobial activity with molecular docking interaction pattern [56]. The desired compounds were prepared in multisteps. Initially α-acetophenone was brominated using glacial acetic acid and Br_2 to produce phenacyl bromide, which was further grounded with substituted aniline using anhydrous K_2CO_3, KSCN, and *p*-TsOH using a mortar and pestle. Finally, this reaction mixture was heated at 80–90°C to furnish imidazole derivatives (Scheme 5.11). An in vitro antimicrobial study was performed against bacterial strains (*S. aureus*, *P. aeruginosa*, and *E. coli*) and fungal strains (*A. niger* and *C. albicans*). Most of the compounds displayed moderate pathogenic inhibition against the reference drug. Compounds **14k**, **14q**, and **14r** exhibited the highest antibacterial activity and compounds **14g** and **14h** showed antifungal activity

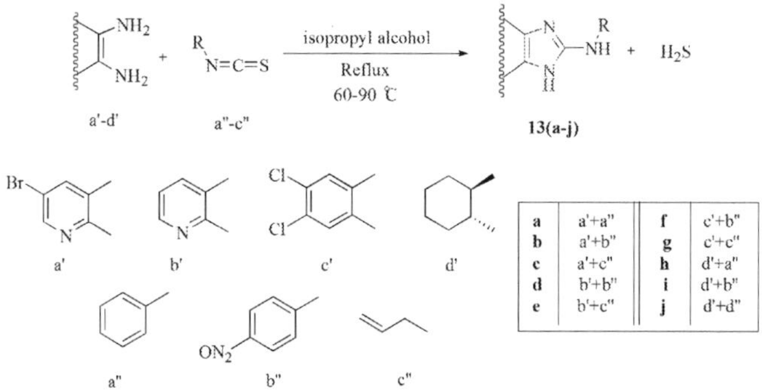

a	a'+a"	f	c'+b"
b	a'+b"	g	c'+c"
c	a'+c"	h	d'+a"
d	b'+b"	i	d'+b"
e	b'+c"	j	d'+d"

SCHEME 5.10 Synthesis of substituted imidazole derivatives.

TABLE 5.3 In vitro study data of some imidazole derivatives MIC value (μg/mL).

Compound	*S. aureus*	*B. subtilis*	*K. pneumoniae*	*E. coli*
13a	10.12	7.8	11.11	9.78
13c	9.16	10.11	12.28	14.28
13d	6.6	8.12	10.13	18.14
13e	9.18	13.46	13.17	20.22
13f	8.17	10.64	16.72	21.42
13i	8.16	8.14	16.37	12.66
Ciprofloxacin	16.18	14.66	18.86	22.16

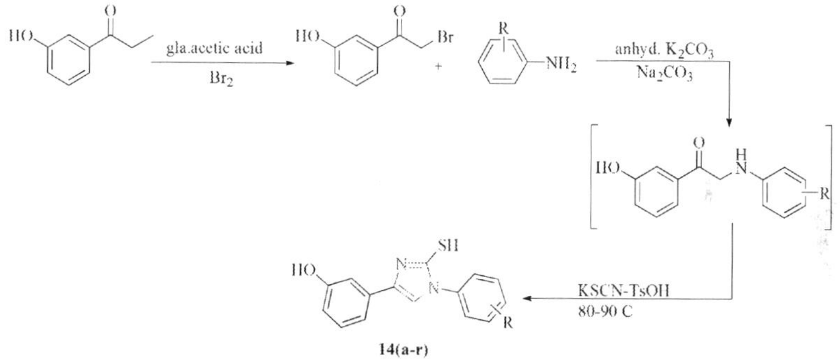

14(a-r)

H, o-CH_3, p-CH_3, o-OH, m-OH, p-OH, p-OCH_3, o-NO_2, m-NO_2, p-NO_2, o,p-diNO_2, m-Cl, p-Cl, o,m-diCl, o,p-diCl, m,p-diCl, p-F, p-Br

SCHEME 5.11 Synthesis of 1,4-diaryl-2-mercaptoimidazole derivatives.

due to the presence of the *p*-fluoro, *p*-bromo, *o,p*-dinitro, and hydroxy group, which enhanced their interaction efficiency. An in silico study indicated that all compounds showed good affinity and minimum binding energy for the active site of protein.

Imidazol-5-one derivatives were synthesized and screened for antibacterial activity against gram-(+ve) (*S. mutans* and *S. aureus*) and gram-(−ve) (*E. coli*, *K. pneumonia*, and *P. aeruginosa*) bacteria via an agar well diffusion procedure using ampicillin and gentamicin as reference drugs [57]. The synthesis of desired compounds proceeded via multistep synthesis; initially benzoylglycine and 4-formyl arylbenzoates were condensed in Ac_2O, AcONa at 100°C to form **15**, which further reacted with different types of amines in glacial acetic acid, AcONa at 100°C to produce respective imidazoles (Scheme 5.12). Among

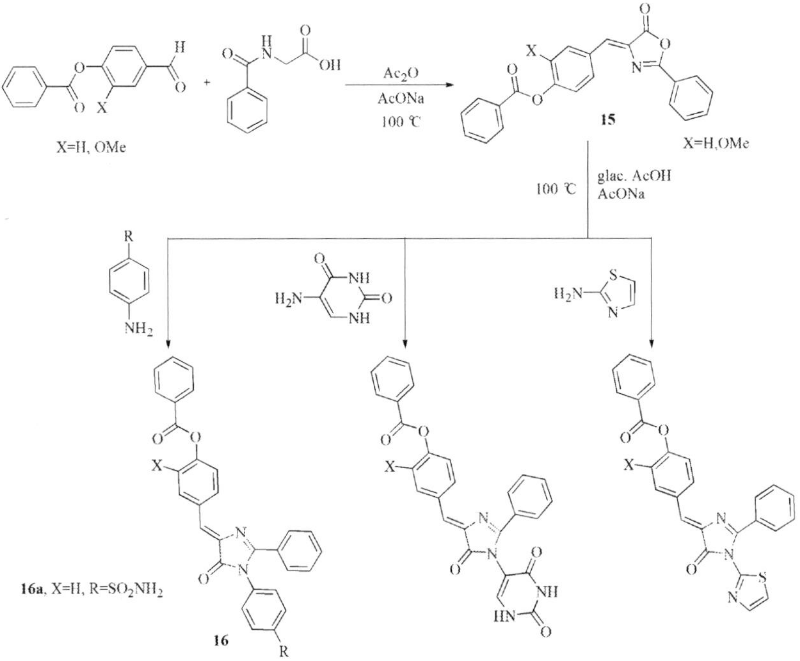

SCHEME 5.12 Synthesis of bio-active imidazolone derivatives comprising a benzoate moiety.

all the selected compounds, compound **16a** (X=H, R=SO_2NH_2) exhibited the highest toxicity (40.6±0.6, 184%) and an MIC (62.5 μg/mL) value against *S. aureus* and good inhibition activity for *E. coli.*

Recently, Raghu and companions reported MoS2-supported-calix[4]arene (MoS2-CA4) catalyzed one-pot synthesis of 2,4,5-trisubstituted imidazoles and evaluated their activity against *M. tuberculosis* (Mtb H37Rv), and the docking study revealed their protein-ligand interaction [58]. A heterogeneous catalyst supported condensation of substituted 1,2-diones, and aldehydes with NH_4OAc under solvent-free conditions yielded desired products in excellent yields with good recyclability of (MoS2-CA4). In vitro antitubercular activity results concluded that all the synthesized compounds showed higher inhibition activity than the reference drugs ciprofloxacin, ethambutol, and pyrazinamide, and the nature of substitutions affected their antibacterial potency. The presence of 2-methylfuran carboxamide, other amide groups, and ERG improved their pathogen-inhibition tendency. The docking results demonstrated that most of the compounds formed hydrogen, electrostatic, and hydrophobic interaction with the target protein.

Very recently, novel dihydro-1*H*-imidazole-2-yl)-[1,1′-biphenyl]-2-carboxamides were synthesized and studied for their utility as antimicrobial agents against broad spectrum pathogens comprising *E. coli*, *P. aeruginosa*,

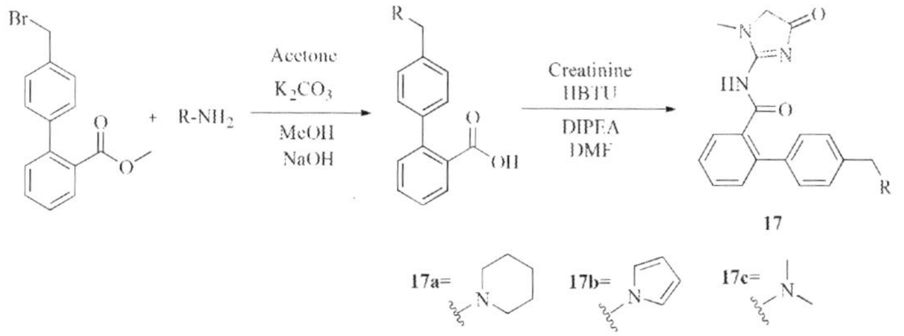

SCHEME 5.13 Synthesis of dihydro-1*H*-imidazole-2-yl)-[1,1′-biphenyl]-2-carboxamides.

K. pneumoniae, *S. aureus*, *C. diphtheria*, and *S. pyogenes* strains [59]. The synthesis of desired imidazoles proceeded via a multistep process, where initially, methyl 4′-(bromomethyl)-[1,1′-biphenyl]-2-carboxylate was alkylated using substituted amine, then refluxed in methanol and NaOH to prepare acid derivatives, which further reacted with creatinine, HBTU, DIPEA, and DMF solvent to obtain 70%–80% yields (Scheme 5.13). Among all the derivatives, compounds **17a** (0.75 μg/mL), **17b** (1.05 μg/mL), and **17c** (1.25 μg/mL) showed higher inhibition against most of the bacterial strains than ciprofloxacin (1.25 μg/mL), the standard drug.

2.1.1.3 Imidazole derivatives

In 2011, the anticandida potency of the 1-[(3-substituted-3-phenyl)propyl]-1*H*-imidazole derivatives (**18** and **19**) was evaluated via an agar diffusion technique, and compared with tioconazole and miconazole [60] (Fig. 5.4). The halogen derivatives of imidazole, **18a** (4-Cl-C_6H_4CO) and **18b** (2,4-Cl_2-C_6H_3CO), were found to be equipotent with the reference tioconazole exhibiting an MIC value of 0.3 μg/mL and more potent than miconazole (0.5 μg/mL) against *C. albicans* and *C. pseudotropicales* and **18c** (4-OCH_3-C_6H_4CO) displayed an MIC value of 0.4 mg/mL.

A range of [2-(substituted phenyl)-imidazole or (benzimidazole)-1-yl]-pyridin-3-yl-methanones (**20a–t**) was synthesized and studied for their inhibition potency against several pathogens, viz. *S. aureus*, *B. subtilis*, *E. coli*, *C. albicans*, and *A. niger* against ciprofloxacin and FCZ via the agar dilution method (Fig. 5.4) [61]. The *in vitro* study results demonstrated that compounds **20 h** and **20p** were equipotent against MTB compared to the reference antibiotic. The antibacterial studies revealed broad-spectrum activity; in particular, compound **20k** was found to be more potent toward pathogenic inhibition against tested strains, while high activity was displayed by derivatives **20c**, **20i**, and **20j** in comparison to standard drugs (Table 5.4).

A range of imidazoles and benzimidazoles **21** was designed for the study of antifungal activity against *C. albicans*, *C. tropicalis*, *C. glabrata*, *C. parapsilosis*, *C. krusei*, *C. dubliniensis*, *A. fumigatus*, *A. flavus*, *M. canis*, *M. gypseum*, *T. mentagrophyte*, and *E. floccosum* via a microdilution method and agar

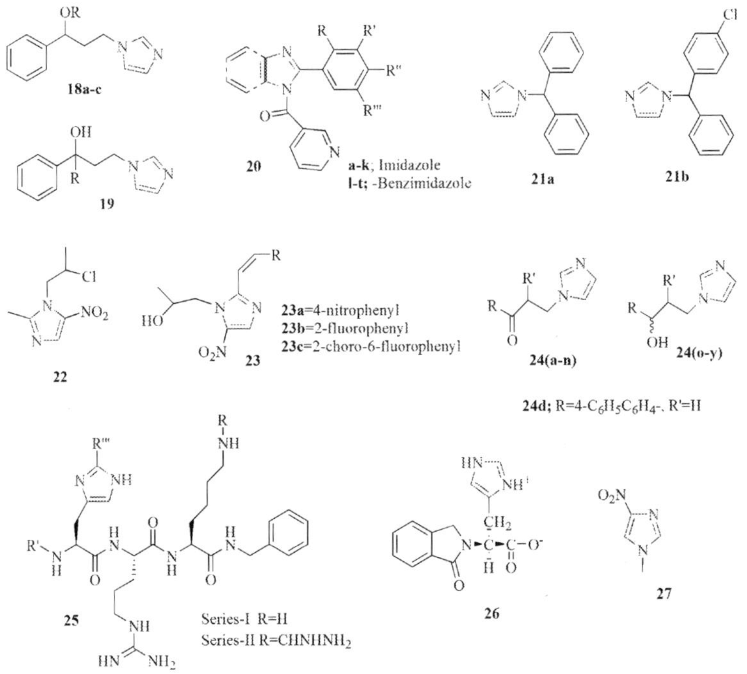

FIG. 5.4 Imidazole-based antimicrobial compounds.

dilution assay (Fig. 5.4) [62]. Most of them exhibited good antifungal potency compared to standard FCZ and itraconazole (ITZ), whereas derivatives **21a** and **21b** displayed broad-spectrum activity against various standard resistant fungi (Table 5.5).

Compound 1-(2-chloropropyl)-2-methyl-5-nitro-1*H*-imidazole **22** was synthesized and screened against *Helicobacter pylori* microbes, and the results revealed that this molecule displayed more potency with $13 \pm 2\,\mu M$ IC_{50} value than that of secnidazole (156 μM) and acetyl-hydroxy acid ($16 \pm 2\,\mu M$) (Fig. 5.4) [63].

Secnidazole is an antibacterial imidazole drug which is more effective than other commonly used IMD drugs. In consequence of this, 20 novel derivatives of 1-(2-hydroxypropyl)-2-styryl-5 nitroimidazole **23** were synthesized and their antimicrobial efficiency was determined against *B. thuringiensis*, *B. subtilis*, *E. coli*, and *P. aeruginosa* as FabH inhibitor using the MTT method (Fig. 5.4) [64]. Most of the compounds displayed selective inhibition among tested pathogens, whereas derivatives possessing nitro substituents displayed broad-spectrum inhibition with low MIC values. Notable growth inhibition was displayed by **23a** (36.8 μg/mL), **23c** (5.7 μg/mL), **23c** (6.1 μg/mL), **23b** (3.4 μg/mL), and **23a** (4.6 μM) derivatives against *E. coli*, *P. aeruginosa*, *B. subtilis*, B. thuringiensis, and *E. coli* FabH, respectively, compared to the reference, kanamycin.

TABLE 5.4 Antimicrobial and antimycobacterial activities of selected [2-(substituted phenyl)-imidazol-1-yl/benzimidazole-1-yl]-pyridin-3-yl-methanones.

	Substituents				MIC (μM/mL)					MIC (μg/mL)
Compound	*R*	*R′*	*R″*	*R‴*	*S.a.*	*B.s.*	*E.c.*	*C.a.*	*A.n.*	*MTB*
20c	CH_3	H	H	CH_3	NT	NT	NT	0.005	0.005	25
20h	H	H	NO_2	H	0.010	0.010	0.040	0.020	0.040	3.13
20i	NO_2	H	H	H	0.040	0.002	0.040	0.020	0.040	6.25
20j	H	NO_2	H	H	0.013	0.005	0.002	0.025	0.025	6.25
20k	COOH	H	H	H	0.002	0.002	0.002	0.020	0.040	>25.0
20p	H	H	NO_2	H	–	–	–	–	–	1.56
Ciprofloxacin					0.004	0.004	0.004	–	–	3.13
FCZ					–	–	–	0.005	0.005	–
Isoniazid/ethambutol					–	–	–	–	–	0.10/1.56

Note: *NT* = not tested.

TABLE 5.5 In vitro antifungal study of some bio-active imidazole derivatives against various fungal strains.

	Candida sp. (MIC 90% value)								Aspergillus, dermatophytes (μg/mL)				
Compound	*C.a.*	*C.t.*	*C.g.*	*C.p.*	*C.k.*	*C.d.*	*C.a.*	*C.t.*	*A.f.*	*A. flavus*	*T.m.*	*M.c.*	*M.g.*
21a	1	1	0.5	0.5	0.5	0.5	0.5	64	0.5	0.5	1	1	0.5
21b	0.5	2	0.5	8	0.5	0.5	8	4	4	0.5	0.5	1	8
Fluconazole	>256	32	4	2	16	1	R	R	4	NT	NT	NT	NT
Itraconazole	>256	>256	0.06	0.03	0.03	>256	R	R	–	–	–	–	–
Griseofulvin	–	–	–	–	–	–	–	–	NT	NT	5	0.6	8

Notes: *R* = resistant, *NT* = not tested.

A range of 3-(1*H*-imidazol-1-yl)-1-(4-biphenylyl)-1-propanone (**24**) was prepared and screened as broad-spectrum antifungal drug candidates. An in vitro antifungal study was performed against 16 candida species through a broth microdilution technique using fluconazole as a standard drug (Fig. 5.4) [65]. Most of the synthesized derivatives were highly potent, and among them, compound **24d** displayed broad activity against all the tested strains with MIC values from 1 to 8 mg/L. A SAR study explained that compounds with carbonyl functional groups were more active than their corresponding alcohol derivatives.

Two series of tripeptide-based imidazole derivatives (**25**) were designed, synthesized, and scrutinized for antimicrobial activity against *C. albicans*, *C. kyfer*, *C. neoformans*, *A. niger*, and *N. crassa* pathogens, compared with amphotericin B antibiotic (Fig. 5.4) [66]. Significant microbe growth inhibition was displayed by most of the derivatives, although the highest activity was exhibited against *C. neoformans*. The SAR study adduced that peptides might be involved in the disintegration of the cell membrane, which finally caused the microbe's cell death.

2-(1-Oxo-1*H*-2,3-dihydroisoindol-2-yl)-3-imidazolyl-L-lactamic acid (**26**) was prepared from the condensation of L-histidine and *o*-phthalaldehyde and displayed moderate pathogenic growth inhibition against *C. albicans*, *S. aureus*, *E. coli*, and *P. aeruginosa* strains (Fig. 5.4) [67].

Shahid and companions introduced two new methods for the synthesis of 1-methyl- 4-nitro-1*H*-imidazole (**27**) with regio-specificity and demonstrated their antimicrobial potency against variety of pathogens (Fig. 5.4) [68]. The in vitro study was done via the microdilution method and revealed moderate growth inhibition against all the tested strains (Table 5.6).

A range of Schiff's base containing imidazole derivatives was synthesized as microbe growth inhibitors against *E. coli*, *P. aeruginosa*, *B. subtilis*, *S. aureus*, and *E. coli* FabH [69]. The molecular interaction and binding energy were examined by an in silico docking study. Good antimicrobial activity was displayed by all the synthesized derivatives and compound **28a** was found to be highly active against all strains, including *E. coli* FabH inhibition with an IC_{50} value of 2.6883 μM compared with standard DCCP (IC50 = 3.1542) (Fig. 5.5). The docking study of bonding of molecules with active site of receptors presented five π-π interactions and one hydrogen bonding.

A variety of novel biphenyl imidazole derivatives **29(a–v)** and **30(a–b)** were prepared via multistep synthesis and were studied as antifungal agents for the inhibition of *C. albicans* and *C. neoformans*, *C. tropicalis*, *A. fumigatus* against FCZ and ITZ; the inhibition potency of cytochrome P450 (CYP) enzyme (five isoforms) was also evaluated, which played a key role in drug metabolism (Fig. 5.5) [70]. Compounds **29a–b**, **29f–g**, and **30a–b** displayed high growth inhibition with 0.03125–2 μg/mL MIC value against *C. albicans* and *C. neoformans*; however, moderate inhibition was exhibited against CYP isoform.

Novel imidazole derivatives were synthesized and studied for antifungal activity against *C. krusei*, *C. glabrata*, *C. parapsilosis*, and *C. albicans* using

TABLE 5.6 In vitro antimicrobial data of 1-methyl- 4-nitro-1*H*-imidazole (27) against various microbes [MIC in μg/mL].

Bacterial strains				Fungal strains	
Bacteria	MIC	Bacteria	MIC	Fungal strains	MIC
M. leutus	180	*P. vulgaris*	280	*C. albicans*	240
S. aures	220	*Morganella morganii*	200	*C. galbrata*	180
P. aeruginosa	120	*P. mirabilis*	100	*C. kurzi*	200

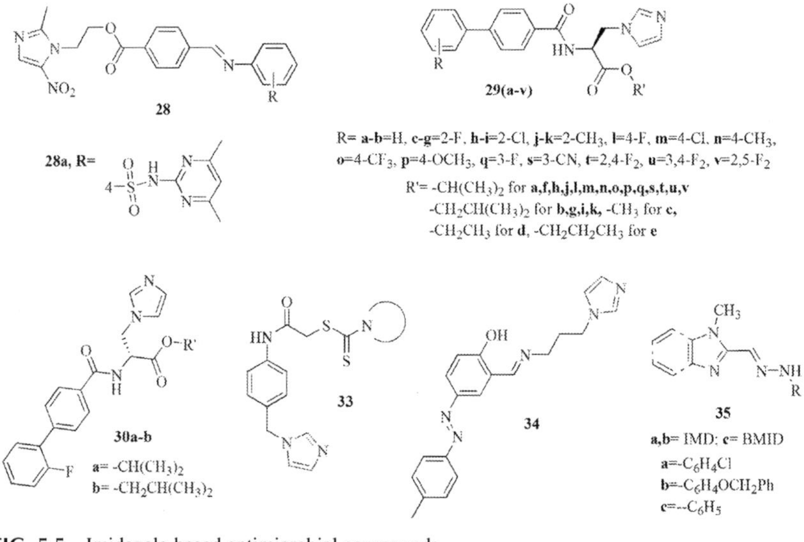

FIG. 5.5 Imidazole-based antimicrobial compounds.

FCZ and ketoconazole as standard drugs via the broth microdilution method [71]. The reaction of 2-chloro-*N*-[4-(1*H*-imidazole-1-yl)phenyl]acetamide with (benz)azolethiol derivatives or dithiocarbamate sodium salts furnished substituted imidazole compounds (**31,32**) (Scheme 5.14). Most of the compounds displayed moderate anticandidal effects and compound **31a** (4-methoxybenzyl piperazine derivative) showed the highest inhibition with MIC values of 1.56 and 0.78 μg/mL for *C. albicans* and *C. krusei*, respectively. The docking study of **31a** revealed a bonding pattern with 14-α-sterol demethylase, two nitrogen atoms of imidazole, amino and carbonyl groups of amide and first nitrogen of piperazine and phenyl ring via different type of interaction, further strengthening their inhibition potency.

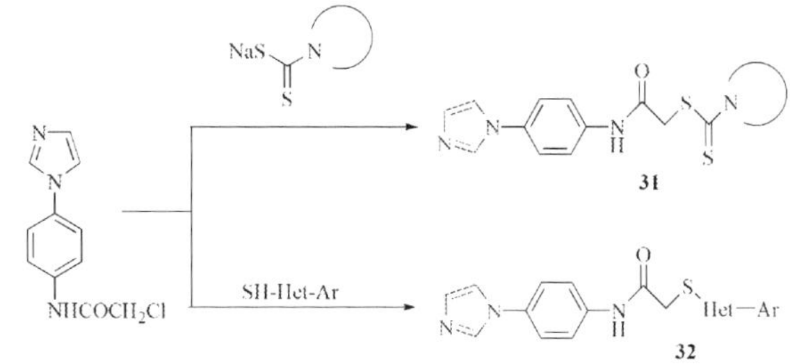

SCHEME 5.14 Synthesis of novel antifungal benzimidazole derivatives.

After this success, a range of 2-(substituted-dithiocarbamoyl)-*N*-[4-((1*H*-imidazol-1-yl)methyl)phenyl] acetamide derivatives were synthesized and screened for antifungal activity against four strains, *C. glabrata*, *C. krusei*, *C. parapsilosis*, and *C. albicans*, using FCZ and ketoconazole as reference drugs (Fig. 5.5) [72]. Most of the compounds displayed good activity against *C. albicans* and *C. krusei* with MIC_{50} values in the range 12.5–25 1g/mL. Lanosterol-14-α-demethylase is a vital enzyme of sterol biosynthesis in fungal strains for life functions, and was used for the study of interaction of compounds via a molecular docking study. Compound **33** displayed metal coordination, hydrogen bonding, and p-p interaction with the amino, carbonyl group, and imidazole ring, respectively.

The interesting properties of Schiff bases make them important centers of pharmaceutical study; therefore, Schiff base-based imidazole ligand (**34**) was synthesized and screened for in vitro antibacterial activities for the inhibition of *S. aureus*, *P. putida*, *K. pneumoniae*, and *E. coli* via a diffusion agar technique using chloramphenicol as a standard (Fig. 5.5) [73]. The results revealed that the ligand showed high ZOI and significant activity against *S. aureus* with an MIC value of 28.84 mg/mL.

Substituted (benz)imidazole-hydrazones were synthesized at ambient temperature from the reaction of carbaldehyde derivatives of imidazole with hydrazine and their potency as antibacterial agents was evaluated against *Fusarium oxysporum* fungal strains using carbendazim as a positive control (Fig. 5.5) [74]. The in vitro study results depicted that most of the compounds displayed good inhibition. Among them, molecules **35a** (IC50 = 268 mM), **35b** (IC50 = 198.4 ± 1.7 mM), and **35c** (IC50 = 203.6 ± 0.9 mM) showed high inhibition potency with inhibition rates of 47.5%, 57.7%, and 50.7%, and it was concluded that imidazole derivatives was more responsive than benzimidazole derivatives for pathogen inhibition.

2.1.2 Benzimidazole

2.1.2.1 Ring formation

An isomeric range of substituted 1-alkyl-2-trifluoromethyl benzimidazole compounds (**36**) was prepared and screened against four bacterial strains, *B. subtilis*,

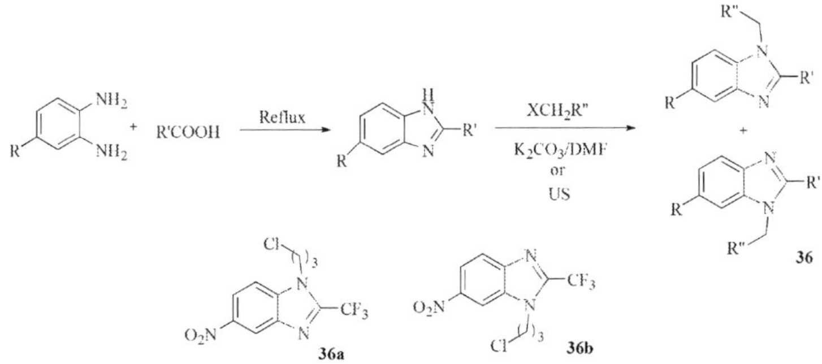

SCHEME 5.15 Synthesis of novel antimicrobial benzimidazole derivatives.

S. aureus, *E. coli*, and *P. aeruginosa*, and four fungal strains, *C. albicans*, *A. niger*, *R. oryzae*, and *S. cerevisiae* (Scheme 5.15). According to the in vitro study data, most of them displayed promising antimicrobial potency, and compounds **36a** and **36b** showed the highest inhibition against standard antibiotic ciprofloxacin and FCZ [75].

A range of 2-chloromethyl-1*H*-benzimidazole derivatives (**37**, **38**, and **39**) was synthesized in multistep with good yields and their antifungal potency was evaluated against phytopathogens fungi *Cytospora* sp., *C. gloeosporioides*, *B. cinerea*, *A. solani*, and *F. solani* via a mycelium growth rate method (Scheme 5.16) [76]. Nearly all of them showed significant growth inhibition for the studied pathogens. Compounds **37a**, **38a**, and **39a** were found to be most active for the inhibition of selective strains due to the structural parameters like the presence of a Cl atom, benzene ring, and sulfonyl group (Table 5.7).

A greener and eco-benign synthesis of 2-styrylbenzimidazoles **40** and their antimicrobial activity was discovered by Kumar and companions [77]. The reaction of *o*-phenylenediamines and various acid derivatives in the presence

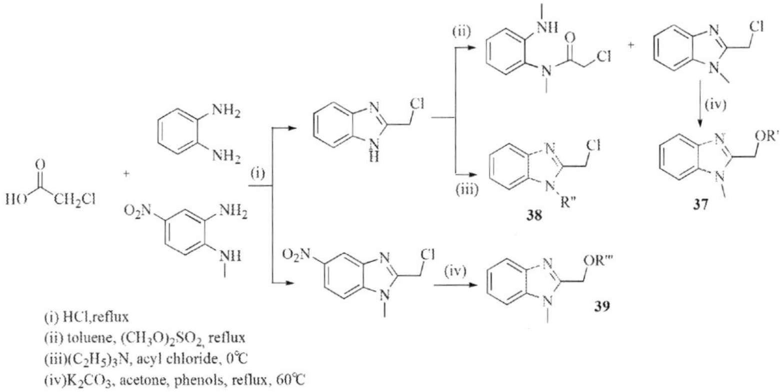

SCHEME 5.16 Synthesis of 2-chloromethyl-1*H*-benzimidazole derivatives.

TABLE 5.7 Antifungal activity of 2-chloromethyl-1*H*-benzimidazole derivatives against five phytopathogens [$IC_{50} \pm SD$(μg/mL)].

Compound	R′/R″/R‴	*C.s.*	*C.g.*	*B.c.*	*A.s.*	*F.s.*
37a	$-C_6H_5$	50.04 ± 1.03	20.76 ± 0.67	> 150	27.58 ± 0.59	18.60 ± 1.10
38a	$-O_2SCH_3$	30.97 ± 1.12	11.38 ± 0.90	57.71 ± 3.51	> 100	40.15 ± 0.76
39a	$-C_6H_4$-Cl (p)	NA	NA	13.36 ± 1.35	NA	NA
Hy	–	> 100	> 100	8.92 ± 0.44	14.03 ± 0.14	38.49 ± 6.95

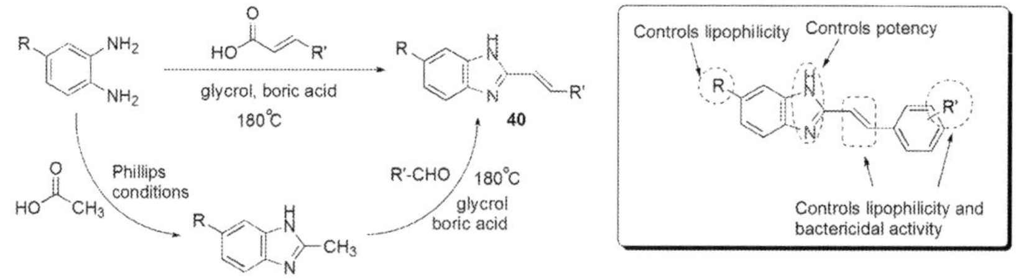

SCHEME 5.17 Synthesis of 2-styrylbenzimidazoles 6(a–x) and their pharmacophoric behavior.

of glycerol and boric acid furnished benzimidazole derivatives in high yields (Scheme 5.17). In vitro antimicrobial activity was studied against eight bacterial strains: *M. luteus*, *S. aureus*, *S. aureus* MLS-16, *B. subtilis*, *E. coli*, *P. aeruginosa*, *K. planticola*, and *C. albicans*. Some of the tested compounds displayed high pathogenic growth inhibition compared to the reference drug, ciprofloxacin (Table 5.8).

A range of 2-substituted fluorinated benzimidazoles **41** was tested for in vitro antimicrobial activity against *E. coli*, *K. pneumonia*, *S. aureus*, *P. aeruginosa*, and *C. albicans* with standards ampicillin, ciprofloxacin, and itoconazole via disk diffusion assay [78]. The synthesis of desired products was completed in multisteps; initially 4-fluoro-3-nitrobenzoic acid was esterified and further reacted with 4-fluorophenyl aniline to produce methyl 4-(4-fluoroaniline)-3-nitrobenzoate (Scheme 5.18). This nitro derivative was reduced to amino compounds and then condensed with aldehydes in three different reaction conditions and yielded final compounds in high amounts. Among all the tested compounds, 2-OH, 4,6-$(OCH_3)_2$ and 2,3,4-$(OH)_3$ substituted benzimidazoles displayed broad-spectrum growth inhibition against all the tested bacterial strains.

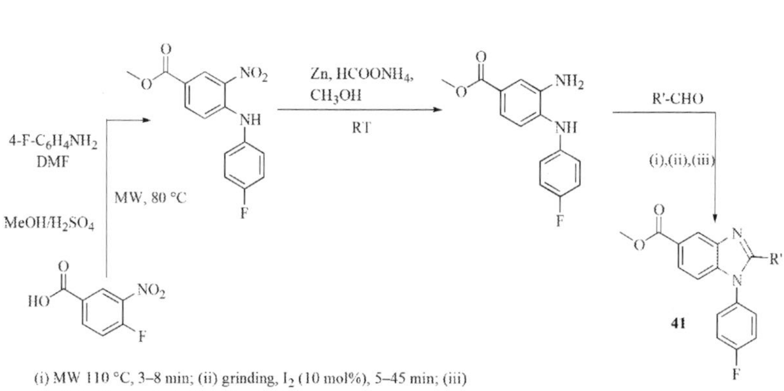

(i) MW 110 °C, 3–8 min; (ii) grinding, I_2 (10 mol%), 5–45 min; (iii) $Na_2S_2O_5$ (20 mol%), EtOH, reflux, 8–18 hr

SCHEME 5.18 Synthesis of 2-substituted fluorinated benzimidazoles.

TABLE 5.8 In vitro study data of some imidazole derivatives MIC value (μg/mL).

Compound	R	R′	*M.l.*	*S.a.*	*S.a.* MLS-16	*B.s.*	*E.c.*	*P.a.*	*K.p.*
40a	H	C_6H_5-p-F	31.2	7.8	3.9	3.9	31.2	31.2	15.6
40b	H	Furan	31.2	31.2	3.9	31.2	62.5	>125.0	62.5
40c	H	Thiophene	62.5	62.5	3.9	7.8	31.2	31.2	62.5
40d	NO_2	C_6H_5-p-F	7.8	1.9	0.9	0.9	7.8	7.8	7.8
40e	NO_2	C_6H_5-p-Cl	7.8	1.9	0.9	0.9	7.8	7.8	7.8
40f	NO_2	1,3-Benzodioxole	15.6	7.8	3.9	3.9	31.2	15.6	31.2
40g	NO_2	Furan	15.6	3.9	3.9	3.9	31.2	31.2	15.6
40h	COC_6H_5	C_6H_5-p-NO_2	62.5	62.5	31.2	7.8	31.2	62.5	>125.0
Ciprofloxacin	–	–	0.9	0.9	0.9	0.9	0.9	0.9	0.9

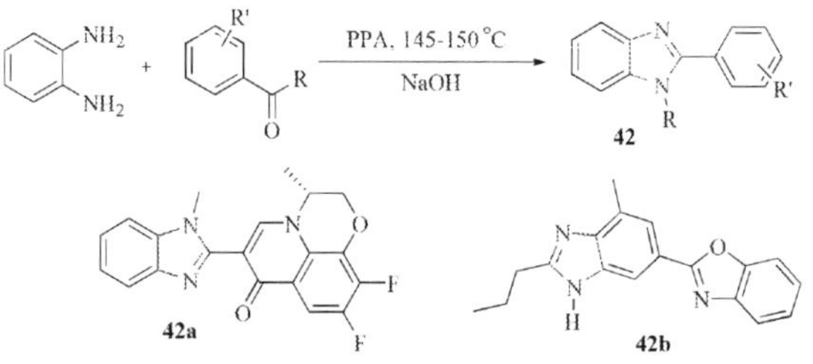

SCHEME 5.19 Synthesis of 2-substituted benzimidazole with highly active derivatives.

A range of 2-substituted benzimidazoles **42** was utilized as potent antimicrobial agents against a panel of bacterial (*E. coli*, *B. subtilis*, *S. aureus*, and *P. aeruginosa*) and fungal strains (*A. niger*, *A. flavus*, and *M. purpureus*) using the cup-plate method, and the in silico study validated their binding efficiency with microbial DNA gyrase [79]. The reaction of substituted *o*-phenylenediamine and aldehydes using polyphosphoric acid and NaOH was performed to furnish high amounts of corresponding benzimidazoles (Scheme 5.19). The MIC results revealed that compounds **42a** (0.625–12.5 μg/mL) and **42b** (0.156–12.5 μg/mL) demonstrated broad-spectrum antibacterial activity and other derivatives exhibited moderate inhibition against the reference ciprofloxacin, miconazole, and FCZ. The docking study showed good binding affinity and binding energies (−49.642 to −39.176 kcal/mol) with docking scores from −8.46 to −7.00, and displayed electrostatic interactions, H-bonding, pi–pi stacking with active sites of enzyme. ADME values also confirmed their drug-likeness, possibilities of activity and efficiency as oral drug candidates.

Three novel benzimidazole derivatives were synthesized and considered as potent antimicrobial drug candidates through an in vitro study against bacterial pathogens (*E. aerogenes*, *M. luteus*, *S. aureus*, and *E. coli*) and fungal strains (*A. flavus*, *F. solani*, and *A. fumigatus*) via the disk diffusion method using kanamycin and terbinafine as positive controls [80]. Among all the derivatives, **43b** displayed broad-spectrum inhibition compared to other derivatives due to their higher lipophilicity, and the presence of ERG (–CH_3) simplified the penetration in the outer layer and stopped bacterial growth by preventing protein synthesis (Table 5.9) (Fig. 5.6).

2.1.2.2 Benzimidazole derivatives

1*H*-Benzimidazole-2-carboxamido derivatives were synthesized **44** and studied for in vitro antimicrobial activities against *S. aureus*, *E. coli*, *B. subtilis*, and *C. albicans* through disk diffusion techniques (Fig. 5.7) [81]. These compounds were more active toward *S. aureus* with 16 mm ZOI compared to standard ampicillin.

A wide range of benzimidazole derivatives **45** was screened for antimicrobial QSAR study against varied bacterial and fungal strains. This study

TABLE 5.9 *In vitro* study data of tested imidazole derivatives MIC values (μg/mL).

Compound	R′	*M. luteus*	*S. aureus*	*E. aerogenes*	*E. coli*	*A. flavus*	*A. fumigatus*	*F. solani*
43a	H	–	50 ± 3.1	200 ± 2.5	12.5 ± 1.2	100 ± 1.6	200 ± 1.1	–
43b	3-CH_3	25 ± 1.5	25 ± 1.7	25 ± 1.0	12.5 ± 2.2	50 ± 1.5	50 ± 0.7	50 ± 2.3
43c	NO_2	–	50 ± 2.5	–	25 ± 1.5	200 ± 2.5	100 ± 1.0	–
Kanamycin		6.25 ± 0.5	12.5 ± 0.4	12.5 ± 0.7	6.25 ± 0.3	–	–	–
Terbinafine		–	–	–	–	6.25 ± 0.4	25 ± 0.1	12.25 ± 0.1

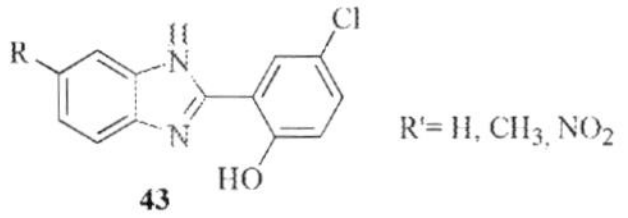

FIG. 5.6 Structure of novel benzimidazole derivatives.

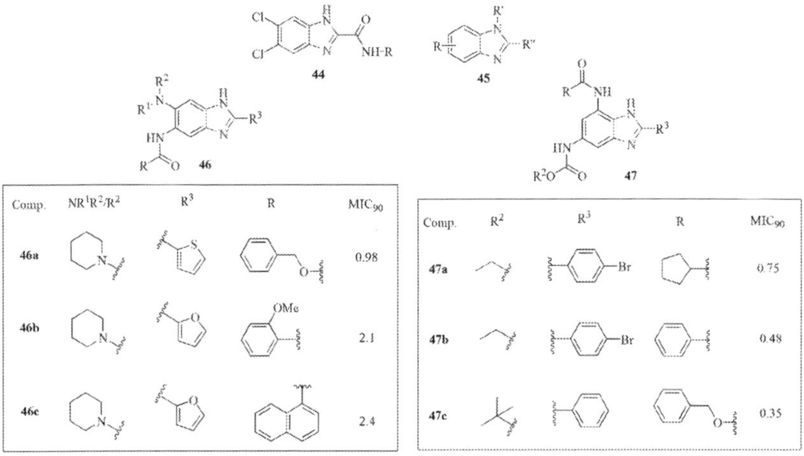

FIG. 5.7 Structure of selected antimicrobial benzimidazole derivatives.

displayed good predictive performances against *E. coli*, *S. aureus*, and MRSA (clinical isolates) with 0.833, 0.8485, and 0.8443 correlation coefficient values and 0.039, 0.2662, and 0.2488 RMSE values, respectively, which suggested that these compounds are good antimicrobial agents (Fig. 5.7) [82].

F. tularensis is one such most virulent gram-negative bacteria which causes most lethal pulmonary infections such as tularemia or "rabbit fever" [83, 84]. To discover newer antibacterial drugs for the treatment of this pathogen, a library of 2,5,6- and 2,5,7-trisubstituted benzimidazoles were screened [85]. After preliminary screening, 23 compounds showed >90% activity at 1 μg/mL and 21 compounds displayed 0.35–48.6 μg/mL MIC_{90} values among all. Compounds **46a**, **47a**, **47b**, and **47c** showed the highest growth inhibition with MIC_{90} values <1.0 μg/mL. For the ex vivo study, compounds **46a**, **46b**, and **46c** exhibited high potency in the concentration range of 10–50 μg/mL against thiabendazole and doxycycline, the reference compounds (Fig. 5.7).

2-[4-(1*H*-Benzimidazol-2-ylmethoxy)phenyl]-2-(4-chlorophenylamino) acetonitrile derivatives were synthesized and screened for microbial inhibition against *S. aureus*, *E. coli*, *S. typhi*, and *C. albicans* species via the broth microdilution method [86]. The in vitro study results revealed that compounds **48d** (3.9 μg/mL) and **48b** (7.8 μg/mL) showed the highest inhibition against *E. coli* and *S. typhi* compared to standard tetracycline and **48a–c** (3.9 μg/mL) and

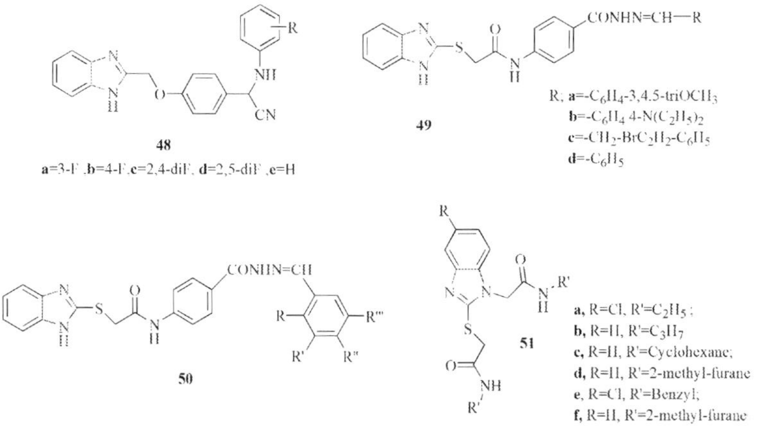

FIG. 5.8 Structure of antimicrobial benzimidazole derivatives.

48e (7.8 μg/mL) depicted the highest antifungal activity compared to standard antibiotic FCZ (Fig. 5.8). However, most of the compounds exhibited moderate to high pathogenic inhibition with good MIC values. The structure–activity relationship indicated that the presence of different substituents influenced their activity, viz. a phenyl ring having a fluorine atom at the ortho and para positions showed high activity and electron-donating substituents (methoxy and ethoxy group) alleviated their activity.

A range of substituted 2-mercaptobenzimidazole Schiff bases was studied as efficient targets for antimicrobial activity against broad-spectrum microbes via tube dilution method using cefadroxil and fluconazole as reference drugs [87]. Among all the synthesized compounds, compounds **49(a–d)** exhibited high pathogenic inhibition against all the tested strains, *S. aureus*, *B. subtilis*, *E. coli*, *P. aeruginosa*, *S. enterica*, *A. niger*, and *C. albicans*. The SAR studies concluded that presence of tri-OCH_3, -$N(C_2H_5)_2$, benzylidene hydrazide enhanced their inhibition activity (Fig. 5.8).

In continuation of previous research, Talhan et al. synthesized 2-mercaptobenzimidazole azomethine derivatives **50** and screened them for antimicrobial evaluation against a range of bacterial (*B. subtilis*, *E. coli*, *P. aeruginosa*, *S. typhi*, and *K. pneumoniae*) and fungal strains (*A. niger* and *C. albicans*) via the serial dilution method [88] (Fig. 5.8). Significant pathogenic inhibition was displayed by all compounds against all the selected strains compared to ofloxacin (antibacterial) and FCZ (antifungal), standard drugs; however, molecules **50(a–g)** showed high antimicrobial activity against specific strains due to presence of different substituents, that indicated the diverse structural requirements for the inhibition of particular pathogenic targets (Table 5.10).

Recently, the in vitro antimicrobial activity of *N*- and *S*-alkylated benzimidazole-2-thiols against various pathogens including *S. aureus*, *M. luteus*, *S. aureus* MLS-16, *B. subtilis*, *E. coli*, and *P. aeruginosa* were studied and their broad

TABLE 5.10 In vitro study data of selected 2-mercaptobenzimidazole azomethine derivatives.

No.	R	R′	R″	R‴	*B. subtilis*	*P. aeruginosa*	*E. coli*	*K. pneumoniae*	*C. albicans*
50a	H	OCH_3	OCH_3	OCH_3	2.41	1.20	2.41	4.81	4.81
50b	H	H	$N(C_2H_5)_2$	H	2.50	0.30	2.50	4.99	4.99
50c	H	OCH_3	OCH_3	H	1.28	1.28	1.28	5.11	5.11
50d	OCH_3	H	H	H	0.68	0.68	2.72	5.44	5.44
50e	H	OCH_3	H	H	1.36	1.36	1.36	5.44	1.36
50f	H	H	OCH_3	H	0.68	1.36	2.72	5.44	5.44
50g	H	H	Br	H	4.92	4.92	2.46	1.23	4.92
Std.					1.73[a]	3.46[a]	3.46[a]	3.46[a]	4.08[b]

[a] *Ofloxacin.*
[b] *Fluconazole.*

spectrum antibiofilm activity was also evaluated [89] (Fig. 5.8). Most of the compounds displayed high inhibition activity with selected strains, whereas compounds **51a–f** exhibited broad-spectrum inhibition with MIC values from 1.9 to 3.9 μg/mL and MBC (minimum bactericidal concentration) from 3.9 to 7.8 μg/mL; they also demonstrated significant antibiofilm properties. The docking study confirmed the interaction between molecules and proteins, which showed their strong H-bonding with N and O atoms and high binding energies. The SAR studies concluded that amide linkage possessing small substituents showed higher inhibition compared to bulky functional groups.

2.2 Based on linked moieties

2.2.1 Amide

In 2014, a novel range of 4-chloro substituted imidazole amides and sulfonamides were synthesized and screened as antimicrobial agents against bacterial strains (*S. aureus*, *E. coli*, and *K. pneumoniae*) and fungal strains (*C. albicans*, *A. flavus*, and *Rhizopus* sp.) with ciprofloxacin (Cfn) and amphotericin B (Am B) antibiotics using the serial plate dilution method (Fig. 5.9) [90]. Most of the synthesized derivatives were highly active against all the tested pathogens. Furthermore, derivatives **52c–52e** exhibited the highest microbial growth inhibition potency compared to standard. The SAR study revealed that the presence of the sulfonamide group might be responsible for enhancing antimicrobial activity (Table 5.11).

2-(5-Methoxy-1*H*-benzo[*d*]imidazol-2-ylthio)-*N*-phenylacetamide derivatives **53** were prepared in multisteps and applied for antimicrobial screening via the microdilution/broth titer method (Fig. 5.9) [91]. The in vitro screening results indicated that compounds **53a**, **53b**, **53d**, and **53f** were selectively more potent, some of the derivatives were equally active compared to the reference, and others displayed significant inhibition against selective strains. According to the data and SAR study, it was perceived that compounds containing EWG (–F, $-NO_2$, –Cl) were comparatively more active than ERG-containing compounds (Table 5.12).

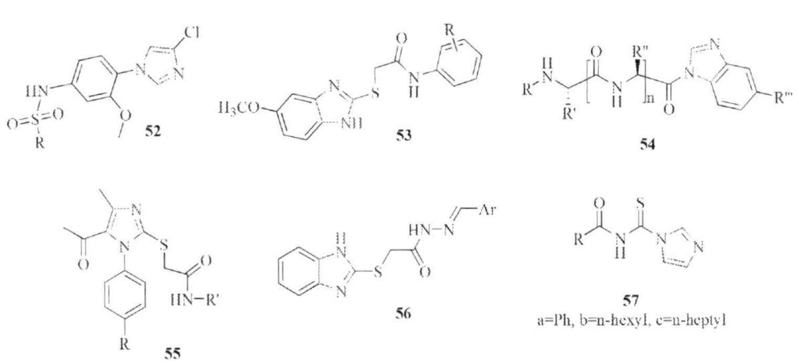

FIG. 5.9 Structure of some microbial active amide-based imidazole compounds.

TABLE 5.11 Antimicrobial activity data of the sulfonamides-imidazole conjugates (MIC in μg/mL).

Compound	R	*S. aureus*	*E. coli*	*K. pneumoniae*	*C. albicans*	*A. flavus*	*Rhizopus* sp.
52a	2,4-diClC_6H_3	8	4	64	128	64	128
52b	-CH_2-3-$CF_3C_6H_4$	4	64	4	64	32	128
52c	3-F,5-$CF_3C_6H_4$	4	4	8	8	8	16
52d	4-$CF_3C_6H_4$	4	16	8	8	16	16
52e	2-Cl,3-CF_3, 6-FC_6H_4	4	4	8	8	16	16
Cfn	–	8	4	16	–	–	–
Am B	–	–	–	–	8	8	16

TABLE 5.12 MIC (μg/mL) values of sulfonamides-imidazole conjugates against various microbes.

Compound	R	*S. aureus*	*E. faecalis*	*E. coli*	*P. aeruginosa*	*C. albicans*	*A. niger*
53a	3-F-4-Cl	125	125	62.5	62.5	125	250
53b	-H	125	125	62.5	125	125	125
53c	4-F	125	150	62.5	125	125	125
53d	2-OCH_3	31.25	125	62.5	125	250	125
53e	3-OCH_3	62.5	125	125	62.5	250	250
53f	2-NO_2	62.5	125	62.5	125	250	250
Flc	–	–	–	–	–	125	62.5
Cfn	–	62.5	125	125	125		

Novel amino acids and dipeptide benzimidazole conjugates **54** were prepared in a single step with retention of the original chirality and their utility as antimicrobial agents was studied against *S. aureus*, *E. faceium* NJ-1, *E. coli*, *P. aeruginosa*, *C. albicans*, and *C. tropicalis* through the MIC method (Fig. 5.9) [92]. All compounds displayed moderate activity with an MIC range of 100–800 μg/mL compared to the reference ampicillin, ceftazidime, and fluconazole.

2-((5-Acetyl-4-methyl-1-phenyl-1*H*-imidazole-2-yl)thio)-*N*-phenylacetamides **55** were efficiently synthesized via conventional heating and MW irradiation and studied for antimicrobial activity against *S. aureus*, *E. coli*, *P. aeruginosa*, *S. pyogenes*, *C. albicans*, *A. niger*, and *A. clavatus* with standard chloroamphenicol, Cfn, ampicillin, and griseofulvin (Fig. 5.9) [93]. Biological activity data revealed that most of the synthesized compounds showed high inhibition against all the tested strains, whereas some derivatives showed selectivity toward various pathogens (Table 5.13). However, the derivatives were less potent against *A. niger* and *A. clavatus*. The molecular docking studies showed good binding affinity and binding energies for most of the compounds. Moreover, SAR studies concluded that the electronic behavior of compounds controlled their bioactivity.

A range of benzimidazole derivatives **56** were prepared and demonstrated as effective antimicrobial agents against *S. aureus*, *B. subtilis*, *E. coli*, *C. albicans*, and *A. niger* and their in vitro and in vivo antitubercular activity was also evaluated (Fig. 5.9) [94]. The in vitro study revealed that all the synthesized compounds were more active growth inhibitors against all the tested strains with MIC values ranging from 0.016 to 0.073 μM compared to the standards, cefadroxil (0.345 μM) and fluconazole (0.40 and 0.82 μM). Most of the compounds worked as equipotent antitubercular agents compared to streptomycin antibiotic, as noted from the ZOI (> 20 mm) and MIC values (12.5 μM). The in vivo results depicted that these molecules were effective and safe for specific TB bacteria without any animal death.

N-Acyl-1*H*-imidazole-1-carbothioamides **57** were synthesized and screened for antimicrobial activity against various bacterial and fungal strains through the agar disk diffusion method using ampicillin and amphotericin B as reference antibiotics [95]. The results indicated that among all the screened derivatives, compounds **57a**, **57b**, and **57c** showed good antibacterial inhibition whereas compound **57b** was highly potent against all the tested fungus, viz. *A. flavus* (9.2 ± 0.54 nm), *A. fumigatus* (11.6 ± 1.5 nm), *A. niger* (9.6 ± 1.37 nm), and *F. solani* (11.7 ± 1.8 nm) with good ZOI (Fig. 5.9).

2.2.2 Triazole

A range of indole–benzimidazole-based 1,2,3-triazole analogs were prepared from microwave-assisted click reaction and utilized as antimicrobial agents against *B. subtilis*, *S. aureus*, *E. coli*, *P. vulgaris*, *A. niger*, and *C. albicans* (Fig. 5.10) [96]. The in vitro screening results displayed good activity of all compounds, and derivatives **58a**, **58b**, and **58c** were highly active against all

TABLE 5.13 Antibacterial screening value of acetamides-imidazole conjugates against various microbes [MIC in μg/mL].

Compound	R	R′	*E. coli*	*P. aeruginosa*	*S. aureus*	*S. pyogenes*	*C. albicans*
55a	Cl	4-FC_6H_4	62.5	125	100	250	500
55b	Cl	4-ClC_6H_4	12.5	50	12.5	50	250
55c	Cl	C_6H_5	25	50	12.5	125	500
55d	Cl	3-OHC_6H_4	25	50	12.5	125	250
55e	Cl	4-$OCH_3C_6H_4$	12.5	50	250	250	1000
55f	Cl	4-BrC_6H_4	125	100	62.5	50	500
55g	Cl	2-Benzothiazole	250	125	62.5	62.5	500
55h	Cl	4-$NO_2C_6H_4$	12.5	125	50	100	250
55i	Cl	2-Pyridine	62.5	62.5	100	62.5	250
55j	F	4-$OCH_3C_6H_4$	12.5	62.5	62.5	62.5	500
55k	H	4-$OCH_3C_6H_4$	125	125	250	500	1000
55l	OCH_3	4-$OCH_3C_6H_4$	125	100	125	250	>1000
55m	NO_2	4-$OCH_3C_6H_4$	250	100	125	250	500
Gentamycine	–	–	0.05	1	0.25	0.5	–
Ampicillin	–	–	100	–	250	100	–
Nystatin		–	–	–	–	–	100

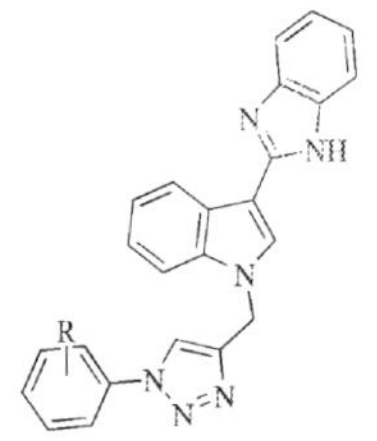

58

R=2-Cl, 4-Cl, 2-NO$_2$, 3-CF$_3$

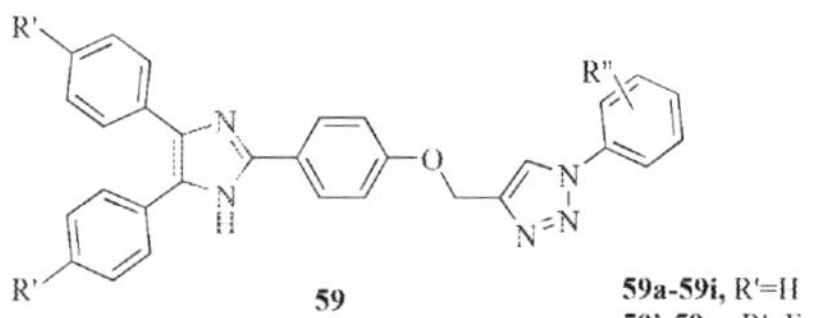

59

59a-59i, R'=H
59j-59r, R'=F

R= a=4-CH$_3$, b=2-F,c=4-OCH$_3$, d=4-F, e=3-Cl, f=4-Cl, g=2-CH$_3$-3-Cl, h=4-Br, i=4-NO$_2$, j=4-CH$_3$, k=2-F, l=4-NO$_2$, m=4-F, n=3-Cl, o=4-Cl, p=2-CH$_3$-3-Cl, q=4-Br, r=4-OCH$_3$

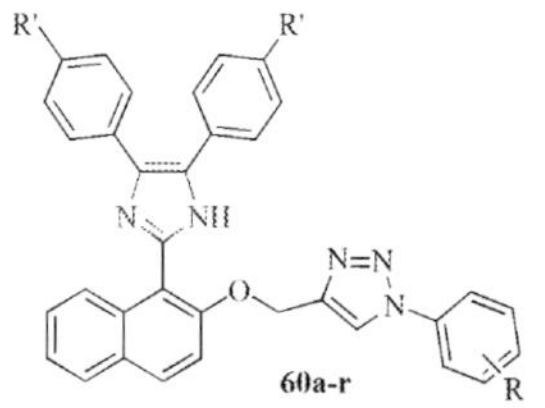

60a-r

60a-60i, R'=H
60j-60r, R'=F

R= a=4-CH$_3$, b=2-CH$_3$-3-Cl, c=3-Cl, d=4-Cl, e=2-NO$_2$, f=2-F, g=4-F, h=4-Br, i=4-OCH$_3$, j=4-CH$_3$, k=3-Cl-2-CH$_3$, l=3-Cl, m=4-Cl, n=2-NO$_2$, o=2-F, p=4-F, q=4-Br, r=4-OCH$_3$

FIG. 5.10 Structure of selective antimicrobial triazole-based imidazole compounds.

strains with MIC values of 3.125–6.25 μg/mL and compounds **58a**, **58c**, and **58d** exhibited significant antitubercular activity having MIC values of 3.125–6.25 μg/mL against *M. tuberculosis* H37Rv strains.

Novel 4-((4-(4,5-diaryl-1*H*-imidazol-2-yl)phenoxy)methyl)-1-substituted-1*H*-1,2,3-triazole derivatives **59(a–r)** were synthesized and their antimicrobial potency were scrutinized against *B. subtilis*, *S. epidermidis*, *E. coli*, *P. aeruginosa*, *A. niger*, and *C. albicans* via a serial dilution method, and the in vitro study data revealed substantial growth inhibition against most of the strains [97]. These derivatives were found to be comparatively more active against fungal strains than that of bacterial strains and the highest inhibition was shown by compounds, **59c–59h** and **59j–59p** against *C. albicans* with MIC values of 0.0057–0.0064 μmol/mL and compound **59p** (0.0113 μmol/mL) against *A. niger* compared to the fluconazole standard with MIC values of 0.0051 and 0.0102 μmol/mL, respectively (Fig. 5.10).

Subsequently, the authors designed and synthesized 2,4,5-trisubstituted-1*H*-imidazole–triazoles and monitored their bioactivity against previously studied microbes and found significant growth inhibition against most of the selected pathogens [98]. The highest activity was displayed by compounds **60c–60i** (0.0052–0.0058 μmol/mL) against *S. epidermidis*, compounds **60c–60f** (0.0055–0.0058 μmol/mL) against *E. coli*, compound **60g** (0.0058 μmol/mL) against *P. aeruginosa*, and compound **60k** (0.0103 μmol/mL) against *A. niger* compared to the reference drugs, ciprofloxacin (0.0047 μmol/mL), and fluconazole (0.0102 μmol/mL), respectively (Fig. 5.10). The docking study of compound **60e** demonstrated good hydrophobic, electrostatic interaction, and H-bonding with active spot of DNA gyrase topoisomerase II of *E. coli*.

2.2.3 Pyrazole

Novel tri-substituted imidazole derivatives **61(a–g)** were synthesized from Radziszewski reaction and studied as antimicrobial agents against various gram-(+ve and −ve) bacterial, fungal, and *M. tuberculosis* H37Rv strains via the serial broth dilution method (Fig. 5.11) [99]. Most of the derivatives exhibited moderate activity against all the strains. However, some of the derivatives were more potent and equally potent compared to the reference, viz. compound **61a** (50 μg/mL) for *E. coli*, **61e** and **61 g** (100 μg/mL) for *P. aeruginosa*, **61d** and **61f** (100 μg/mL) for *S. aureus*, **61d** (50 μg/mL) for *S. pyogenes*, **61d** (200 μg/mL) for *C. albicans*, **61d** (100 μg/mL) for *A. niger*, etc.

A range of biarylpyrazole imidazole hybrids were synthesized and screened as antimycobacterial agents against *M. tuberculosis* H37Rv through Resazurin Microtiter Assay (Fig. 5.11) [100]. The in vitro studies revealed high inhibition activity for all the derivatives and displayed that $-CH_2-$ series (**62**) (MIC values 6.25–25 μg/mL) were more potent than $-C(O)NH(CH_2)_2-$ (**63**) series compared to the clotrimazole (20 μg/mL) and fluconazole (100 μg/mL) reference drugs.

A novel efficient one-pot methodology for the synthesis of pyrazole clubbed imidazo[1,2-*a*]pyrimidine derivatives was introduced and their in vitro antimicrobial activity was evaluated against various pathogens via the disk diffusion method [101]. Most of the derivatives were highly active growth inhibitors against all the strains. All compounds displayed selectivity in high microbe inhibition, viz. compound **64f** against *E. coli*, **64b**, **65b**, and **65e** against *S. typhi*, **64g** against *V. cholera*, **64a**, **64e**, and **65b** against *S. pneumoniae*, **64f** and **65d** against *B. subtilis*, **64b** and **65b** against *C. tetani* possessing an MIC value of 62.5 μg/mL and **65b** against *C. albicans*, and **64b** and **65e** against *A. niger* with

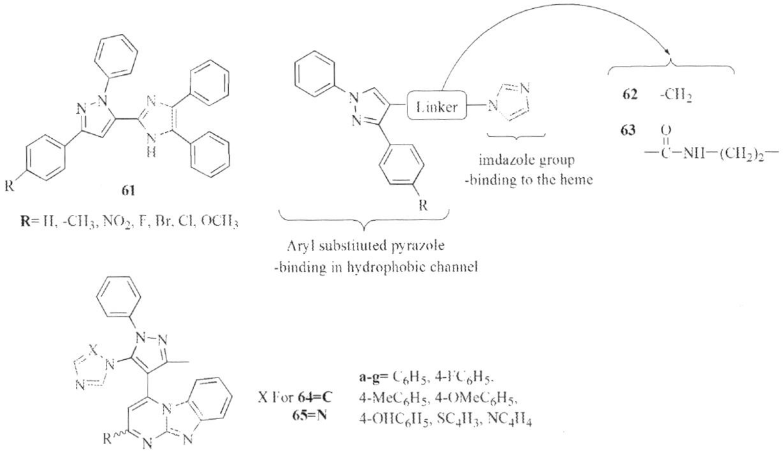

FIG. 5.11 Structure of antimicrobial pyrazole clubbed imidazole derivatives.

an MIC value of 100 μg/mL. Additionally, **64b** and **65e** showed more than 90% inhibition against the *M. tuberculosis* H37Rv strain compared to the standard (Fig. 5.11).

2.2.4 Quinolone/quinoline

In 2011, Desai and companions introduced a facile and efficient microwave-assisted synthesis of *N*-(4-((2-chloroquinolin-3-yl)methylene)-5-oxo-2-phenyl-4,5-dihydro-1*H*-imidazol-1-yl)(aryl)amides and studied their bioactivity against various pathogens through the serial broth dilution method [102]. The in vitro study data indicated that significant growth inhibition was displayed by most of the compounds; additionally, some compounds showed higher activity, viz. **66a** (50 μg/mL) against *E. coli*, **66b** (50 μg/mL) and **66c** (100 μg/mL) against *S. aureus*, and some were equipotent like **66d** and **66e** against *A. niger* with an MIC value of 100 μg/mL against standard drugs (Fig. 5.12). The SAR study revealed that the electronic environment of quinoline-imidazole derivatives affected their selectivity and inhibition activity.

7-Alkyloxy-4,5-dihydro-imidazo[1,2-*a*]quinoline derivatives **67** were synthesized and studied as antibacterial agents against *S. aureus* and *E. coli* against reference drugs, ciprofloxacin and amoxicillin via the broth microdilution method [103]. Compound **67a** was found to be more active with a 0.5 μg/mL MIC value than the reference against *E. coli* and other compounds displayed moderate activity for both pathogens (Fig. 5.12). This most active derivative **67a** further underwent antimicrobial activity against four bacterial strains and exhibited higher potency against *Proteus vulgaris* (16 μg/mL), *Salmonella* (64 μg/mL), and *Pasteurella* (2 μg/mL), and equal inhibition potency against *P. aeruginosa* (16 μg/mL) and *K. pneumoniae* (> 128 μg/mL) compared to the standards.

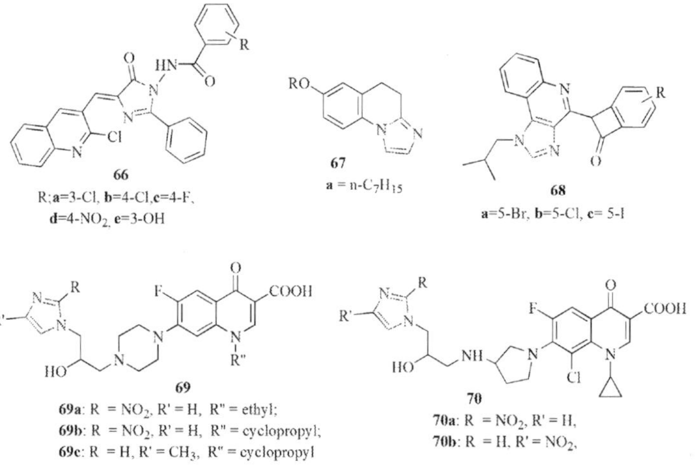

FIG. 5.12 Structure of antimicrobial quinoline clubbed imidazole derivatives.

7-(1-Isobutyl-1*H*-imidazo[4,5-*c*]quinolin-4-yl)-7-azabicyclo[4.2.0]octa-1,3,5-trien-8-ones were synthesized and discovered as antimicrobial agents against various bacterial and fungal strains [104]. All derivatives were good inhibitors and among them, the compounds **68a**, **68b**, and **68c** were found to be equipotent against *S. aureus*, *P. marneffei*, and *T. mentagrophytes* and more potent against *E. coli* with an MIC value of 6.25 μg/mL compared to ampicillin and itraconazole antibiotics (Fig. 5.12).

A wide range of quinolone imidazoles was prepared from multistep synthesis and screened as antimicrobial agents against a panel of microbes via the two-fold broth dilution method [105]. An in vitro microbial study revealed that most of the compounds were highly potent growth inhibitors against the reference antibiotic and less toxic toward human cell lines. Compounds **69a**, **69b**, **69c**, **70a**, and **70b** exhibited higher and equivalent growth inhibition activity, in the MIC value range 0.015–7.74 nmol/L, than that of standards chloramphenicol (8.91–35.66 nmol/L), ciprofloxacin (1.26–8.79 nmol/L), and norfloxacin (1.31–41.79 nmol/L) against most of the pathogens (Fig. 5.12).

2.2.5 Thiadiazole

Imidazo[2,1-*b*][1,3,4]thiadiazole derivatives **71(a–t)** were prepared via multistep synthesis and screened as antibacterial agents against *E. coli*, *K. pneumonia*, *S. typhimurium*, *Y. enterocolitica*, and *P. mirabilis* from an MIC value study [106]. Most of the compounds displayed moderate activity, and compounds **71c**, **71g**, and **71t** were highly active with MIC values of 625 > 5000 μg/mL against selective strains; the results have also been validated from the docking studies (Fig. 5.13).

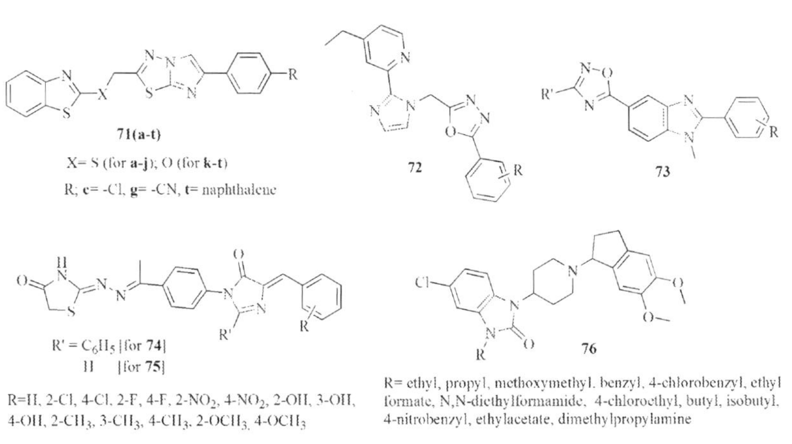

FIG. 5.13 Structure of antimicrobial hybrid imidazole derivatives.

2.2.6 Oxazole/oxadiazole

1,3,4-Oxadiazole clubbed 2-(4-ethyl-2-pyridyl)-1*H*-imidazole derivatives were synthesized and their in vitro antifungal activity was screened against various *Candida* species through a broth microdilution assay (Fig. 5.13) [107]. The results indicated that all the derivatives were highly active, and compounds **72a**, **72b**, and **72c** were the most potent inhibitors against all the strains compared to the reference fluconazole (Table 5.14). The presence and position of various functional groups affected their inhibition potency, viz. oxadiazole, -NO_2 and -NH_2 substituents at the para position enhanced their activity as revealed by SAR. The authors also studied the impact of these molecules on ergosterol biosynthesis, which validated their activity via interaction with active sites of target enzymes.

Hybrid compounds, 2-aryl-5-(3-aryl-[1,2,4]-oxadiazol-5-yl)-1-methyl-1*H*-benzo[*d*]imidazole **73** and their derivatives, were designed and synthesized, and their antimicrobial efficiency was screened against varied pathogens compared to ciprofloxacin and fluconazole (Fig. 5.13) [108]. All derivatives displayed good inhibition against all strains; in particular, derivatives **73a–73e** were highly potent with good ADME properties (Table 5.15).

2.2.7 Thiazolidine

2-((1-(4-(4-(Arylidene)-5-oxo-2-phenyl-4,5-dihydro-1*H*-imidazol-1-yl)phenyl) ethylidene)hydrazono)thiazolidine-4-one derivatives **74(a–o)** were synthesized and utilized as antimicrobial agents against a panel of microbes through a serial broth dilution assay (Fig. 5.13) [109]. Compounds **74n** (100 ± 3.60 μg/mL) against *E. coli*, **74c** (100 ± 1.00 μg/mL) and **74l** (250 ± 1.04 μg/mL) against *P. aeruginosa*, **74l** (100 ± 3.78 μg/mL) and **74o** (50 ± 2.05 μg/mL) against *S. pyogenes*, **74j** (100 ± 1.16 μg/mL) against *C. albicans*, **74g** (100 ± 1.60 μg/mL) against *A. niger*, and **74e** (100 ± 1.50 μg/mL) and **74o** (100 ± 1.00 μg/mL) against *A. clavatus* were found to be the most effective inhibitors against the standards ampicillin and griseofulvin.

After this success, the authors explored their activity via replacing the phenyl group with a hydrogen atom **75(a–o)** and found high antimicrobial potency via the serial broth dilution method against various pathogens such as bacteria and fungi [110]. The *in vitro* antimicrobial study revealed significant growth inhibition, whereas compounds **74f** (100 ± 1.86 μg/mL), **74h** (25 ± 2.18 μg/mL), **74m** (12.5 ± 1.64 μg/mL), **74o** (50 ± 4.16 μg/mL), **74m** (250 ± 2.88 μg/mL), **74h** (50 ± 1.18 μg/mL), and **74n** (25 ± 1.86 μg/mL) were found to be more potent against *E. coli*, *P. aeruginosa*, *S. aureus*, *S. pyogenes*, *C. albicans*, *A. niger*, and *A. clavatus* strains, respectively (Fig. 5.13).

2.2.8 Piperidine

A series of novel 1-(piperidin-4-yl)-1*H*-benzo[*d*]imidazol-2(3*H*)-one derivatives were prepared and the antimicrobial potency was studied against *S. aureus*,

TABLE 5.14 MIC (μg/mL) values of 1,3,4-oxadiazole clubbed imidazole derivatives for fungal strains.

		Laboratory strains		Clinical strains				
				C. albicans	*C. albicans*	*C. albicans*	*C. tropicalis*	*C. tropicalis*
Compound	R	*C. albicans*	*C. tropicalis*	**2323**	**2367**	**2779**	**2029**	**2356**
72a	*p*-NO_2	0.008	0.016	0.016	0.016	0.032	0.016	0.064
72b	*p*-NH_2	0.016	0.016	0.032	0.032	0.064	0.032	0.125
72c	*o*-$OCOCH_3$	0.002	0.004	0.016	0.008	0.016	0.016	0.032
Flc	Cl	0.008	0.016	0.032	0.016	0.032	0.032	0.062

TABLE 5.15 MIC (μg/mL) values of benzimidazole-oxadiazole hybrid against microbes.

Compound	R	R′	*S. aureus*	*E. faecalis*	*E. coli*	*K. pneumoniae*	*C. albicans*	*A. fumigatus*
73a	2-Cl-6-F	4-Chlorophenyl	3.12	1.6	3.12	3.12	1.6	3.12
73b	3-Cl-2-F	4-Chlorophenyl	1.6	1.6	3.12	6.25	0.8	1.6
73c	3-Cl-2-F	3-Chlorobenzyl	1.6	3.12	3.12	12.5	1.6	1.6
73d	2,3-$(Cl)_2$	2-Chlorophenyl	3.12	1.6	3.12	6.25	3.12	6.25
73e	2,3-$(Cl)_2$	3-Chlorobenzyl	6.25	6.25	3.12	3.12	1.6	1.6
Ciprofloxacin			2.0	1.0	2.0	1.0	–	8.0
Flc			–					16.0

S. pyogenes, *E. coli*, *P. aeruginosa*, *C. albicans*, and *A. clavatus* via the disk agar diffusion technique [111]. The growth inhibition study revealed their promising activity; furthermore, compounds **76a**, **76b**, **76k**, and **76e** were found selectively more potent growth inhibitors compared to the standards ampicillin, chloramphenicol, and fluconazole (Fig. 5.13).

2.2.9 Coumarin

Coumarin–imidazole hybrids were synthesized via a one-pot, multicomponent reaction (MCR) and were converted in their acrylates to evaluate them as potent antimicrobial agents against *B. flexus*, *P.* spp., *S.* spp., and *A. tereus* via the broth dilution technique [112]. All of the synthesized compounds exhibited higher growth inhibition activity with MIC values of 0.1–0.7 μg/cm^3 except the compounds **77b** and **78b** for bacterial strains and **77c**, **77d**, **77f**, and **78d** for fungal strains compared to standard ciprofloxacin and nystatin (0.7 μg/cm^3) (Fig. 5.14).

2.2.10 Azepine

3-Biphenyl-*3H*-imidazo[1,2-*a*]azepin-1-ium bromides **80** were synthesized and studied for their inhibition activity against various microbes, i.e., one gram-(+ve) (*S. aureus*), four gram-(−ve) (*E. coli*, *P. aeruginosa*, *K. pneumoniae*, *A. baumannii*) bacteria, and two yeast (*C. albicans*, *C. neoformans*) strains

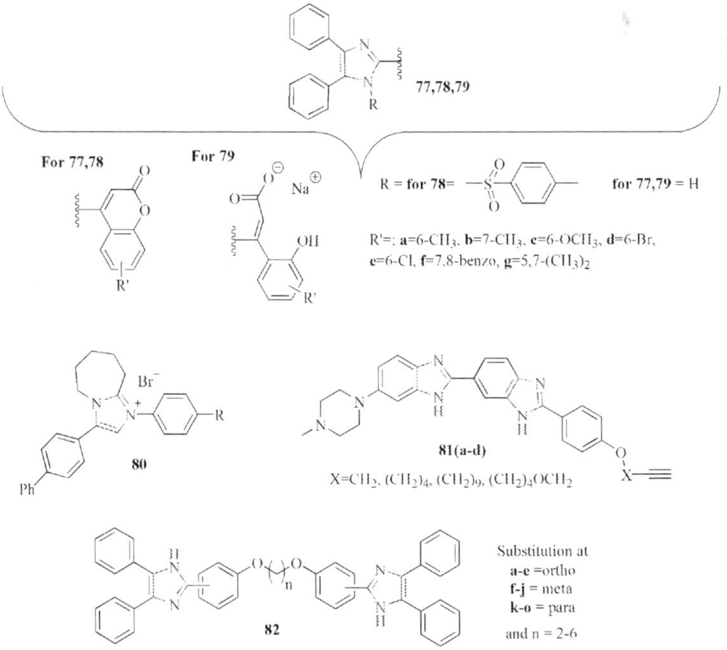

FIG. 5.14 Structure of various antimicrobial clubbed imidazole derivatives.

(Fig. 5.14) [113]. Most of the compounds revealed significant activity against gram-(+ve) bacteria and yeast pathogen, and did not show any hemolysis in human red blood cells (RBCs).

2.2.11 Bis-imidazole and bis-benzimidazole

Hoechst dyes possess a bis-benzimidazole core structure and inhibit the functions of mammalian topoisomerase I and II. By virtue of this, four alkynyl-bisbenzimidazoles derivatives **81** were prepared and studied as bacterial and topoisomerase I inhibitors (Fig. 5.14) [114]. The study on *E. coli* DNA topoisomerase (I and DNA gyrase) and human DNA topoisomerase (I, and II) revealed that all the compounds showed good inhibitory activity against all targets except DNA gyrase. The in vitro antimicrobial study exhibited most of the derivatives as highly active, and compounds **81b** (4 μg/mL) and **81d** (8 μg/mL) displayed better activity than the reference drug against *E. faecalis*.

A range of bis(4,5-diphenylimidazol-2-yl-phenyl)glycols **82(a–o)** were prepared from glycol and benzil/benzoin using iodine/acetic acid and their antimicrobial activity was screened against bacterial strains (*S. aureus*, *B. cereus*, *E. coli*, and *P. vulgaris*) and fungal strains (*A. niger*, *C. albicans*, *A. foetidus*, and *C. rogosa*) via a cup-plate agar diffusion process [115]. All of the compounds were highly active against all strains and compounds **82g**, **82j**, **82m**, and **82o** exhibited higher growth inhibition compared to the reference drugs ampicillin and amphotericin B (Fig. 5.14).

2.2.12 Miscellaneous

A series of amido-linked imidazoles were designed and synthesized, and their utility as antimicrobial agents was studied against various pathogens [116]. The in vitro results depicted moderate activity for all molecules; however, compound **83** was the most active compound against all the tested strains with an MIC value of 12.5 μg for bacterial (*S. aureus*, *B. subtilis*, *P. aeruginosa*, and *K. pneumoniae*) and 25 μg for fungal (*A. niger* and *P. chrysogenum*) strains compared to standard chloramphenicol and ketoconazole (Fig. 5.15).

Substituted 5 benzimidazolyl-2′-deoxyuridine derivatives **84(a–h)** were synthesized via MW-assisted synthesis and their antimicrobial profile was screened against a large panel of microbes using ciprofloxacin and linezolid as reference antibiotics [117]. The *in vitro* results indicated that compounds **84a**, **84b**, **84c**, **84d**, **84g**, and **84h** were highly active growth inhibitors against *S. aureus*, *E. faecalis*, *E. faecium*, and *S. pneumoniae*, having higher than 64 μg/mL MIC values (Fig. 5.15).

A novel range of quinazolinone imidazole derivatives were prepared and scrutinized as antimicrobial drug candidates against various strains, and among all the 33 derivatives, most of them revealed higher or equal potency in comparison to standard antibiotic through *in vitro* antimicrobial assay [118]. Series **85** was comparatively more active than other series and displayed MIC value in the range of 0.02–0.05 μmol/mL for all tested strains except *P. aeruginosa* in

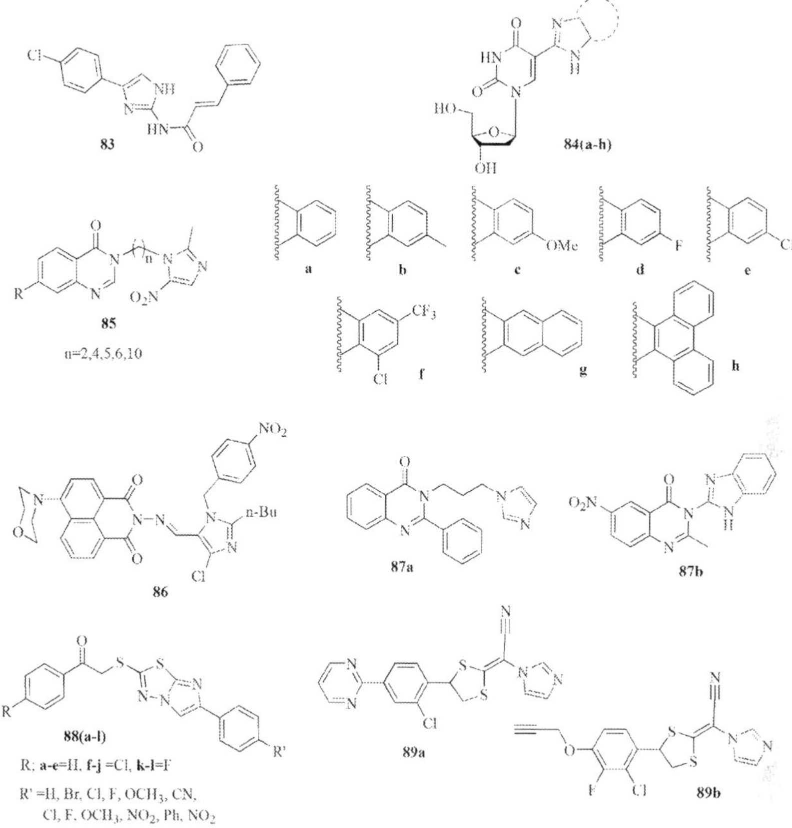

FIG. 5.15 Structure of potent antimicrobial clubbed imidazole compounds.

comparison to fluconazole (0.01–0.84 μmol/mL) and chloromycin (MIC = 0.02–0.10 μmol/mL) (Fig. 5.15).

Schiff base-linked imidazole naphthalimide derivatives were prepared and their antimicrobial behavior was screened against 15 microbes via the microbroth dilution method [119]. The in vitro assessment results illustrated moderate activity of all compounds, whereas compounds **86** exhibited similar activity against *P. aeruginosa* (0.005 μmol/mL) and *C. albicans* (0.003 μmol/mL) in comparison with the standards, norfloxacin and fluconazole. Various parameter studies demonstrated that compounds **86** effectively permeated the microbes' membranes, easily killed selected strains, and good interaction with calf thymus DNA, which validated their anti-MRSA and good therapeutic activity with low drug-resistance (Fig. 5.15).

A series of disubstituted-quinazolin-4(3*H*)-one derivatives has been synthesized and the antimicrobial activity was evaluated against various microbes through the disk diffusion method. Most of the synthesized compounds were good inhibitors against most of the pathogens, although comparatively high activity was displayed against the *S. aureus*, *C. albicans*, *A. niger*, *C. gullerimondii*, and *A. flavus* strains [120]. Compound **87a** showed high inhibition against *S. aureus* (32 µg/mL), *E. coli* (64 µg/mL), and *K. pneumoniae* (16 µg/mL) compared to the standard drug, ciprofloxacin. Moreover, compound **87b** displayed good inhibition with MIC values of 64 and 16 µg/mL against the resistance strains, *S. aureus* and *P. aeruginosa*, respectively (Fig. 5.15).

A range of phenyl substituted imidazo[2,1-*b*][1,3,4]thiadiazole derivatives **88** was prepared and their antimicrobial assay was studied against *S. aureus*, *B. subtilis*, and *E. coli*. The assessment results depicted 10 times more activity of compounds **88a**, **88b**, **88d**, **88g**, **88h**, and **88l** against *S. aureus* and *B. subtilis* with 0.03 µg/mL MIC in comparison to standard chloramphenicol (0.4 and 0.85 µg/mL, respectively) (Fig. 5.15) [121].

Novel ketene dithioacetals-based imidazole derivatives were designed, synthesized, and screened as agrochemical fungicides against *A. solani*, *B. fuckeliana*, *E. necator*, and *Z. tritici* via whole-plant assays [122]. The data revealed that compounds **89a** and **89b** displayed more than 90% efficiency at 60 ppm against all the tested phytopathogens (Fig. 5.15).

2.3 Based on metal complex

2.3.1 Imidazole complexes

In 2011, a novel series of symmetrical and nonsymmetrical *N*-heterocyclic carbenes (NHCs) and their silver acetate complexes were designed, synthesized, and analyzed for antimicrobial activity against *S. aureus* and *E. coli* via the Kirby–Bauer disk-diffusion method [123]. The in vitro study results concluded that complexes **90** were more active toward *S. aureus* than *E. coli* growth inhibition and as the complex concentration was increased, their activity also improved (Fig. 5.16). These complexes displayed high activity compared to their precursors due to chelation, and increased lipophilicity, which controlled the penetration in the cell membrane to deactivate the enzymes of pathogens.

In 2013, McCann and coauthors designed imidazole-based silver(I) complexes and anticipated their biological activity [124]. The results of in vitro antimicrobial study confirmed that the complexes $[Ag_4(DMSO)_4(9\text{-aca})_4]n$ (1.92 µM), $[Ag(2\text{-Me-imH})_2(9\text{-aca})]$ (1.58 µM), $[Ag(4\text{-Ph-imH})_2(9\text{-aca})]$ (2.53 µM), $[Ag(2\text{-Mebenz-imH})_2(9\text{-aca})]$ (2.62 µM), and [Ag(4,5-CN-im)] (1.83 µM) (where 2-Me-imH = 2-methylimidazole; 4-Ph-imH = 4-phenylimidazole; 2-MebenzimH = 2-methylbenzimidazole; Benz-imH = benzimidazole; 2-Ph-imH = 2-phenylbenzimidazole; 4,5-CN-imH = 4,5-dicyanoimidazole) were exceedingly active against *C. albicans* compared to standard ketoconazole (4.70 µM). A number of the complexes offered a good prognosis for *G. mellonella* infected with a lethal dose of *C. albicans* cells.

In 2015, Tabrizi and colleagues demonstrated the synthesis of sodium valproate-based Cd(II), Co(II), Mn(II) imidazole complexes and evaluated their

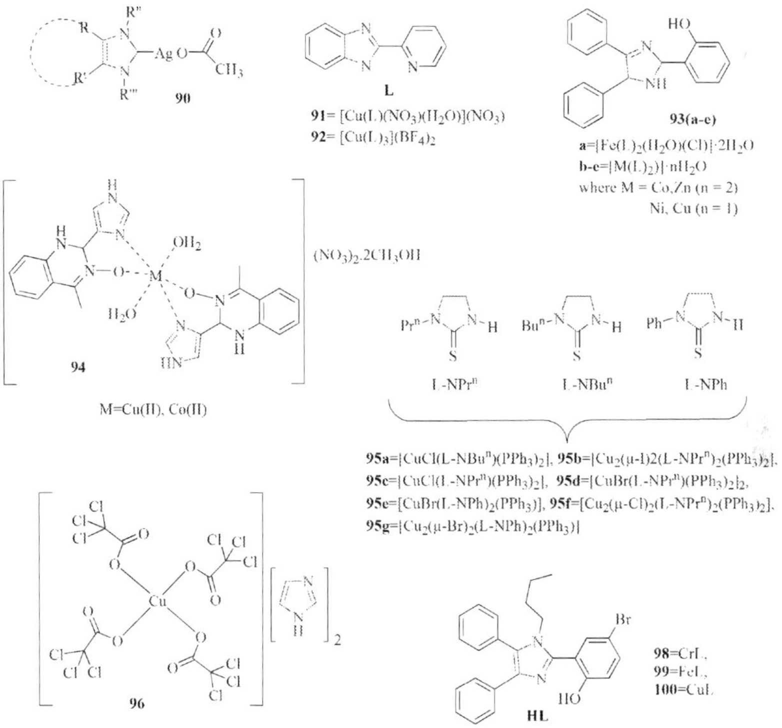

FIG. 5.16 Structure of various antimicrobial imidazole-based metal complexes.

antibacterial activity using an in vitro agar diffusion method against various pathogens [125]. Among the synthesized complexes, Cd(II) and Co(II) complexes displayed excellent activity against *S. aureus* and *B. subtilis*, and good activity against *S. typhi*, *E. coli*, and *P. aeruginosa*.

In 2016, the imidazole-based silver metal complex $(Ag(C_4H_6N_2)_2]_2Cr_2O_7)$ **90** was synthesized and examined for antibacterial activity against *E. coli*, *P. aeruginosa*, *B. subtilis*, and *S. aureus* [126]. The antibacterial activity results demonstrated that the complex reacted with the negatively charged groups of the cell membrane and outer membrane of the examined bacteria. It disrupted the membrane potential of the affected bacteria. In addition to this, N atoms of the 2-methylimidazole rings reduced the (+ve) charge on the Ag(I) ion, which in turn led to the interactions between the (−ve) charges of bacterial cytoplasmic membrane and metal ions. An increase in the number of silver (I), Cr (IV) atoms increased the number of electrostatic interactions with the (−ve) charge of the bacterial cytoplasmic membrane. Thus, the antibacterial activity of the complexes was found to be slightly higher than that of Ag_2CrO_4.

Lewis and coworkers synthesized copper-based imidazole complexes **91** and **92** and assessed their biological activity via MIC value using the agar disk diffusion method (Fig. 5.16) [127]. The complexes demonstrated the highest antibacterial activity; nevertheless, there was no noteworthy difference between the mono- and tri-substituted complexes. There is a significant difference in the antibacterial activity of both complexes, i.e., **91** > **92** against *S. aureus*, but they showed similar activity against *E. coli*.

In 2017, Li and coworkers synthesized two novel Ag(I) complexes based on propyl-substituted imidazole-4,5-dicarboxylic acid, $\{[Ag_2(Hpimdc)(H_2O)](H_2O)\}_n$ and $[Ag_2(Hpimdc)(pda)](2H_2O)\}n$ (where H_3pimdc = 2-propyl-1H-imidazole-4,5-dicarboxylic acid; pda = 1,2-propanediamine) [128]. The antimicrobial results revealed that complex $[Ag_2(Hpimdc)(pda)](2H_2O)\}n$ with MIC being 0.19 mM was more potent against *E. coli* than complex $\{[Ag_2(Hpimdc)(H_2O)](H_2O)\}n$ with MIC being 0.28 mM. The complexes have somewhat diverse effects on the growth of gram-positive bacteria, *S. aureus*, with MICs of 0.28 and 0.24 mM, respectively.

In 2018, Ismael and coworkers demonstrated the synthesis of five metal complexes, $[Fe(L)_2(H_2O)(Cl)]\cdot 2H_2O$ and $[M(L)_2)]\cdot nH_2O$ **93(a–e)** (Fig. 5.16) [129]. The ligand and its metal-complexes were examined against the growth of pathogenic microbes. The results of in vitro activity demonstrated that all the synthesized complexes were more potent in comparison to the ligand. In particular, Zn complexes exhibited 1.3, 1.4, and 1.5 times greater activity than the free ligand against *E. coli*, *B. cereus*, and *A. fumigatus*, respectively. The utmost activity was found for Cu complex with 1.6 and 1.7 times for *E. coli* and *A. fumigatus*, respectively, and the Fe complex was found to show the highest antibacterial activity against *B. cereus*.

In 2018, Abdolmaleki and coworkers synthesized four different coordination complexes using proton transfer compounds like pyridine-2,6-dicarboxylic acid ($pydcH_2$) with 2-methylimidazole (2-mim) and imidazole (imi) [130]. The synthesized complexes $[2\text{-mimH}]_2[Co(pydc)_2][ClO_4]. H_2O$, $[2\text{-mimH}]_2[Zr(pydc)_3]$, $[imiH]_2,[Zr(pydc)_3].4H_2O$, and $[Ca(pydcH_2)2(H_2O)]2[Co(pydc)_2]_2.6H_2O$ were anticipated for antimicrobial activity via agar dilution and diffusion pathways against the pathogenic microbes *S. aureus*, *B. cereus*, *E. coli* and *K. oxytoca*. The highest antibacterial activity was exhibited by complex $[2\text{-mimH}]_2[Zr(pydc)_3]$ at MIC values in the range of 75–175 μg/mL and demonstrated greater effects than gentamicin (standard drug) toward *K. oxytoca*.

In 2018, Chai and coworkers devised an imidazole-based copper and cobalt complexes, $[M(L)_2]$ $(NO_3)_2\cdot 2CH_3OH$ **94** and examined their antibacterial activity and found greater activity due to chelation of metal ion with the ligand (Fig. 5.16) [131]. In comparison to Co (II) complex, Cu (II) complex showed more potent inhibitory activities, which was explained by chelation, i.e., the positive charge of copper and cobalt ions in chelated complex was partly shared with the ligand's donor atoms so that there is an electron delocalization over the entire chelate ring. Furthermore, this increased the lipophilic character of the metal. The antimicrobial potency could be related to the interruption of structural enzymes that played a crucial role in metabolic mechanism of the microorganisms.

In the same year, SKA and coworkers derived a complex having imidazole-4-acetate anion (Iaa), with formula $[Ni(Iaa)_2(H_2O)_2]\cdot H_2O$ [132]. The biophysical studies explained that the complex interacted with DNA without degradation and with BSA as a model protein. This complex, belonging to azoles, might inhibit fungal sterol 14α-demethylase which removed the 14α-methyl group from the lanosterol during ergosterol biosynthesis.

Also in 2018, Lobana and coworkers designed different mono-nuclear and halogen bridge di-nuclear imidazole copper complexes **95(a–g)**, and examined their antimicrobial activity against several pathogens like bacteria (*S. aureus*, *S. epidermidis*, *E. faecalis*, *S. flexneri*, and *E. coli*) and yeast (*C. tropicalis*) (Fig. 5.16) [133]. Some of them were found to be more potent inhibitors than reference antibiotic, viz. complexes **95a** and **95b** showed MIC values, 7 µg/mL against *E. faecalis*, and complexes **95c–95g** displayed 5–7 µg/mL MIC values against *S. epidermidis* compared to standard gentamicin (30 µg/mL).

In 2019, Anastasiadou and coauthors designed bi- and tri-nuclear copper-based metal complexes [134]. The antimicrobial activity of complexes was evaluated by determining their respective IC_{50} and MIC values toward *E. coli*, *X. campestris*, *S. aureus*, and *B. subtilis* bacterial species. The MIC values of the complexes were in the range of 1.11–631.9 µg/mL. The interaction of the complexes with DNA depends on the structure and stability of the complexes as well as the nature of their ligands. The interaction of metal complexes with DNA might take place via formation of covalent bonds or in a noncovalent manner (including intercalation, groove-binding, electrostatic interactions) caused cleavage of the DNA-helix.

In 2019, Ishak and coworkers designed four different Ni(II) complexes having imidazole (im), benzimidazole (bz), and tridentate Schiff bases as ligands [135]. The results of antibacterial activity of the compounds showed that the PT24D Schiff base, Ni(MT24D)im, Ni(MT24D)bz, and Ni(PT24D)bz (where 24D =2.4-dihydroxybenzaldehyde, M=methyl, P=phenyl) complexes have specific and selective activity against *S. aureus*, *B. subtilis*, *P. acnes*, and *E. aerogenes*. The mixed-ligand Ni(II) complexes were found to have better antibacterial activity compared to the Schiff bases; this was due to the reduced polarity of the compounds through the partial distribution of (+ve) charge with the donor atoms of the Schiff base.

In 2019, Obaleye and coworkers devised a Cu(II) complex with trichloroacetic acid and imidazole [136]. The imidazole scaffold was not directly bound to Cu but H-bonding existed between the imidazole and the Cu(II) tetratrichloroacetato complex. The antimicrobial activity was tested on *M. luteus*, *S. aureus*, *E. coli*, *S. typhi*, *K. pneumoniae*, and *C. albicans* using the agar-well diffusion method. The *in vitro* studies demonstrated that complex **96** was capable of hindering the growth of four bacterial strains and effectively showed higher potency against *S. typhi* compared to the parent ligands (Fig. 5.16). Thus, *S. typhi* was found to be greatly susceptible to the copper(II) complex followed by *K. pneumoniae*. This suggested that the binding of the copper ions to the surface

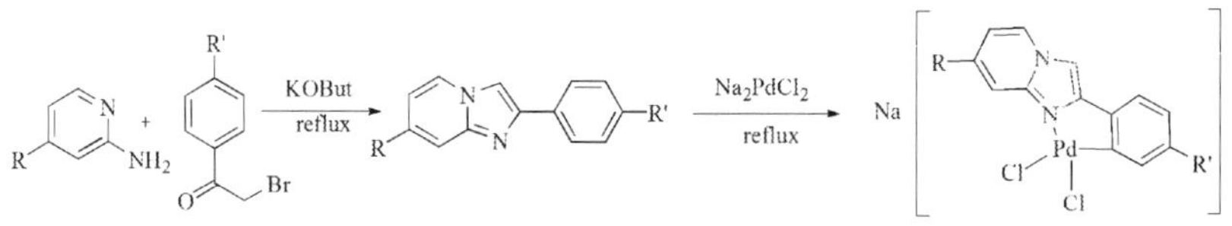

97(a-f)

Compd.	a	b	c	d	e	f
R	H	CH_3	H	CH_3	H	CH_3
R'	H	H	CH_3	CH_3	Cl	Cl

SCHEME 5.20 Synthesis of antimicrobial Pd(II)-based complexes

of bacterial cells affected the respiratory processes of the cells, which played a crucial role in their bactericidal activity.

In 2019, Thakor and coworkers synthesized Pd(II)-based organometallic complexes Na[Pd(3n)Cl_2] **97(a–f)** (where 3n = imidazole ligands) and scrutinized them against *E. coli*, *S. marcescens*, *P. aeruginosa*, *B. subtilis*, and *S. aureus* for the antibacterial activity in terms of MIC using the broth dilution method (Scheme 5.20) [137]. The complexes showed MIC values in the range of 30–125 μM.

Abdel Rahman et al. synthesized novel imidazole-based complexes **98–100** via bonding of 2-(1-butyl-4,5-diphenyl-1*H*-imidazole-2-yl) (4-bromophenol) (HL) ligand with different metals, Cr(III), Fe(III), and Cu(II), and evaluated their antimicrobial activity against bacterial strains (*S. aureus*, *E. coli*, and *P. aeruginosa*) and fungal strains (*A. flavus*, *C. albicans*, and *T. rubrumin*) (Fig. 5.16) [138]. Most of the compounds displayed good antimicrobial potency against selected microbes, and Cu(LH) exhibited highest antimicrobial activity among all the compounds. The authors also studied the interaction of these complexes with CT-DNA and found the intercalative binding mode.

Abdalla et al. designed and synthesized salen-based ligand (S) and their binary (MS) and ternary (MSI) complexes **101** with various metals like Cd, Cu, Ni, Co, etc. and further evaluated their microbial inhibition potency [139]. The reaction of salicylaldehyde and ethylenediamine in ethanol (50 mL) yielded 2,20-{1,2-ethanediylbis [nitrilo(E)methylylidene]}diphenol (salen) and their binary compounds, which coordinated with metal salt (Cd(II), Cu(II), Ni(II) and Co(II), Al(III)) in an aqueous ethanoic solution at 70°C (Scheme 5.21). The antimicrobial efficacy for the MS complexes followed the order: CdS > gentamicin > CuS > NiS > CoS > LaS > AlS > S, whereas MSI complexes potencies are ordered as CuSI > gentamicin > CdSI > NiSI > CoSI > LaSI > AlSI > S.

2.3.2 Benzimidazole complexes

In 2010, Cu(II)- and Fe(II)-based complexes **102** having -(5-Cl/NO_2-1H-benzimidazol-2-yl)-4-Br/NO_2-phenols as a ligand have been synthesized and examined for antimicrobial activity (Fig. 5.17) [140]. The complexes were

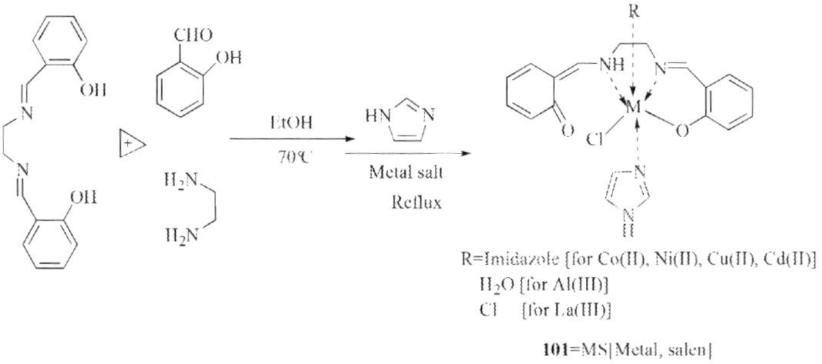

SCHEME 5.21 Synthesis of salen-based metal-complexes by Abdalla et al.

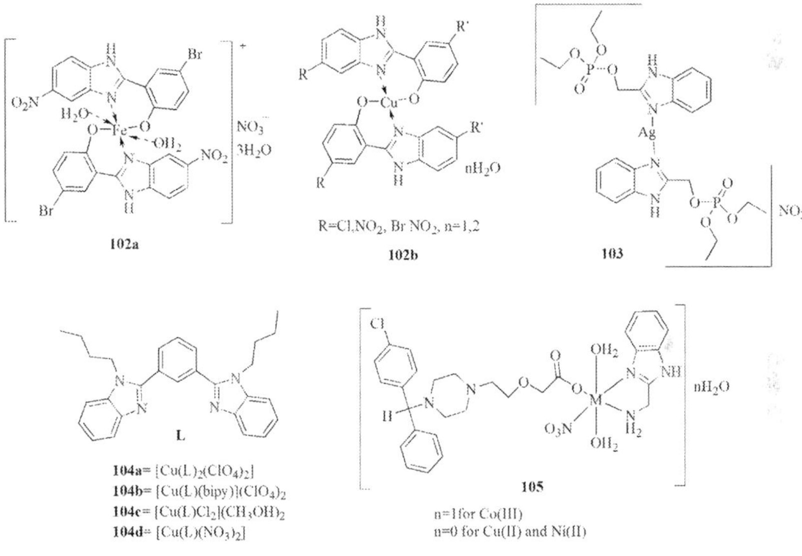

FIG. 5.17 Structure of various antimicrobial benzimidazole-based metal complexes.

anticipated against *S. aureus*, *S. epidermidis*, *E. coli*, *K. pneumoniae*, *P. aeruginosa*, and *P. mirabilis*. Bromo and nitro substituted complexes (**102a** and **102b**) showed good activity against *S. aureus* and *S. epidermidis* with MIC values of 39 μg/mL compared to the reference drug ciprofloxacin.

Furthermore, silver and copper complexes of benzimidazole were synthesized and examined for antimicrobial activity against six strains of gram-(+ve) and gram-(−ve) bacteria and two strains of Candida albicans, yeast [141]. The silver complexes exhibited a broad spectrum of activity against different organisms and the antifungal activity was in the range of 18.8–21.3 μM. The complex **103** was found to be the most active against *P. aeruginosa* and *E. coli* with an MIC value of 37.6 μM (Fig. 5.17).

In 2014, Kose and coworkers designed copper-based metal complexes $[Cu(L)_2(ClO_4)_2]$ and $[Cu(L)(bipy)](ClO_4)_2$ (**104a** and **104b**) and anticipated their potency against different bacterial strains, *B. subtilis*, *B. cereus*, *K. pneumoniae*, *E. coli*, and *E. aerogenes*, and three yeasts, *C. albicans*, *C. utilis*, and *S. cerevisiae*, using the diffusion agar method (Fig. 5.17) [142]. Both the complexes were equally potent toward the bacteria and showed greater activity compared to ampicillin (10μg). The complexes also showed potent antifungal activity against *S. cerevisiae*, *C. utilis*, and *C. albicans* compared with Nystatin 100U (antifungal).

In the same year, the authors followed their work and synthesized *N*-alkylated benzimidazole-based mononuclear copper(II) complexes, $[Cu(L)Cl_2](CH_3OH)_2$ and $[Cu(L)(NO_3)_2]$ (**104c** and **104d**) (Fig. 5.17) [143]. The complexes were examined for in vitro antibacterial and antifungal activities. The complexes exhibited moderate antifungal activities against *S. cerevisiae*, *C. albicans*, and C. utilis. To determine the MIC, microdilution broth susceptibility assay was used, and to estimate the microbial growth, the turbidimetric method was used. The complexes did not possess antibacterial activity against *B. subtilis*, *B. cereus*, and *E. aerogenes*. These complexes were more effective against *S. cerevisiae* than the other bacterial and fungal strains with MIC values of 75 and 100μgmL^{-1}.

In 2015, cetirizine-modified imidazole-based bivalent metal complexes **105** were synthesized and examined for antimicrobial activity against different gram-(+ve) and gram-(−ve) bacteria (Fig. 5.17) [144]. The complexes were more active against gram-(+ve) bacteria compared to gram-(−ve) bacteria. The activity of complexes followed the order Cu(II) > Ni(II) > Co(II).

A few years later, in 2018, Mahmood and coworkers designed different Schiff bases of benzimidazole and fabricated their metal complexes with Zn(II), Pd(II), Ni(II), and Cu(II) (Fig. 5.18) [145]. These metal complexes **106(a–d)** exhibited interactions with the minor groove of DNA. A UV–vis spectroscopy technique was used to examine the binding ability of small molecules of DNA. The results of the experimental study revealed that complexes **106a**, **106c**, and **106d** interacted with DNA since the distinct hypo-chromic effect proposed the formation of DNA-ligand complexes. The Pd(II) complex showed lack of DNA binding due to the large atomic radius of the metal core, which obstructed the potential intercalation or minor groove binding. The Zn(II) complex stuck to the DNA by intercalation due to the planarity of the Schiff base that triggered the intramolecular H-bond formation between the N atom of imino and the OH of the phenolic group. Cu(II) complexes provoked the reactive oxygen species (ROS), thereby damaging DNA. The MIC values were in the range of 7.5–166.7μg/mL.

Also in 2018, Casanova and coworkers synthesized metal(II) complexes **107** with a 1*H*-anthra[1,2-*d*]imidazol-6,11-dione-2-[2-hydroxyphenyl] ligand that has the capability to form di-nuclear metal species via oxygen atoms that acted as a bridge between two metal atoms (Fig. 5.18) [146]. This was possible because of the presence of weakly acidic N–H and O–H groups. The in vitro

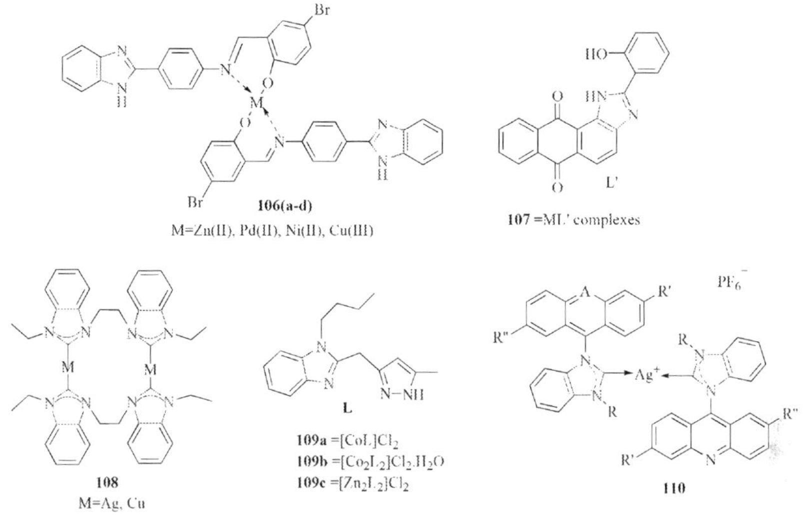

FIG. 5.18 Structure of selected antimicrobial benzimidazole-based metal complex.

antimicrobial activity of these metal complexes and ligands was examined against various bacteria and fungi. $[NiL(bpy)]_2$ and all tested Cd(II) complexes were the most active compounds, showing the highest inhibitory effect against bacilli (MIC 1.5–3 μg/mL) and *Sarcina*, *Streptococci*, and Haemophilus influenz*ae*, bacterial strains (MIC 12–50 μg/mL), while almost no antifungal properties were observed.

In the same year, Dileepan and coworkers devised a novel benzimidazole ligand and synthesized Ag(I) and Cu(I) N-heterocyclic carbene complexes **108** and utilized them as antimicrobial agents (Fig. 5.18) [147]. Among the synthesized complexes and ligands, Ag–NHC strongly hampered the growth of fungal strains in MIC ranging from 46 to 750 μg/mL. A distinct reduction in cell growth was found with *C. albicans* when treated with Ag–NHC complex. This complex caused severe membrane damage and accumulation of intracellular reactive oxygen species (ROS).

In 2020, Chkirate and coauthors devised a pyrazolyl-benzimidazole complexes of Co(II) and Zn (II). Three mono-nuclear coordination complexes $[CoL]Cl_2$, $[Co_2L_2]Cl_2$. H_2O and $[Zn_2L_2]Cl_2$ **109(a–c)** with ligand 1-butyl-2-((5-methyl-1*H*-pyrazol-3-yl)methyl)-1*H*-benzimidazole (L) were synthesized (Fig. 5.18) [148]. The complex and ligands were examined against *E. coli*, *P. aeruginosa*, and *S. aureus*. According to the study, the metal complexes were found to be more active than the ligand (Table 5.16). The increase in activity of the complexes can be demonstrated with Overtone's concept on cell permeability and Tweedy's chelation theory. In accordance with Overtone, the lipid membrane that neighbors the cell only favors the transition of lipid-soluble materials, and is responsible for the antifungal activity. Upon chelation, there is a

TABLE 5.16 MIC value (μg/mL) of pyrazolyl-benzimidazole complexes of Co(II) and Zn(II).

Compound	*S. aureus*	*E. coli*	*P. aeruginosa*
L	12.5	50	50
109a	6.25	25	50
109b	6.25	6.25	50
109c	12.5	12.5	50
Chloramphenicol	12.5	6.25	6.25

reduction in the polarity of metal ions due to the overlap of the ligand orbital and partial sharing of the positive charge of the metal ions with donor groups. Moreover, it increased the delocalization of p-electrons upon the entire chelate ring and improved the lipophilicity of the complexes. This augmented lipophilicity improved the penetration of the complexes into the lipid membrane and restricted further multiplicity of the microorganisms.

A range of 9-substituted acridine-based *N*-alkylated benzimidazolium salts and their Ag(I) carbene complexes **110** were synthesized and screened against *E. coli*, *S. aureus*, *S. epidermis*, *P. aeruginosa*, and *S. typhi* using ampicillin as a reference drug (Fig. 5.18) [149]. Moderate activity was displayed by all complexes, and the complexes displayed comparatively higher inhibition activity than their ligands.

2.4 Based on imidazolium salt or ionic liquids

Imidazolium-based ionic liquids and salt demonstrated wide-ranging applications in the field of medicinal chemistry and drug-designing due to their unique structural properties like ionic nature, solubility, lipophilicity, and ease of functionality [150, 151].

Cationic imidazolium-based ethoxyether amphiphilic ionic liquids **111** displayed superior antimicrobial activity against various pathogens (Fig. 5.19) [152]. This polymer break of the bioenvelope of infectious microbes and longer alkyl chain polymer displayed comparatively higher inhibition potency. The molecule **111c** exhibited MIC values of 5 μM, 89 μM, and 13 μM against *A. baumannii*, *P. aeruginosa*, and *E. coli*, respectively, and 7 μM against *S. aureus* and *S. agalactiae*.

Imidazolium chiral ionic liquids **112** were assessed for antimicrobial activity against a wide range of microorganisms through a different method (Fig. 5.19) [153]. Higher antimicrobial potency was exhibited by ionic liquids which comprised long alkyl chains. The molecules inhibited the action of microorganisms through cell membrane destruction.

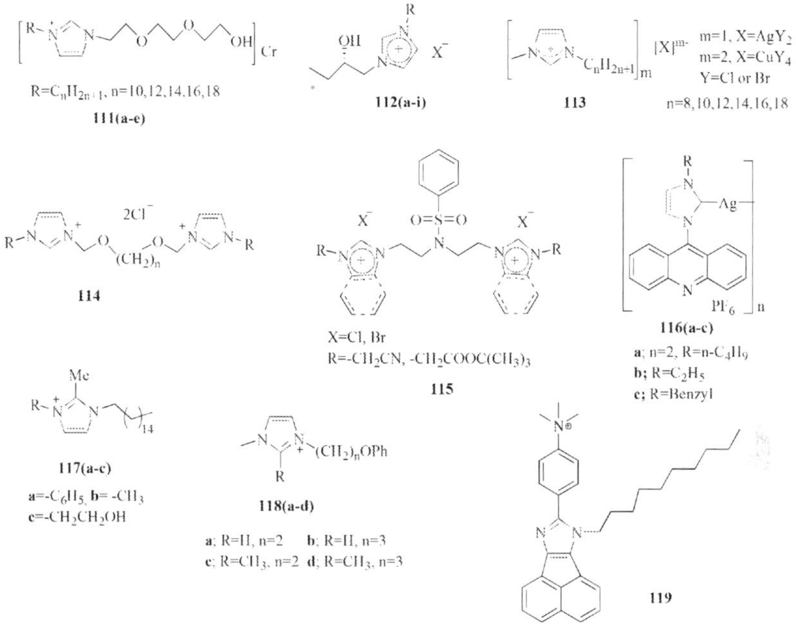

FIG. 5.19 Structure of antimicrobial imidazolium salt.

In 2013, a range of Ag- and Cu-containing 1-alkyl-3-methylimidazolium ionic liquids **113** were screened as antimicrobial agents against a wide range of microbes (Fig. 5.19) [154]. [C_{16}mim][$AgBr_2$] and [C_{14}mim][$AgBr_2$] were most potent against gram-(+ve) and gram-(−ve) bacteria, respectively, and Ag(I) anionic ionic liquids were more active than other ionic liquids in growth inhibition. [C_{16}mim][$AgBr_2$] demonstrated 3.61 μmol/L MIC and 7.23 μmol/L MBC values against *S. epidermidis* and *S. aureus*, respectively. Pałkowski and coworkers introduced the efficient synthesis of novel 70 gemini imidazolium-based chloride ionic liquids **114** and studied their antimicrobial activity and SAR data (Fig. 5.19) [155]. The in vitro data revealed that all of the synthesized molecules displayed broad-spectrum growth inhibition and their activity was affected by surface active properties, type of substituents, their position, and length of alkyl substituent.

Furthermore, bis-imidazolium and benzimidazolium-based gemini-type ionic liquids **115** were prepared and their antibacterial activity were screened against 10 bacterial strains via micro-broth dilution bioassay using amoxicillin and kanamycin as standard antibiotics [156]. Acetonitrile substituents showed highest growth inhibition, i.e., 0.05 mg/mL MIC against *B. subtilis*, *E. coli*, and *A. calcoaceticus*. Among all the tested strains, gram-(+ve) bacteria were more sensitive compared to gram-(−ve) bacteria (Fig. 5.19).

The acridine-imidazolium silver oxide salt-based coordination polymers **116(a–c)** were assessed for antimicrobial activity against a panel of pathogens [157]. The *in vitro* study data revealed that **116a**, **116b**, and **116c** have 16, 8 and 32 mg/mL MIC values, respectively, against most of the tested strains (Fig. 5.19).

The design, synthesis, and antimicrobial activity of 30 diverse NH125 (an imidazolium cation with a 16-membered fatty tail) analogs **117** were studied against a wide range of microbes [158]. Compounds **117a**, **117b**, and **117c** displayed high microbe growth inhibition, excellent biofilm eradication, and more rapid MRSA persister cell death compared to the standards (Fig. 5.19).

A range of novel imidazolium salt and ionic liquids **118** was prepared via MW-assisted synthesis and they were evaluated as potent antimicrobial agents [159]. Most of the compounds exhibited high inhibition; however, **118(b–d)** were equipotent against *S. aureus* with 16 μg/mL MIC value, and **118c** were equipotent against *A. baumannii* with 32 μg/mL MIC value compared to standard ampicillin and rifampicin, respectively (Fig. 5.19).

Novel highly substituted fluorescent imidazoles **119** were prepared which displayed high inhibition potency against a range of bacterial strains, including both gram-(+ve) and-(−ve) strains, and especially methicillin-resistant *S. aureus* (MRSA) (MIC = 8 μg/mL) compared to standard kanamycin and ampicillin antibiotic drugs using a broth microdilution process (Fig. 5.19) [160]. The design of compounds displayed special properties which were responsible for their antibacterial activity; for example, the imidazole ring enhanced the possibilities of diversified substitution, the long alkyl chain elevated the hydrophobicity, membrane permeability, and cellular uptake, the ionic structure enhanced the interaction with bacteria and genomic DNA via electrostatic forces, and the acenaphthene motif made feasible the observation of cell uptake and antibacterial mode of action from fluorescence monitoring.

2.5 Based on polymer

In past decades, antimicrobial polymers have gained much interest due to their special features such as eco-friendliness, low toxicity, long lifetime, efficiency, and low resistance compared to the smaller molecules. Imidazole-based polymers displayed various interactions in living organisms like electrostatic, self-assembly, and aggregation to inhibit the growth of microorganisms [161, 162].

Vinyl imidazole (VI)-based silica polymer nanocomposites (SBA/VI) were screened for antibacterial activity against different bacterial strains [163]. The in vitro study results revealed their good bioactivity and the activity increased with increasing concentration of VI and was found to be maximum at SBA/ VI_{55} (5% of styrene and 5% of VI) with 20 mm ZOI. The authors explained the mechanistic pathway of inhibition through electrostatic interactions among vinyl imidazole's positive charge and the bacterial membrane's negative charge and hydrophobic interactions in alkyl chain of polymer and cytoplasmic film of pathogen, which caused agglomeration and cell death.

The antibacterial potential of the imidazolium ion-fabricated electrospun membrane (Bmim-PU) was studied against two types of bacteria and improved inhibition was found compared to that without fabrication [164]. The Bmim-PU431 membrane demonstrated 91.26% and 88.14% inhibition against *E. coli* and *S. aureus*, respectively, through breaking of the cell wall via electrostatic interaction and accumulation. In 2017, five novel Ag(I)-based coordination polymers were synthesized from bis-(imidazole-2-thione) ligands; they were screened for antimicrobial efficiency and exhibited broad-spectrum activity [165]. The presence of both positive and negative charge centers increased their interaction with microbes' cell membranes and showed higher inhibition compared to ligands, and the activity was also affected by the number of Ag atoms, polymer structure, and size. After this success, recently, three new Ag(I)-based 3D supra-molecular coordination imidazole polymers were synthesized and they showed higher antibacterial activity against tested strains compared to their salts with 0.5–4 mg/mL MBC and MIC values [166].

3. Conclusion and future perspectives

Infectious diseases directly affect millions of people each year and have caused significant increases in morbidity and mortality rates in living organisms. This chapter has thrown light upon the global account of antimicrobial activity of imidazole compounds based on their structural variations like central ring formation, hybrid with other heterocycles, various metal-complexes, polymers, imidazolium ionic liquids, and salts, which show the versatility of present pharmacophores in the context of microbial treatment. The in silico study and SAR have been discussed due to their different molecular targets, which provides a detailed insight on molecular interactions with an active site of the protein and is also useful for future drug development. The abovementioned findings and impressive development validate the higher and equal antimicrobial efficiency of these motifs, compared with the standard antibiotics available in the market. The present study provides a low-height flying bird's eye view of imidazole-based antimicrobial agents to medicinal chemists and drug designers via combining chemical and biological information. This molecule compiles unique characteristics and versatile structural features, yet several protocols and methodologies are in progress to prepare efficient drugs. Further efforts for the design and development of novel, safe, and less resistant drugs are still ongoing, and breakthroughs are eagerly awaited. The path of drug discovery is arduous, and dexterous researchers face various hurdles to combat ailments. Some challenges for future scientists are summarized as follows:

- Discover novel, safe, economical, and eco-friendly pathways to develop future drugs because, for a drug molecule, purity and stereochemistry are the most acute parameters in reference to human beings and medicinal industrial perspectives.

- Discover hybrid pharmacophores that will be effective against different targets for broad-spectrum activity, high tolerability with good pharmacokinetics, and high stability [167].
- Draw attention to in silico and other computational studies to find various binding patterns to discover novel sites of interaction.
- Focus more on experimental studies in animal models and in vivo assay to make available drug candidates for the use of humankind.
- Place more emphasis on the study of the toxicity of synthesized drugs to abolish their adverse effects.
- Study the various aspects of resistance like intrinsic, acquired and evolved for better understanding of mutation in microbes' genotype [168].
- The most attention should be devoted to the generation of multidrug-resistant pharmacophores, as the WHO has also published a list of 12 drug-resistant pathogens that urgently require new drug formations to ensure the safety of public health [169–171].
- Investigate the novel mechanism of action triggered by modified imidazole compounds.
- Most importantly, spread public awareness about the improper use of antibiotics by academia, nongovernmental organizations (NGOs), industries, and government committees.

Acknowledgments

The authors are grateful to the Department of Chemistry, Mohan Lal Sukhadia University, Udaipur (Raj.), India, for providing necessary library facilities for carrying out the work. N.S. is deeply grateful to the Council for Scientific and Industrial Research (CSIR) (file no. 09/172(0088)2018-EMR-I), New Delhi, for providing a Senior Research Fellowship as financial support. A.S. is grateful to University Grants Commission (UGC) for providing a fellowship to carry out this work.

Conflict of interest

The authors declare no conflict of interest, financial or otherwise.

Funding source

This work was supported by the CSIR [file no. 09/172(0088)2018-EMR-I], UGC-MANF (201819MANF-2018-19-RAJ-91971).

References

[1] Wright AJ. The penicillins. Mayo Clin Proc 1999;74(3):290–307. https://doi.org/10.4065/74.3.290.

[2] Gligić BD, Djuran MI. Gold complexes as antimicrobial agents: an overview of different biological activities in relation to the oxidation state of the gold ion and the ligand structure. Dalton Trans 2014;43:5950–69. https://doi.org/10.1039/c4dt00022f.

[3] Ansari A, Ali A, Asif M, Shamsuzzaman. Review: biologically active pyrazole derivatives. New J Chem 2017;41:16–41. https://doi.org/10.1039/c6nj 03181a.
[4] Hiremathad A, Patil MR, Chethana KR, Chand K, Santos MA, Keri RS. Benzofuran: an emerging scaffold for antimicrobial agents. RSC Adv 2015;5:96809–28.
[5] Baul TSB. Antimicrobial activity of organotin(IV) compounds: a review. Appl Organomet Chem 2008;22:195–204. https://doi.org/10.1002/aoc.1378.
[6] Goel S, Parihar PS. Plant-derived quinones as a source of antibacterial and anticancer agents. In: Meshram V, Singh J, Meshram V, Gupta M, editors. Bioactive natural products in drug discovery. Singapore: Springer; 2020. https://doi.org/10.1007/978-981-15-1394-7_6.
[7] Górniak I, Bartoszewski R, Króliczewski J. Comprehensive review of antimicrobial activities of plant flavonoids. Phytochem Rev 2019;18:241–72. https://doi.org/10.1007/s11101-018-9591-z.
[8] Cushnie TPT, Cushnie B, Lamb AJ. Alkaloids: an overview of their antibacterial, antibiotic-enhancing and antivirulence activities. Int J Antimicrob Agents 2014;44(5):377–86.
[9] Kathiravan MK, Salake AB, Chothe AS, Dudhe PB, Watode RP, Mukta MS, Gadhwe S. The biology and chemistry of antifungal agents: a review. Bioorg Med Chem 2012;20:5678–98.
[10] Gaba M, Mohan C. Development of drugs based on imidazole and benzimidazole bioactive heterocycles: recent advances and future directions. Med Chem Res 2016;25:173–210. https://doi.org/10.1007/s00044-015-1495-5.
[11] Bambeke FV. Glycopeptides and glycodepsipeptides in clinical development: a comparative review of their antibacterial spectrum, pharmacokinetics and clinical efficacy. Curr Opin Investig Drugs 2006;7(8):740–9.
[12] Bush K, Bradford PA. β-Lactams and β-lactamase inhibitors: an overview. Cold Spring Harb Perspect Med 2016;6(8):a025247. https://doi.org/10.1101/cshperspect.a025247.
[13] Krause KM, Serio AW, Kane TR, Connolly LE. Aminoglycosides: an overview. Cold Spring Harb Perspect Med 2016;6(6):a027029. https://doi.org/10.1101/cshperspect.a027029.
[14] Nelson ML, Levy SB. The history of the tetracyclines. Ann N Y Acad Sci 2011;1241:17–32. https://doi.org/10.1111/j.1749-6632.2011.06354.x.
[15] Tenson T, Lovmar M, Ehrenberg M. The mechanism of action of macrolides, lincosamides and streptogramin B reveals the nascent peptide exit path in the ribosome. J Mol Biol 2003;330(5):1005–14. https://doi.org/10.1016/S0022-2836(03)00662-4.
[16] Heeb S, Fletcher MP, Chhabra SR, Diggle SP, Williams P, Cámara M. Quinolones: from antibiotics to autoinducers. FEMS Microbiol Rev 2011;35(2):247–74. https://doi.org/10.1111/j.1574-6976.2010.00247.x.
[17] Campoy S, Adrio JL. Antifungals. Biochem Pharmacol 2017;133:86–96. https://doi.org/10.1016/j.bcp.2016.11.019.
[18] Shafii M, Peyton L, Hashemzadeh M, Foroumadi A. History of the development of antifungal azoles: a review on structures, SAR, and mechanism of action. Bioorg Chem 2020;104:104240. https://doi.org/10.1016/j.bioorg.2020.104240.
[19] Aminov R. History of antimicrobial drug discovery – major classes and health impact. Biochem Pharmacol 2017;133:4–19. https://doi.org/10.1016/j.bcp.2016.10.001.
[20] Deak D, Outterson K, Powers JH, Kesselheim AS. Progress in the fight against multidrug-resistant bacteria? A review of US Food and Drug Administration approved antibiotics, 2010-2015. Ann Intern Med 2016;165:363–72.
[21] Burnham CAD, Leeds J, Nordmann P, O'Grady J, Patel J. Diagnosing antimicrobial resistance. Nat Rev Microbiol 2017;15(11):697–703.
[22] Baym M, Stone LK, Kishony R. Multidrug evolutionary strategies to reverse antibiotic resistance. Science 2016;351(6268):aad3292.

[23] Piddock LJV. Understanding drug resistance will improve the treatment of bacterial infections. Nat Rev Microbiol 2017;15(11):639–40.

[24] Marson CM. Saturated heterocycles with applications in medicinal chemistry. Adv Heterocycl Chem 2017;121:13–33.

[25] Zhuang J, Ma S. Recent development of pyrimidine containing antimicrobial agents. Chem Med Chem 2020;15(20):1875–86. https://doi.org/10.1002/cmdc.202000378.

[26] Henary M, Kananda C, Rotolo L, Savino B, Owensab EA, Cravotto G. Benefits and applications of microwave-assisted synthesis of nitrogen containing heterocycles in medicinal chemistry. RSC Adv 2020;10:14170. https://doi.org/10.1039/d0ra01378a.

[27] Kumaria A, Singh RK. Medicinal chemistry of indole derivatives: current to future therapeutic prospectives. Bioorg Chem 2019;89:103021. https://doi.org/10.1016/j.bioorg.2019.103021.

[28] Sahiba N, Sethiya A, Soni J, Agarwal DK, Agarwal S. Saturated five-membered thiazolidines and their derivatives: from synthesis to biological applications. Top Curr Chem (Z) 2020;378:34. https://doi.org/10.1007/s41061-020-0298-4.

[29] Sahiba N, Agarwal S. Recent advances in the synthesis of perimidines and their applications. Top Curr Chem (Z) 2020;378:44. https://doi.org/10.1007/s41061-020-00307-5.

[30] Wright JB. The chemistry of the benzimidazoles. Chem Rev 1951;48(3):397–541.

[31] Arrang JM, Garbarg M, Lancelot JC, Lecomte JM, Pollard H, Robba M, Schunack W, Schwartz JC. Highly potent and selective ligands for a new class H3 of histamine receptor. Invest Radiol 1988;23:S130–2.

[32] Sugumaran M, Kumar MY. Synthesis and biological activity of novel 2, 5-disubstituted benzimidazole derivatives. Int J Pharm Sci Drug Res 2012;4(1):80–3.

[33] Verma A, Joshi S, Singh D. Imidazole: having versatile biological activities. J Chem 2013;1:2013.

[34] Massal S, Santol RD, Retico A, Artico M, Simonett N, Fabrizi PG, Lamba D. Antifungal agents. 1. Synthesis and antifungal activities of estrogen-like imidazole and triazole derivatives. Eur J Med Chem 1992;27:495–502.

[35] Elagamey AGA, Harbb AFA, Khodeir MN, SZA S. Synthesis of hnidazole derivatives containing biologically active units. Arch Pharm Res 1987;10(3):153–7.

[36] Melaiye A, Sun Z, Hindi K, Milsted A, Ely D, Reneker DH, Tessier CA, Youngs WJ. Silver(I)-imidazole cyclophane gem-diol complexes encapsulated by electrospun tecophilic nanofibers: formation of nanosilver particles and antimicrobial activity. J Am Chem Soc 2005;127:2285–91.

[37] Sharma D, Narasimhan B, Kumar P, Judge V, Narang R, Clercq ED, Balzarini J. Synthesis, antimicrobial and antiviral evaluation of substituted imidazole derivatives. Eur J Med Chem 2009;44:2347–53.

[38] Tullio V, Cuffini AM, Carlone NA. In vitro antifungal activities of new imidazole salts towards dermatophytes. Mycoses 1990;33(5):257–63.

[39] Uchida K, Nishiyama Y, Tanaka T, Yamaguchi H. In vitro activity of novel imidazole antifungal agent NND-502 against Malassezia species. Int J Antimicrob Agents 2003;21:234–8.

[40] Pye GW, Marriot MS. Inhibition of sterol C14 demethylation by imidazole-containing antifungals. Sabouraudia 1982;20(4):325–9.

[41] Zampieri D, Mamolo MG, Vio L, Banfi E, Scialino G, Fermeglia M, Ferronec M, Pricl S. Synthesis, antifungal and antimycobacterial activities of new bis-imidazole derivatives, and prediction of their binding to P45014DM by molecular docking and MM/PBSA method. Bioorg Med Chem 2007;15:7444–58.

[42] Rodrigues AD, Gibson GG, Ioannides C, Parke DV. Ideations of imidazole antifungal agents with purified cytochrome P-450 proteins. Biochem Pharmacol 1987;36(24):4277–81.

[43] Khalafi-Nezhad A, Rad MNS, Mohabatkar H, Asraria Z, Hemmateenejad B. Design, synthesis, antibacterial and QSAR studies of benzimidazole and imidazole chloroaryloxyalkyl derivatives. Bioorg Med Chem 2005;13:1931–8.

[44] Liang J, She J, He H, Fan Z, Chen S, Li J, Liu B. A new approach to fabricate polyimidazolium salt (PIMS) coatings with effient antifouling and antibacterial properties. Appl Surf Sci 2019;478:770–8. https://doi.org/10.1016/j.apsusc.2019.01.228.

[45] Rokon Ul Karim M, Harunari E, Oku N, Akasaka K, Igarashi Y. Bulbimidazoles A-C, antimicrobial and cytotoxic alkanoyl imidazoles from a marine gammaproteobacterium microbulbifer species. J Nat Prod 2020;83(4):1295–9. https://doi.org/10.1021/acs.jnatprod.0c00082.

[46] Jayabharathi J, Thanikachalam V, Rajendraprasath N, Saravanan K, Perumal MV. Antioxidant potential and antimicrobial screening of some novel imidazole derivatives: greenway efficient one pot synthesis. Med Chem Res 2012;21:1850–60. https://doi.org/10.1007/s00044-011-9702-5.

[47] Desai NC, Pandya MR, Rajpara KM, Joshi VV, Vaghani HV, Satodiya HM. Synthesis and antimicrobial screening of novel series of imidazo-[1,2-a]pyridine derivatives. Med Chem Res 2012;21:4437–46. https://doi.org/10.1007/s00044-012-9988-y.

[48] Gupta GK, Saini V, Khare R, Kumar V. 1,4-Diaryl-2-mercaptoimidazoles derivatives as a novel class of antimicrobial agents: design, synthesis, and computational studies. Med Chem Res 2014;23:4209–20. https://doi.org/10.1007/s00044-014-0994-0.

[49] Desai NC, Vaghani HV, Rajpara KM, Joshi VV, Satodiya HM. Novel approach for synthesis of potent antimicrobial hybrid molecules containing pyrimidine-based imidazole scaffolds. Med Chem Res 2014;23:4395–403. https://doi.org/10.1007/s00044-014-1005-1.

[50] Kalhor M, Salehifar M, Nikokar I. Synthesis, characterization, and antibacterial activities of some novel N,N'-disubstituted thiourea, 2-amino thiazole, and imidazole-2-thione derivatives. Med Chem Res 2014;23:2947–54. https://doi.org/10.1007/s00044-013-0883-y.

[51] Girish YR, Sharathkumar KS, Thimmaiah KN, Rangappa KS, Shashikanth S. ZrO_2-β-cyclodextrin catalyzed synthesis of 2,4,5-trisubstituted imidazoles and 1,2-disubstituted benzimidazoles under solvent free conditions and evaluation of their antibacterial study. RSC Adv 2015;5:75533–46. https://doi.org/10.1039/C5RA13891D.

[52] Nural Y, Gemili M, Seferoglu N, Sahin E, Ulger M, Sari H. Synthesis, crystal structure, DFT studies, acid dissociation constant, and antimicrobial activity of methyl 2-(4-chlorophenyl)-7a-((4-chlorophenyl)carbamothioyl)-1-oxo-5,5-diphenyl-3-thioxo-hexahydro-1H-pyrrolo[1,2-e]imidazole-6-carboxylate. J Mol Struct 2018;1160:375–82. https://doi.org/10.1016/j.molstruc.2018.01.099.

[53] de NR E, Macı'as M, Castillo JC, Sortino M, Svetaz L, Zacchino S, Portilla J. Synthesis and in vitro antifungal evaluation of novel n-substituted 4-aryl-2-methylimidazoles. ChemistrySelect 2018;3:5220–7.

[54] Salhi L, Achouche-Bouzroura S, Nechak R, Nedjar-Kolli B, Rabia C, Merazig H, Poulain-Martini S, Dunach E. Synthesis of functionalized dihydroimidazo[1,2- A]pyridines and 4-thiazolidinone derivatives from maleimide, as new class of antimicrobial agents. Synth Commun 2020;50(3):412–22. https://doi.org/10.1080/00397911.2019.1699933.

[55] Basha SM, Varalakshmi M, Basha ST, Venkataramaiah C, Shafi SS, Raju CN, Rajendra W. Synthesis, spectral characterization, and biological evaluation of imidazole-substituted pyridine- 2-amine and benzo-substituted imidazole-2-amine. J Chin Chem Soc 2019;66(12):1700–7. https://doi.org/10.1002/jccs.201800461.

[56] Rani N, Singh R. Molecular modelling studies of 1,4-Diaryl-2-Mercaptoimidazole derivatives for antimicrobial potency. Curr Comput Aided Drug Des 2019;15(4):409–20.

[57] Metwally NH, Mohamed MS. New imidazolone derivatives comprising a benzoate or sulfonamide moiety as anti-inflammatory and antibacterial inhibitors: design, synthesis, selective COX-2, DHFR and molecular-modeling study. Bioorg Chem 2020;99:103438. https://doi.org/10.1016/j.bioorg.2019.103438.
[58] Raghu MS, Kumar CBP, Prasad KNN, Prashanth MK, Kumarswamy YK, Chandrasekhar S, Veeresh B. MoS2-calix[4]arene catalyzed synthesis and molecular docking study of 2,4,5-trisubstituted imidazoles as potent inhibitors of mycobacterium tuberculosis. ACS Comb Sci 2020;22(10):509–18. https://doi.org/10.1021/acscombsci.0c00038.
[59] Ashok SR, Shivananda MK, Prakash MS, Sreenivasa S, Manikandan A. Synthesis and discovery of N-(1-methyl-4-oxo-4,5-dihydro-1H-imidazol-2-yl)-[1,10-biphenyl]-2-carboxamide derivatives as antimicrobial agents. J Heterocycl Chem 2020;57(3):1211–9. https://doi.org/10.1002/jhet.3857.
[60] Aboul-Enein MN, El-Azzouny AAE-S, Attia MI, Saleh OA, Kansoh AL. Synthesis and anti-Candida potential of certain novel 1-[(3-Substituted-3-phenyl)propyl]-1 H-imidazoles. Arch Pharm Chem Life Sci 2011;344:794–801.
[61] Narasimhan B, Sharma D, Kumar P, Yogeeswari P, Sriram D. Synthesis, antimicrobial and antimycobacterial evaluation of [2-(substituted phenyl)-imidazol-1-yl]-pyridin-3-ylmethanones. J Enzyme Inhib Med Chem 2011;26(5):720–7.
[62] Rezaei Z, Khabnadideh S, Zomorodian K, Pakshir K, Kashi G, Sanagoei N, Gholami S. Design, synthesis and antifungal activity of some new imidazole and triazole derivatives. Arch Pharm Chem Life Sci 2011;344:658–65.
[63] Huang XS, Wang LS, Yin Y, Li W-M, Duan M, Ran W, Zhu H-L. Synthesis, characterization and bioactivity research of a derivative of secnidazole: 1-(2-chloropropyl)-2-methyl-5-nitro-1H-imidazole. J Chem Crystallogr 2011;41:1360–4. https://doi.org/10.1007/s10870-011-0104-9.
[64] Wang Z-C, Duan YT, Qin Y-J, Wang P-F, Luo Y, Wen Q, Yang Y-A, Sun J, Hu Y, Sang Y-L, Zhu H-L. Potentiating 1-(2-hydroxypropyl)-2-styryl-5-nitroimidazole derivatives against antibacterial agents: design, synthesis and biology analysis. Eur J Med Chem 2013;65:456–63. https://doi.org/10.1016/j.ejmech.2013.05.004.
[65] Roman G, Mares M, Năstasa V. A novel antifungal agent with broad spectrum: 1-(4-Biphenylyl)-3-(1 H-imidazol-1-yl)-1-propanone. Arch Pharm Chem Life Sci 2013;346:110–8.
[66] Mittal S, Kaur S, Swami A, Maurya IK, Jain R, Wangoo N, Sharma RK. Alkylated histidine based short cationic antifungal peptides: synthesis, biological evaluation and mechanistic investigations. RSC Adv 2016;6:41951–61. https://doi.org/10.1039/C6RA05883C.
[67] Jia T, Zhang W-L, Chen Y, Cai S-L, Yi H-B. Synthesis, structure, optical properties, antifungal and antibacterial activities of 2-(1-oxo-1H-2,3-dihydroisoindol-2-yl)-3-imidazolylL-lactamic acid. J Mol Struct 2013;1050:211–5.
[68] Shahid HA, Jahangir S, Yousuf S, Hanif M, Sherwani SK. Synthesis, crystal structure, structural characterization and in vitro antimicrobial activities of 1-methyl-4-nitro-1H-imidazole. Arab J Chem 2016;9(5):668–75. https://doi.org/10.1016/j.arabjc.2014.11.001.
[69] Zhang X, Sangani CB, Jia L-X, Gong P-X, Wang F, Wang J-F, Zhu H-L. Synthesis and antibacterial evaluation of novel Schiff's base derivatives of nitroimidazole nuclei as potent E. coli FabH inhibitors. RSC Adv 2014;4:54217–25.
[70] Zhao D, Zhao S, Zhao L, Zhang X, Wei P, Liu C, Hao C, Sun B, Su X, Cheng M. Discovery of biphenyl imidazole derivatives as potent antifungal agents: design, synthesis, and structure-activity relationship studies. Bioorg Med Chem 2017;25(2):750–8. https://doi.org/10.1016/j.bmc.2016.11.051.

[71] Işık A, Çevik UA, Sağlık BN, Özkay Y. Synthesis, characterization, and molecular docking study of some novel imidazole derivatives as potential antifungal agents. J Heterocycl Chem 2019;56(1):142–52.

[72] Altındağ FD, Sağlık BN, Çevik UA, Işıkdağ İ, Özkay Y, Gençer HK. Novel imidazole derivatives as antifungal agents: synthesis, biological evaluation, ADME prediction and molecular docking studies. Phosphorus Sulfur Silicon Relat Elem 2019;194(9):887–94. https://doi.org/10.1080/10426507.2019.1565761.

[73] Slassi S, Aarjane M, Yamni K, Amine A. Synthesis, crystal structure, DFT calculations, Hirshfeld surfaces, and antibacterial activities of schiff base based on imidazole. J Mol Struct 2019;1197:547–54.

[74] Khodja IA, Boulebd H, Bensouici C, Belfaitah A. Design, synthesis, biological evaluation, molecular docking, DFT calculations and in silico ADME analysis of (benz)imidazole-hydrazone derivatives as promising antioxidant, antifungal, and antiacetylcholinesterase agents. J Mol Struct 2020;1218:128527. https://doi.org/10.1016/j.molstruc.2020.128527.

[75] Sathaiah G, Kumar AR, Shekhar C, Raju K, Rao PS, Narsaiah B, Reddy AR, Lakshmi D, Sridhar B. Design and synthesis of positional isomers of 1-alkyl-2-trifluoromethyl-5 or 6-substituted benzimidazoles and their antimicrobial activity. Med Chem Res 2013;22:1229–37. https://doi.org/10.1007/s00044-012-0131-x.

[76] Bai Y-B, Zhang A-L, Tang J-J, Gao J-M. Synthesis and antifungal activity of 2-chloromethyl-1hbenzimidazole derivatives against phytopathogenic fungi in vitro. J Agric Food Chem 2013;61:2789–95. https://doi.org/10.1021/jf3053934.

[77] Kumar TA, Babu PNK, Devi BR. Syntheses of arylcinnamic acids, using Alum-Cs_2CO_3 as precursors of new 2-heterostyrylbenzimidazoles and their antimicrobial evaluation. Med Chem Res 2015;24:1351–64. https://doi.org/10.1007/s00044-014-1208-5.

[78] Shintre SA, Ramjugernath D, Singh P, Mocktar C, Koorbanally NA. Microwave synthesis, biological evaluation and docking studies of 2-substituted methyl 1-(4-fl uorophenyl)-1H-benzimidazole-5-carboxylates. Med Chem Res 2020;10:11615–23. https://doi.org/10.1007/s00044-016-1763-z.

[79] Kashid BB, Ghanwat AA, Khedkar VM, Dongare BB, Shaikh MH, Deshpande PP, Wakchaure YB. Design, synthesis, in vitro antimicrobial, antioxidant evaluation, and molecular docking study of novel benzimidazole and benzoxazole derivatives. J Heterocycl Chem 2019;56(3):895–908.

[80] Mahmood K, Akhter Z, Asghar MA, Mirza B, Ismail H, Liaqat F, Kalsoom S, Ashraf AR, Shabbir M, Qayyum MA, McKee V. Synthesis, characterization and biological evaluation of novel benzimidazole derivatives. J Biomol Struct Dyn 2019;38(6):1670–82. https://doi.org/10.1080/07391102.2019.16177.

[81] ÖzdenS,UstaF,AltanlarN,GökeraH.Synthesisofsomenew1h-benzimidazole-2-carboxamido derivatives and their antimicrobial activitiy. J Heterocycl Chem 2011;48:1317.

[82] Worachartcheewan A, Nantasenamat C, Isarankura-Na-Ayudhya C, Prachayasittikul V. Predicting antimicrobial activities of benzimidazole derivatives. Med Chem Res 2013;22:5418–30. https://doi.org/10.1007/s00044-013-0539-y.

[83] Oyston PCF, Sjoestedt A, Titball RW. Tularaemia: bioterrorism defence renews interest in Francisella tularensis. Nat Rev Microbiol 2004;2(12):967–78.

[84] Ryan KJ, Ray CG. Sherris medical microbiology: an introduction to infectious diseases. McGraw Hill; 2004.

[85] Kumar K, Awasthi D, Lee S-Y, Cummings JE, Knudson SE, Slayden RA, Ojima I. Benzimidazole- based antibacterial agents against F. tularensis. Bioorg Med Chem 2013;21(11):3318–26. https://doi.org/10.1016/j.bmc.2013.02.059.

[86] Shaikh IN, Hosamani KM, Kurjogi MM. Design, synthesis, and evaluation of new α-aminonitrile-based benzimidazole biomolecules as potent antimicrobial and antitubercular agents. Arch Pharm Chem Life Sci 2018;351(2). https://doi.org/10.1002/ardp.201700205, e1700205.

[87] Tahlan S, Narasimhan B, Lim SM, Ramasamy K, Mani V, Shah SAA. 2-Mercaptobenzimidazole schiff bases: design, synthesis, antimicrobial studies and anticancer activity on HCT-116 cell line. Mini Rev Med Chem 2019;19:1080–92.

[88] Tahlan S, Narasimhan B, Lim SM, Ramasamy K, Mani V, Shah SAA. Design, synthesis, SAR study, antimicrobial and anticancer evaluation of novel 2-mercaptobenzimidazole azomethine derivatives. Mini Rev Med Chem 2020;20:1559–71.

[89] Singu PS, Kanugala S, Dhawale SA, Kumar CG, Kumbhare RM. Synthesis and pharmacological evaluation of some amide functionalized 1 H-benzo[d]imidazole-2-thiol derivatives as antimicrobial agents. ChemistrySelect 2020;5:117–23.

[90] Ranjith PK, Pakkath R, Haridas KR, Kumari SN. Synthesis and characterization of new N-(4-(4-chloro-1 H-imidazol-1-yl)-3-methoxyphenyl)amide/sulfonamide derivatives as possible antimicrobial and antitubercular agents. Eur J Med Chem 2014;71:354–65.

[91] Joshi D, Parikh K. Synthesis and evaluation of novel benzimidazole derivatives as antimicrobial agents. Med Chem Res 2014;23:1290–9. https://doi.org/10.1007/s00044-013-0732-z.

[92] Buğday N, Küçükbay FZ, Apohan E, Küçükbay H, Serindağ A, Yeşilada O. Synthesis and evaluation of novel benzimidazole conjugates incorporating amino acids and dipeptide moieties. Lett Org Chem 2017;14:198–206.

[93] Daraji DG, Patel KD, Patel HD, Rajani DP. Synthesis, in vitro biological screening, and in silico computational studies of some novel imidazole-2-thiol derivatives. J Heterocycl Chem 2019;56(2):539–51.

[94] Yadav S, Narasimhan B, Lim SM, Ramasamy K, Vasudevan M, Shah SAA, Mathur A. Synthesis and evaluation of antimicrobial, antitubercular and anticancer activities of benzimidazole derivatives. Egypt J Basic Appl Sci 2018;5(1):100–9. https://doi.org/10.1016/j.ejbas.2017.11.001.

[95] Aziz H, Saeed A, Khan MA, Afridi S, Jabeen F, Ur Rehman A, Hashim M. Novel N-acyl-1H-imidazole-1-carbothioamides: design, synthesis, biological and computational studies. Chem Biodivers 2020;17(3). https://doi.org/10.1002/cbdv.201900509, e1900509.

[96] Ashok D, Gundu S, Aamate VK, Devulapally MG. Conventional and microwave-assisted synthesis of new indole-tethered benzimidazole-based 1,2,3-triazoles and evaluation of their antimycobacterial, antioxidant and antimicrobial activities. Mol Divers 2018;22:769–78. https://doi.org/10.1007/s11030-018-9828-1.

[97] Chauhan S, Verma V, Kumar D, Kumar A. Facile synthesis, antimicrobial activity and molecular docking of novel 2,4,5-trisubstituted-1 H-imidazole–triazole hybrid compounds. J Heterocycl Chem 2019;56(9):2571–9.

[98] Chauhan S, Verma V, Kumar D, Kumar A. Synthesis, antimicrobial evaluation and docking study of triazole containing triaryl-1 H-imidazole. Synth Commun 2019;49(11):1427–35. https://doi.org/10.1080/00397911.2019.1600192.

[99] B'Bhatt H, Sharma S. Synthesis, characterization, and biological evaluation of some trisubstituted imidazole/thiazole derivatives. J Heterocycl Chem 2014;52(4):1126–31.

[100] Taban IM, Elshihawy HEAE, Torun B, Zucchini B, Williamson CJ, Altuwairigi D, Ngu AST, McLean KJ, Levy SW, Sood S, Marino LB, Munro AW, de Carvalho LPS, Simons C. Novel aryl substituted pyrazoles as small molecule inhibitors of cytochrome P450 CYP121A1: synthesis and antimycobacterial evaluation. J Med Chem 2017;60:10257–67.

[101] Prasad P, Kalola AG, Patel MHP. Microwave assisted one-pot synthetic route to imidazo[1,2-a]pyrimidine derivatives of imidazo/triazole clubbed pyrazole and their pharmacological screening. New J Chem 2018;42:12666–76. https://doi.org/10.1039/c8nj00670a.

[102] Desai NC, Maheta AS, Rajpara KM, Joshi VV, Vaghani HV, Satodiya HM. Green synthesis of novel quinoline basedimidazole derivatives and evaluation of their antimicrobial activity. J Saudi Chem Soc 2014;18(6):963–71.

[103] Sun XY, Wu R, Wen X, Guo L, Zhou C-P, Li J, Quan Z-S, Bao J. Synthesis and evaluation of antibacterial activity of 7-alkyloxy-4,5-dihydro-imidazo[1,2-a]quinoline derivatives. Eur J Med Chem 2013;60:451–5.

[104] Kayarmar R, Nagaraja GK, Bhat M, Naik P, Rajesh KP, Shetty S, Arulmoli T. Synthesis of azabicyclo[4.2.0]octa-1,3,5-trien-8-one analogues of 1H-imidazo[4,5-c]quinoline and evaluation of their antimicrobial and anticancer activities. Med Chem Res 2014;23:2964–75. https://doi.org/10.1007/s00044-013-0885-9.

[105] Zhang L, Kumar KV, Rasheed S, Geng R-X, Zhou C-H. Design, synthesis and antimicrobial evaluation of novel quinolone imidazoles and interactions with MRSA DNA. Chem Biol Drug Des 2015;86(4):648–55.

[106] Er M, Ozer€ A, Direkel S, Karakurt T, Tahtaci H. Novel substituted benzothiazole and imidaz o[2,1-b][1,3,4]thiadiazole derivatives: synthesis, characterization, molecular docking study, and investigation of their in vitro antileishmanial and antibacterial activities. J Mol Struct 2019;1194:284–96.

[107] Wani MY, Ahmad A, Shiekh RA, AlGhamdi KJ, Sobral AJFN. Imidazole clubbed 1,3,4-oxadiazole derivatives as potential antifungal agents. Bioorg Med Chem 2015;23(15):4172–80. https://doi.org/10.1016/j.bmc.2015.06.053.

[108] Shruthi N, Poojary B, Kumar V, Hussain MM, Rai VM, Pai VR, Bhata M, Revannasiddappa BC. Novel benzimidazole-oxadiazole hybrid molecules as promising antimicrobial agents. RSC Adv 2016;6:8303–16.

[109] Desai NC, Joshi VV, Rajpara KM, Makwana AH. A new synthetic approach and in vitro antimicrobial evaluation of novel imidazole incorporated 4-thiazolidinone motifs. Arab J Chem 2017;10:s589–99. https://doi.org/10.1016/j.arabjc.2012.10.020.

[110] Desai NC, Joshi VV, Rajpara KM, Vaghani HV, Satodiya HM. Microwave-assisted synthesis and antimicrobial screening of new imidazole derivatives bearing 4-thiazolidinone nucleus. Med Chem Res 2013;22:1893–908. https://doi.org/10.1007/s00044-012-0190-z.

[111] Patel V, Bhatt N, Bhatt P, Joshi HD. Synthesis and pharmacological evaluation of novel 1-(piperidin-4-yl)-1H-benzo[d]imidazol-2(3H)-one derivatives as potential antimicrobial agents. Med Chem Res 2014;23:2133–9. https://doi.org/10.1007/s00044-013-0799-6.

[112] Holiyachi M, Samundeeswari S, Chougala BM, Naik NS, Madar J, Shastri LA, Joshi SD, Dixit SR, Dodamani S, Jalalpure S, Sunagar VA. Design and synthesis of coumarin–imidazole hybrid and phenylimidazoloacrylates as potent antimicrobial and anti-inflammatory agents. Monatsh Chem 2018;149:595–609. https://doi.org/10.1007/s00706-017-2079-5.

[113] Demchenko S, Lesyk R, Zuegg J, Elliott AG, Fedchenkova Y, Suvorova Z, Demchenko A. Antibacterial and antifungal activity of new 3-biphenyl-3H-[1,2-a]azepin-1-ium bromides. Eur J Med Chem 2020;201:112477. https://doi.org/10.1016/j.ejmech.2020.112477.

[114] Ranjan N, Fulcrand G, King A, Brown J, Jiang X, Leng F, Arya DP. Selective inhibition of bacterial topoisomerase I by alkynyl-bisbenzimidazoles. Med Chem Comm 2014;5(6):816–25.

[115] Goud GL, Ramesh S, Ashok D, Reddy VP. One-pot synthesis of bis(4,5-diphenylimidazol-2-yl-phenyl)glycols and evaluation of their antimicrobial activity. Russ J Gen Chem 2015;85(3):673–8.

[116] Padmavathi V, Kumara CP, Venkatesh BC, Padmaja A. Synthesis and antimicrobial activity of amido linked pyrrolyl and pyrazolyl-oxazoles, thiazoles and imidazoles. Eur J Med Chem 2011;46:5317–26.

[117] Krim J, Grünewald C, Taourirte M, Engels JW. Efficient microwave-assisted synthesis, antibacterial activity and high fluorescence of 5 benzimidazolyl-2′-deoxyuridines. Bioorg Med Chem 2012;20:480–6.

[118] Peng L-P, Nagarajan S, Rasheed S, Zhou C-H. Synthesis and biological evaluation of a new class of quinazolinone azoles as potential antimicrobial agents and their interactions with calf thymus DNA and human serum albumin. Med Chem Commun 2015;6:222–9.

[119] Gong H-H, Baathulaa K, Lv J-S, Cai G-X, Zhou C-H. Synthesis and biological evaluation of Schiff base-linked imidazolyl naphthalimides as novel potential anti-MRSA agents. Med Chem Commun 2016;7:924–31. https://doi.org/10.1039/C5MD00574D.

[120] Patil DA, Surana SJ. Synthesis, biological evaluation of 2,3-disubstituted-imidazolyl/benzimidazolyl-quinazolin-4(3H)-one derivatives. Med Chem Res 2016;25:1125–39. https://doi.org/10.1007/s00044-016-1552-8.

[121] Tahtaci H, Karacık H, Ece A, Er M, Şeker MG. Design, synthesis, SAR and molecular modeling studies of novel imidazo[2,1-b][1,3,4]thiadiazole derivatives as highly potent antimicrobial agents. Mol Inform 2017;36:1700083.

[122] Jeanmart S, Gagnepain J, Maity P, Lamberth C, Cederbaum F, Rajan R, Jacob O, Blum M, Bieri S. Synthesis and fungicidal activity of novel imidazole-based ketene dithioacetals. Bioorg Med Chem 2018;26:2009–16.

[123] Patil S, Deally A, Gleeson B, Müller-Bunz H, Paradisi F, Tacke M. Novel benzyl-substituted N-heterocyclic carbene–silver acetate complexes: synthesis, cytotoxicity and antibacterial studies. Metallomics 2011;3:74–88. https://doi.org/10.1039/c0mt00034e.

[124] McCann M, Curran R, Ben-Shoshan M, McKee V, Devereux M, Kavanagh K, Kellett A. Synthesis, structure and biological activity of silver(I) complexes of substituted imidazoles. Polyhedron 2013;56:180–8.

[125] Tabrizi L, McArdle P, Ektefan M, Chiniforoshan H. Synthesis, crystal structure, spectroscopic and biological properties of mixed ligand complexes of cadmium(II), cobalt(II) and manganese(II) valproate with 1,10-phenanthroline and imidazole. Inorg Chim Acta 2016;439:138–44. https://doi.org/10.1016/j.ica.2015.10.015.

[126] Beheshti A, Hashemi F, Monavvar MF, Khorrmdin R, Abrahams CT, Motamedi H, Shakerzadeh E. Synthesis, structural characterization, antibacterial activity, DNA binding and computational studies of bis(2-methyl-1H-imidazole κN3)silver(I)dichromate(VI). J Mol Struct 2017;1133:591–606. https://doi.org/10.1016/j.molstruc.2016.11.064.

[127] Lewis A, McDonald M, Scharbach S, Hamaway S, Plooster M, Peters K, Fox KM, Cassimeris L, Tanski JM, Tyler LA. The chemical biology of Cu(II) complexes with imidazole or thiazole containing ligands: synthesis, crystal structures and comparative biological activity. J Inorg Biochem 2016;157:152–61. https://doi.org/10.1016/j.jinorgbio.2016.01.014.

[128] Li Y, Lu X, Jing H, Wang Q, Cai Y. Synthesis, structures and antimicrobial activities of silver(I) complexes derived from 2-propyl-1H-imidazole-4,5-dicarboxylic acid. Inorg Chim Acta 2017. https://doi.org/10.1016/j.ica.2017.07.070.

[129] Ismael M, Abdou A, Abdel-Mawgoud A-M. Synthesis, characterization, modeling, and antimicrobial activity of FeIII, CoII, NiII, CuII, and ZnII complexes based on trisubstituted imidazole ligand. Z Anorg Allg Chem 2018;644(20):1203–14. https://doi.org/10.1002/zaac.201800230.

[130] Abdolmaleki S, Ghadermazi M, Ashengroph M, Saffari A, Sabzkohi SM. Cobalt (II), zirconium(IV), calcium(II) complexes with dipicolinic acid and imidazole derivatives: X-ray studies, thermal analyses, evaluation as in vitro antibacterial and cytotoxic agents. Inorg Chim Acta 2018;480:70–82.

[131] Chai LQ, Zhou L, Zhang KY, Zhang HS. Structural character izations, spectroscopi c, electrochemical proper ties, and antibacterial activities of copper (II) and cobalt (II) complexes containing imidazole ring. Appl Organometal Chem 2018;e4576. https://doi.org/10.1002/aoc.4576.

[132] SKA KG, Kurdziel K, Ciepluch K, Rachuna J, Kowalska M, Madej Ł, Węgierek-Ciuk A, Lankoffd A, Arabski M. Synthesis, physicochemical and biological characterization of Ni(II) complex with imidazole-4-acetate anion as new antifungal agent. J Chem Sci 2018;130:169. https://doi.org/10.1007/s12039-018-1574-5.

[133] Lobana TS, Aulakh JK, Sood H, Arora DS, Garcia-Santos I, Kaur M, Duff CE, Jasinski JP. Synthesis, structures and antimicrobial activity of copper derivatives of N-substituted imidazolidine-2-thiones: unusual bio-activity against Staphylococcus epidermidis and enterococcus faecalis. New J Chem 2018;42:9886–900. https://doi.org/10.1039/C8NJ00206A.

[134] Anastasiadoua D, Psomasa G, Kalogiannis S, Geromichalos G, Hatzidimitrioua AG, Aslanidisa P. Bi- and trinuclear copper(I) compounds of 2,2,5,5-tetramethylimidazolidine-4-thione and 1,2-bis(diphenylphosphano)ethane: synthesis, crystal structures, in vitro and in silico study of antibacterial activity and interaction with DNA and albumins. J Inorg Biochem 2019;198:110750. https://doi.org/10.1016/j.jinorgbio.2019.110750.

[135] Ishak NNM, Jamsari J, Ismail AZ, Tahir MIM, Tiekink ERT, Veerakumarasivam A, Ravoof TBSA. Synthesis, characterisation and biological studies of mixed-ligand nickel (II) complexes containing imidazole derivatives and thiosemicarbazide Schiff bases. J Mol Struct 2019;1198:126888.

[136] Obaleye JA, Ajibola AA, Bernardus VB, Hosten EC. Synthesis, X-ray crystallography, spectroscopic and in vitro antimicrobial studies of a new Cu(II) complex of trichloroacetic acid and imidazole. J Mol Struct 2020;1203:127435. https://doi.org/10.1016/j.molstruc.2019.127435.

[137] Thakor KP, Lunagariya MV, Bhatt BS, Patel MN. Fluorescence and absorption titrations of bio-relevant imidazole based organometallic Pd(II) complexes with DNA: synthesis, characterization, DNA interaction, antimicrobial, cytotoxic and molecular docking studies. J Inorg Organomet Polym Mater 2019;29:2262–73. https://doi.org/10.1007/s10904-019-01184-2.

[138] Abdel-Rahman LH, Abdelhamid AA, Abu-Dief AM, Shehata MR, Bakheet MA. Facile synthesis, X-Ray structure of new multi-substituted aryl imidazole ligand, biological screening and DNA binding of its Cr(III), Fe(III) and Cu(II) coordination compounds as potential antibiotic and anticancer drugs. J Mol Struct 2020;1200:127034.

[139] Abdalla EM, Rahman LHA, Abdelhamid AA, Shehata MR, Alothman AA, Nafady A. Synthesis, characterization, theoretical studies, and antimicrobial/antitumor potencies of salen and salen/imidazole complexes of Co (II), Ni (II), Cu (II), Cd (II), Al (III) and La (III). Appl Organomet Chem 2020;34(11), e5912.

[140] Tavman A, Boz I, Birteksoz AS, Cinarli A. Spectral characterization and antimicrobial activity of cu(II) and Fe(III) complexes of 2-(5-Cl/NO_2-1H-benzimidazol-2-yl)-4-Br/NO_2-phenols. J Coord Chem 2010;63(8):1398–410. https://doi.org/10.1080/00958971003789835.

[141] Kalinowska-Lis U, Szewczyk EM, Chęcińska L, Wojciechowski JM, Wolf WM, Ochocki J. Synthesis, characterization, and antimicrobial activity of silver(I) and copper(II) complexes of phosphate derivatives of pyridine and benzimidazole. ChemMedChem 2014;9(1):169–76.

[142] Kose M, Digrak M, Gonul I, Mckee V. Two Cu(II) complexes from an N-alkylated benzimidazole: synthesis, structural characterization, and biological properties. J Coord Chem 2014;67(10):1746–59. https://doi.org/10.1080/00958972.2014.920502.

[143] Kose M. Synthesis, characterization, and antimicrobial properties of two Cu(II) complexes derived from a benzimidazole ligand. J Coord Chem 2014;67(14):2377–92. https://doi.org/10.1080/00958972.2014.940924.

[144] Abdelkarim AT, Al-Shomrani MM, Rayan AM, El-Sherif AA. Mixed ligand complex formation of cetirizine drug with bivalent transition metal(II) ions in the presence of 2-aminomethylbenzimidazole: synthesis, structural, biological, pH-metric and thermodynamic studies. J Solution Chem 2015;44:1673–704. https://doi.org/10.1007/s10953-015-0362-9.

[145] Mahmood K, Hashmi W, Ismail H, Mirza B, Twamley B, Akhter Z, Rozas I, Baker RJ. Synthesis, DNA binding and anti-bacterial activity of metal(II) complexes of a benzimidazole Schiff base. Polyhedron 2019;157:326–34. https://doi.org/10.1016/j.poly.2018.10.020.

[146] Casanova I, Durán ML, Viqueira J, Sousa-Pedrares A, Zani F, Realc JA, García-Vázquez JA. Metal complexes of a novel heterocyclic benzimidazole ligand formed by re-arrangement cyclization of the corresponding Schiff base. Electrosynthesis, structural characterization and antimicrobial activity. Dalton Trans 2018;47:4325–40. https://doi.org/10.1039/c8dt00532j.

[147] Dileepan AGB, Kumar AG, Mathumidha R, Rajaram R, Rajam S. Dinuclear rectangular-shaped assemblies of bis-benzimidazolydine salt coordinated to Ag(I) and Cu(I) N-heterocyclic carbene complexes and their biological applications. Chem Zvesti 2018;72:3017–31. https://doi.org/10.1007/s11696-018-0538-z.

[148] Chkirate K, Karrouchi K, Dege N, Sebbar NK, Ejjoummany A, Radi S, Adarsh NN, Talbaoui A, Ferbinteanu M, Essassia EM, Garcia Y. Co(II) and Zn(II) pyrazolyl-benzimidazole complexes with remarkable antibacterial activity. New J Chem 2020;44:2210–21. https://doi.org/10.1039/c9nj05913j.

[149] Heidelberg S, Hashim NM, Al-Madhagi WM, Ali HM. Benzimidazolium-acridine-based silver N-heterocyclic carbene complexes as potential anti-bacterial and anti-cancer drug. Inorg Chim Acta 2020;504:119462.

[150] Riduan SN, Zhang Y. Imidazolium salts and their polymeric materials for biological applications. Chem Soc Rev 2013;42(23):9055–70. https://doi.org/10.1039/c3cs60169b.

[151] Gravel J, Schmitzer AR. Imidazolium and benzimidazolium-containing compounds: from simple toxic salts to highly bioactive drugs. Org Biomol Chem 2017;15:1051–71. https://doi.org/10.1039/C6OB02293F.

[152] Huang RTW, Peng KC, Shih HN, Lin GH, Chang TF, Hsu SJ, Hsua TST, Lin IJB. Antimicrobial properties of ethoxyether-functionalized imidazolium salts. Soft Matter 2011;7:8392–400.

[153] Borowiecki P, Milner-Krawczyk M, Brzezin'ska D, Wielechowska M, Plenkiewicz J. Synthesis and antimicrobial activity of imidazolium and triazolium chiral ionic liquids. Eur J Org Chem 2013;712–20. https://doi.org/10.1002/ejoc.201201245.

[154] Gilmore BF, Andrews GP, Borberly G, Earle MJ, Gilea MA, Gorman SP, Lowry AF, Mc Laughlina M, Seddon KR. Enhanced antimicrobial activities of 1-alkyl-3-methylimidazolium ionic liquids based on silver or copper containing anions. New J Chem 2013;37:873–6.

[155] Pałkowski Ł, Błaszczynski J, Skrzypczak A, Błaszczak J, Kozakowska K, Wroblewska J, Kozuszko S, Gospodarek E, Krysinski J, Słowinski R. Antimicrobial activity and SAR study of new gemini imidazolium-based chlorides. Chem Biol Drug Des 2014;83:78–288.

[156] Al-Mohammed NN, Alias Y, Abdullah Z. Bis-imidazolium and benzimidazolium based gemini-type ionic liquids structure: synthesis and antibacterial evaluation. RSC Adv 2015;5:92602–17.

[157] He Z, Huang K, Xiong F, Zhang S-F, Xue J-R, Liang Y, Jing L-H, Qin D-B. Self-assembly of imidazoliums salts based on acridine with silver oxide as coordination polymers: synthesis, fluorescence and antibacterial activity. J Organometal Chem 2015;797:67–75.

[158] Basak A, Abouelhassan Y, Zuo R, Yousaf H, Ding Y, Huigens RW. Antimicrobial peptide-inspired NH125 analogues: bacterial and fungal biofilm-eradicating agents and rapid killers of MRSA persisters. Org Biomol Chem 2017;15:5503–12. https://doi.org/10.1039/c7ob01028a.

[159] Aljuhani A, El-Sayed WS, Sahu PK, Rezki N, Aouad MR, Salghi R, Messali M. Microwave-assisted synthesis of novel imidazolium, pyridinium and pyridazinium-based ionic liquids and/or salts and prediction of physico-chemical properties for their toxicity and antibacterial activity. J Mol Liq 2018;249:747–53.

[160] Bulut O, Oktem HA, Yilmaz MD. A highly substituted and fluorescent aromatic-fused imidazole derivative that shows enhanced antibacterial activity against methicillin-resistant Staphylococcus aureus (MRSA). J Hazard Mater 2020;399:122902. https://doi.org/10.1016/j.jhazmat.2020.122902.

[161] Jain A, Duvvuri LS, Farah S, Beyth N, Domb AJ, Khan W. Antimicrobial polymers. Adv Healthc Mater 2014;3(12):1969–85. https://doi.org/10.1002/adhm.201400418.

[162] Anderson EB, Long TE. Imidazole- and imidazolium-containing polymers for biology and material science applications. Polymer 2010;51:2447–54.

[163] Sharma A, Wilson GR, Dubey A. Antibacterial activity of vinyl imidazole(VI) functionalized silica polymer nanocomposites (SBA/VI) against gram negative and gram positive bacteria. New J Chem 2016;40:764–9. https://doi.org/10.1039/c5nj01536g.

[164] Zhu L, Dai J, Chen L, Chen J, Na H, Zhu J. Design and fabrication of imidazolium ion-immobilized electrospun polyurethane membranes with antibacterial activity. J Mater Sci 2017;52:2473–83.

[165] Beheshti A, Soleymani-Babadi S, Mayer P, Abrahams CT, Motamedi H, Trzybinski D, Wozniak K. Design, synthesis, and antibacterial assessment of silver(i)-based coordination polymers with variable counterions and unprecedented structures by the tuning spacer length and binding mode of flexible bis(imidazole-2-thiones) ligands. Cryst Growth Des 2017;17(10):5249–62. https://doi.org/10.1021/acs.cgd.7b00784.

[166] Beheshtia A, Panahia F, Soleymani-Babadia S, Mayerb P, Lipkowskic J, Motamedid H, Samiee S. Rational synthesis, structural characterization, theoretical studies, antibacterial activity and selective dye absorption of new silver coordination polymers generated from a flexible bis (imidazole-2-thione) ligand. Inorg Chim Acta 2020;504:119406.

[167] Klahn P, Bronstrup M. Bifunctional antimicrobial conjugates and hybrid antimicrobials. Nat Prod Rep 2017;34:832–85.

[168] Theuretzbacher U, Bush K, Harbarth S, Paul M, Rex JH, Tacconelli E, Thwaites GE. Critical analysis of antibacterial agents in clinical development. Nat Rev Microbiol 2020;18(5):286–98. https://doi.org/10.1038/s41579-020-0340-0.

[169] Brown ED, Wright GD. Antibacterial drug discovery in the resistance era. Nature 2016;529:336–43. https://doi.org/10.1038/nature17042.

[170] Bassetti M, Ginocchio F, Mikulska M, Taramasso L, Giacobbe DR. Will new antimicrobials overcome resistance among gram-negatives? Expert Rev Anti Infect Ther 2011;9(10):909–22.

[171] Global priority list of antibiotic-resistant bacteria to guide research, discovery, and development of new antibiotics by WHO report- https://www.who.int/medicines/publications/global-priority-list-antibiotic-resistant-bacteria/en/ [Assessed 27 November 2020].

Chapter 6

Imidazole containing heterocycles as antioxidants

Nusrat Sahiba, Ayushi Sethiya, Pankaj Teli, and Shikha Agarwal
Synthetic Organic Chemistry Laboratory, Department of Chemistry, Mohanlal Sukhadia University, Udaipur, Rajasthan, India

Abbreviations

ADME properties	absorption, distribution, metabolism and excretion (pharmcokintetic properties)
Am B	amphotericin B
Cfn	ciprofloxacin
ERG	electron releasing group
EWG	electron withdrawing group
FCZ	fluconazole
IC_{50} value	half maximal inhibitory concentration
ITZ	itraconazole
MBC	minimum bactericidal concentration
MIC	minimum inhibitory concentration
MW	microwave
QSAR	quantitative SAR
RMSE	root mean square error
ROS	reactive oxygen species
SAR	structure-activity relationship
TB bacteria	mycobacterium tuberculosis
ZOI	zone of inhibition

1. Introduction

In the present era, degenerative diseases such as hypertension, cancer, heart failure, and arthritis are global concerns, causing high rates of morbidity and mortality. The main reason behind this issue is free-radical generation [1]. This is a natural process in aerobic oxidation, which is generally produced as a consequence of adenosine triphosphate production by the mitochondria. Several other mechanisms like endogenous and environmental factors are also responsible for their production. Endogenous sources involve respiratory burst, autoxidation reactions, and enzyme reactions, and environmental factors implicate tobacco smoke, ultraviolet radiation, ionizing radiation, xenobiotics, other pollutants, etc.

Imidazole-Based Drug Discovery. https://doi.org/10.1016/B978-0-323-85479-5.00007-1
Copyright © 2022 Elsevier Inc. All rights reserved.

Free radicals are highly reactive and unstable species that contains unpaired electrons and are capable of independent existence. Free radicals can either donate or receive an electron from other molecules and reduce or oxidize the molecules in living cells. Oxygen- and nitrogen-centered species like peroxyl ($ROO^{\cdot}$), superoxide ($O_2^{\cdot}-$), alkoxyl ($RO^{\cdot}$), hydroxyl ($HO^{\cdot}$), peroxynitrite ($^{\cdot}ONOO$), and nitric oxide ($NO^{\cdot}$) are common examples of free radicals [2]. These free radicals play a significant role in several bioactive processes of the cells like immune function and cellular response, but become toxic at higher concentrations. This situation is known as oxidative stress, which adversely alters DNA, RNA, lipids, and proteins and trigger various diseases like Alzheimer's, cancer, arthritis, schizophrenia, cataract, Parkinson's, emphysema, neurodegenerative diseases, and so on [3]. The human body has several settings and defense mechanisms to counteract the negative impact of free radicals by producing antioxidants, either via in situ natural synthesis or delivered by food and supplements [4].

Antioxidants are components that control free-radical generation, and prevent and neutralize their adverse effect on cells [5, 6]. Antioxidants can be classified on the basis of various parameters like their nature, bioavailability, mode of action, and working profile [7]. Generally, antioxidants work via three phases, *viz.* antioxidant enzymes (catalase, ceruloplasmin, superoxide dismutases), chain breaking antioxidants (tocopherols, ubiquinol, glutathione), and transition metal binding proteins (lactoferrin, ferritin, transferrin) [8, 9].

In the present time, various natural and synthetic antioxidants are available for the treatment of antagonistic and deleterious effects of free radicals on cellular activities [10–12]. Nitrogen-containing heterocyclic compounds occupy an indispensable position in medicinal chemistry and have substantial importance in the search of new bioactive molecules. Imidazole, a five-membered heterocycle, has been established as an important moiety for the development of therapeutic molecules of pharmaceutical importance [13, 14]. The potency and wide applicability of this pharmacophore in the field of antioxidant activity can be attributed to its hydrogen bond donor-acceptor capability as well as its high binding affinity for metals. Tremendous research has been done in the field of imidazole-based antioxidant derivatives, fused analogs, and metal complexes. This chapter gives detail insight on the recent advancements in the field of imidazole-based antioxidant molecules like various derivatives, hybrids, and complexes, and discusses their potential role in preventing and repairing damage caused by free radicals to enrich the knowledge of active scientists in this field.

2. Imidazole-based antioxidant compounds

For the ease of understanding of readers, antioxidant active imidazole derivatives have been classified on the basis of their structural modifications:

1. Ring generation/formation;
2. Imidazole-based hybrids; and
3. Imidazole metal complexes

2.1 Imidazole

Neurological disorders, cerebral ischemia, involve reperfusion and free radical generation in brain tissues. Regarding this issue, in 2006, Sorrenti et al. [15] introduced imidazole compounds **1** as antioxidants and NOS (nitric oxide synthase) inhibitors via DPPH assay (Fig. 6.1). The results revealed that all compounds showed dose-dependent DPPH quenching; however, the presence of the keto group increased the antioxidant efficiency. Compounds possessing the $-NO_2$ group at the phenyl ring and the $-CH_3/-CH(CH_3)_2$ group at the imidazole ring were the most active compared to the reference trolox. Most of the tested compounds demonstrated superoxide anion scavenger ability and inhibited *in vitro* lipid peroxidation.

In 2007, Soujanya and Sastry [16] studied the antioxidant mechanism of 1,3-dihydro-1-methyl-2*H*-imidazole-2-selenol (MSeI) **2** via density functional theory at B3LYP/6-31G(d) level and gave detailed information about intermediates, transition structures, and the energy profile of catalytic cycle. The molecule demonstrated GPx-like antioxidant activity (Fig. 6.1).

Salerno and companions [17] studied imidazoles as effective inhibitors of both neuronal nitric oxide synthase (nNOS) and endothelial nitric oxide synthase (eNOS). For the evaluation of imidazole biological potency, the authors studied three NOS isoforms via enzymatic and DPPH assay. Among all the screened compounds, compound **3** was the most potent and selective inhibitor against nNOS. Compound **4** exhibited high selectivity against eNOS, nNos, iNOS, and cytochrome P450, and displayed good antioxidant properties. These molecules also efficiently reduced lipid peroxidation (Fig. 6.1).

2.1.1 Ring generation/formation

The multistep synthesis for a range of 2-(2-methylbenzylthio)-1*H*-imidazole-4,5-dicarboxylic acid (**5**,**6**) and their antioxidant activity was demonstrated by Maddila and coauthors [18] (Scheme 6.1). The results of nitric oxide methods

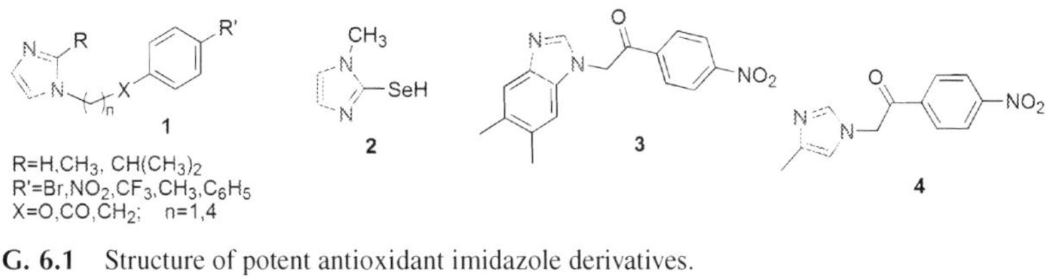

FIG. 6.1 Structure of potent antioxidant imidazole derivatives.

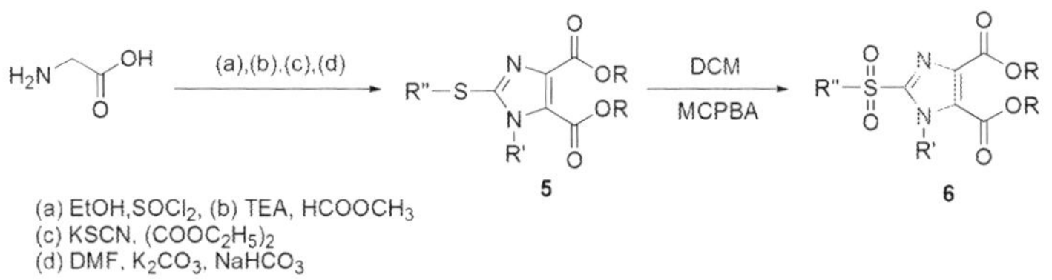

SCHEME 6.1 Synthesis of imidazole derivatives.

and DPPH assay concluded that all the synthesized compounds were highly potent for free-radical inhibition at 100 μM concentration (Table 6.1).

Jayabharathi and companions [19] reported a facile, efficient, and greener route for the synthesis of imidazole derivatives from 1,2-diketone, NH_4OAc, and aromatic aldehydes using molecular iodine at ambient temperature through a mechanochemical process. These synthesized molecules were screened for the antioxidant activity via DPPH (1,1-diphenyl-2-picrylhydrazyl) superoxide anion, and hydroxyl radical scavenging methods. All the synthesized compounds were highly potent radical inhibitors, and compounds having dimethoxyphenyl at N3 and fluorophenyl at the C2 position of the imidazole ring displayed the highest DPPH radical scavenging potency. The presence of dimethoxyphenyl at the N3 position revealed the highest superoxide anion and hydroxy radical scavenging ability.

A range of 4,5-diaryl-1*H*-imidazole-2(3*H*)-thiones was prepared and screened as potent free radical scavengers against the reference drug, nordihydroguaiaretic acid. Among all the synthesized ones, compound **7** displayed the highest free radical scavenging activity with an IC_{50} value of 14 μM via DPPH colorimetric assay [20] (Fig. 6.2).

In 2014, synthesis of substituted imidazoles **8** and their antioxidant properties were demonstrated by Nayak and his group [21]. The reaction of itaconic anhydride and ortho-phenylenediamine in the presence of PTSA yielded substituted imidazole compounds in good yields. The in vitro antioxidant results revealed that both compounds showed noticeable DPPH radical scavenging activity, nitric oxide scavenging activity, and reducing power capacity compared to standard glutathione. The molecular docking study with tyrosinase enzyme displayed good binding pattern with −213.96 (**8a**) and −221.41 (**8b**) energy values (Scheme 6.2).

Guda et al. [22] introduced 2,4,5-trisubstituted imidazole derivatives **9** and **10** as antioxidant agents against the standard, ascorbic acid via DPPH

TABLE 6.1 In vitro antioxidant study data of some imidazole derivatives.

				% Inhibition at 100 μM	
Compound	R	R′	R″	Nitric oxide method	DPPH method
5a	H	$CH_2CH(CH_3)_2$	o-CH_3-$CH_2C_6H_4$	82.25	84.74
5b	H	H	$C_{10}H_{11}CH_2$	91.18	91.18
6a	Et	COOEt	o-CH_3-$CH_2C_6H_4$	96.42	94.38
6b	H	$CH_2C_6H_5$	$CH_2C_6H_5$	90.15	85.26
6c	H	$CH_2CH(CH_3)_2$	$CH_2CH(CH_3)_2$	80.65	82.45

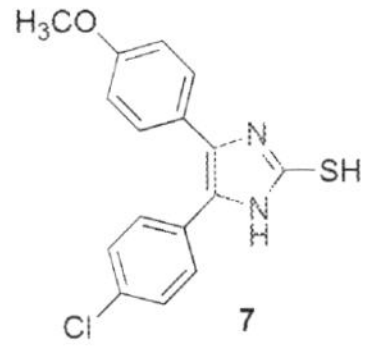

FIG. 6.2 Structure of 4,5-diaryl-1*H*-imidazole-2(3*H*)-thione.

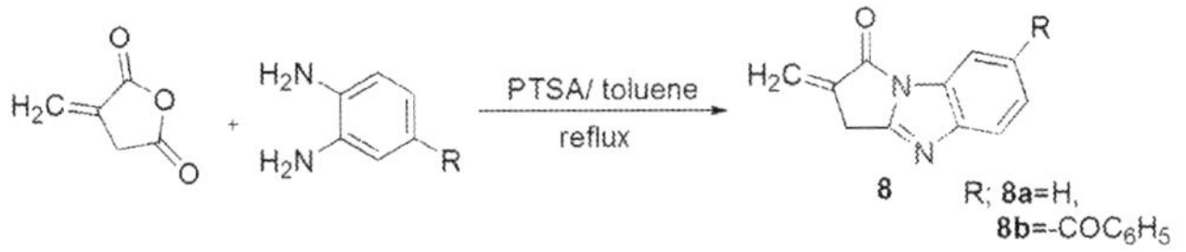

SCHEME 6.2 Synthesis of benzimidazol-1-ones.

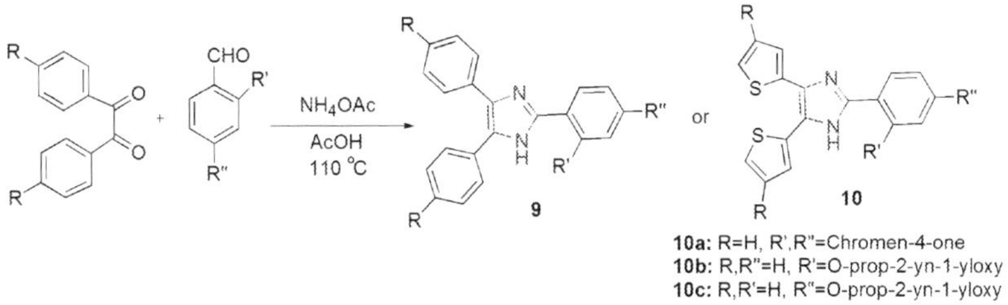

SCHEME 6.3 Synthesis of 2,4,5-trisubstituted imidazole derivatives.

radical scavenging assay. Substituted benzimidazoles were prepared from the reaction of benzil or thenil with different aldehydes and NH_4OAc at 110 °C to obtain good yields (Scheme 6.3). Compounds **10a**, **10b**, and **10c** displayed $4.95 \pm 1.81\,\mu M$, $5.87 \pm 1.73\,\mu M$, and $6.29 \pm 1.27\,\mu M$ IC_{50} values, which showed their high antioxidant potency. The interaction mechanisms of molecules with various protein receptors, namely EGFR and HER2, were studied from molecular docking studies. The results concluded that most of the derivatives showed good binding energies in the range of −7.63 kcal/mol to −11.15 kcal/mol compared to the standard.

Ghorbani and companions [23] introduced ionic liquid-functionalized magnetic nanoparticles catalyzed synthesis of imidazoles and studied their antioxidant activity. The reaction of benzoin and substituted aldehydes with amine and ammonium acetate in the presence of $H_2PW_{12}O_{40}/Fe_3O_4@SiO_2$–Pr–Pi catalyst yielded corresponding imidazoles **11** with high efficiency via an in situ oxidation-condensation process. The results of DPPH radical scavenger assay displayed that among all the synthesized compounds, compound **11a** ($IC_{50} = 0.12$) was found to be highly potent compared to the reference, ascorbic acid (Scheme 6.4).

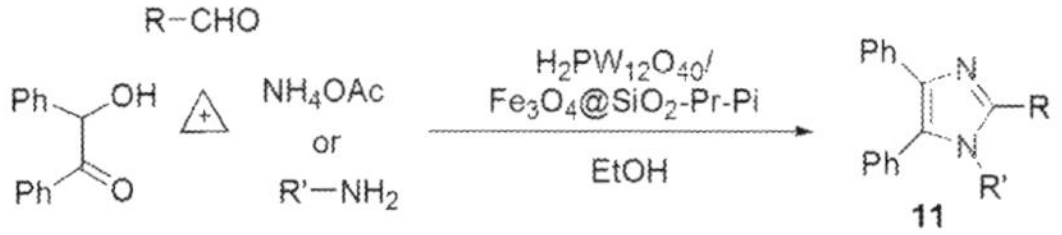

11

11a; R= thiophene-2-yl
R'=4-$CH_3C_6H_4$

SCHEME 6.4 Synthesis of imidazole using $H_2PW_{12}O_{40}$/Fe_3O_4@SiO_2–Pr–Pi catalyst.

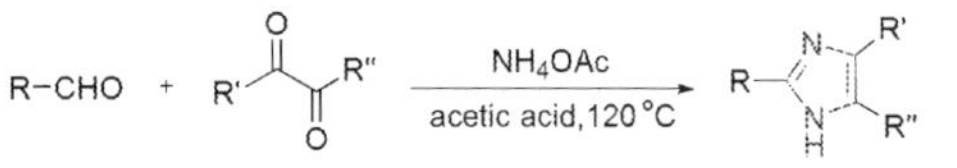

12

12(a-c)R=2-naphthyl
12a; R',R''=4-$CH_3C_6H_4$, **12b**;R',R''=C_6H_5
12c;R=2-naphthyl, R'=3-furfural, R''=2-furfural

SCHEME 6.5 Synthesis of naphthalene-containing 2,4,5-trisubstituted imidazoles.

Somashekara and coauthors [24] synthesized imidazole derivatives and revealed their antioxidant properties through DPPH free radical scavenging activity. A mixture of substituted benzil, naphthaldehyde, and NH_4OAc was refluxed in the presence of acetic acid at 120°C to obtain moderate to high yields of 2,4,5-trisubstituted imidazoles. From the DPPH assay it was concluded that all compounds showed good to moderate activity at 100μg/mL. Among all the derivatives, compounds **12d**, **12g**, and **12k** presented good antioxidant activity due to the presence of 4-methyl-phenyl, phenyl, and furan substituents at the 4,5 positions, respectively (Scheme 6.5).

Abdelhamid et al. [25] utilized pyrrolidinium hydrogen sulfate (PHS) as an efficient recyclable catalyst for the synthesis of novel imidazole derivatives and studied their in vivo antioxidant activity on rats. In the present protocol, the reaction of aldehydes, ethyl glycinate hydrochloride, 1,2-diphenylethane-1,2-dione, and NH_4OAc using PHS, afforded corresponding imidazoles in excellent yields and of high purity with short reaction time. The in vivo antioxidant activity determined from various methods comprised liver reduced glutathione content (GSH), malondialdehyde level (MDA), and nitric oxide level (NO). Remarkable activity was demonstrated by all compounds and the highest value was shown by **13b** for GSH level, **13a** for MDA level, and **13b** and **14** for NO level (Scheme 6.6).

Recently, Noriega-Iribe et al. [26] screened 2,4,5-trisubstituted imidazole compounds as potent antioxidant drug candidates through in vitro DPPH and ABTS assay and in silico studies. For this purpose, a range of imidazole derivatives were prepared from the reaction of benzil, aromatic aldehydes, and ammonium acetate in acetic acid via Radziszewski reaction. The results of DPPH assay indicated that the synthesized molecules showed moderate to high antioxidant activity, and compounds **15a**, **15b**, **15c**, and **15d** were found to be the most active with EC_{50} values of 0.141, 0.174, 0.341, and 1.389mg/mL, respectively. Similar results were displayed by 2–2′-azino-bis-(3-ethylbenzothiazoline-6-sulfonate)

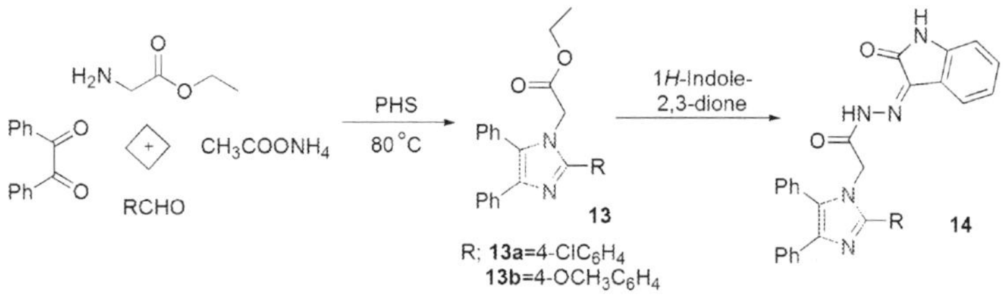

SCHEME 6.6 Synthesis of 1,2,4,5-tetrasubstituted imidazoles using ionic-liquids.

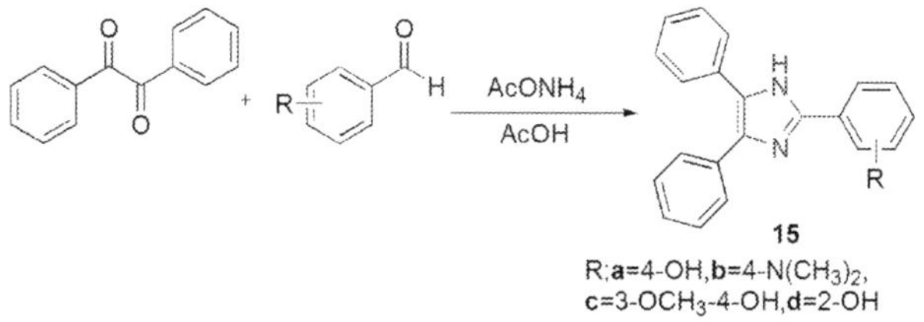

SCHEME 6.7 Synthesis of antioxidant 2,4,5-triphenyl-1*H*-imidazole derivatives.

(ABTS) assay, and molecules **15b**, **15a**, **15d**, and **15c** were found to be the most active and presented EC_{50} values of 0.162, 0.168, 0.188, and 0.199 mg/mL, respectively (Scheme 6.7). The data revealed that the presence of ERG like –OH, $–N(CH_3)_2$ on the aromatic ring increased the antioxidant potency of imidazole compounds. These results were also validated by in silico study and ADME analysis.

2.1.2 Imidazole hybrids

A range of 3-chloro-1-(5*H*-dibenz[*b,f*]azepine-5yl)propan-1-one derivatives **16(a–k)** were synthesized by *N*-acylation of 5*H*-dibenz[*b,f*]azepine followed by condensation of **16** with various amino acids via base condensation reaction [27]. The whole range of derivatives was checked for their antioxidant activity through the inhibition of lipid peroxidation by linoleic acid and β-carotene assay. The presence of different functional groups on amino acid conjugates of **16** enhanced antioxidant activity. Among all the synthesized compounds, the imidazole derivative **16k** presented good antioxidant activities with 84.96% and 86.76% antioxidant activity at 10 and 25 μM, respectively (Scheme 6.8).

Five imidazole alkaloids (lepidines **17a**, **17b**, **17c**, **17d**, and **17e**) were studied for their antioxidant activity against hydroperoxyl radicals via single-electron transfer (SET) reactions [28]. Hydroxyperoxyl radicals are produced within living organisms and are associated with the oxidation of lipoproteins or biological membranes. The lepidines with a phenolic ring, i.e., lepidines **17d**, **17b**, and **17e**, were found to be more effective to deactivate hydroperoxyl free radicals in an aqueous solution. The acid-base reaction played a pivotal role in their activity (Fig. 6.3).

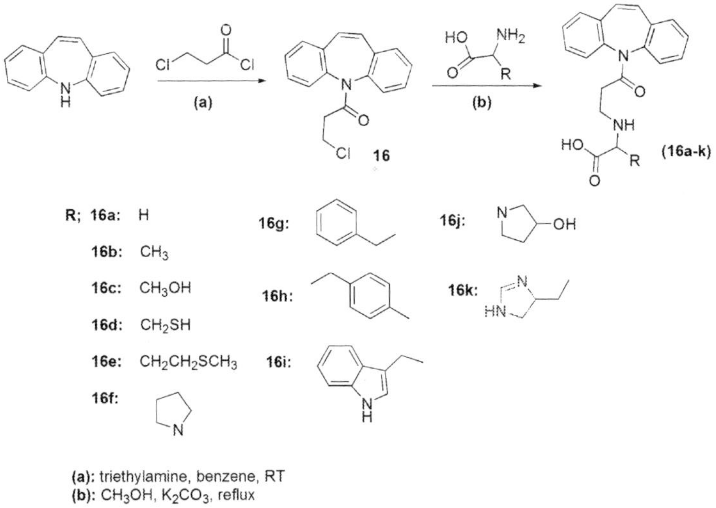

SCHEME 6.8 Synthetic pathway for 3-chloro-1-(5H-dibenz[*b,f*]azepine-5yl)propan-1-one derivatives.

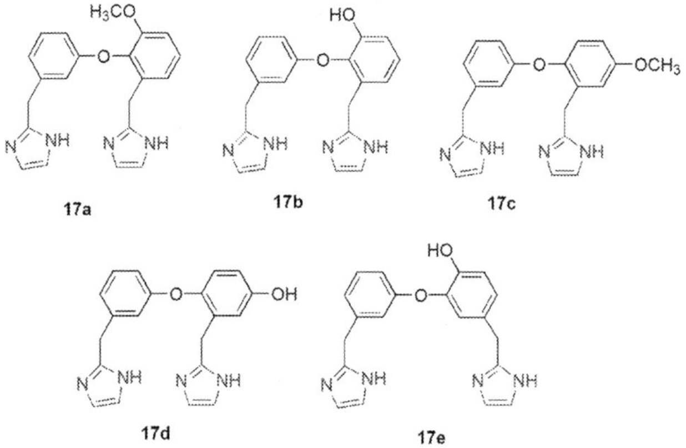

FIG. 6.3 Structure of some imidazole alkaloids (lepidines).

Three inulin derivatives named 2-(1-methylimidazole)acetyl inulin chloride (MIAIL), 2-imidazoleacetyl inulin chloride (IAIL), and 2-imidazoleacetyl inulin chloride (IAIL) were synthesized through the reaction of chloracetyl inulin (CAIL) with tertiary amine and their antioxidant activity was evaluated against superoxide and hydroxyl radicals [29]. The study has shown that IAIL exhibited the most promising scavenging activity on superoxide and hydroxyl radicals with 67.8% and 86.7% at 1.6 mg/mL, respectively (Scheme 6.9).

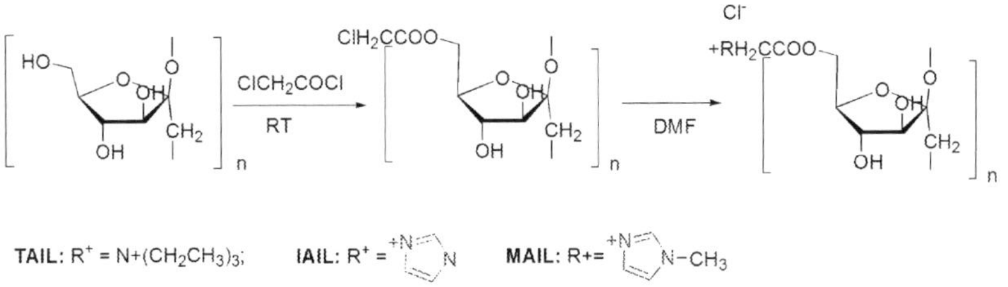

SCHEME 6.9 Synthetic protocol for the inulin derivatives.

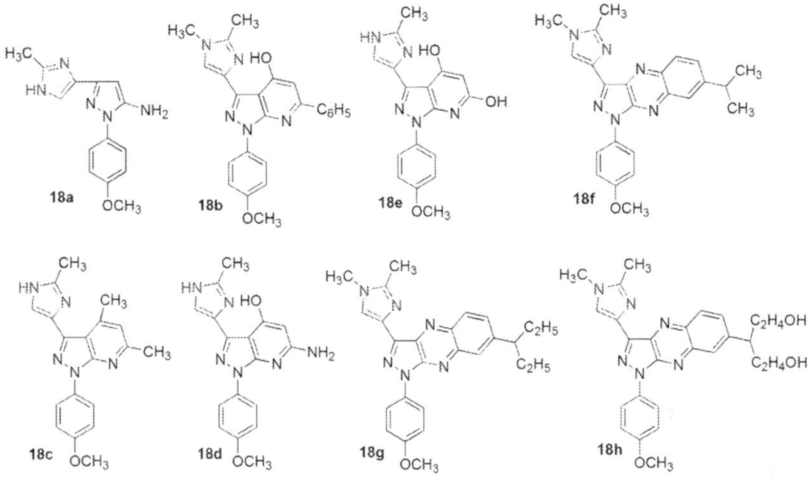

FIG. 6.4 Illustration of antioxidant compounds of N-fused heterocycles bearing imidazole moiety.

A class of N-fused heterocylic compounds bearing an imidazolyl nucleus, i.e., imidazolyl pyrazolopyridines, and imidazolyl pyrazoloquinoxaline were synthesized using 5-amino-1-*p*-anisyl-3-(1,2-disubstituted-imidazol-4-yl) pyrazole as a precursor under microwave irradiations, and DPPH assay was performed to check the antioxidant activity of compounds [30]. Compounds **18a** and **18(f–g)** showed higher radical scavenging activity than the reference molecule (ascorbic acid) with IC_{50} values of 35.64, 36.46, 37.68, and 32.88 μg/mL, respectively. Moreover, compounds **18b** (IC_{50} = 49.33 μg/mL), **18c** (IC_{50} = 57.81 μg/mL), **18d** (IC_{50} = 50.13 μg/mL), and **18f** (IC_{50} = 62.33 μg/mL) presented excellent DPPH scavenging activity (Fig. 6.4).

For the exploration of efficient and potent antioxidant agents, several pyrrolo[1,2-c]imidazoles were synthesized by the reaction of aldehydes, malononitrile, and chiral thiohydantoins using NEt_3 as a catalyst [31]. Among all the synthesized derivatives, compound **19l** presented the highest antioxidant activity (90%) in DPPH radical scavenging assay (Scheme 6.10).

Different imidazole-linked 1,2,3-triazole hybrid compounds were synthesized by click chemistry followed by multicomponent reaction in microwave condition,

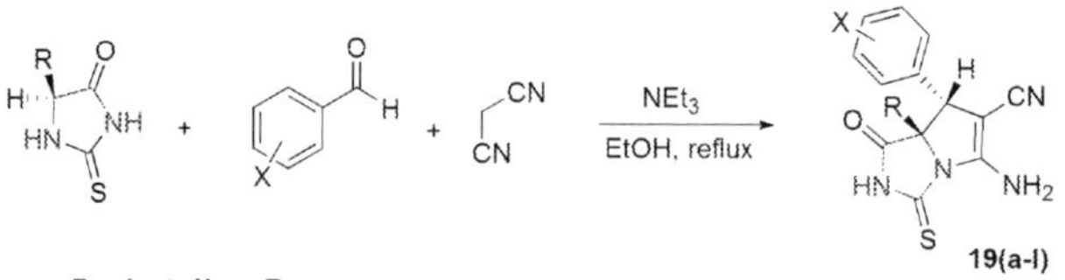

Product	X	R	Product	X	R
19a	4-Cl	CH_3	**19g**	4-Cl	$CH_2CH_2(CH_3)_2$
19b	2-Cl	CH_3	**19h**	2-Cl	$CH_2CH_2(CH_3)_2$
19c	4-Br	CH_3	**19i**	4-Br	$CH_2CH_2(CH_3)_2$
19d	4-Cl	$CH_2(CH_3)_2$	**19j**	4-Cl	CH_2Ph
19e	2-Cl	$CH_2(CH_3)_2$	**19k**	2-Cl	CH_2Ph
19f	4-Br	$CH_2(CH_3)_2$	**19l**	4-Br	CH_2Ph

SCHEME 6.10 Synthesis of pyrrolo[1,2-*c*]imidazoles.

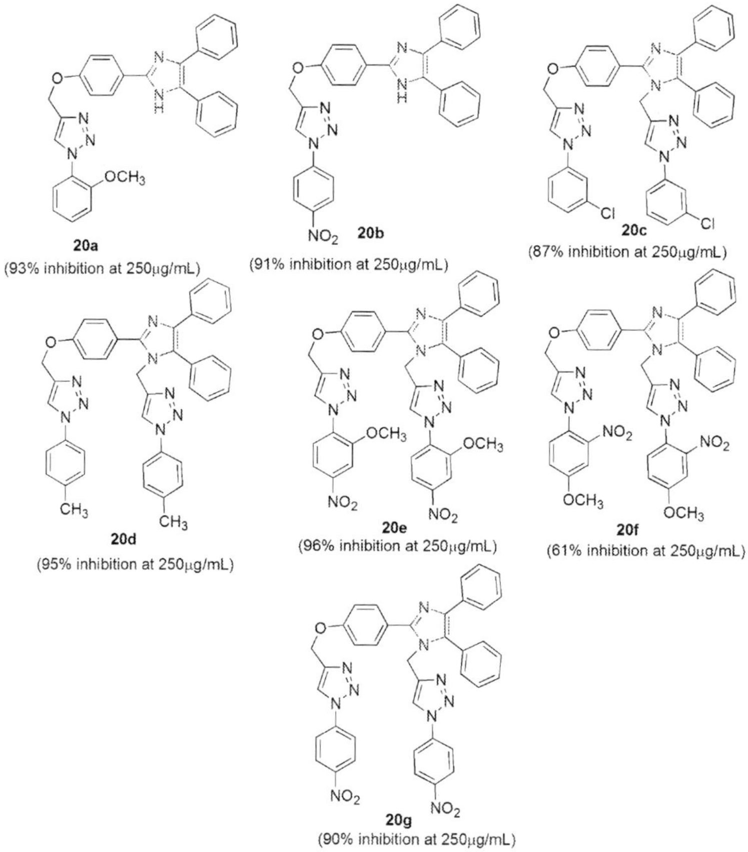

FIG. 6.5 Structural design of imidazole-linked 1,2,3-triazole hybrid compounds.

and the in vitro antioxidant activity of synthesized compounds was also examined. Most of the compounds exhibited antioxidant activity, but a few of them exhibited significant antioxidant activity, particularly **20(a–g)** (Fig. 6.5) [32].

A series of *N*-acyl-1*H*-imidazole-1-carbothioamide derivatives were synthesized and accessed for the antioxidant activity via molecular docking study [33].

The derivative **21d** showed the most promising reducing capacity of 170.19 ± 1.9 μg ascorbic acid equivalent per mg and compound **21f** exhibited significant free radical scavenging activity with 91.74% at 200 μg/mL having an IC_{50} of 93.89 μg/mL with high biocompatibility (Scheme 6.11).

A sequence of derivatives bearing 6-O-imidazole-based quaternary ammonium salts **22(a–i)** were synthesized and their antioxidant activities were evaluated against DPPH radicals, superoxide radicals, and hydroxyl radicals via an in vitro technique [34]. Compounds **22d** and **22h** exhibited excellent antioxidant activity with more than 90% scavenging indices at 1.6 mg/mL (Fig. 6.6). Moreover, these derivatives possessed effective antifungal activity against *Gibberella zeae* and *Botrytis cinerea*.

A library of 7-O-b-D-glucopyranosyloxy-3-(4,5-disubstituted imidazol-2-yl)-4*H*-chromen-4-ones were synthesized and assessed for their *in vitro* antioxidant and antimicrobial activity [35]. The *in vitro* study has shown that the novel glucosides of 7-hydroxy-3-imidazolyl-4*H*-chromen-4-ones **23(a–e)** have good antioxidant activity (75.67%–91.15% inhibition of DPPH radical) (Fig. 6.7).

A library of 1-(furan-2-ylmethyl)-2,4,5-triphenyl-*1H*-imidazole derivatives were synthesized by a four-component reaction of benzil, ammonium acetate, furfurylamine, and benzaldehyde in the ethanol medium using sulfated yatria (SO_4^{2-}/Y_2O_3) as a catalyst and their antioxidant activity was assessed through DPPH assay [36]. Among all the derivatives, compounds **24b**, **24c**, and **24i**

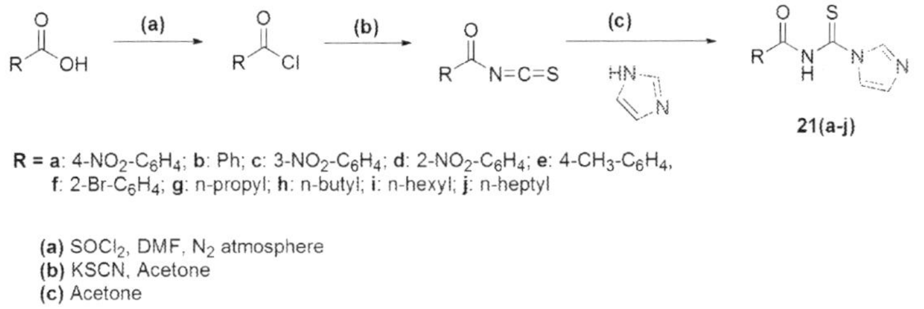

SCHEME 6.11 Synthesis of N-acyl-1H-imidazole-1-carbothioamide derivatives.

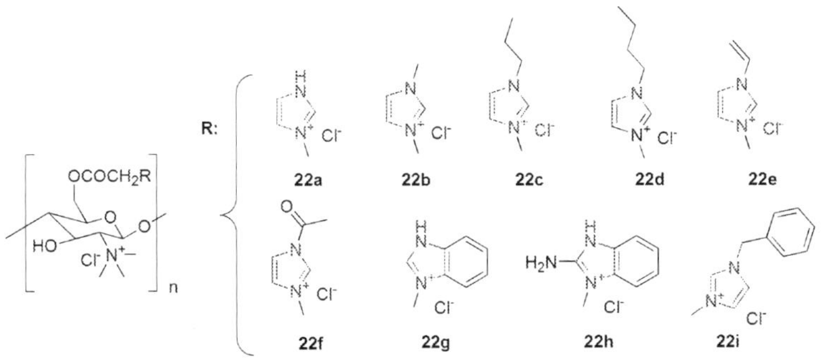

FIG. 6.6 Derivatives bearing 6-O-imidazole-based quaternary ammonium salts.

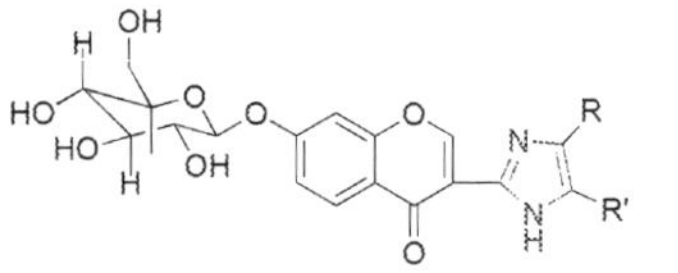

	23a	23b	23c	23d	23e
R:	H	CH_3	C_6H_5	C_6H_5	2-ClC_6H_4
R':	H	CH_3	H	4-$OCH_3C_6H_4$	2-ClC_6H_4

FIG. 6.7 Presentation of 7-hydroxy-3-imidazolyl-4H-chromen-4-ones.

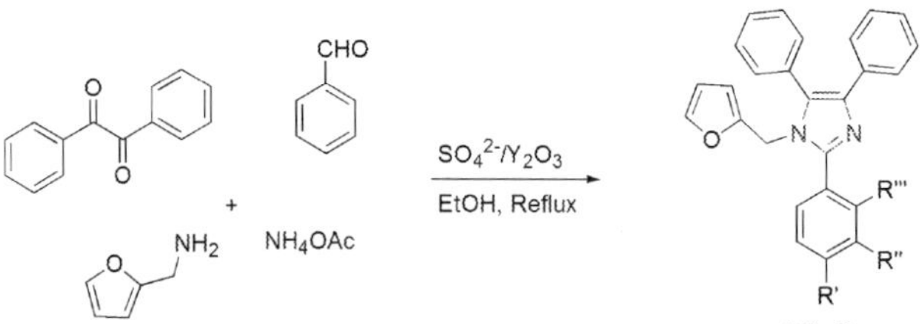

24(a-j)

	24a	24b	24c	24d	24e	24f	24g	24h	24i	24j
R'	H	4-CH_3	4-OCH_3	4-F	4-Cl	H	4-Br	H	4-OH	4-$N(CH_3)_2$
R''	H	H	H	H	H	H	H	3-NO_2	H	H
R'''	H	H	H	H	H	2-Cl	H	H	H	H

SCHEME 6.12 Synthesis of 1-(furan-2-ylmethyl)-2,4,5-triphenyl-1H-imidazole derivatives.

exhibited the most promising antioxidant activity with IC_{50} values of 0.18, 0.13, and 0.21 μg/mL, respectively (Scheme 6.12).

Four novel phenanthroimidazole-imine derivatives **25(a–d)** were synthesized by N-alkylation of compound **25** followed by reduction with Pd/C and coupling with different aldehydes [37]. The antioxidant activity of these compounds was assessed with respect to ascorbic acid as a standard molecule using DPPH method. Compound **25c** showed weak antioxidant activity (0.2% DPPH inhibition value at 100 μg/mL) while compound **25d** exhibited significant activity (52% DPPH inhibition value 100 μg/mL) with respect to ascorbic acid (Scheme 6.13).

Compound **26a** was synthesized by the reaction of 1-(5-methyl-2-phenyl-1H-imidazol-4-yl)ethanone with thiosemicarbazide and was used as a precursor in the synthesis of novel coumarin, thiazole and arylidiene derivatives via reaction with 2,3-dichloroquinoxaline, hydrazonoyl chlorides, and phenacyl bromides, respectively [38]. The antioxidant activity of compounds was assessed and it was found that compounds **26b**, **26c**, and **26d** (IC_{50} value = 6.25 mg/mL each) showed the most promising antioxidant property through the DPPH method (Scheme 6.14).

García et al. [39] introduced novel losartan-hydrocaffeic linked imidazole hybrids as potent antioxidant drug candidates. The results of the ABTS

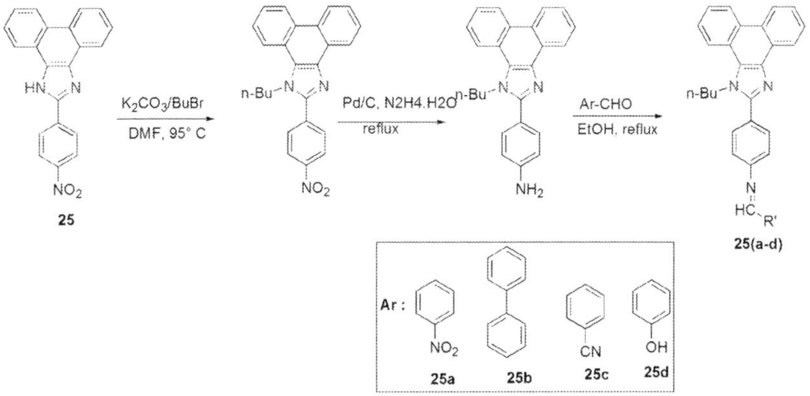

SCHEME 6.13 Synthesis of phenanthroimidazole-imine derivatives.

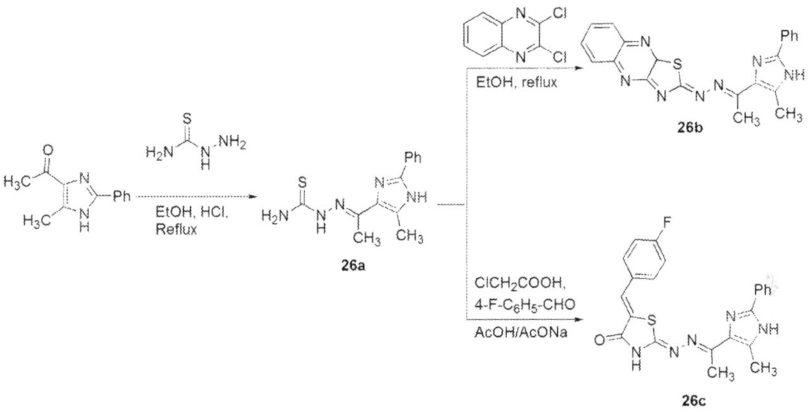

SCHEME 6.14 Synthesis of various antioxidant active derivatives of imidazole-based heterocycles.

method confirmed that all the hybrids displayed three to nine times higher antioxidant activity than the reference losartan and showed 88%–95% inhibition. Compound **27** was found to be the best antioxidant hybrid which revealed extra tissue protection properties (Scheme 6.15). A structure-activity relationship study disclosed the structural necessity of imidazole hybrids (**28–31**) to work as good antioxidants: first, phenols must have at least two hydroxy groups, among them one has to be present in the p-position with respect to the spacer chain; second is the presence of at least the methylene group in the spacer; and third is the presence of amide moiety (Table 6.2).

Brahmbhatt and coauthors [40] synthesized pyrazole-fused tri-substituted imidazole derivatives and screened them for antioxidant activity. The reaction of substituted 1*H*-pyrazole-4-carbaldehyde and benzil with NH_4OAc yielded corresponding imidazoles in good yields via Radziszewski reaction (Scheme 6.16).

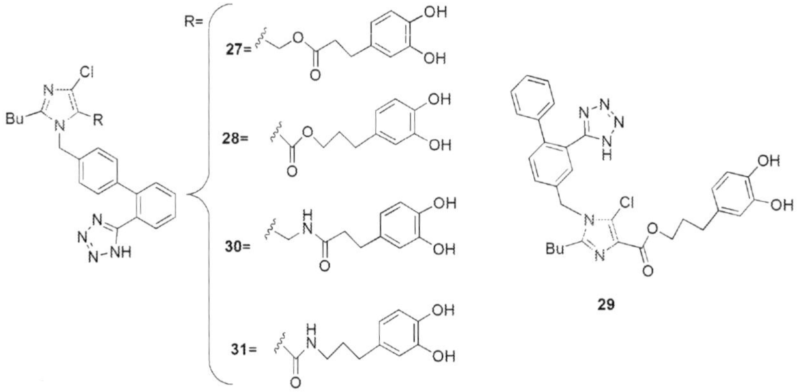

SCHEME 6.15 Structure of losartan-hydrocaffeic linked imidazole hybrids.

TABLE 6.2 Antioxidant ability and inhibition of cellular contraction of the new hybrids.

Compound	27	28	29	30	31
TEAC (Mm; $n=8$)	0.31 ± 0.03	0.10 ± 0.02	0.24 ± 0.03	0.27 ± 0.04	0.17 ± 0.02
Cell conc. Inhibition (%; $n=5$)	94	88	95	90	93

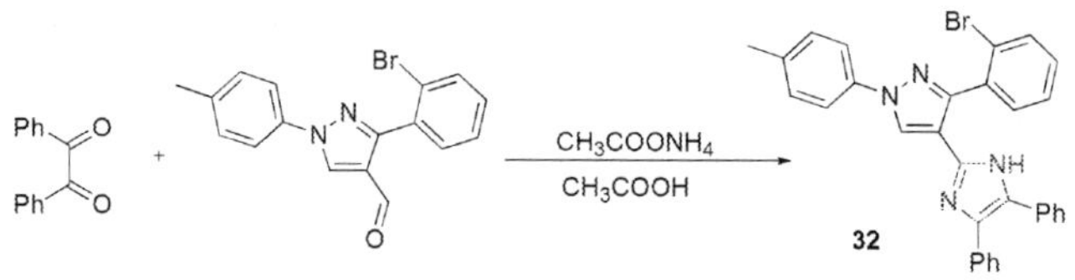

SCHEME 6.16 Synthesis of pyrazole-fused tri-substituted imidazole.

From DPPH scavenging data it was concluded that among all the synthesized compounds, only compound **32** displayed moderate activity and the others were not active inhibitors.

2.1.3 Imidazole metal complex

Recently, Sasahara et al. [41] synthesized two nitro-imidazole ruthenium complexes **33(a,b)** and screened them for in vitro antioxidant activity via intracellular reactive oxygen species (ROS) levels, superoxide anion production, and lipid peroxidation study. Both the *cis*-$[Ru(NO_2)(bpy)_2(5NIM)]PF_6$ (**33a**) and *cis*-[RuCl(bpy)$_2$(MTZ)]PF$_6$ (**33b**) complexes showed high antioxidant activity, and decreased intracellular ROS levels, superoxide anion production (0.22 ± 0.01 and 0.26 ± 0.03, respectively), and lipid peroxidation (**33a** = 0.2 ± 0.06 nmol/mg tissue) compared to α-tocopherol and dexamethasone, the reference drugs (Fig. 6.8).

Tolfenamic acid- and mefenamic acid-based Cu(II) and Zn(II) imidazole complexes **34(a,b)** and **35** were synthesized and screened as potent antioxidant molecules [42]. The antioxidant properties were studied through various processes like DPPH and ABTS radical scavenging assay, iron chelating ability of the compounds, and hydroxyl radical scavenging potential. The study data revealed that all three complexes exhibited good antioxidant activity, and complex **34b** was found to be the most active compared to standard BHT, as depicted in Table 6.3. The biological potency of complexes increased with elevation in their concentrations (Fig. 6.9).

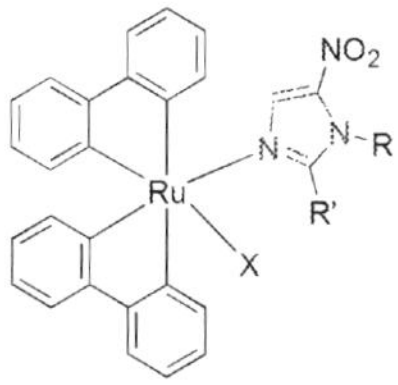

33(a,b)
X;**a**=-NO$_2$, **b**=Cl
R;**a**=H, **b**=-CH$_2$CH$_2$OH
R';**a**=H, **b**=CH$_3$

FIG. 6.8 Structure of nitro-imidazole ruthenium complexes.

TABLE 6.3 Antioxidant activity data of complex 34(a,b) and 35.

	IC$_{50}$ (μg/mL)			
Parameters	34a	34b	35	Butylated hydroxyl toluene (BHT)
DPPH	1702.72 ± 1.04	1199.36 ± 9.63	6612.51 ± 0.49	2313.33 ± 8.43
Iron chelating	1041.24 ± 9.85	516.07 ± 9.37	7551.05 ± 6.35	110.01 ± 3.14
Hydroxyl radical	330.08 ± 8.04	454.73 ± 6.53	4669.03 ± 2.92	6990.55 ± 5.54
ABTS	2023.90 ± 5.33	535.48 ± 4.61	7348.64 ± 3.55	4689.93 ± 9.23

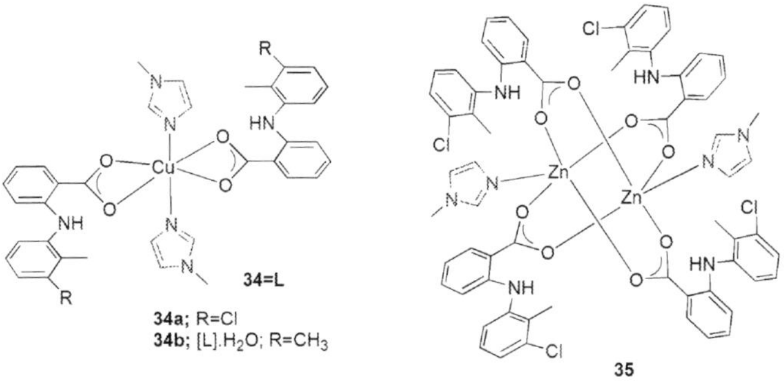

FIG. 6.9 Structure of Cu(II) and Zn(II) imidazole complexes.

2.2 Benzimidazole

2.2.1 Ring generation/formation

In 2016, Dogan et al. [43] synthesized novel series of 2-substituted benzyl-4(7)-phenyl-1*H*-benzo[*d*]imidazole derivatives **36(a–k)** (Scheme 6.17) and examined their tyrosinase inhibitory activity and antioxidant activity. The DPPH radical scavenging activity was used and SC_{50} values in the range of 0.504 ± 0.009 to 0.549 ± 0.014 mM have been estimated. Among the synthesized compounds, compound **36j** exhibited the lowest SC_{50} values with 0.504 ± 0.009 mM followed by **36h** (0.508 ± 0.009 mM) and **36g** (0.519 ± 0.016 mM), respectively. The investigation clearly confirmed that these derivatives have low scavenging activities compared to the standard gallic acid. The ferric reducing antioxidant potency was also determined and the derivatives possessed moderate activity compared to the standard drug, with the following order: **36j** < **36h** < **36g** < **36a** < **36c** < **36e** < **36b** < **36d** < **36k** < **36i** < **36f**.

In the same year, Nayak and coworkers [21] synthesized benzimidazoles **37(a–b)** from itaconic anhydride and substituted o-phenylenediamines (Scheme 6.18). The derivatives were investigated for antioxidant activity by determining DPPH radical scavenging activity, ferric reducing antioxidant power (FRAP) assay, and nitric oxide scavenging assay. Among the compounds **37a** and **37b**, compound **37a** showed moderate DPPH scavenging activity and FRAP assay, and both compounds possessed nitric oxide scavenging ability at a concentration of 1 mg/mL compared with the standard glutathione.

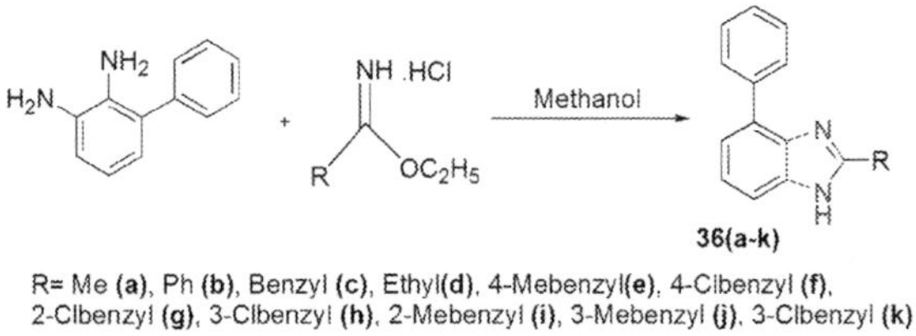

SCHEME 6.17 Synthesis of Benzyl-4(7)-phenyl-1H-benzo[*d*]imidazole derivatives.

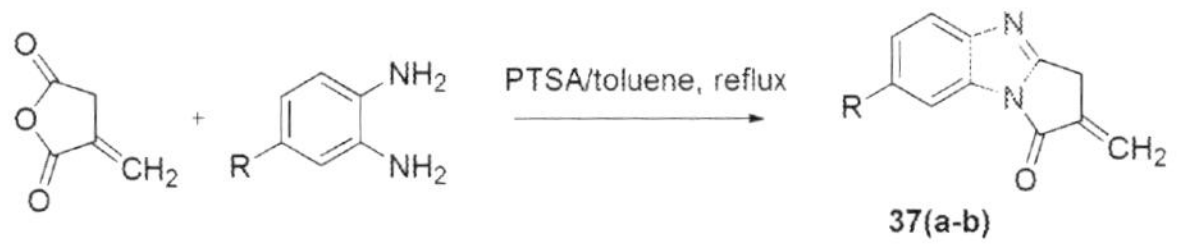

SCHEME 6.18 Synthesis of benz-imidazoles.

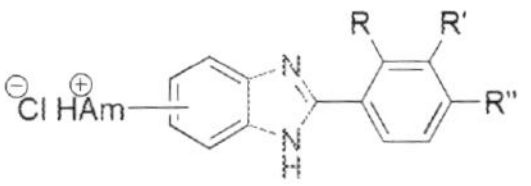

FIG. 6.10 Structure of amidino-substituted benzimidazole derivatives.

Racane and coworkers [44] developed an efficient and facile protocol for the synthesis of novel amidino-substituted benzimidazole derivatives **38(a–h)** (Fig. 6.10) and anticipated their antioxidant activity using three biological assays: DPPH, ABTS, and FRAP. The most prominent antioxidative activity was exhibited by compounds having trihydroxy substituents, **38d** and **38h**. It was observed that the unsubstituted amidino group induced more pronounced activity compared to compounds having the 2-imidazolinyl group.

In the same year, Bellam et al. [45] synthesized N-substituted pyrazole-containing benzimidazoles **39(a–l)** (Scheme 6.19) using 1,2-diaminobenzene and pyrazole-4-carbaldehyde in good yields. The antioxidant activity of all the synthesized compounds was examined using DPPH and H_2O_2 assays. Compound **39f** possessed the highest antioxidant activity with IC_{50} values of 49.48 ± 0.87 mg/mL (DPPH) and 79.95 ± 0.59 mg/mL (H_2O_2 activity).

Shankar and coworkers [46] designed a library of 2-(6-alkyl-pyrazin-2-yl)-1*H*-benzo[*d*]imidazole analogs **40(a–j)** (Scheme 6.20) from benzo[*d*]imidazole analogs by reacting with different substituted cyclic and acyclic secondary amines in the presence of a Pd-PEPPSI-Mes catalyst. All the prepared derivatives were examined for antioxidant activity. All the derivatives showed good percentages of inhibition in DPPH radical scavenging in the range of

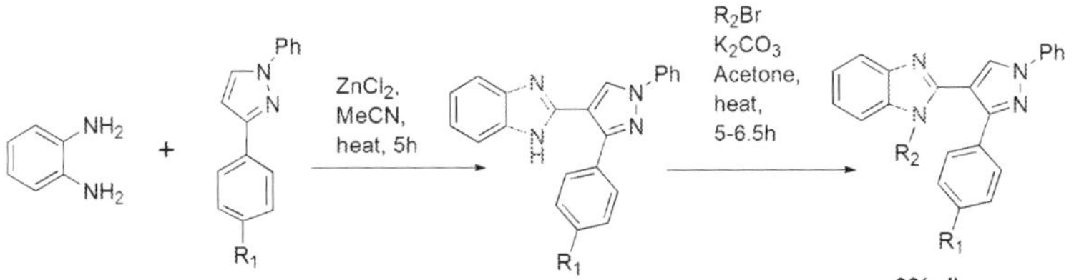

SCHEME 6.19 Structure of N-substituted pyrazole-containing benzimidazole derivatives.

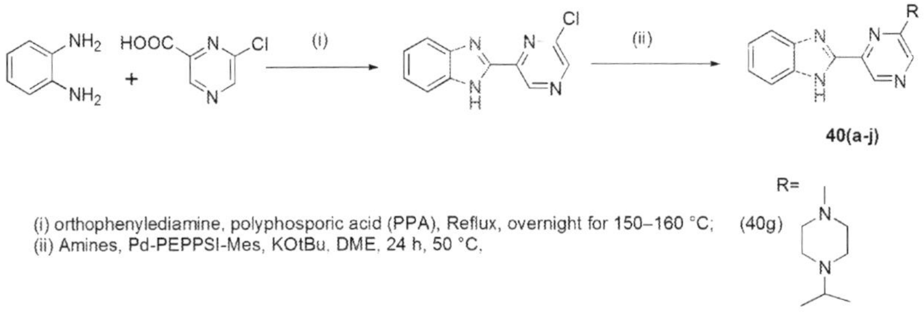

SCHEME 6.20 Synthesis of 2-(6-alkyl-pyrazin-2-yl)-1Hbenz[d] imidazole derivatives.

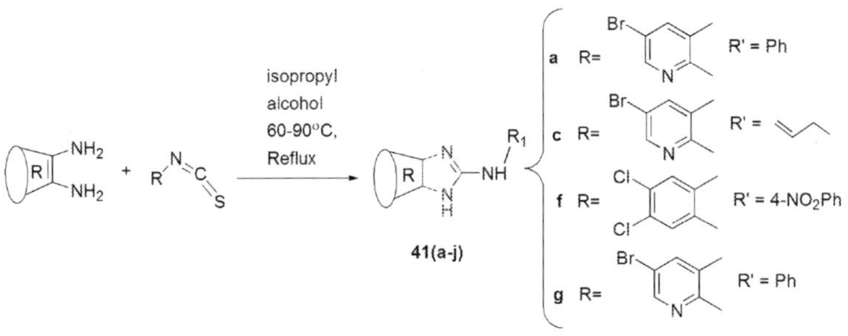

SCHEME 6.21 Imidazole-substituted pyridine-2-amine and benzo-substituted imidazol-2-amine analogs.

33.60 ± 0.87 to 82.40 ± 0.65 for 0.1 mM concentration. Compound **40g** was found to be the most active with an IC_{50} value of $21.7 \pm 0.45\,\mu M$.

Later on in 2018, imidazole-substituted pyridine-2-amine and benzo-substituted imidazol-2-amine compounds **41(a–j)** (Scheme 6.21) were synthesized using o-phenylenediamine and isothiocyanate in the presence of isopropyl alcohol at 60–90 °C by Basha and coauthors [47]. The DPPH radical scavenging activity and its IC_{50} values suggested that compound **41a** ($IC_{50} = 38.56\,\mu g/mL$) was the most active compound compared to ascorbic acid. The H_2O_2 scavenging activity was also determined and it was observed that compounds **41c**, **41f**, and **41g** have potent activity.

In 2019, Baldisserotto and coworkers [48] synthesized 2-substituted benzimidazole derivatives **42(a–ao)**, (Scheme 6.22) (43 derivatives). The synthesized compounds were examined for their antioxidant potency using three biological assays, namely DPPH, FRAP, and ORAC methods. The results of antioxidant activity showed that the nature of substituents, position of substituents, and number of substituents present on the derivatives affected the potency of compounds like the presence of OH, NMe_2 group increased activity compared to Cl, OEt substituents. As per the DPPH and FRAP assay, compounds **42ao**, **42ak**, and **42al** were found to be the most active whereas in the ORAC assay, compound **42ad** showed better activity.

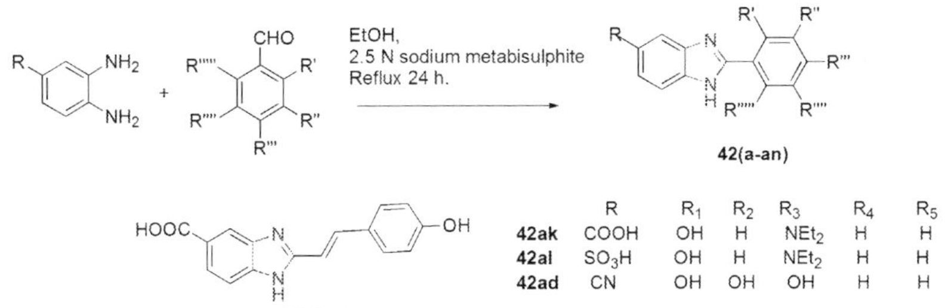

	R	R_1	R_2	R_3	R_4	R_5
42ak	COOH	OH	H	NEt_2	H	H
42al	SO_3H	OH	H	NEt_2	H	H
42ad	CN	OH	OH	OH	H	H

SCHEME 6.22 Structure of 2-substituted benzimidazole derivatives.

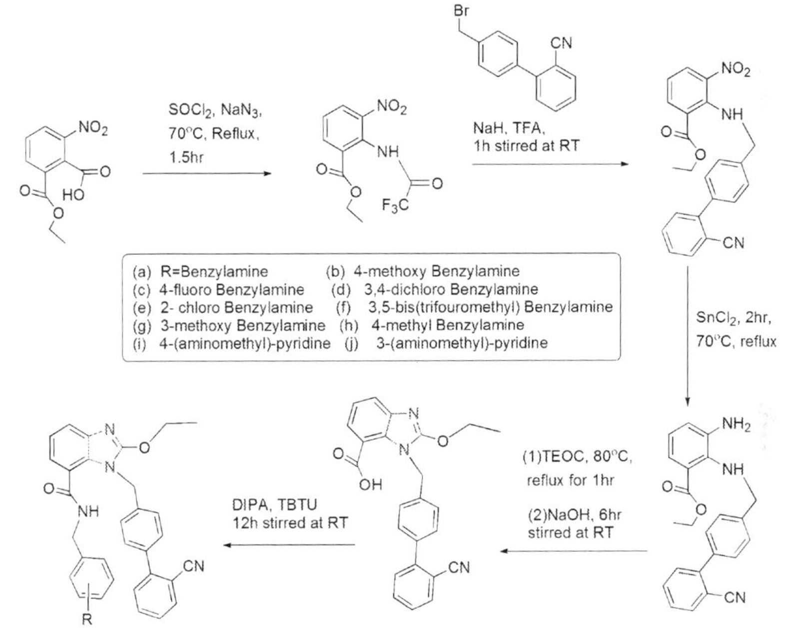

43(a-j)

SCHEME 6.23 Synthesis of benzimidazole carboxamide derivatives.

Prabhu and coworkers [49] developed a facile approach for the synthesis of new benzimidazole carboxamide derivatives **43(a–j)** by a multicomponent multistep synthesis (Scheme 6.23). The prepared derivatives were examined for antioxidant activity. Compounds **43a**, **43d**, and **43f** showed good activity and the rest of the compounds exhibited moderate activity in comparison to ascorbic acid in DPPH assay. The higher antioxidant potency of compounds can be elucidated by the existence of the EWG, *viz.* Cl, F, that determines stabilization of free radicals of N atoms in DPPH radicals. The increased stability of the radical structure is due to the conjugation between the free radical of the N atom and the π electrons of the benzyl ring system.

2.2.2 Benzimidazole hybrid

In 2013, Ravindranath et al. [50] devised a one-pot MCR strategy for the synthesis of library of novel benzo[*d*]imidazolyl tetrahydropyridine carboxylate derivatives **9(a–n)** (Scheme 6.24). The prepared compounds showed a concentration-dependent DPPH radical scavenging. The IC_{50} values of derivatives **44(a–n)** were in the range of 3.8–40 μM in comparison to standard ascorbic acid. The results of the antioxidant activity revealed that compounds **44b** (IC_{50} = 6.6 μM), **44i** (IC_{50} = 4.2 μM), and **44j** (IC_{50} = 3.8 μM), having 2-OH, 2-OMe, and 4-NO_2 functional groups on the benzene ring, respectively, exhibited better activity against DPPH free radicals. The presence of either ERG (Me, OH) or EWG (Cl, NO_2, Br) on the benzene ring augmented the antioxidant activity.

In view of this, Ashok et al. [51] designed microwave-induced click chemistry-inspired synthesis of indole-tethered benzimidazole-based 1,2,3-triazole derivatives **45(a–j)** (Scheme 6.25). The DPPH and H_2O_2 assays were used to determine the antioxidant activity. Among the synthesized compounds, derivatives **45e**, **45f**, and **45g** with 4-methyl, 2-methoxy, and 4-methoxy substitution on the phenyl ring showed outstanding radical scavenging activity with IC_{50} values of 10.04 ± 1.10, 8.52 ± 0.57, and 9.25 ± 0.85 μg/mL, respectively. These results were more promising compared to the reference compounds, ascorbic acid and butylated hydroxyl toluene (BHT), with 13.62 ± 0.69 and 17.16 ± 0.25 μg/mL, respectively.

Recently, Yildiz et al. [52] synthesized thiol bridged bis-benzimidazole derivatives and its dicationic analogs **46(a–b)** from 2-mercaptobenzimidazole derivatives (Fig. 6.11). The antioxidant activity was determined with the help of DPPH and H_2O_2 radical scavenging activity, and it was observed that compounds **46a** and **46b** have IC_{50} values of 14.5 μM and 57.5 μM, respectively, for DPPH assay. The IC_{50} values in hydrogen peroxide scavenging assay for **46a** and **46b** were 638.6 μg/mL and 398.9 μg/mL, respectively. The structural

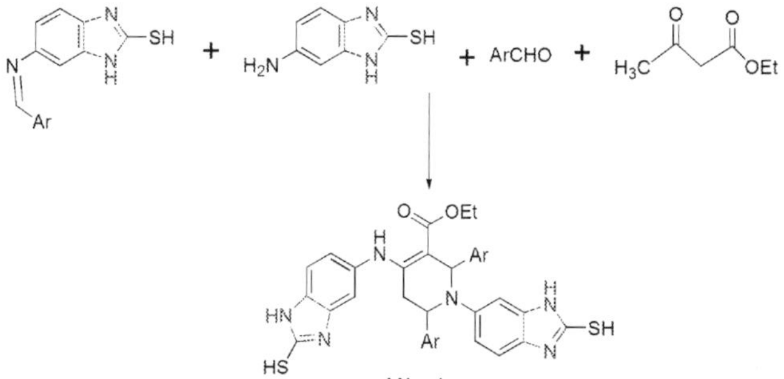

44(a-n)

(a) Ar = C_6H_5, Ar' = C_6H_5 (b) Ar = C_6H_5, Ar' = 2-OHC_6H_4 (c) Ar = C_6H_5, Ar' = 2-ClC_6H_4
(d) Ar = C_6H_5, Ar' = 4-ClC_6H_4 (e) Ar = C_6H_5, Ar' = 2-BrC_6H_4 (f) Ar = C_6H_5, Ar' = 4-BrC_6H_4
(g) Ar = C_6H_5, Ar' = 4-$CH_3C_6H_4$ (h) Ar = C_6H_5, Ar' = 4-$OCH_3C_6H_4$ (i) Ar = C_6H_5, Ar' = 2-$OCH_3C_6H_4$
(j) Ar = C_6H_5, Ar' = 4-$NO_2C_6H_4$ (k) Ar = 2-OHC_6H_4, Ar' = 2-ClC_6H_4 (l) Ar = 2-OHC_6H_4, Ar' = 2- BrC_6H_4
(m) Ar = 2-ClC_6H_4, Ar' = 2-OHC_6H_4 (n) Ar = 2-ClC_6H_4, Ar' = 2-$OCH_3C_6H_4$

SCHEME 6.24 Synthesis of benzo[*d*]imidazolyl tetrahydropyridine carboxylates derivatives.

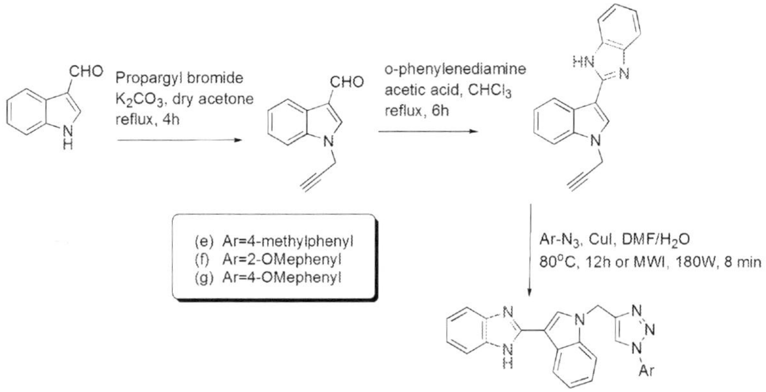

45(a-j)

SCHEME 6.25 Synthesis of indole-tethered benzimidazole-based 1,2,3-triazoles derivatives.

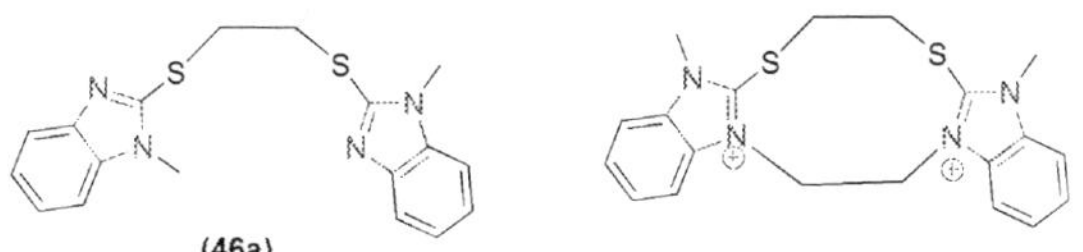

(46a) **(46b)**

FIG. 6.11 Structure of bis-benzimidazole analogs.

analogs did not contain any functional group, therefore they did not damage the DNA. They could defend the DNA structure from the adverse effects of radicals.

Recently, (benz)imidazole-hydrazone analogs **47(a–j)** (Fig. 6.12) were synthesized using 1-methylimidazole-2-carbaldehyde, 1-methylbenzimidazole-2-carbaldehyde, and phenylhydrazine derivatives in ethanol at room temperature to afford the desired products in high yields [53]. The DPPH radical scavenging activity results showed that all the synthesized compounds possessed IC_{50} values in the range of $22.4 \pm 1.2\,\mu M$ to $82.4 \pm 3.4\,\mu M$ in comparison to the references BHT ($70.8 \pm 6.6\,\mu M$) and BHA ($26.0 \pm 1.9\,\mu M$). The highest activity was shown by compound **47c** with IC_{50} $22.4 \pm 1.2\,\mu M$. In ABTS assay, all the examined compounds demonstrated high antioxidant activity and the benzimidazole derivatives **47(g–j)** were found to be the best antioxidant agents ($IC_{50} \leq 3.6 \pm 0.1\,\mu M$), followed by compound **47d** ($IC_{50} = 7.3 \pm 0.8\,\mu M$), compared to the standards BHA ($8.2 \pm 0.4\,\mu M$) and BHT (IC_{50}: $7.2 \pm 1.7\,\mu M$). In CUPRAC assay, all the synthesized hydrazones showed comparable antioxidant activity ($A_{0.5}$ in the range of 17.8 ± 0.4 and $49.5 \pm 0.4\,\mu M$) to that of the standard BHT ($53.4 \pm 4.8\,\mu M$). Finally, in GOR assay, all the compounds exhibited high antioxidant activity with IC_{50} values between 25.2 ± 0.5 and $54.0 \pm 1.0\,\mu M$. The best result was obtained with compound **47i** having an IC_{50} value of $25.2 \pm 0.5\,\mu M$, less than that of the standard BHA ($48.4 \pm 0.1\,\mu M$) and BHT ($29.3 \pm 0.2\,\mu M$).

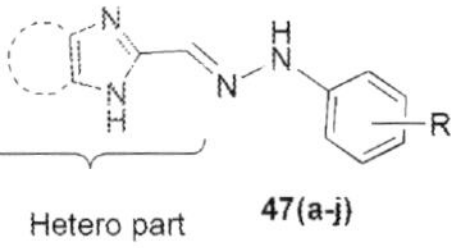

47(a-j)

Hetero d=imidazole, g=benzimidazole
R; d=4-Cl, g=4-OMe; h=3,4-$(Me)_2$; i =4-Cl
j=4-OCH_2Ph

FIG. 6.12 (Benz)imidazole-hydrazone analogs.

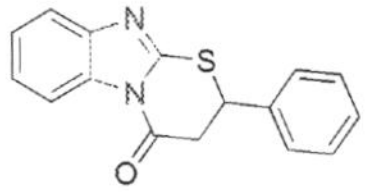

(48)

FIG. 6.13 Structure of 2-aryl-2,3-dihydro-4H-[1,3]thiazino[3,2-a]benzimidazol-4-one.

In view of this, Rodríguez and coworkers designed an efficient methodology for the synthesis of 2-aryl-2,3-dihydro-4*H*-[1,3]thiazino[3,2-a]benzimidazol-4-one (**48**) (Fig. 6.13) [54]. The antioxidant activity of compound **48** was determined using DPPH and ABTS assay. The mechanism of antioxidant activity was explained through the HAT (H12) mechanism. The H atom of position 12 was transferred to stabilize the radical and form DPPH. There was a conjugation of charge between the C12 of the thiazine ring and the aromatic ring. The same was also confirmed by H function calculation. Compound **48** showed an inhibition of $99 \pm 1.18\%$ in comparison to the reference, ascorbic acid ($98 \pm 0.34\%$), which specified a high scavenging capacity. In DPPH radical scavenging activity, inhibition of $73 \pm 2.42\%$ was shown in comparison to the reference antioxidant trolox ($70 \pm 0.35\%$).

2.2.3 Benzimidazole metal complex

Benzimidazole-based heterocycles have been used as ligands in metal complexes and possess different biological applications. In 2010, Liu and coauthors [55] synthesized a novel ligand DBHIP (2-(3,5-dibromo-4-hydroxyphenyl) imidazo[4,5-f][1,10]phenanthroline) and prepared two ruthenium(II) complexes: $[Ru(dmb)_2(DBHIP)](ClO_4)_2$ (**49a**) and $[Ru(dmp)_2(DBHIP)](ClO_4)_2$ (**49b**) (Fig. 6.14). These compounds also possessed good antioxidant activity against hydroxyl radicals ($OH^{\cdot}$). The ligand DBHIP and metal complexes **49a** and **49b** were examined, and it was demonstrated that the antioxidant activity against $OH^{\cdot}$ radical for complex **49a** ($IC_{50} = 0.79\,\mu M$) was higher than that of complex **49b** ($IC_{50} = 0.93\,\mu M$) and ligand DBHIP ($IC_{50} = 1.11\,\mu M$) under the same experimental conditions. These outcomes confirmed that the formation of metal complexes increased their antioxidant activity.

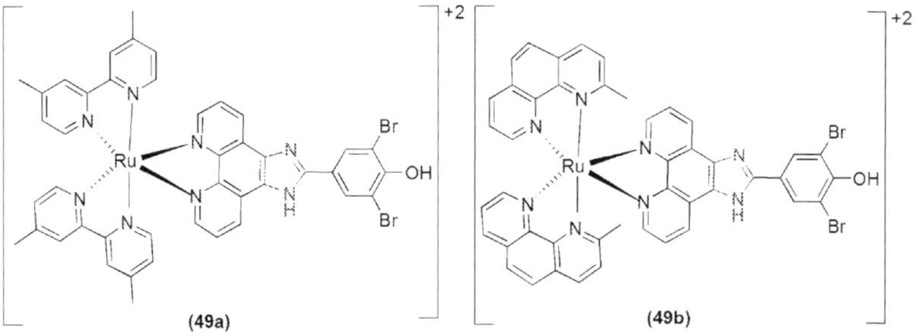

FIG. 6.14 Ruthenium(II) complexes with DBHIP ligand.

After a few years, in 2015, Singla et al. [56] synthesized copper(II) nano-coordination complexes [LCuCl]Cl (**50a**), [LCu(SCN)]SCN (**50b**), and [LCu(NO_3)]NO_3 (**50c**) capped by L = bis-benzimidazole diamide ligand N^1,N^5-bis{(1H-benzo[*d*]imidazole-2-yl)methyl}glutarimide. The metal complex inhibited release of nitric oxide production, which acted as an antioxidant and showed *in vitro* cytotoxicity and antitumor activity. The bis-benzimidazole diamide ligand was coordinated with metallic salts, and they played a crucial role by acting as an antioxidant and an antitumor agent. The presence of imine nitrogen and carbonyl groups might be functionalized by the H-bonding network, which would further inhibit NO release, thereby acting as an antitumor agent. Compound (**50b**) was found to be the best antitumor agent and (**50c**) was the least cytotoxic.

3. Conclusion and future perspectives

This chapter demonstrated in detail that imidazole and their analogs have achieved a cardinal place in the armory of biochemists who are involved in the synthesis of antioxidant imidazole molecules over the past decades. Synthesis of antioxidant imidazole compounds comprising diversified substitution, fusion with various other moieties, and complex formation with different metals have gained significant interest from future organic-chemists and pharmacologists. However, many problems are still unresolved to make this structure safer and more effective, such as detailed mechanistic study at cellular level, side effects, cost-efficiency, and selection of suitable assay [57, 58]. This study provided a comprehensive outlook of the biochemistry of these advantageous heterocycles in the context of antioxidant activity. In the coming years, many innovations are expected to uncover several exciting aspects of imidazole, and it is anticipated that this chapter will be of assistance in this area.

Conflict of interest

The authors declare no conflict of interest, financial or otherwise.

Acknowledgments

The authors are grateful to the Department of Chemistry, Mohan Lal Sukhadia University, Udaipur (Raj.), India, for providing the necessary library facilities for carrying out the work. N. Sahiba is deeply grateful to the Council for Scientific and Industrial Research (CSIR) (file no. 09/172(0088)2018-EMR-I), New Delhi, for providing a Senior Research Fellowship as financial support. A. Sethiya and P. Teli are also grateful to UGC-MANAF and the CSIR, India, respectively for providing a fellowship to carry out this work.

Funding source

This work was supported by the CSIR [file no. 09/172(0088)2018-EMR-I], [file no. 09/172(0099)2019-EMR-I], UGC-MANF (201819MANF-2018-19-RAJ-91971).

References

[1] Huang D, Boxin O, Prior RL. The chemistry behind antioxidant capacity assays. J Agric Food Chem 2005;53:1841–56.

[2] Lobo V, Patil A, Phatak A, Chandra N. Free radicals, antioxidants and functional foods: impact on human health. Pharmacogn Rev 2010;4(8):118–26.

[3] Pham-Huy LA, He H, Pham-Huy C. Free radicals, antioxidants in disease and health. Int J Biomed Sci 2008;4(2):89–96.

[4] Yadav A, Kumari R, Yadav A, Mishra JP, Srivatva S, Prabha S. Antioxidants and its functions in human body-a review. Res Environ Life Sci 2016;9(11):1328–31.

[5] Stanner S, Weichselbaum E. Antioxidants. Encyclopedia of human nutrition. vol. 1; 2013. https://doi.org/10.1016/B978-0-12-375083-9.00013-1. [S. Stanner, E. Weichselbaum, Antioxidants, Editor(s): Benjamin Caballero, Encyclopedia of Human Nutrition (Third Edition), Academic Press, 2013, Pages 88–99, ISBN 9780123848857].

[6] Kurutas EB. The importance of antioxidants which play the role in cellular response against oxidative/nitrosative stress: current state. Nutr J 2016;15:71.

[7] Sindhi V, Gupta V, Sharma K, Bhatnagar S, Kumari R, Dhaka N. Potential applications of antioxidants: a review. J Pharm Res 2013;7:828–35.

[8] Young IS, Woodside JV. Antioxidants in health and disease. J Clin Pathol 2001;54:176–86.

[9] Kattappagari KK, Teja CSR, Kommalapati RK, Poosarla C, Gontu SR, Reddy BV. Role of antioxidants in facilitating the body functions: a review. J Orofac Sci 2015;7:71–5.

[10] Yashin A, Yashin Y, Xia X, Nemzer B. Antioxidant activity of spices and their impact on human health: a review. Antioxidants 2017;6(3):70.

[11] Liu R, Xing L, Fu Q, Zhou G-H, Zhang W-G. A review of antioxidant peptides derived from meat muscle and by-products. Antioxidants 2016;5(3):32.

[12] Pietta P-G. Flavonoids as antioxidants. J Nat Prod 2000;63:1035–42.

[13] Kerru N, Gummidi L, Maddila S, Gangu KK, Jonnalagadda SB. A review on recent advances in nitrogen-containing molecules and their biological applications. Molecules 2020;25:1909.

[14] Zhang L, Peng X-M, Damu GLV, Geng RX, Zhou CH-H. Comprehensive review in current developments of imidazole-based medicinal chemistry. Med Res Rev 2014;34(2):340–437.

[15] Sorrenti V, Salerno L, Di Giacomo C, Acquaviva R, Siracusa MA, Vanella A. Imidazole derivatives as antioxidants and selective inhibitors of nNOS. Nitric Oxide 2006;14:45–50.

[16] Soujanya Y, Sastry GN. Theoretical elucidation of the antioxidant mechanism of 1,3-dihydro-1-methyl-2 H-imidazole-2-selenol (MSeI). Tetrahedron Lett 2007;48:2109–12.

[17] Salerno L, Modica MN, Romeo G, Pittalà V, Siracusa MA, Amato ME, Acquaviva R, Di Giacomo C, Sorrenti V. Novel inhibitors of nitric oxide synthase with antioxidant properties. Eur J Med Chem 2012;49:118–26.

[18] Maddila S, Palakondu L, Chunduri VR. Synthesis, antibacterial, antifungal and antioxidant activity studies on 2-benzylthio- and 2-benzylsulfonyl-1H-imidazoles. Pharm Lett 2010;2(4):393–402.

[19] Jayabharathi J, Thanikachalam V, Rajendraprasath N, Saravanan K, Perumal MV. Antioxidant potential and antimicrobial screening of some novel imidazole derivatives: greenway efficient one pot synthesis. Med Chem Res 2012;21:1850–60.

[20] Assadieskandar A, Amini M, Salehi M, Sadeghian H, Alimardani M, Sakhteman A, Nadri H, Shafiee A. Synthesis and SAR study of 4,5-diaryl-1H-imidazole-2(3H)-thione derivatives as potent 15-lipoxygenase inhibitors. Bioorg Med Chem 2012;20:7160–6.

[21] Nayak PS, Narayana B, Sarojini BK, Sheik S, Shashidhara KS, Chandrashekar KR. Design, synthesis, molecular docking and biological evaluation of imides, pyridazines, imidazoles derived from itaconic anhydride for potential antioxidant and antimicrobial activities. J Taibah Univ Sci 2016;10(6):823–38.

[22] Guda R, Kumar G, Korra R, Balaji S, Dayakar G, Palabindela R, Myadaraveni P, Yellu NR, Kasula M. EGFR, HER2 target based molecular docking analysis, in vitro screening of 2, 4, 5-trisubstituted imidazole derivatives as potential anti-oxidant and cytotoxic agents. J Photochem Photobiol B Biol 2017;176:69–80.

[23] Ghorbani-Vaghei R, Izadkhah V, Mahmoodi J, Karamian R, Khoei MA. The synthesis of imidazoles and evaluation of their antioxidant and antifungal activities. Monatsh Chem 2018;149:1447–52.

[24] Somashekara B, Thippeswamy B, Vijayakumar GR. Synthesis antioxidant and α-amylase inhibition activity of naphthalene-containing 2,4,5-trisubstituted imidazole derivatives. J Chem Sci 2019;131(7):62.

[25] Abdelhamid AA, Salah HA, Marzouk AA. Synthesis of imidazole derivatives: ester and hydrazide compounds with antioxidant activity using ionic liquid as an efficient catalyst. J Heterocycl Chem 2020;57(2):676–85.

[26] Noriega-Iribe E, Díaz-Rubio L, Estolano-Cobián A, Barajas-Carrillo VW, Padrón JM, Salazar-Aranda R, Díaz-Molina R, García-González V, Chávez-Santoscoy RA, Chávez D, Córdova-Guerrero I. In vitro and in silico screening of 2,4,5-trisubstituted imidazole derivatives as potential xanthine oxidase and acetylcholinesterase inhibitors, antioxidant, and antiproliferative agents. Appl Sci 2020;10:2889.

[27] Kumar HV, Kumar CK, Naik N. Synthesis of novel 3-chloro-1-(5H-dibenz[b,f]azepine-5yl) propan-1-one derivatives with antioxidant activity. Med Chem Res 2011;20:101–8.

[28] Pérez-González A, García-Hernández E, Chigo-Anota E. The antioxidant capacity of an imidazole alkaloids family through single-electron transfer reactions. J Mol Model 2020;6:321. https://doi.org/10.1007/s00894-020-04583-2.

[29] Chen Y, Zhang J, Tan W, Wang G, Dong F, Li Q, Guo Z. Antioxidant activity of inulin derivatives with quaternary ammonium. Starch 2017;69(11-12):1700046. https://doi.org/10.1002/star.201700046.

[30] Ibrahim SA, Rizk HF, El-Borai MA, Sadek ME. Green routes for the synthesis of new pyrazole bearing biologically active imidiazolyl, pyridine and quinoxaline derivatives as promising antimicrobial and antioxidant agents. J Iran Chem Soc 2021. https://doi.org/10.1007/s13738-020-02119-2.

[31] Mollanejad K, Asghari S, Jadidi K. Diastereoselective synthesis of pyrrolo[1, 2-c]imidazoles using chiral thiohydantoins, malononitrile, and aldehydes and evaluation of their antioxidant and antibacterial activities. J Heterocycl Chem 2020;57(2):556–64.

[32] Subhashini NJP, Praveenkumar E, Gurrapu N, Yerragunta V. Design and synthesis of imidazolo-1, 2,3-triazoles hybrid compounds by microwave-assisted method: evaluation as an antioxidant and antimicrobial agents and molecular docking studies. J Mol Struct 2019;1180:618–28.
[33] Aziz H, Saeed A, Khan MA, Afridi S, Jabeen F, Rehman AU, Hashim M. Novel N-acyl-1H-imidazole-1-carbothioamides: design, synthesis, biological and computational studies. Chem Biodivers 2020;17(3). https://doi.org/10.1002/cbdv.201900509, e1900509.
[34] Wei L, Li Q, Chen Y, Zhang J, Mi Y, Dong F, Lei C, Guo Z. Enhanced antioxidant and antifungal activity of chitosan derivatives bearing 6-O-imidazole-based quaternary ammonium salts. Carbohydr Polym 2018;206:493–503.
[35] Hatzade K, Sheikh J, Taile V, Ghatole A, Ingle V, Genc M, Lahsasni S, Hadda TB. Antimicrobial/antioxidant activity and POM analyses of novel 7-O-b-D-glucopyranosyloxy-3-(4,5-disubstitutedimidazol-2-yl)-4H-chromen-4-ones. Med Chem Res 2015;24:2679–93.
[36] Rajaraman D, Sundararajan G, Rajkumar R, Bharanidharan S, Krishnasamy K. Synthesis, crystal structure investigation, DFT studies and DPPH radical scavenging activity of 1-(furan-2-ylmethyl)-2,4,5-triphenyl-1H-imidazole derivatives. J Mol Struct 2016;1108:698–707.
[37] Gülle S, Erbaş SC, Uzel A. Synthesis and spectroscopic studies of phenanthroimidazole-imine derivatives and evaluation of their antioxidant activity. J Fluoresc 2018;28:217–23.
[38] Abdel-Wahab BF, Awad GEA, Badria FA. Synthesis, antimicrobial, antioxidant, anti-hemolytic and cytotoxic evaluation of new imidazole-based heterocycles. Eur J Med Chem 2011;46(5):1505–11.
[39] García G, Serrano I, Sánchez-Alonso P, Rodríguez-Puyol M, Alajarín R, Griera M, Vaquero JJ, Rodríguez-Puyol D, Álvarez-Builla J, Díez-Marqués ML. New losartan-hydrocaffeic acid hybrids as antihypertensive-antioxidant dual drugs: ester, amide and amine linkers. Eur J Med Chem 2012;50:90–101.
[40] Brahmbhatt H, Molnar M, Pavi V. Pyrazole nucleus fused tri-substituted imidazole derivatives as antioxidant and antibacterial agents. Karbala Int J Mod Sci 2018;4:200–6.
[41] Sasahara GL, Júnior FSG, de Oliveira Rodrigues R, de Souza Zampieri D, da Cruz Fonseca SG, de Cássia Ribeiro Gonçalves R, Athaydesd BR, Kitagawad RR, Santose FA, de Sousab EHS, Diasa ATN, de França Lopesb LG. Nitro-imidazole-based ruthenium complexes with antioxidant and anti-inflammatory activities. J Inorg Biochem 2020;206:111048.
[42] Shamle NJ, Tella AC, Obaleye JA, Balogun FO, Ashafa AOT, Ajibade PA. Synthesis, characterization, antioxidant and antidiabetic studies of Cu(II) and Zn(II) complexes of tolfenamic acid/mefenamic acid with 1-methylimidazole. Inorganica Chim Acta 2020;513:119942.
[43] Dogan IS, O€zel A, Birinci Z, Barut B, Sellitepe HE, Kahveci B. Synthesis of some novel 2-substitutedbenzyl-(4)7-phenyl-1hbenzo[d]imidazoles in mild conditions as potent anti-tyrosinase and antioxidant agents. Arch Pharm Chem Life Sci 2016;349:1–8.
[44] Racané L, Cindrić M, Perin N, Roškarić P, Starčević K, Mašek T, Maurić M, Dogan J, Karminski-Zamola G. Synthesis and anti-oxidative potency of novel amidino substituted benzimidazole and benzothiazole derivatives. Croat Chem Acta 2017;90(2):187–95. https://doi.org/10.5562/cca3146.
[45] Bellam M, Gundluru M, Sarva S, Chadive S, Netala VR, Tartte V, Cirandur SR. Synthesis and antioxidant activity of some new *N*-alkylated pyrazole-containing benzimidazoles. Chem Heterocycl Compd 2017;53(2):173–8.
[46] Shankar B, Jalapathi P, Anil Valeru, Kumar AK, Saikrishna B, Karunakar ao Kudle KR. Synthesis and biological evaluation of new 2-(6-alkyl-pyrazin-2-yl)-1H-benz[d]imidazoles as potent anti-inflammatory and antioxidant agents. Med Chem Res 2017;26:1835–46. https://doi.org/10.1007/s00044-017-1897-7.

[47] Basha MS, Varalakshmi M, Basha ST, Venkataramaiah C, Shafi SS, Raju CN, Rajendra W. Synthesis, spectral characterization, and biological evaluation of imidazole-substituted pyridine-2-amine and benzo-substitutedimidazole-2-amine. J Chin Chem Soc 2019;66(12):1700–7.

[48] Baldisserotto A, Demurtas M, Lampronti I, Tacchini M, Moi D, Balboni G, Pacifico S, Vertuani S, Manfredini S, Onnis V. Synthesis and evaluation of antioxidant and antiproliferative activity of 2-arylbenzimidazoles. Bioorg Chem 2020;94:103396.

[49] Prabu DSD, Lakshmanan S, Ramalakshmi N, Thirumurugan K, Govindaraj D, Antony SA. Synthesis, characterization of benzimidazole carboxamide derivatives as potent anaplastic lymphoma kinase inhibitor and antioxidant activity. Synth Commun 2019;49(2):266–78. https://doi.org/10.1080/00397911.2018.1554144.

[50] Ravindernath A, Reddy MS. Synthesis and evaluation of anti-inflammatory, antioxidant and antimicrobial activities of densely functionalized novel benzo [d] imidazolyl tetrahydropyridine carboxylates. Arab J Chem 2017;10:S1172–9.

[51] Ashok D, Gundu S, Aamate VK, Devulapally MG. Conventional and microwave-assisted synthesis of new indole-tethered benzimidazole-based 1,2,3-triazoles and evaluation of their antimycobacterial, antioxidant and antimicrobial activities. Mol Diver 2018;22:769–78. https://doi.org/10.1007/s11030-018-9828-1.

[52] Yildiz U. Anti-oxidant and DNA damage protecting activities of newly synthesized thiol bridged bis-benzimidazole derivative and its dicationic analogue. J Heterocycl Chem 2020;57(11):4007–12.

[53] Khodja IA, Boulebd H, Bensouici C, Belfaitah A. Design, synthesis, biological evaluation, molecular docking, DFT calculations and *in silico* ADME analysis of (benz)imidazole-hydrazone derivatives as promising antioxidant, antifungal, and anti-acetylcholinesterase agents. J Mol Struct 2020;1218:128527. https://doi.org/10.1016/j.molstruc.2020.128527.

[54] Rodríguez OAR, Vergara NEM, Sanchez JPM, Martínez MTS, Sandoval ZG, Cruz A, Organillo AR. Synthesis, crystal structure, antioxidant activity and DFT study of 2-aryl-2,3-dihydro-4H-[1,3]thiazino[3,2-a]benzimidazol-4-one. J Mol Struct 2020;1199:127036.

[55] Liu YJ, Liang Z-H, Li Z-Z, Zeng C-H, Yao J-H, Huang H-L, Wu F-H. 2-(3,5-Dibromo-4-hydroxyphenyl)imidazo[4,5f][1,10]phenanthrolinoruthenium(II) complexes: synthesis, characterization, cytotoxicity, apoptosis, DNA-binding and antioxidant activity. Biometals 2010;23:739–52. https://doi.org/10.1007/s10534-010-9340-2.

[56] Singla M, Ranjan R, Mahiya K, Mohapatra SC, Ahmade S. Nitric oxide inhibition, antioxidant, and anti-tumour activities of novel copper(II) bis-benzimidazole diamide nanocoordination complexes. New J Chem 2015;39:4316–27. https://doi.org/10.1039/c4nj02147a.

[57] Apak R. Current issues in antioxidant measurement. J Agric Food Chem 2019;67(33):9187–202.

[58] Apak R, Özyürek M, Güçlü K, Çapanoğlu E. Chemical principles, mechanisms and electron transfer (ET)-based assays. J Agric Food Chem 2016;64(5):997–1027.

Chapter 7

Miscellaneous biological activity profile of imidazole-based compounds: An aspirational goal for medicinal chemistry

Nusrat Sahiba[a], Pankaj Teli[a], Dinesh K. Agarwal[b], and Shikha Agarwal[a]
[a]*Synthetic Organic Chemistry Laboratory, Department of Chemistry, Mohanlal Sukhadia University, Udaipur, Rajasthan, India,* [b]*Department of Pharmacy, PAHER University, Udaipur, Rajasthan, India*

1. Introduction

Nitrogen-containing heterocycles are among the most utilized effective pharmacophores to combat lethal diseases. Within the last two decades, the imidazole nucleus has established itself as an integral part of synthetic-medicinal chemistry [1–3]. The growth in the field of imidazole-based compounds arises from inherent properties, viz. its structural features, viz. the presence of heteroatom, wide substitution possibilities, interaction ability with various proteins, and good binding affinity with vital elements of living organisms [4–6]. Its diversified structure, mode of action, and mechanistic pathways have opened new possibilities for drug designing.

Imidazole-based compounds have well-known biological activities (Fig. 7.1) such as antiparasitic [7], antihypertensive [8], antidiabetic [9], anti-Alzheimer [10], antinociceptive [11] anticonvulsant [12–14], antiinflammatory [15,16], and also work as opioid receptor agonists for gastrointestinal disorders [17], H3-receptor antagonists [18,19], transient receptor potential vanilloid 1 antagonists [20], cholecystokinin 1 receptor (CCK1R) agonists [21], selective CB1 cannabinoid receptor antagonists [22], and others. These derivatives are involved in a number of enzymatic inhibition processes like inhibition of retinoic acid 4-hydroxylase (CYP26) [23], glutaminyl cyclase [24], human liver glycogen phosphorylase [25], 5-lipoxygenase [26], 15-lipoxygenase [27], thromboxane synthase [28], p38 MAP (Mitogen-Activated Protein) kinase [26,29–32], casein kinase 1δ [33], gastric H^+/K^+-ATPase [34], isocitric dehydrogenase 1 (IDH1) [35], cytochrome P450 enzyme [36], aldosterone synthase (CYP11B2) [37], angiotensin-converting enzyme (ACE) [38], and so on. Moreover, they possess

Imidazole-Based Drug Discovery. https://doi.org/10.1016/B978-0-323-85479-5.00008-3
Copyright © 2022 Elsevier Inc. All rights reserved.

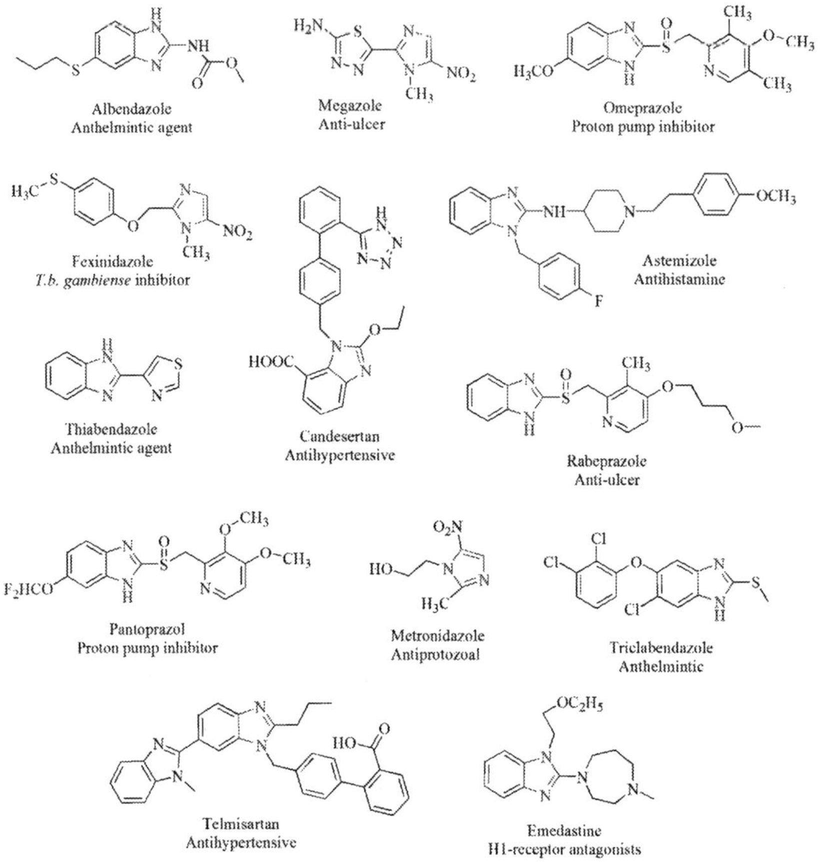

FIG. 7.1 Various imidazole-based drugs available on the market.

bioactivity against anticholinesterase intoxication [39], proprotein convertase subtilisin/Kexin 9 (PCSK9) inhibition [40], and work as ligands for telomeric G-quadruplex DNA [41], R1â2γ2S GABAA receptors, and F1 GABAC receptors (GABA, γ-aminobutyric acid) [42].

A variety of imidazole analogs has also been synthesized and used for lowering intraocular pressure (IOP) [43,44], calcium channel blocker [45], forming caseoperoxidase (a mixed β-casein-SDS-hemin imidazole complex) that works as a nanoartificial enzyme [46], sequence-specific DNA alkylating agents [47], and in hydrogel formation in biomedical applications [48], etc. The reactivity, properties, and high efficacy have been emphasized through several mechanistic investigations, based on experimental in vitro and in vivo studies, computational studies such as molecular modeling, and in silico studies. This chapter intends to give an overview of the applications of the imidazole-based analogs in terms of their bioactivity against several noxious targets. The present study outlines the state-of-the-art innovation in the field of imidazole motifs as antiparasitic, anti-Alzheimer's, antiinflammatory, antidiabetic, antihypertensive, antidepressant, anticonvulsant, antihelminthic, anticoagulant, antiprotozoal, and antiulcer activities along with a critical discussion on drug design and development.

2. Imidazole-based bioactive compounds

Biologically active imidazole derivatives have been categorized on the basis of their work and mode of action on particular diseases (Fig. 7.2).

2.1 Antiparasitic activity

Malaria, caused by the parasite *Plasmodium falciparum*, is a common parasitic infection all over the world. According to a WHO report, 229 million people are still at risk of malaria. In 2019, there were 409,000 deaths and most of them were young people [49]. Astemizole **1** is a recently screened drug that inhibits the proliferation of three parasitic strains of *Plasmodium falciparum*. Inspired by its mode of action, Roman and his companions synthesized imidazole derivatives **2** with several structural modifications in an astemizole skeleton to improve antiplasmodium activity [50]. The in vitro evaluation revealed that all the synthesized compounds showed significant inhibition and high toxicity toward the parasite over CHO (Chinese ovarian hamster) with remarkable selectivity. From the data it was concluded that either the presence of substituted phenylalkyl group or the absence of substituents at these two regions favored antiplasmodium activity. Moreover, the presence of the secondary amine group at the second position exhibited high selectivity toward the parasite (Fig. 7.3).

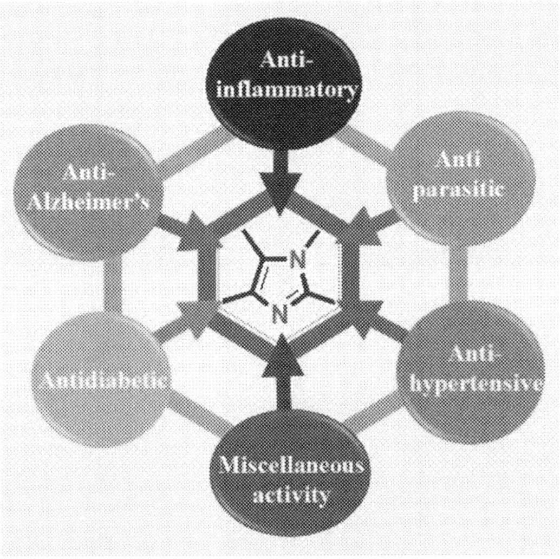

FIG. 7.2 Diverse biological activities of imidazole scaffold.

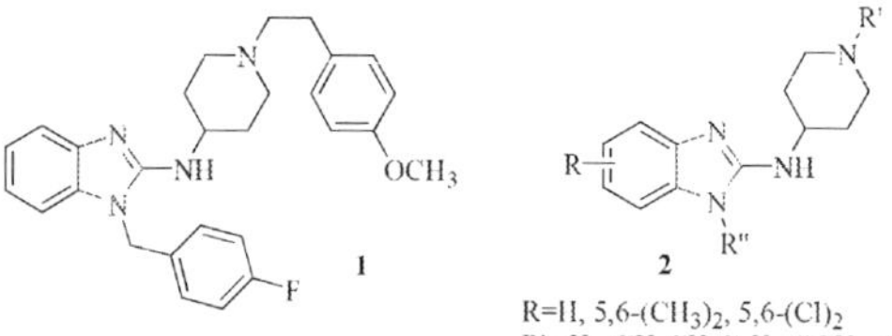

1 2

R=H, 5,6-$(CH_3)_2$, 5,6-$(Cl)_2$
R'= H, $-CH_2CH_2C_6H_4-OCH_3$, $COOC_2H_5$
R''= H, $-CH_2C_6H_4F$, $-CH_2C_6H_4CF_3$

FIG. 7.3 Astemizole-related benzimidazole analogs for antiplasmodium activity.

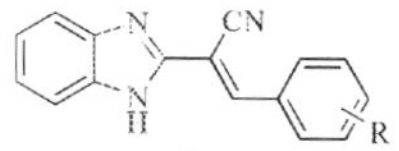

3

FIG. 7.4 Benzimidazole acrylonitrile derivatives.

Drug resistance reduces the effectiveness of a medicine for curing any disease and may emerge as a serious problem in the case of any infection. Meanwhile, benzimidazole acrylonitrile derivatives **3** were prepared and screened against hemozoin and falcipain-2 formation [51] (Fig. 7.4). Some of the synthesized compounds were highly potent toward antimalarial activity with acceptable cytotoxicity limits. Three derivatives, **3a**, **3b**, and **3c**, were the most active inhibitors of *Plasmodium falciparum*, which were further validated by docking studies and MD simulations (Table 7.1).

Subsequently, a range of 4-aminoquinoline-imidazoles was synthesized via one-pot multicomponent reaction (MCR) and evaluated against two strains, K1 (chloroquine resistant) and 3D7 (chloroquine sensitive) of *Plasmodium falciparum* [52]. The synthesized compounds showed moderate activity against both strains, and compounds **4a**, **4b**, **4c**, and **4d** exhibited good inhibition potency toward the K1 strain with IC_{50} values of 0.46, 0.45, 0.29, and 0.34 μM, respectively, compared to chloroquine (0.25 μM). These compounds showed high selectivity for parasite and inhibited β-hematin polymerization via formation of a

TABLE 7.1 In vitro antimalarial activity data.

		IC_{50} (μM)				
Compound	R	3D7	SI	RKL9	SI	cLogP
3a	3-OCH_3	00.69	34.2	3.41	7.00	2.799
3b	3,5-OCH_3	01.60	23.2	14.32	6.94	2.888
3c	3-OH,4-OCH_3	01.61	23.6	5.46	6.97	2.06
Chloroquine	–	01.53	–	7.12	–	–

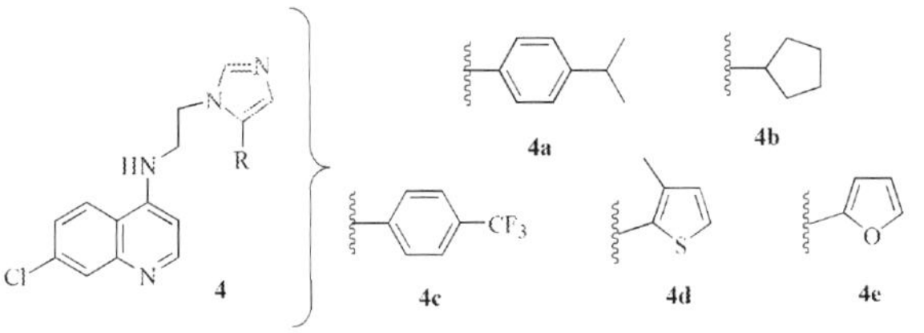

FIG. 7.5 4-Aminoquinoline-imidazole derivatives.

strong complex with hematin. Among all analogs, compounds **4c** and **4e** exhibited significant log k values, 5.93 and 5.71, respectively (Fig. 7.5).

Leishmaniasis is a parasitic disease caused by the Leishmania parasite, with 20 known species. This disease has shown high rates of morbidity and mortality globally, and endemics have occurred in 98 countries. Despite several drug discoveries, this disease is spreading very fast due to several reasons like lack of effective vaccine, highly toxic, long and expensive treatments, drug-resistance, etc. In this regard, Pandey and companions demonstrated virtual screening of 5-nitroimidazole analogs for the inhibition of *Leishmania donovani* trypanothione reductase [53]. The authors docked 984 analogs with a reference, clomipramine. Among them, 147 analogs showed better docking properties compared to the standard and were selected for further studies. After ADMET, QikProp, and prime MM-GBSA studies, 24 imidazole analogs were selected. Afterward, two molecules were chosen from MM-GBSA dG bind data, which underwent further molecular dynamic simulation to find a binding pattern with protein. Finally, compound, ethyl 2-acetyl-5-[4-butyl-2-(3-hydroxypentyl)-5-nitro-1*H*-imidazol-1-yl]pent-2-enoate **5** was determined as the most active inhibitor of VL in the Indian subcontinent with a docking score of −7.63 (Fig. 7.6).

The synthesis and antileishmanial activity of 1,2-substituted-1*H*-benzo[*d*] imidazole derivatives was reported by De Luca et al. [54] These derivatives showed good affinity toward *L. mexicana* CPB2.8ΔCTE in the submicromolar range ($Ki = 0.15–0.69\,\mu M$). Among all the screened compounds, derivative **6d** displayed the best activity against intracellular amastigotes of *L. infantum* with an IC_{50} value of 6.8 μM, though with some toxicity. The in silico molecular docking and ADME-Tox results indicated that derivative **6d** possessed good oral availability with nonmutagen and noncarcinogenic properties (Fig. 7.7).

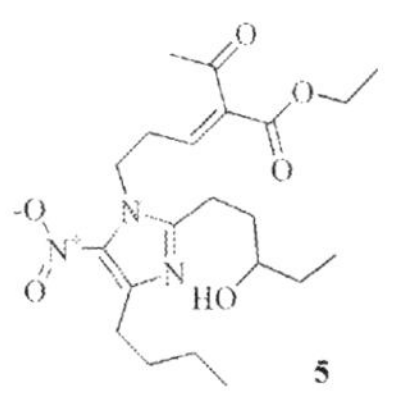

FIG. 7.6 5-Nitroimidazole analogs.

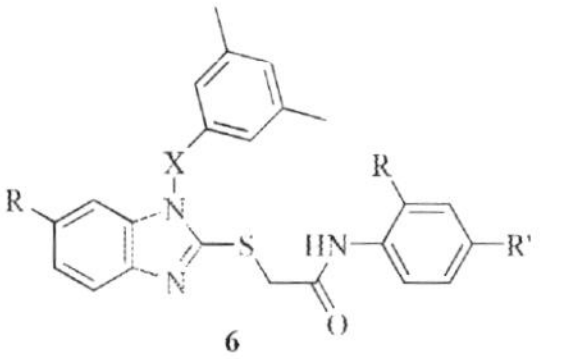

6

Compd.	X	R	R'
6a	$-SO_2$	H	H
6b	$-SO_2$	H	CO_2Me
6c	$-SO_2$	Cl	CO_2Me
6d	$-CH_2$	Cl	CO_2Me

FIG. 7.7 1,2-Substituted-1*H*-benzo[*d*]imidazole derivatives.

Recently, Revuelto and companions worked on the proteomimetic approach-based design and synthesis of nonpeptidic small molecule Li-TryR dimerization disruptors [55]. For this purpose, a range of 5-6-5 imidazole-phenylthiazole α-helix-mimetic motifs were suitably designed with substituents that could mimic three key residues, K, Q, and I, of the linear peptide prototype (PKIIQSVGIS-Nle-K-Nle). All compounds were screened for enzyme inhibition and antileishmanial activity. Compounds **7e** (5.1 μM) and **7f** (8.6 μM) were the most active for killing intracellular and extracellular parasites in cell culture and Li-TryR enzyme inhibition at low micromolar concentrations (Fig. 7.8).

A varied range of imidazoles, viz. bis-imidazole **8**, phenyl substituted imidazole **9**, **10**, thiophene-imidazole **11**, etc., was prepared from the reaction of benzil, NH_4OAc, and 2-OH carbonyl compounds in different environmental conditions and evaluated for antiparasitic activity against *T. gondii* [56]. The parasitic inhibition activity of synthesized compounds followed the order: thiopene-imidazoles (>91%), phenyl-substituted 1*H*-imidazoles (>90%), and bis-imidazoles (>75%). The in vitro study results indicated that compounds **9(a–d)**, **11a**, and **11b** showed high selectivity >1176 to >27,666 against the parasite versus the host cells (Fig. 7.9).

In 2018, Saccoliti et al. synthesized a broad range of imidazole compounds **12–14** as effective antiprotozoal agents [57]. Most of the screened derivatives displayed high to moderate inhibition activity in the micromolar and submicromolar ranges against *T. b. rhodesiense*, *L. donovani*, and *P. falciparum* with high selectivity index. Compounds **12a** (0.035 μM) and **12b–d** (0.002 μM, 0.021 μM, 0.034 μM, respectively) were the most potent against *T. cruzi* in comparison with benznidazole (1.65 μM) and the mouse model showed that compound **12a** (98.5%) reduced parasitemia in *T. cruzi* without acute toxicity (Fig. 7.10).

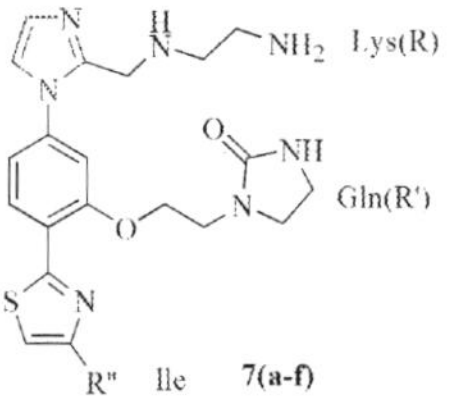

7(a-f)

R'': a=isobutyl, b=neopentyl, c=CH_2Ph, d=$(CH_2)_2Ph$, e=$(CH_2)_2$naphthyl, f=$(CH_2)_2$biphenyl

FIG. 7.8 Structure of imidazole-phenylthiazole α-helix-mimetic motifs.

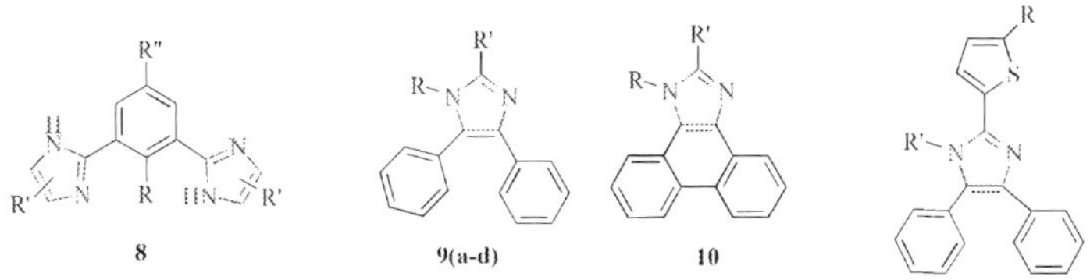

8 9(a-d) 10 11(a,b)

a; R=C_6H_5 R'=2-naphthol, b;R=C_6H_5 R'=2-OH-5-OCH_3-C_6H_3, c;R=4-OCH_3-C_6H_4 R'=2-OH-C_6H_4, d;R=3,4-$(OCH_3)_2$-C_6H_3 R'=2-OH-3-OCH_3-C_6H_3,

R;a=CH_3, b=Br
R'=H

FIG. 7.9 Structure of antiparasitic active imidazole analogs.

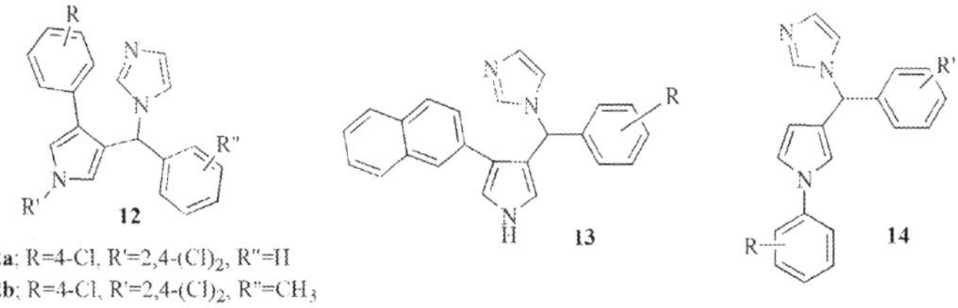

12 13 14

12a; R=4-Cl, R'=2,4-$(Cl)_2$, R"=H
12b; R=4-Cl, R'=2,4-$(Cl)_2$, R"=CH_3
12c; R=3,4-$(Cl)_2$, R'=2,4-$(Cl)_2$, R"=H
12d; R=2,4-$(Cl)_2$, R'=2,4-$(Cl)_2$, R"=H

FIG. 7.10 Antiprotozoal imidazole scaffolds.

Consequently, a range of imidazole-based compounds **15–19** was synthesized and screened for antiprotozoal activity against *T. b. rhodesiense*, *T. cruzi*, *L. donovani*, and *P. falciparum* [58]. Most of the derivatives showed good activity with IC_{50} values in the nanomolar range against *T. cruzi* and *P. falciparum*, single-digit micromolar against *L. donovani*, and double-digit micromolar against *T. b. rhodesiense.* The authors also studied the role of stereochemistry of molecules in the inhibition activity and revealed high selectivity ($373 < SI < 4030$). The binding affinity and inhibitory mechanism of the most active compounds were assessed with recombinant *T. cruzi* CYP51. Derivatives **15**, **19b**, and **17b** were found to be the most potent through in vitro inhibition of human CYPs; they underwent further in vivo inhibition studies and it was found that derivatives **17a** and **17b** decreased parasitemia by 99.5 and 99.4%, respectively (Fig. 7.11, Table 7.2).

According to the WHO, the third most widespread tropical disease is Chagas disease, caused by kinetoplastid protozoan *Trypanosoma cruzi*, which is endemic throughout Latin America. The disease is transmitted by triatomine bugs and shown in two phases, acute and chronic. In 2010, Sanchez-Moreno and companions demonstrated imidazole-based benzo[*g*]phthalazine derivatives for the treatment of both phases of Chagas disease [59]. After that, Papadopoulou et al. introduced nitro(triazole/imidazole)-based heteroarylamides/sulfonamides **20(a–d)** as effective inhibitors against *T. cruzi* with <1 mM IC_{50} values [60]. The inhibitor efficiency of the compounds followed the order: nitrotriazoles > 2-nitroimidazole > 4-nitroimidazoles. The nitrotriazoles and

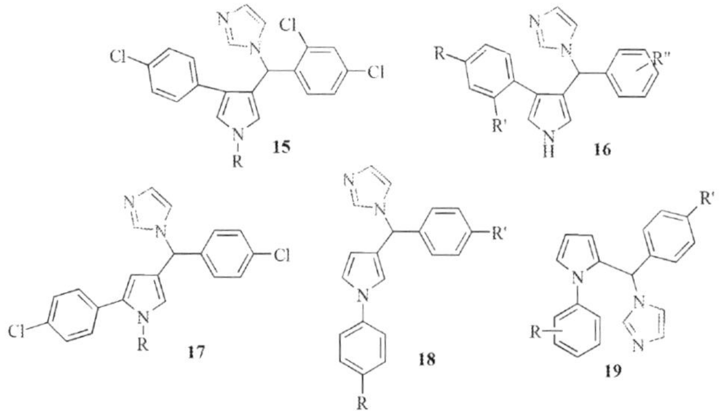

FIG. 7.11 Antiprotozoal imidazole derivatives.

2-nitroimidazole-based derivatives displayed >1400-fold and <50 selectivity toward the parasite *T. cruzi* (Fig. 7.12).

Trunzandcompanionsreportedthesynthesisof1-aryl-4-nitro-1*H*-imidazoles**21** (43 compounds) and assessed them as antitrypanosomal drug candidates [61]. Among all the synthesized compounds, 28 derivatives displayed good inhibition activity with IC_{50} values in the micromolar range against *T. b. rhodesiense*. Compounds **21a** (0.16μM) and **21b** (0.10μM) were found to be the most active compounds compared to megazol (0.10μM) and were further evaluated via mouse models of HAT. The SAR study demonstrated that the presence of F, Cl, Br, CF_3, and NO_2 in the 3′ (5′) position, and Cl, CF_3, and OCF_3 substituents in the 4′ position increased the antiparasitic activity. In this study, both compounds were beneficial for 100% cure in stage I with oral dose 25 or 50mg/kg/day of **21c** and **21a** administered, respectively, for 4days. In stage II, a chronic CNS model showed 100% cure with 50 or 100mg/kg of **21c** and **21a** administered, respectively, for 5days (twice in a day). Compounds **21c** (2.7μM) and **21d** (7.0μM) were found to be the most potent against *T. cruzi* and both (*T. cruzi and T. b. rhodesiense*), respectively (Fig. 7.13).

In 2019, a novel range of 4-nitroimidazole compounds was synthesized and screened for trypanocidal activity [62]. Among all the synthesized derivatives, compound **22** was found to be an active inhibitor of both trypomastigote and amastigote forms of *T. cruzi* with IC_{50} values of 5.4 and 12.0μM, respectively, compared to the standard benznidazole. The docking study showed the improved interaction due to the presence of the hydrophobic group (Fig. 7.14).

de Souza et al. demonstrated a structure-based drug design approach for imidazole-based molecules for the generation of novel and potent cruzain inhibitors and trypanocidal active drug candidates [63]. The cruzain enzyme is a major cysteine protease present in the entire life cycle of *T. cruzi* and vital for several necessary physiological processes. All compounds displayed high to moderate inhibition of cruzain and *T. cruzi* via in vitro and in vivo assay without any cytotoxicity. Among the synthesized compounds, derivatives **23a** (1.1μM),

TABLE 7.2 Antiprotozoal activity of imidazole compounds.

				IC_{50} (μM)						
Compound	**R**	**R′**	**R″**	**Tb**	**Tc**	**SI**	**Ld**	**SI**	**Pf**	**SI**
16a	2-Naphthyl	–	2,4-Cl_2	15.58	0.018	363	6.41	1	0.34	19
17a	H	–	–	7.39	0.003	2217	3.01	2	0.78	9
17b	CH_3	–	–	40.02	0.003	4030	5.70	2	1.41	9
17c	CH_2CHCH_2	–	–	38.69	0.002	3880	2.16	4	0.63	12
19a	4-Cl	2,4-Cl_2	–	37.0	0.005	2776	5.34	3	0.18	77
(S)(+)-**15**	Cl	H	2,4-Cl_2	12.19	0.017	327	5.02	1	0.11	51
(R)(−)-**15**	Cl	H	2,4-Cl_2	9.21	0.0009	2844	3.77	–	0.092	28
(±)-**15**	Cl	H	2,4-Cl_2	17.33	0.035	275	3.77	3	0.16	60
Standard				MEL	Bz		MF		CHQ	
S. value				0.0088	1.65	>233.3	0.56	244.6	0.174	533.3

Notes: *CHQ*, chloroquine; *MEL*, melarsoprol; *MF*, miltefosine.

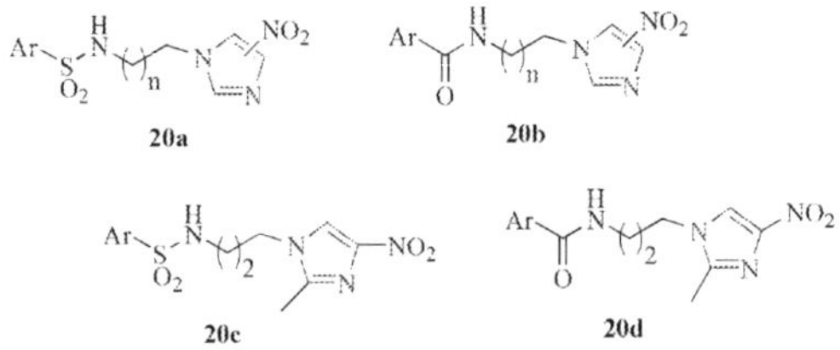

FIG. 7.12 Imidazole compounds as chagasic inhibitors.

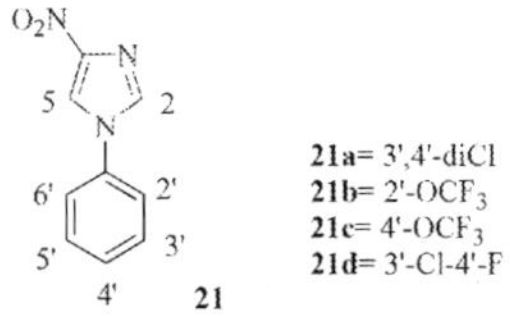

FIG. 7.13 Antitrypanosomal active 1-aryl-4-nitro-1H-imidazoles.

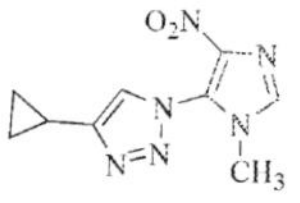

FIG. 7.14 Active anti-*T. cruzi* 4-nitroimidazole molecule.

and **23g** (1.7 μM) displayed threefold more trypanocidal activity than standard **BZ** (IC_{50} = 3.3 μM). Molecules **23e** (2.0 μM), **23b** (2.3 μM), **23d** (2.9 μM), and **23c** (3.4 μM) exhibited equipotency with reference against *T. cruzi*. The cruzain inhibition study revealed high inhibition activity of compounds **23f** and **23h**, with IC_{50} values of of 7.5 and 6 μM, respectively. Compounds **23a**, **23g**, **23e**, **23b**, and **23d** were found to be the most selective analogs for *T. cruzi* with SI values of 86, 58, 50, 42, and 33, respectively (Fig. 7.15).

2.2 Anti-Alzheimer's activity

LOX (lipoxygenase) enzymes catalyze the hydroperoxidation of polyunsaturated fatty acids and are associated with neurodegenerative disease. Quinoline-based imidazole-fused systems were synthesized and utilized as soybean 15-LOX

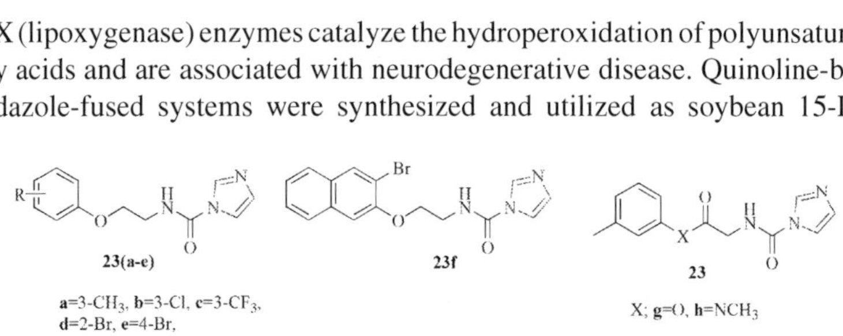

FIG. 7.15 Active imidazole derivatives as cruzain inhibitors with trypanocidal activity.

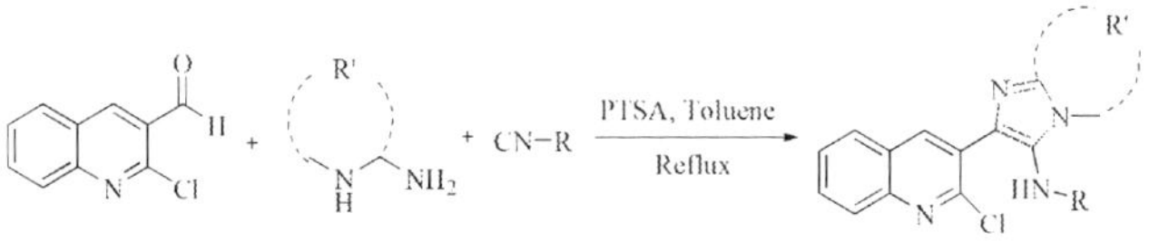

Compd	R'	R
24a	2-aminothiazole	tert-butyl
24b	2-amino-4-methyl	pyridine cyclohexyl
24c	2-amino-4-methyl	pyridine 1,1,3,3-tetramethylbutyl

SCHEME 7.1 Synthesis of LOX-inhibitor imidazole analogs.

inhibitors [27]. One-pot reaction of heteroaromatic amidine, 2-chloroquinoline-3-carbaldehyde, and alkyl isocyanides produced quinoline-based imidazole compounds (Scheme 7.1). All compounds displayed significant anti-15-LOX activity and compounds **24a**, **24b**, and **24c** were found to be the most potent with IC_{50} values of 11.5, 15.3, and 14.1 μM, respectively. The docking study showed that the synthesized compounds interacted with ligands via π-cation interaction and hydrophobic interactions. The SAR study results demonstrated that the presence of different substituents on alkylamine side chain affected the biological activity.

In Alzheimer's disease (AD), Aβ peptides and tau proteins are responsible for the spread of disease via enlarged fabrication of amyloid precursor protein (APP) and Aβ plaque deposition. The production of APP and Aβ plaque increased in neurons via binding between lipoxygenase (5-LOX) and cyclooxygenase (COX) derived eicosanoid and their related receptors. For this purpose, a range of 1,5-diarylimidazoles **25** was synthesized and studied as microtubule stabilizing agents and 5-LOX and COX pathways inhibitors through QBI MT assay, RBL-1 assay, and 5-LOX-derived LTB4, COX-derived PGD2 and PGE2 studies [64]. The structure activity relationship (SAR) revealed that the presence of different functional groups affected their activity profiles in eicosanoid and microtubule synthesis (Fig. 7.16).

A series of o/p-(3-substitutedpropoxy)phenyl]-1*H*-benzimidazole was designed, synthesized and screened against cholinesterase (ChE) inhibitory activity for the treatment of Alzheimer's disease [65]. The study of cell viability on SH-SY5Y cells lines, metal chelating ability, partition coefficient, and docking study revealed that most of the compounds showed good biological activity, and molecules **26a** and **26b** exhibited high potency toward AChE (acetylcholinesterase) and BuChE (butyrylcholinesterase) inhibition and high neuroprotection activity (Fig. 7.17).

In Alzheimer's disease, beta amyloid cleavage enzyme-1 (BACE1) works as a key enzyme in Aβ peptide synthesis. Previously, the relationship between Aβ effect and oxidative stress in neurodegeneration process was also demonstrated by researchers. In this context, in vitro lipid peroxidation assay and in silico ADME analysis were carried out for imidazole-based molecules, and the results showed that all the derivatives were moderately active for BACE1 inhibition

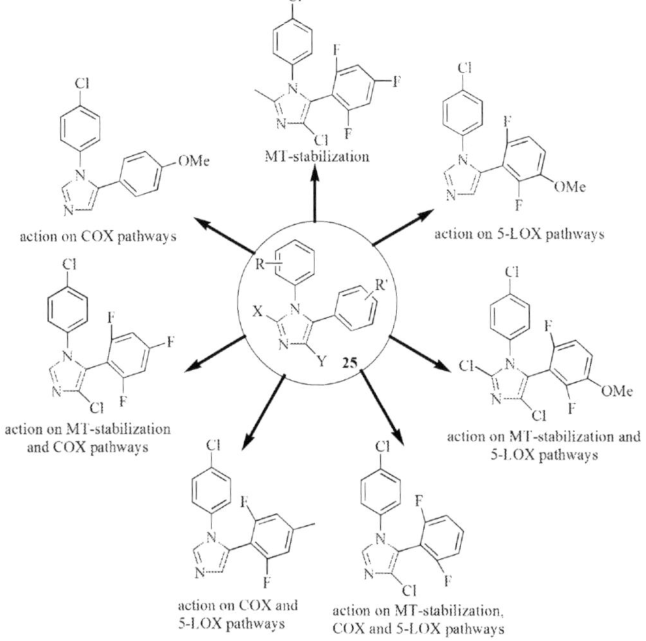

FIG. 7.16 Different activity profiles exhibited by 1,5-diarylimidazoles.

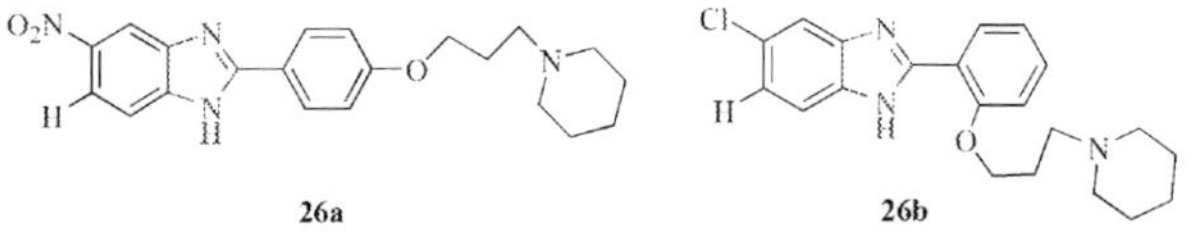

26a **26b**

compd. \ IC_{50}	AchE	BuChE
26a	0.14 ± 0.021	2.47 ± 0.315
26b	6.20 ± 0.071	0.22 ± 0.005

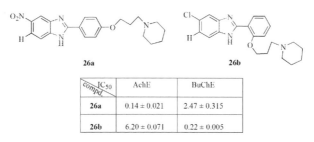

FIG.7.17 Alzheimer'sinhibitionactivityofo/p-(3-substitutedpropoxy)phenyl]-1H-benzimidazole.

and neuroprotection; however, derivatives **27a** and **27b** were the most potent with memory retention in mice [66] (Fig. 7.18).

Previous investigations have shown that various factors like deposition of amyloid β (Aβ) peptide, oxidative stress, neurofibrillary tangles (p-Tau), and acetylcholine (ACh) level in the cortex and hippocampus region of brain were responsible for Alzheimer's disease [67]. Here, a novel range of 1,3-dimethylbenzimidazolinone was designed, synthesized, and evaluated for in vitro cholinesterase (ChEs) inhibition activity, and most of the compounds displayed effective ChEs inhibition in the μ molar range [68]. Among all derivatives, compounds **28a** and **28b** showed high potency toward AChE and BChE inhibition, neuroprotective effect

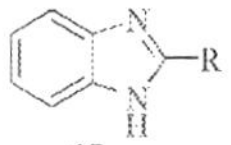

27

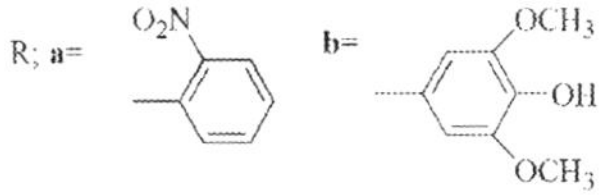

FIG. 7.18 Structure of BACE1 active benzimidazole motif.

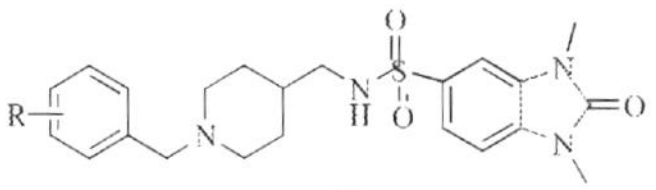

28

compd. \ IC_{50}	R	AchE	BuChE
28a	2-CH_3	0.39 ± 0.11	0.66 ± 0.16
28b	3-Br	0.39 ± 0.15	0.16 ± 0.04

FIG. 7.19 Structure and acetylcholine inhibition activity of 1,3-dimethylbenzimidazolinone.

against H_2O_2-induced oxidative damage on PC12 cells, antioxidant activity, and low hepatotoxicity in comparison to tacrine and donepezil. Moreover, derivatives **28a** and **28b** did not display any considerable change in ALT and AST levels, and significantly ameliorated the cognition impairment in scopolamine-induced mouse models (Fig. 7.19).

Glycogen synthase kinase-3β (GSK3β) and p38α mitogen-activated protein (MAP) kinase is a serine/threonine kinase that is allied with the pathways leading to Alzheimer's disease. Therefore, 39 novel pyridinylimidazoles **29** were designed, synthesized, and studied for dual inhibition of GSK3β and p38α MAP kinase [69]. Enzyme-linked immunosorbent assay (ELISA), molecular modeling, and docking study results demonstrated that most of the compounds were highly potent inhibitors; among them, *N*-(4-(2-ethyl-4-(4-fluorophenyl)-1*H*-imidazol-5-yl)pyridin-2-yl) cyclopropane carboxamide (**29a**) showed IC_{50} values of 16 and 35 nM for p38α MAP kinase and GSK3β, respectively, and demonstrated excellent metabolic stability. Compound **29a** displayed 8.5-fold isoform selectivity toward GSK3β, which supported pharmacokinetic properties and demonstrated 43% inhibition of human ether-a-go-go-related gene (hERG) channel and 0.155 mg/mL (threefold improved) water solubility (Fig. 7.20).

Benzimidazole-2-thiols were synthesized and anti-Alzheimer's activity was demonstrated by Latif and companions [70]. All compounds were screened for the in vitro inhibition of both AChE and BuChE against galantamine, the reference drug. Among all the synthesized compounds, **30a** and **30b** exhibited significant inhibition against the standard drugs. Molecular docking was further performed to investigate the mechanism of anticholinesterase activity.

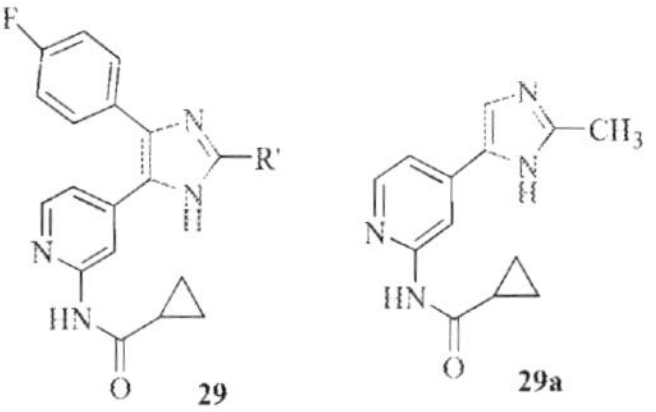

29 29a

FIG. 7.20 Pyridinylimidazole derivatives.

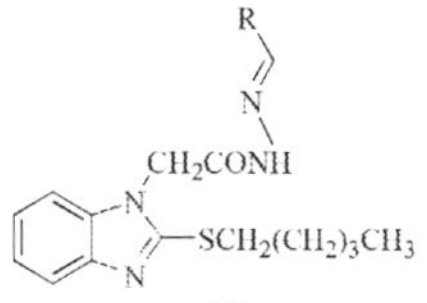

30

Compd.	R	AChE	BuChE
30a	C_7H_{15}	121.2	38.3
30b	4-$CH_3C_6H_4$	180.3	35.4

FIG. 7.21 Anticholinesterase activity of potent derivatives of benzimidazole-2-thiol (IC_{50} value in μM).

Compounds **30a** and **30b** showed various binding patterns such as π-alkyl, π–sulfur, and π–π stacked interaction via docking study (Fig. 7.21).

2.3 Antiinflammatory activity

Inflammatory drugs display their action via inhibition of cyclooxygenase enzymes (COX-1, COX-2, and COX-3) [71]. For this purpose, a range of imidazole derivatives was synthesized and applied as antiinflammatory agents and cyclooxygenase inhibitors. All of the synthesized compounds were found to be active toward nonselective COX-1 and COX-2 enzyme inhibition, and derivatives **31a** and **31b** were highly potent against COX-2 inhibition. Compound **31a** showed the highest in vivo antiinflammatory activity relative to diclofenac and lower ulcerogenic effect (UI = 12.20 and 12.81, respectively) compared to indomethacin. The SAR study concluded that the presence of unsaturated moieties like allyl group and aryl groups in both series enhanced cyclooxygenase inhibition and the presence of benzyl group decreased inhibition (Fig. 7.22).

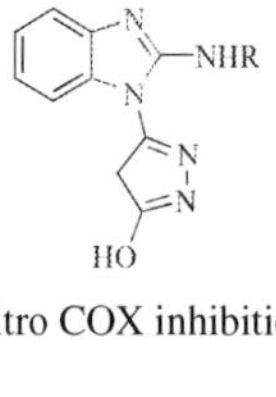

compd. \ IC_{50}	R	COX-1(nM)	COX-2(nM)
31a	n-C_4H_9	0.2272	0.0469
31b	3-$CH_3C_6H_4$	0.1664	0.0370

FIG. 7.22 In vitro COX inhibition of most active compounds.

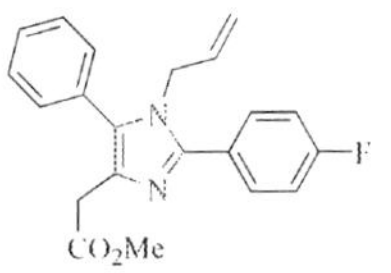

32

FIG. 7.23 Structure of antiinflammatory active imidazole derivative.

In 2018, eight novel imidazole molecules were synthesized and studied for antiinflammatory activity through *in vitro* and in vivo studies [72]. The data of inflammatory activity concluded that all compounds displayed high to moderate antiinflammatory activity, and compound **32** exhibited high inhibition potency toward NOx, TNF-α, IL-6, and IL-1β production, and inflammatory activity in J774 macrophages and inhibited nuclear factor kappa B (NF-κB) transmigration. Compound **32** reduced leukocyte migration and exudate formation in the pleural cavity of carrageenan's mice via lowering superoxide dismutase, glutathione S-transferase, catalase, and myeloperoxidase activity. The study revealed that oxidative stress and NF-κB play important roles in inflammatory activity (Fig. 7.23).

Holiyachi et al. demonstrated a facile and efficient one-pot MCR of coumarin and 1,2-diketone with NH_4OAc and an acid catalyst to produce coumarin–imidazole hybrid compounds **33–35**; they further converted them into phenyl-imidazole acrylates [73]. All the synthesized derivatives were employed as antiinflammatory active compounds through the gelatin zymography method. For the study of inflammation and autoimmunity stats, two types of matrixes, metalloproteinases (MMP), MMP-2 and MMP-9, were selected. All compounds showed high antiinflammatory potency, and derivatives **35(a–f)** were found to be more active against both MMPs due to the hydrophilic nature of the molecules (Scheme 7.2).

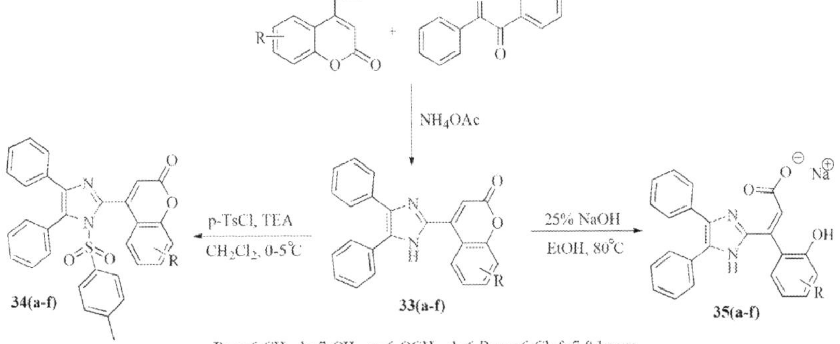

SCHEME 7.2 Synthesis of coumarin–imidazole and phenylimidazoloacrylates hybrid as antiinflammatory agents.

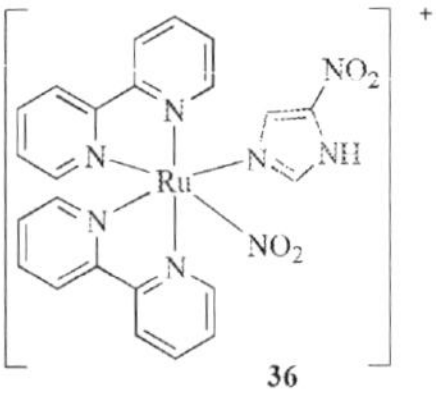

36

FIG. 7.24 Nitro-imidazoles-containing ruthenium complexes.

Recently, nitro-imidazole-containing ruthenium complexes cis-[Ru(NO2)(bpy)2(5NIM)]PF6 (**36**) were synthesized, where bpy=2,2′-bipyridine, 5NIM=5-nitroimidazole, and MTZ=metronidazole [74]. The complex displayed antiinflammatory properties equivalent to dexamethasone via down-regulation of key pro-inflammatory mediators like inhibiting *IL-6* and *TNF*-α production, and *COX-2* and *IL-1*β gene expression in LPS-stimulated RAW264.7 macrophages (Fig. 7.24).

2.4 Antidiabetic activity

In 2016, a range of amino acid conjugates of chromene-3-imidazoles based on natural isoflavonoids was designed and synthesized for the inhibition study of aldose reductase (ALR) [75]. ALR2 showed plasticity in the active site vicinities and possible shifts in the neighboring two alpha helices in the diabetic activity. All tested compounds showed ALR2 inhibition activity with an IC_{50} value range of 0.031 ± 0.082 mM to 4.29 ± 0.55 mM. Compound **37a** showed the highest inhibition activity through biochemical and in silico study with high selective index against ALR1. Compound **37a** decreased blood glucose levels and delayed cataract progression in STZ-induced rats in a dose-dependent manner (Fig. 7.25).

In 2018, N^{α}-arylsulfonyl histamines were prepared from the extract of *Urtica urens* L and further modified by the reaction from benzenesulfonyl chloride [76]. The semisynthetic compounds were applied for the inhibition of β glucosidase. The in vitro study results revealed that among all derivatives of **38**

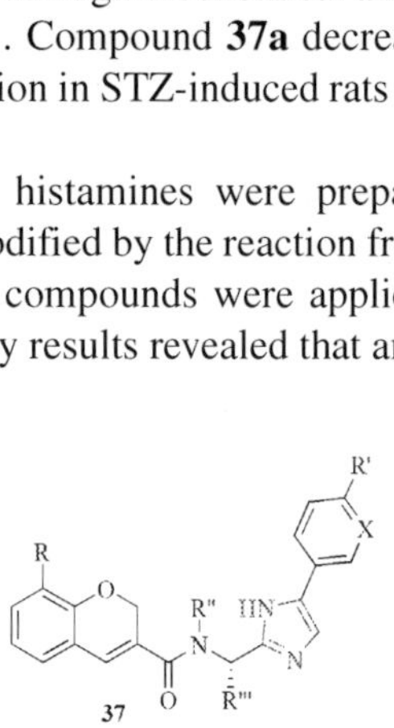

37

					IC50 values (mM)	
Compd.	R	R'	R''/R'''	X	ALR2	ALR1
37a	OCH_3	Cl	$-CH_2-CH_2-CH_2-$	C	0.031 ± 0.082	3.422 ± 0.081

FIG. 7.25 Amino acid conjugates of chromene-3-imidazoles.

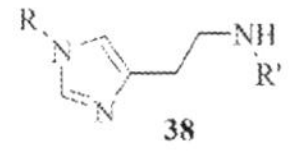

38 series I		I-a	I-b	I-c	I-d	I-e	I-f	I-g	I-h	I-i
R=H.	R'	m	n	o	p	q	r	s	t	u
38 series II		II-a	II-b	II-c	II-d	II-e	II-f	II-g	II-h	II-i
	R=R'	m	n	o	p	q	r	s	t	u

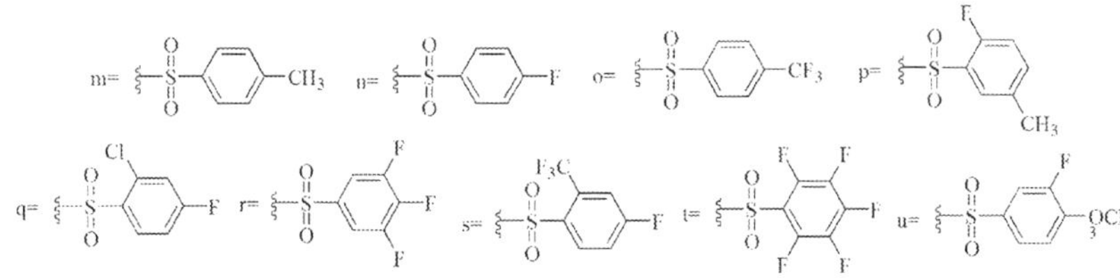

FIG. 7.26 Arylsulfonyl histamine molecules.

series I, derivatives I-c, I-d, I-g, and I-h were found to be the most active b-Glc inhibitors with IC_{50} values of 65.08, 79.50, 71.43, and 72.69 μM, respectively, in comparison to the standard 1-DNJ. Here, series I was found to be more active toward b-Glc inhibition than series II (Fig. 7.26).

Recently, a range of imidazo[1,2-*b*]pyrazole derivatives was synthesized and screened as effective α-glucosidase inhibitors via in vitro, kinetic, and docking study [77] (Scheme 7.3). The in vitro study data revealed that all derivatives were highly potent inhibitors in the range of IC_{50} values 95–372 μM compared to reference drug, acarbose (IC_{50} = 750 μM), and also displayed noncytotoxicity. Among all the derivatives, derivative **39c** exhibited eight-fold more inhibition activity compared to the standard. From the kinetic study, compound **39c** showed 90 μM *Ki* values, which indicated their competitive inhibition with acarbose. The docking studies indicated that compounds **39a**, **39b**, **39c**, and **39d** showed H-bonding, π-ion, and other hydrophobic interaction with amino acids, and more negative binding energies were found compared to the standard.

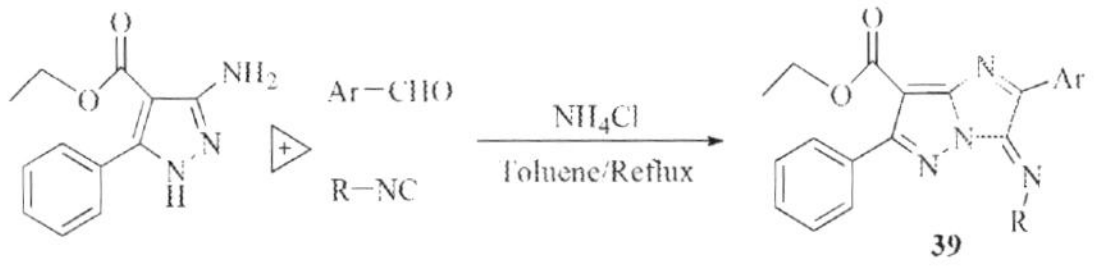

Compd.	R	Ar	IC_{50}(μM)
39a	t-Bu	4-OH-3-OCH_3-C_6H_3	164.5
39b	cyclohexyl	-C_6H_5	111.3
39c	cyclohexyl	4-OH-3-OCH_3-C_6H_3	95.0
39d	cyclohexyl	thiophene	141.0

SCHEME 7.3 Synthesis and in vitro inhibitory activities of imidazole compounds against α-glucosidase.

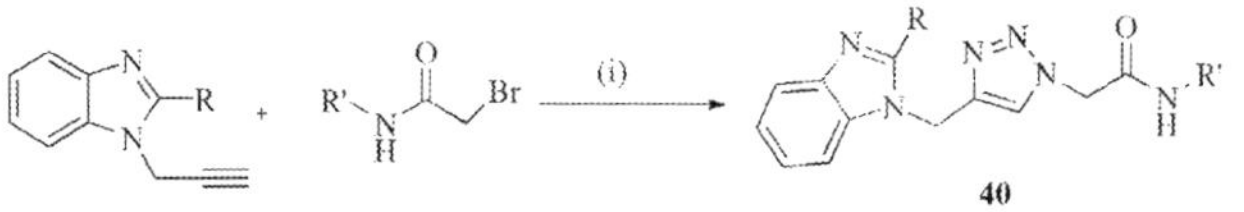

40

			IC_{50} value (μg/ml).	
Compd.	R	R'	α-Amylase inhibition	α-Glucosidase inhibition
40e	$-CH_3$	$-3-F-C_6H_4$	19.44	5.304
40g	$-CH_3$	$-4-OCH_3-C_6H_4$	16.88	5.8
40h	$-CH_3$	$-4-F-C_6H_4$	18.64	9.76
40n	$-SCH_3$	$-3-F-C_6H_4$	16.24	6.44
40p	$-SCH_3$	$-4-OCH_3-C_6H_4$	17.59	8.18
Acarbose			15.31	4.12

(i) NaN_3, $CuSO_4.5H_2O$, sodium ascorbate, DMF/H_2O (8:2), 50°C, stirring 5–6 hr

SCHEME 7.4 Synthesis and antidiabetic activity of benzimidazole-tethered 1,2,3-triazole derivatives.

The click reaction between substituted 1*H*-benzo[*d*]imidazole and in situ azide furnished benzimidazole-tethered 1,2,3-triazole derivatives, and these were screened for antidiabetic activity [78] (Scheme 7.4). All compounds displayed high to moderate inhibition activity and compounds **40e**, **40g**, **40h**, **40n**, and **40p** were the most active for α-glucosidase and α-amylase inhibition activity, with IC_{50} values ranging from 0.0146 to 0.0732 μmol/mL and 0.0410 to 0.0916 μmol/mL, respectively. These biochemical study results were supported by ADME properties study and the docking study via halogen-bonding, H-bonding, and T-shaped π–π interactions with active sites of residue.

2.5 Antihypertensive activity

The renin-angiotensin–aldosterone system (RAAS) plays a significant role in pathophysiology of hypertension and cardiovascular diseases. In the human body, angiotensin I is present in an inert form which converts into an active form, angiotensin II, by enzymatic activity in lungs and other organs. N-substituted 5-butylimidazole derivatives were designed, synthesized, and screened as angiotensin II AT1 receptor blockers [79]. Among all compounds, **41f** (3.0 nM) was found to be more active than the standard, losartan (6.0 nM), via in vitro studies. Moreover, from rat uterotonic test results, compound **41f** (pA2 = 7.83) was found to be the most active compared to the reference in respect to good binding affinity with protein (Fig. 7.27).

A range of novel NO-releasing benzimidazole analogs was synthesized and employed as effective antihypertensive candidates [80]. The in vitro study data showed that all compounds displayed good biological activity, and among them, compounds **42e** (pA2 = 8.033) and **42j** (pA2 = 8.099) were found to be more potent toward antihypertensive activity compared to losartan (pA2 = 7.904). The study revealed that the presence of the NO donor with different linkers significantly increased their Ang II inhibition activity (Fig. 7.28).

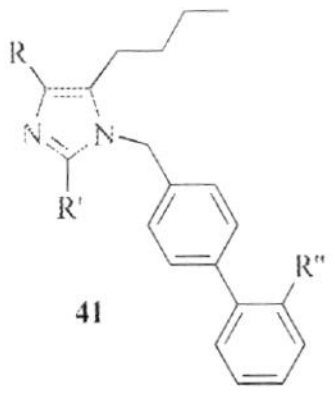

41

41a; R=H, R'=CH_2OH, R''=CN_4H
41b; R=Cl, R'=CH_2OH, R''=CN_4H
41c; R=Br, R'=CH_2OH, R''=CN_4H
41d; R=I, R'=CH_2OH, R''=CN_4H
41e; R=H, R'=CHO, R''=CN_4H
41f; R=H, R'=COOH, R''=CN_4H
41g; R=H, R'=CH_2OH, R''=COOH

FIG. 7.27 N-substituted 5-butylimidazole as angiotensin II blocker.

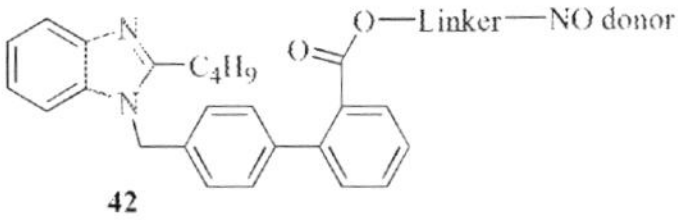

42

Comp.	NO donor	Linker	Comp.	NO donor	Linker
42a		$-CH_2CH_2-$	**42f**		$-CH_2CH_2-$
42b		$-CH_2CH_2CH_2-$	**42g**		$-CH_2CH_2CH_2-$
42c	ONO_2	$-CH_2CH_2CH_2CH_2-$	**42h**	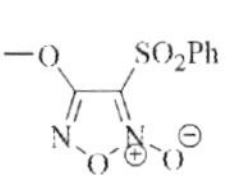	$-CH_2CH_2CH_2CH_2-$
42d		$-(CH_2)_5-$	**42i**		$-CH_2CH_2CHCH_3-$
42e		$-CH_2PhCH_2-$	**42j**		$-CH_2PhCH_2-$
			42k		$-CH_2CH_2OCH_2CH_3$
			42l		$-CH_2CCCH_2-$

FIG. 7.28 Structure of antihypertensive benzimidazole molecules.

Some researchers demonstrated structural modifications of imidazole-based compounds for synthesizing efficient antihypertensive drug candidates. Sharma et al. displayed structural insights for substituted acyl sulfonamides/sulfamides of imidazoles **43** via a molecular modeling approach [81]. The authors applied PLSR and kNN-MFA analysis techniques to develop 2D, G-QSAR, and 3D-QSAR models. Fig. 7.29 illustrates the structural features of biologically active imidazole moiety, i.e., the presence of these substituents favored the antihypertensive activity. This information and proposed models will be helpful to design novel compounds due to their good predicting ability (Fig. 7.29).

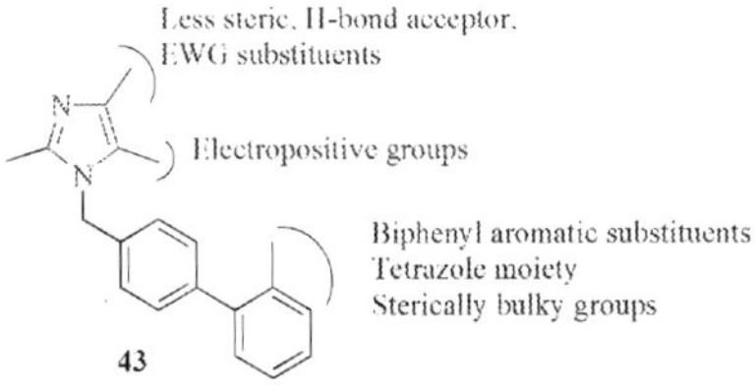

43

FIG. 7.29 Favored conditions of imidazole motif as Ang II inhibition via molecular modeling studies.

2.6 Antidepressant activity

Panneerselvam et al. synthesized novel benzimidazole derivatives and screened them as antidepressant agents via in silico modeling and in vitro studies [82]. The in silico modeling study indicated that compounds **44d** and **44e** possessed significant activity against the standard, flibanserin. The antidepressant activity data revealed that all derivatives showed high to moderate activity at the dose level of 50 mg/kg with reduction in the immobility time. Compounds possessing *p*-NO_2 (− 86.66), *o*-Br (− 75.83), and *m*-OCH_3 (− 70.62) groups were found to be the most active antidepressant agents against the reference drug, clomipramine (− 88.63) (Fig. 7.30).

2.7 Miscellaneous activities

Kenchappa and coauthors introduced the synthesis of benzofuran attached thiazolo[3,2-*a*]benzimidazole analogs and assessed them as anthelmintic drug candidates [83] (Scheme 7.5). The *in vitro* anthelmintic activity data showed that most of the synthesized compounds were effective against earthworm *Pheretima posthuma* in a dose-dependent manner. Among all the synthesized compounds, **45a**, **45b**, and **47i** were the most active for anthelmintic activity and exhibited high binding affinity with β-tubulin through molecular docking studies (Table 7.3). The SAR study revealed that compounds of series **45** and **47** were more active due to the presence of aliphatic methylene linkage compared

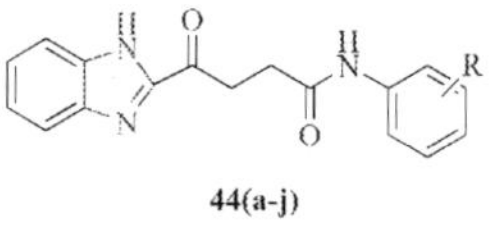

44(a-j)

R; a=H, b=2-NO_2, c=3-NO_2, d=4-NO_2, e=2-Br, f=3-Br, g=4-Br, h=2-OCH_3, i=3-OCH_3, j=4-OCH_3

FIG. 7.30 Structure of 4-(1*H*-benzo [*d*] imidazol-2-yl)-*N*-(substituted phenyl)-4-oxobutanamide.

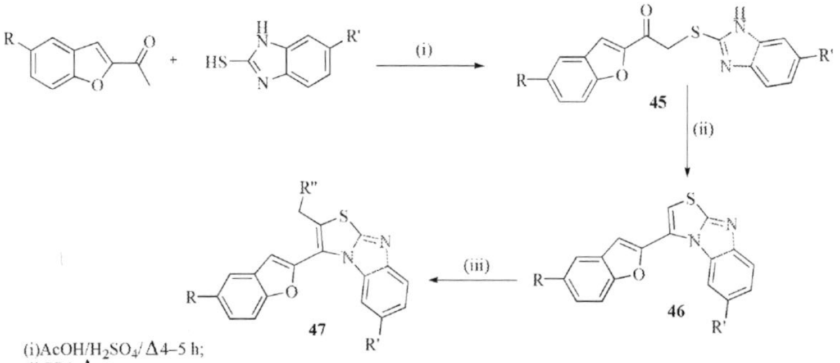

(i)AcOH/H_2SO_4/Δ4–5 h;
(ii)PPA,Δ
(iii)Sec amines/HCHO, EtOH/AcOH

SCHEME 7.5 Synthesis of anthelmintic imidazole compounds.

TABLE 7.3 Anthelmintic activity data of most active compounds.

Concentration (%) compound	R/R′/R″	Paralytic time (min)			Death time (min)		
		1	2	3	1	2	3
45a	H/H/–	12.17 ± 1.08	11.05 ± 1.28	10.30 ± 1.02	22.15 ± 0.75	20.25 ± 0.38	18.25 ± 0.64
45b	H/H/–	12.52 ± 1.03	11.40 ± 1.21	11.05 ± 0.78	23.20 ± 0.39	21.38 ± 1.12	19.35 ± 0.82
47i	H/H/ morpholino	13.58 ± 0.58	12.56 ± 0.63	11.48 ± 0.84	23.38 ± 0.67	22.00 ± 0.91	21.46 ± 0.49

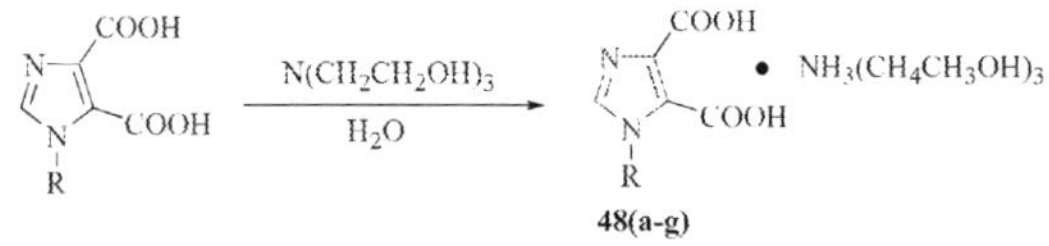

SCHEME 7.6 Synthesis of water-soluble imidazole-based dicarboxylic acid.

to series **46**. In series **47**, the presence of the morpholine group increased their bioactivity compared to pyrrolidine and piperidine.

In 2018, Brusina et al. converted imidazole dicarboxylic acid into its water-soluble salts **48(a–g)** to evaluate their anticonvulsant activity via an NMDA-induced convulsion model [84] (Scheme 7.6). Here, the salts displayed good anticonvulsant activity although the acid was also a convulsant. The alteration in biological activity occurred due to the formation of ion pairs and associates in polar environment, and their presence was confirmed by conductivity measurements.

A range of tetraarylimidazoles **49** was synthesized from the one-pot MCR of 2-arylindole-3-carbaldehydes, benzyl-substituted anilines, and NH_4OAc in acetic acid in high yields [85]. The compounds were screened for antiurease and antilipoxygenase activity. All synthesized derivatives were more active compared to standard thiourea ($21.25 \pm 0.15\,\mu M$) in IC_{50} values, ranging from 0.12 ± 0.06 to $12.13 \pm 0.02\,\mu M$ except for one derivative. However, low LOX inhibition activity was displayed by all the synthesized compounds (Scheme 7.7).

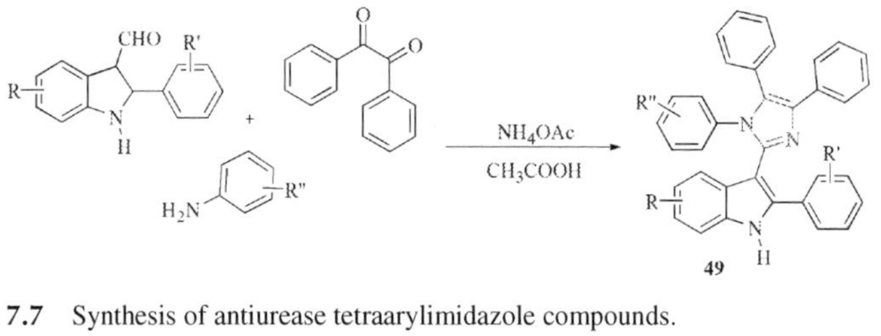

SCHEME 7.7 Synthesis of antiurease tetraarylimidazole compounds.

Ren and his companions docked imidazole derivatives via molecular docking and studied their thrombin inhibitor activity [86]. The high docked score derivatives were selected for synthesis and further screened for in vitro antithrombin activity against the standard, argatroban (IC_{50} 9.88 nM). Molecule **50a** was found to be the most potent inhibitor with an IC_{50} value of 3.11 nM compared to the reference. The results were also validated by the molecular modeling studies (Fig. 7.31).

The design, synthesis, and the inhibitory sirtuin activities were reported by Yoon et al. [87]. The in vitro sirtuin activity was performed for SIRT-I and SIRT-2 via fluorimetric drug discovery. Compounds **51(a–c)** were more potent inhibitors in comparison to the standard, tenovin-6. The other derivatives

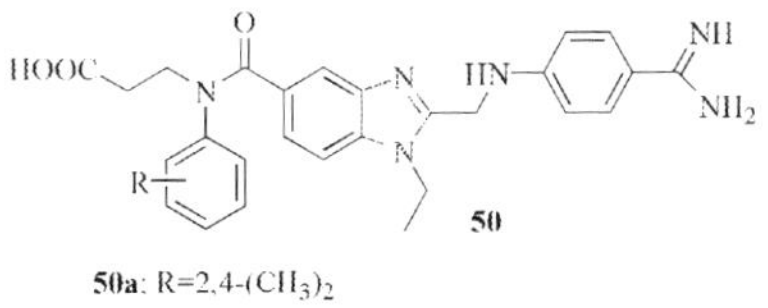

FIG. 7.31 Imidazole scaffolds as thrombin inhibitors.

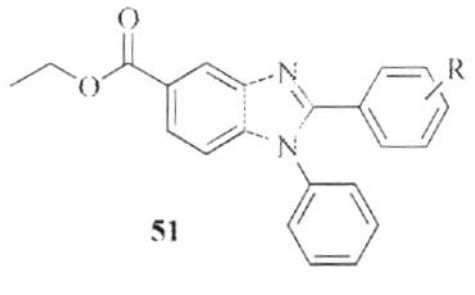

FIG. 7.32 Sirtuin-inhibitor benzimidazole derivatives.

did not display significant inhibition (Fig. 7.32). From a comparative study, it was observed that all compounds were more active toward SIRT-1 compared to SIRT-2 inhibition (Table 7.4).

Substituted methoxybenzyl-sulfonyl-1H-benzo[*d*]imidazole derivatives **52** were designed for the investigation of antiulcer activity [88] (Fig. 7.33). For this purpose, preliminary H^+/K^+-ATPase binding properties and HRBC membrane stabilization activity were evaluated, which are involved in ulcer development. Among all the synthesized analogs, derivatives **52(a–d)** were the most active inhibitors against diclofenac with high HRBC protection (Table 7.5).

Recently, tetrahydrothienopyridine derivatives **53** were synthesized via eco-benign one-pot MCR and assessed for antiplatelet activity [89]. Among all compounds, derivatives **53a** (56.48%) and **53b** (66.38%) were the most active for antiplatelet activity in comparison to aspirin. Both derivatives effectively worked as proton pump inhibitors and reduced gastrointestinal bleeding (Fig. 7.34).

TABLE 7.4 SIRT-1- and SIRT-2-inhibitory activities of benzimidazole derivatives.

	SIRT-1 inhibition		SIRT-2 inhibition	
Compound	(%)	IC_{50} (μM)	(%)	IC_{50} (μM)
51a	60.59	49.12	71.69	26.85
51b	44.05	62.03	50.04	55.90
51c	67.30	35.98	88.57	9.6
Tenovin-6	44.00	58.10	66.20	29.73

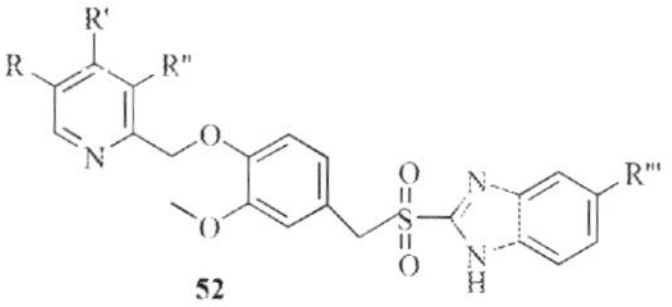

52

FIG. 7.33 Sirtuin-inhibitor benzimidazole derivatives.

TABLE 7.5 In vitro HRBC protection values.

Compound	Mean absorbance	Average % protection	IC_{50} (μM)
52a	0.4428	91.12	0.212
52b	0.4641	94.11	0.186
52c	0.3781	88.45	0.355
52d	0.5278	96.42	0.142
Diclofenac	0.4212	82.24	0.942

Benzimidazole-linked 4-arylamido 5-methylisoxazole derivatives **54** and **55** were designed and synthesized via conformational restriction of an in-house type II FMS inhibitor [90]. Type II FMS inhibitors possess three parts, a central phenyl ring, an H-bonding hinge, and a secondary hydrophobic aromatic ring, which simplify binding to the DFG pocket. Amide or urea linkages connect the middle phenyl ring and secondary hydrophobic aromatic ring. The synthesized compounds were screened against FLT3 and FMS, and found to be selective for FLT3 inhibition. Compound **54a** was the most potent toward FLT3 with an IC_{50} value of 0.495 nM in comparison with the standard, staurosporine (1.13 nM). The SAR study revealed that the presence of the 6-amide group displayed more bioactivity compared to the 7-amide group in the benzimidazole ring. For the determination of bioactivity of compound **54a**, the enzymatic inhibition study was performed against 34 different kinases at a single dose of 10 mM in duplicate, and it was concluded that their strong kinase inhibitor activity was associated with acute myelogenous leukemia (AML) (Fig. 7.35).

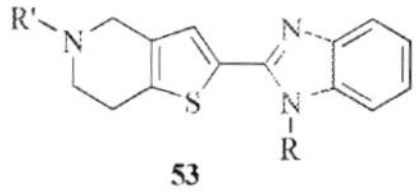

53

a; R=CH_3, R'=-$COCH_3$
b; R=$CH_2C_6H_5F$, R'=-$COC_6H_3FNO_2$

FIG. 7.34 Tetrahydrothienopyridine molecules.

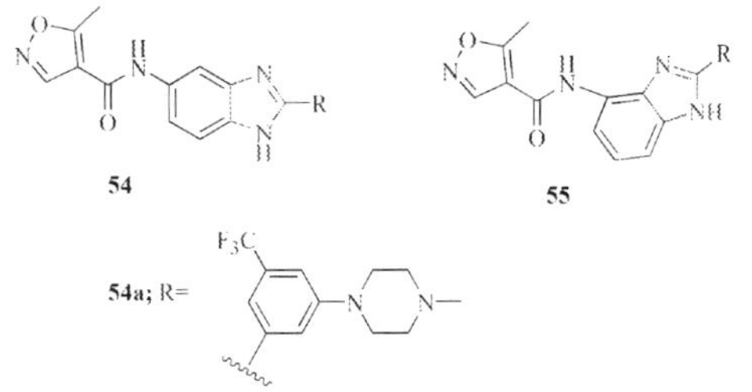

FIG. 7.35 Structure of benzimidazole-based 4-arylamido 5-methylisoxazole derivatives.

3. Conclusion and future perspectives

The chapter demonstrated the evolution in the biological activities of imidazole scaffolds in diverse areas since 2011. The upsurge in its biological applications may be credited to the improved well-established procedures, and investment of scientists, academics, and industries in the regeneration of green economical and sustainable synthetic approaches to furnish imidazole analogs as valuable and effective pharmacophores. Enormous growth has occurred in imidazole drug development, although many challenges in pharmaceutical chemistry still need critical analysis for their solution. Looking to the future, four major goals have been identified to inspire the phase of advancement of bio-active imidazoles: (1) accurate, predictive identification of new derivatives via computational studies (in silico, ADME analysis); (2) introduction of eco-benign and green synthetic protocols for drug development; (3) successful laboratory operations like preclinical and clinical research, viz. in vitro and in vivo studies; and (4) their applicability at industrial level. Presently, many compounds have not completed the drug development process and fulfilled the special requirement of inhibition of particular diseases.

Parasitic diseases are endemic in most countries, responsible for thousands of deaths per year, but unfortunately there is no effective vaccine against most of them, and available vaccines have drawbacks including expensive and long treatment, painful intravenous injections, side effects, toxicity, drug resistance, need of constant medical care, and lack of efficacy. Antiinflammatory drugs get restored after completing their action in living cells, and affect the tissues that lead to autoimmune diseases, chronic infectious diseases, cancer, and cardiovascular side effects. In Alzheimer's disease (AD), only four pharmaceutical ingredients have been approved by the Food and Drug Administration (FDA), but no one is able to stop and cure this disease completely. The main reason for the low clinical success is low understanding of the pathophysiology of AD and most of the drug design is based on Aβ amyloidosis. In diabetic treatment, most of the chemical entities have not shown sufficient efficiency and selectivity against targets. The aforementioned discussion concluded that researchers should increase their knowledge about the work mechanism of particular diseases, correlation between structure and reactivity, and designing chemical entities that are safe, less toxic, and affordable with high efficiency. More efforts should be made toward achieving the synthesis of hybrid, chiral, and versatile pharmacophores via efficient pathways for large-scale synthesis. It is fully anticipated that new techniques and methodologies will be executed in the development of imidazole compounds and applied as fruitful applications in diverse areas in due course.

Conflict of interest

The authors declare no conflict of interest, financial or otherwise.

Acknowledgments

The authors are grateful to the Department of Chemistry, Mohan Lal Sukhadia University, Udaipur (Raj.), India, for providing necessary library facilities for carrying out the work. N. Sahiba and P. Teli are deeply grateful to the Council for Scientific and Industrial Research (CSIR) (file no. 09/172(0088)2018-EMR-I) and (file no. 09/172(0099)2019-EMR-I), New Delhi, for providing a Senior Research Fellowship as financial support.

Funding source

This work was supported by the CSIR [file no. 09/172(0088)2018-EMR-I], [file no. 09/172(0099)2019-EMR-I].

References

[1] Keri RS, Rajappa CK, Patil SA, Nagaraja BM. Benzimidazole-core as an antimycobacterial agent. Pharmacol Rep 2016;68(6):1254–65.

[2] Riduan SN, Zhang Y. Imidazolium salts and their polymeric materials for biological applications. Chem Soc Rev 2013;42(23):9055–70.

[3] Gaba M, Mohan C. Development of drugs based on imidazole and benzimidazole bioactive heterocycles: recent advances and future directions. Med Chem Res 2016;25:173–210.

[4] Akhtar W, Khan MF, Verma G, Shaquiquzzaman M, Rizvi MA, Mehdi SH, Akhter M, Alam MM. Therapeutic evolution of benzimidazole derivatives in the last quinquennial period. Eur J Med Chem 2017;126:705–53.

[5] Devi N, Rawal RK, Singh V. Diversity-oriented synthesis of fused-imidazole derivatives via Groebkee-Blackburne-Bienayme reaction: a review. Tetrahedron 2014;71(2):183–232.

[6] Beltran-Hortelano I, Alcolea V, Font M, Pérez-Silanes S. The role of imidazole and benzimidazole heterocycles in Chagas disease: a review. Eur J Med Chem 2020;206:112692.

[7] Hernández-Núñez E, Tlahuext H, Moo-Puc R, Torres-Gómez H, Reyes-Martínez R, Cedillo-Rivera R, Nava-Zuazo C, Navarrete-Vazquez G. Synthesis and in vitro trichomonicidal, giardicidal and amebicidal activity of N-acetamide(sulfonamide)-2-methyl-4-nitro-1H-imidazoles. Eur J Med Chem 2009;44(7):2975–84.

[8] Prasad J, Pathak MB, Panday SK. An efficient and straight forward synthesis of (5S)-1-benzyl-5-(1H-imidazol-1-ylmethyl)-2-pyrrolidinone (MM1): a novel antihypertensive agent. Med Chem Res 2012;21:321–4.

[9] Shingalapur RV, Hosamani KM, Keri RS, Hugar MH. Derivatives of benzimidazole pharmacophore: synthesis, anticonvulsant, antidiabetic and DNA cleavage studies. Eur J Med Chem 2010;45(5):1753–9.

[10] Yawson GK, Huffman SE, Fisher SS, Bothwell PJ, Platt DC, Jones MA, Ferrence GM, Hamaker CG, Webb MI. Ruthenium(III) complexes with imidazole ligands that modulate the aggregation of the amyloid-β peptide via hydrophobic interactions. J Inorg Biochem 2021;214:111303.

[11] Gentili F, Bousquet P, Brasili L, Caretto M, Carrieri A, Dontenwill M, Giannella M, Marucci G, Perfumi M, Piergentili A, Quaglia W, Rascente C, Pigini M. Alpha2-adrenoreceptors profile modulation and high antinociceptive activity of (S)-(-)-2-[1-(biphenyl-2-yloxy) ethyl]-4,5-dihydro-1H-imidazole. J Med Chem 2002;45(1):32–40.

[12] Robertson DW, Beedle EE, Lawson R, Leander JD. Imidazole anticonvulsants: structure-activity relationships of [(biphenylyloxy)alkyl]imidazoles. J Med Chem 1987;30(5):939–43.

[13] Robertson DW, Krushinski JH, Beedle EE, Leander JD, Wong DT, Rathbun RC. Structure-activity relationships of (arylalkyl)imidazole anticonvulsants: comparison of the (fluorenyl-alkyl)imidazoles with nafimidone and denzimol. J Med Chem 1986;29(9):1577–86.

[14] Walker KA, Wallach MB, Hirschfeld DR. 1-(Naphthylalkyl)-1H-imidazole derivatives, a new class of anticonvulsant agents. J Med Chem 1981;24(1):67–74.

[15] Tuyen TN, Sin KS, Kim HP, Park H. Synthesis and antiinflammatory activity of 1,5-diarylimidazoles. Arch Pharm Res 2005;28(9):1013–8.

[16] Tseng CH, Lin CS, Shih PK, Tsao LT, Wang JP, Cheng CM, Tzeng CC, Chen YL. Furo[3′,2′:3,4]naphtho[1,2-d]imidazole derivatives as potential inhibitors of inflammatory factors in sepsis. Bioorg Med Chem 2009;17(18):6773–9.

[17] Breslin HJ, Miskowski TA, Rafferty BM, Coutinho SV, Palmer JM, Wallace NH, Schneider CR, Kimball ES, Zhang SP, Li J, Colburn RW, Stone DJ, Martinez RP, He W. Rationale, design, and synthesis of novel phenyl imidazoles as opioid receptor agonists for gastrointestinal disorders. J Med Chem 2004;47(21):5009–20.

[18] Mor M, Bordi F, Silva C, Rivara S, Crivori P, Plazzi PV, Ballabeni V, Caretta A, Barocelli E, Impicciatore M, Carrupt PA, Testa B. H3-receptor antagonists: synthesis and structure-activity relationships of para- and meta-substituted 4(5)-phenyl-2-[[2-[4(5)-imidazolyl]ethyl]thio] imidazoles. J Med Chem 1997;40(16):2571–8.

[19] Vacondio F, Mor M, Silva C, Zuliani V, Rivara M, Rivara S, Bordi F, Plazzi PV, Magnanini F, Bertoni S, Ballabeni V, Barocelli E, Carrupt PA, Testa B. Imidazole H3-antagonists: relationship between structure and ex vivo binding to rat brain H3-receptors. Eur J Pharm Sci 2004;23(1):89–98.

[20] Parsons WH, Calvo RR, Cheung W, Lee YK, Patel S, Liu J, Youngman MA, Dax SL, Stone D, Qin N, Hutchinson T, Lubin ML, Zhang SP, Finley M, Liu Y, Brandt MR, Flores CM, Player MR. Benzo[d]imidazole transient receptor potential vanilloid 1 antagonists for the treatment of pain: discovery of trans-2-(2-{2-[2-(4-trifluoromethyl-phenyl)-vinyl]-1H-benzimidazol-5-yl}-phenyl)-propan-2-ol (mavatrep). J Med Chem 2015;58(9):3859–74.

[21] Zhu C, Hansen AR, Bateman T, Chen Z, Holt TG, Hubert JA, Karanam BV, Lee SJ, Pan J, Qian S, Reddy VB, Reitman ML, Strack AM, Tong V, Weingarth DT, Wolff MS, MacNeil DJ, Weber AE, Duffy JL, Edmondson SD. Discovery of imidazole carboxamides as potent and selective CCK1R agonists. Bioorg Med Chem Lett 2008;18(15):4393–6.

[22] Lange JH, van Stuivenberg HH, Coolen HK, Adolfs TJ, McCreary AC, Keizer HG, Wals HC, Veerman W, Borst AJ, de Looff W, Verveer PC, Kruse CG. Bioisosteric replacements of the pyrazole moiety of rimonabant: synthesis, biological properties, and molecular modeling investigations of thiazoles, triazoles, and imidazoles as potent and selective CB1 cannabinoid receptor antagonists. J Med Chem 2005;48(6):1823–38.

[23] Gomaa MS, Bridgens CE, Aboraia AS, Veal GJ, Redfern CP, Brancale A, Armstrong JL, Simons C. Small molecule inhibitors of retinoic acid 4-hydroxylase (CYP26): synthesis and biological evaluation of imidazole methyl 3-(4-(aryl-2-ylamino)phenyl)propanoates. J Med Chem 2011;54(8):2778–91.

[24] Kumar V, Gupta MK, Singh G, Prabhakar YS. CP-MLR/PLS directed QSAR study on the glutaminyl cyclase inhibitory activity of imidazoles: rationales to advance the understanding of activity profile. J Enzyme Inhib Med Chem 2013;28(3):515–22.

[25] Kantsadi AL, Bokor É, Kun S, Stravodimos GA, Chatzileontiadou DSM, Leonidas DD, Juhász-Tóth É, Szakács A, Batta G, Docsa T, Gergely P, Somsák L. Synthetic, enzyme kinetic, and protein crystallographic studies of C-β-d-glucopyranosyl pyrroles and imidazoles reveal and explain low nanomolar inhibition of human liver glycogen phosphorylase. Eur J Med Chem 2016;123:737–45.

[26] Mano T, Stevens RW, Ando K, Nakao K, Okumura Y, Sakakibara M, Okumura T, Tamura T, Miyamoto K. Novel imidazole compounds as a new series of potent, orally active inhibitors of 5-lipoxygenase. Bioorg Med Chem 2003;11(18):3879–87.

[27] Dianat S, Moghimi S, Mahdavi M, Nadri H, Moradi A, Firoozpour L, Emami S, Mouradzadegun A, Shafiee A, Foroumadi A. Quinoline-based imidazole-fused heterocycles as new inhibitors of 15-lipoxygenase. J Enzyme Inhib Med Chem 2016;31(suppl 3):205–9.

[28] Manley PW, Allanson NM, Booth RF, Buckle PE, Kuzniar EJ, Lad N, Lai SM, Lunt DO, Tuffin DP. Structure-activity relationships in an imidazole-based series of thromboxane synthase inhibitors. J Med Chem 1987;30(9):1588–95.

[29] Magnus NA, Diseroad WD, Nevill Jr CR, Wepsiec JP. Synthesis of imidazole based p38 MAP (Mitogen-Activated Protein) kinase inhibitors under buffered conditions. Org Process Res Dev 2006;10:556–60.

[30] Laufer SA, Hauser DR, Domeyer DM, Kinkel K, Liedtke AJ. Design, synthesis, and biological evaluation of novel tri- and tetrasubstituted imidazoles as highly potent and specific ATP-mimetic inhibitors of p38 MAP kinase: focus on optimized interactions with the enzyme's surface-exposed front region. J Med Chem 2008;51(14):4122–49.

[31] Liverton NJ, Butcher JW, Claiborne CF, Claremon DA, Libby BE, Nguyen KT, Pitzenberger SM, Selnick HG, Smith GR, Tebben A, Vacca JP, Varga SL, Agarwal L, Dancheck K, Forsyth AJ, Fletcher DS, Frantz B, Hanlon WA, Harper CF, Hofsess SJ, Kostura M, Lin J, Luell S, O'Neill EA, O'Keefe SJ, et al. Design and synthesis of potent, selective, and orally bioavailable tetrasubstituted imidazole inhibitors of p38 mitogen-activated protein kinase. J Med Chem 1999;42(12):2180–90.

[32] Laufer SA, Zimmermann W, Ruff KJ. Tetrasubstituted imidazole inhibitors of cytokine release: probing substituents in the N-1 position. J Med Chem 2004;47(25):6311–25.

[33] Peifer C, Abadleh M, Bischof J, Hauser D, Schattel V, Hirner H, Knippschild U, Laufer S. 3,4-Diaryl-isoxazoles and -imidazoles as potent dual inhibitors of p38alpha mitogen activated protein kinase and casein kinase 1delta. J Med Chem 2009;52(23):7618–30.

[34] Yamada M, Yura T, Morimoto M, Harada T, Yamada K, Honma Y, Kinoshita M, Sugiura M. 2-[(2-Aminobenzyl)sulfinyl]-1-(2-pyridyl)-1,4,5,6-tetrahydrocyclopent[d]imidazoles as a novel class of gastric H+/K+-ATPase inhibitors. J Med Chem 1996;39(2):596–604.

[35] Zheng Q, Chen Z, Wan H, Tang S, Ye Y, Xu Y, Jiang L, Ding J, Geng M, Huang M, Huang Y. Discovery and structure-activity-relationship study of novel imidazole cyclopropyl amine analogues for mutant isocitric dehydrogenase 1 (IDH1) inhibitors. Bioorg Med Chem Lett 2018;28(23–24):3808–12.

[36] Verras A, Kuntz ID, Ortiz de Montellano PR. Computer-assisted design of selective imidazole inhibitors for cytochrome p450 enzymes. J Med Chem 2004;47(14):3572–9.

[37] Roumen L, Peeters JW, Emmen JM, Beugels IP, Custers EM, de Gooyer M, Plate R, Pieterse K, Hilbers PA, Smits JF, Vekemans JA, Leysen D, Ottenheijm HC, Janssen HM, Hermans JJ. Synthesis, biological evaluation, and molecular modeling of 1-benzyl-1H-imidazoles as selective inhibitors of aldosterone synthase (CYP11B2). J Med Chem 2010;53(4):1712–25.

[38] Jallapally A, Addla D, Bagul P, Sridhar B, Banerjee SK, Kantevari S. Design, synthesis and evaluation of novel 2-butyl-4-chloroimidazole derived peptidomimetics as Angiotensin Converting Enzyme (ACE) inhibitors. Bioorg Med Chem 2015;23(13):3526–33.

[39] Goff DA, Koolpe GA, Kelson AB, Vu HM, Taylor DL, Bedford CD, Musallam HA, Koplovitz I, Harris 3rd RN. Quaternary salts of 2-[(hydroxyimino)methyl]imidazole. 4. Effect of various side-chain substituents on therapeutic activity against anticholinesterase intoxication. J Med Chem 1991;34(4):1363–8.

[40] Lammi C, Sgrignani J, Arnoldi A, Lesma G, Spatti C, Silvani A, Grazioso G. Computationally driven structure optimization, synthesis, and biological evaluation of imidazole-based proprotein convertase subtilisin/Kexin 9 (PCSK9) inhibitors. J Med Chem 2019;62(13):6163–74.

[41] Chen SB, Tan JH, Ou TM, Huang SL, An LK, Luo HB, Li D, Gu LQ, Huang ZS. Pharmacophore-based discovery of triaryl-substituted imidazole as new telomeric G-quadruplex ligand. Bioorg Med Chem Lett 2011;21(3):1004–9.

[42] Madsen C, Jensen AA, Liljefors T, Kristiansen U, Nielsen B, Hansen CP, Larsen M, Ebert B, Bang-Andersen B, Krogsgaard-Larsen P, Frølund B. 5-Substituted imidazole-4-acetic acid analogues: synthesis, modeling, and pharmacological characterization of a series of novel gamma-aminobutyric acid(C) receptor agonists. J Med Chem 2007;50(17):4147–61.

[43] Oresmaa L, Kotikoski H, Haukka M, Salminen J, Oksala O, Pohjala E, Moilanen E, Vapaatalo H, Vainiotalo P, Aulaskari P. Synthesis and ocular effects of imidazole nitrolic acids. J Med Chem 2005;48(13):4231–6.

[44] Oresmaa L, Kotikoski H, Haukka M, Oksala O, Pohjala E, Vapaatalo H, Moilanen E, Vainiotalo P, Aulaskari P. Synthesis, ocular effects, and nitric oxide donation of imidazole amidoximes. Eur J Med Chem 2006;41(9):1073–9.

[45] Iman M, Davood A, Nematollahi AR, Dehpoor AR, Shafiee A. Design and synthesis of new 1,4-dihydropyridines containing 4(5)-chloro-5(4)-imidazolyl substituent as a novel calcium channel blocker. Arch Pharm Res 2011;34(9):1417–26.

[46] Moosavi-Movahedi Z, Gharibi H, Hadi-Alijanvand H, Akbarzadeh M, Esmaili M, Atri MS, Sefidbakht Y, Bohlooli M, Nazari K, Javadian S, Hong J, Saboury AA, Sheibani N, Moosavi-Movahedi AA. Caseoperoxidase, mixed β-casein-SDS-hemin-imidazole complex: a nano artificial enzyme. J Biomol Struct Dyn 2015;33(12):2619–32.

[47] Tao Z-F, Fujiwara T, Saito I, Sugiyama H. Rational design of sequence-specific DNA alkylating agents based on Duocarmycin A and pyrrole-imidazole hairpin polyamides. J Am Chem Soc 1999;121:4961–7.

[48] Singh B, Kumar A. Radiation-induced graft copolymerization of N-vinyl imidazole onto moringa gum polysaccharide for making hydrogels for biomedical applications. Int J Biol Macromol 2018;120(pt B):1369–78.

[49] World Health Organization. World Malaria Report 2020: 20 years of global progress and challenges. Geneva: World Health Organization; 2020.

[50] Roman G, Crandall IE, Szarek WA. Synthesis and anti-Plasmodium activity of benzimidazole analogues structurally related to astemizole. ChemMedChem 2013;8(11):1795–804.

[51] Sharma K, Shrivastava A, Mehra RN, Deora GS, Alam MM, Zaman MS, Akhter M. Synthesis of novel benzimidazole acrylonitriles for inhibition of *Plasmodium falciparum* growth by dual target inhibition. Arch Pharm Chem Life Sci 2017. https://doi.org/10.1002/ardp.201700251, e1700251.

[52] Kondaparla S, Manhas A, Dola VR, Srivastava K, Puri SK, Katti SB. Design, synthesis and antiplasmodial activity of novel imidazole derivatives based on 7-chloro-4-aminoquinoline. Bioorg Chem 2018;80:204–11.

[53] Pandey RK, Sharma D, Bhatt TK, Sundar S, Prajapati VK. Developing imidazole analogues as potential inhibitor for Leishmania donovani trypanothione reductase: virtual screening, molecular docking, dynamics and ADMET approach. J Biomol Struct Dyn 2015;33(12):2541–53.

[54] De Luca L, Ferro S, Buemi MR, Monforte AM, Gitto R, Schirmeister T, Maes L, Rescifina A, Micale N. Discovery of benzimidazole-based *Leishmania mexicana* cysteine protease CPB2.8ΔCTE inhibitors as potential therapeutics for leishmaniasis. Chem Biol Drug Des 2018;92(3):1585–96.

[55] Revuelto A, Ruiz-Santaquiteria M, de Lucio H, Gamo A, Carriles AA, Gutiérrez KJ, Sánchez-Murcia PA, Hermoso JA, Gago F, Camarasa MJ, Jiménez-Ruiz A, Velázquez S. Pyrrolopyrimidine vs imidazole-phenyl-thiazole scaffolds in nonpeptidic dimerization inhibitors of leishmania infantum trypanothione reductase. ACS Infect Dis 2019;5(6):873–91.

[56] Adeyemi OS, Eseola AO, Plass W, Atolani O, Sugi T, Han Y, Batiha GE, Kato K, Awakan OJ, Olaolu TD, Nwonuma CO, Alejolowo O, Owolabi A, Rotimi D, Kayode OT. Imidazole derivatives as antiparasitic agents and use of molecular modeling to investigate the structure-activity relationship. Parasitol Res 2020;119(6):1925–41.

[57] Saccoliti F, Madia VN, Tudino V, De Leo A, Pescatori L, Messore A, De Vita D, Scipione L, Brun R, Kaiser M, Mäser P, Calvet CM, Jennings GK, Podust LM, Costi R, Di Santo R. Biological evaluation and structure-activity relationships of imidazole-based compounds as antiprotozoal agents. Eur J Med Chem 2018;156:53–60.

[58] Saccoliti F, Madia VN, Tudino V, De Leo A, Pescatori L, Messore A, De Vita D, Scipione L, Brun R, Kaiser M, Mäser P, Calvet CM, Jennings GK, Podust LM, Pepe G, Cirilli R, Faggi C, Di Marco A, Battista MR, Summa V, Costi R, Di Santo R. Design, synthesis, and biological evaluation of new 1-(aryl-1 h-pyrrolyl)(phenyl)methyl-1H-imidazole derivatives as antiprotozoal agents. J Med Chem 2019;62(3):1330–47.

[59] Sánchez-Moreno M, Sanz AM, Gómez-Contreras F, Navarro P, Marín C, Ramírez-Macias I, Rosales MJ, Olmo F, Garcia-Aranda I, Campayo L, Cano C, Arrebola F, Yunta MJ. In vivo trypanosomicidal activity of imidazole- or pyrazole-based benzo[g] phthalazine derivatives against acute and chronic phases of Chagas disease. J Med Chem 2011;54(4):970–9.

[60] Papadopoulou MV, Bloomer WD, Rosenzweig HS, Wilkinson SR, Kaiser M. Novel nitro(triazole/imidazole)-based heteroarylamides/sulfonamides as potential antitrypanosomal agents. Eur J Med Chem 2014;87:79–88.

[61] Trunz BB, Jędrysiak R, Tweats D, Brun R, Kaiser M, Suwiński J, Torreele E. 1-Aryl-4-nitro-1H-imidazoles, a new promising series for the treatment of human African trypanosomiasis. Eur J Med Chem 2011;46(5):1524–35.

[62] Mello FdVCE, Castro Salomão Quaresma BM, Resende Pitombeira MC, Araújo de Brito M, Farias PP, Lisboa de Castro S, Salomão K, Silva de Carvalho A, Oliveira de Paula JI, de Brito Nascimento S, Peixoto Cupello M, Paes MC, Boechat N, Felzenszwalb I. Novel nitroimidazole derivatives evaluated for their trypanocidal, cytotoxic, and genotoxic activities. Eur J Med Chem 2020;186:111887.

[63] de Souza ML, de Oliveira Rezende Junior C, Ferreira RS, Espinoza Chávez RM, LLG F, Slafer BW, Magalhães LG, Krogh R, Oliva G, Cruz FC, Dias LC, Andricopulo AD. Discovery of potent, reversible, and competitive cruzain inhibitors with trypanocidal activity: a structure-based drug design approach. J Chem Inf Model 2020;60(2):1028–41.

[64] Cornec AS, Monti L, Kovalevich J, Makani V, James MJ, Vijayendran KG, Oukoloff K, Yao Y, Lee VM, Trojanowski JQ, Smith 3rd AB, Brunden KR, Ballatore C. Multitargeted imidazoles: potential therapeutic leads for Alzheimer's and other neurodegenerative diseases. J Med Chem 2017;60(12):5120–45.

[65] Sarıkaya G, Çoban G, Parlar S, Tarikogullari AH, Armagan G, Erdoğan MA, Alptüzün V, Alpan AS. Multifunctional cholinesterase inhibitors for Alzheimer's disease: synthesis, biological evaluations, and docking studies of o/p-propoxyphenylsubstituted-1H-benzimidazole derivatives. Arch Pharm (Weinheim) 2018. https://doi.org/10.1002/ardp.201800076, e1800076.

[66] Gurjar AS, Solanki VS, Meshram AR, Vishwakarma SS. Exploring beta amyloid cleavage enzyme-1 inhibition and neuroprotective role of benzimidazole analogues as anti-alzheimer agents. J Chin Chem Soc 2019;1–10. https://doi.org/10.1002/jccs.201900200.

[67] (a) Goedert M, Spillantini MG. A century of Alzheimer's disease. Science 2006;314(5800):777–81. (b) Cavalli A, Bolognesi ML, Minarini A, Rosini M, Tumiatti V, Recanatini M, Melchiorre C. Multi-target-directed ligands to combat neurodegenerative diseases. J Med Chem 2008;51(3):347–72.

[68] Mo J, Chen T, Yang H, Guo Y, Li Q, Qiao Y, Lin H, Feng F, Liu W, Chen Y, Liu Z, Sun H. Design, synthesis, *in vitro* and *in vivo* evaluation of benzylpiperidine-linked 1,3-dimethylbenzimidazolinones as cholinesterase inhibitors against Alzheimer's disease. J Enzyme Inhib Med Chem 2020;35(1):330–43.

[69] Heider F, Ansideri F, Tesch R, Pantsar T, Haun U, Döring E, Kudolo M, Poso A, Albrecht W, Laufer SA, Koch P. Pyridinylimidazoles as dual glycogen synthase kinase 3β/p38α mitogen-activated protein kinase inhibitors. Eur J Med Chem 2019;175:309–29.

[70] Latif A, Bibi S, Ali S, Ammara A, Ahmad M, Khan A, Al-Harrasi A, Ullah F, Ali M. New multitarget directed benzimidazole-2-thiol-based heterocycles as prospective anti-radical and anti-Alzheimer's agents. Drug Dev Res 2021;82(2):207–16.

[71] Moneer AA, Mohammed KO, El-Nassan HB. Synthesis of novel substituted thiourea and benzimidazole derivatives containing a pyrazolone ring as anti-inflammatory agents. Chem Biol Drug Des 2016;87(5):784–93.

[72] Nascimento MVPS, Munhoz ACM, Theindl LC, Mohr ETB, Saleh N, Parisotto EB, Rossa TA, Zamoner A, Creczynski-Pasa TB, Filippin-Monteiro FB, Sá MM, Dalmarco EM. A novel tetrasubstituted imidazole as a prototype for the development of anti-inflammatory drugs. Inflammation 2018;41(4):1334–48.

[73] Holiyachi M, Samundeeswari S, Chougala BM, Naik NS, Madar J, Shastri LA, Joshi SD, Dixit SR, Dodamani S, Jalalpure S, Sunagar VA. Design and synthesis of coumarin–imidazole hybrid and phenyl-imidazoloacrylates as potent antimicrobial and antiinflammatory agents. Monatshe Chem 2018;149(3):595–609.

[74] Sasahara GL, Gouveia Júnior FS, Rodrigues RO, Zampieri DS, Fonseca SGDC, Gonçalves RCR, Athaydes BR, Kitagawa RR, Santos FA, Sousa EHS, Nagao-Dias AT, Lopes LGF. Nitro-imidazole-based ruthenium complexes with antioxidant and anti-inflammatory activities. J Inorg Biochem 2020;206:111048.

[75] Gopinath G, Sankeshi V, Perugu S, Alaparthi MD, Bandaru S, Pasala VK, Chittineni PR, Krupadanam GLD, Sagurthi SR. Design and synthesis of chiral 2H-chromene-N-imidazolo-amino acid conjugates as aldose reductase inhibitors. Eur J Med Chem 2016;124:750–62.

[76] Salazar MO, Osella MI, Ramallo IA, Furlan RLE. N^{α}-arylsulfonyl histamines as selective β-glucosidase inhibitors. RSC Adv 2018;8:36209–18.

[77] Peytam F, Adib M, Shourgeshty R, Mohammadi-Khanaposhtani M, Jahani M, Imanparast S, Faramarzi MA, Mahdavi M, Moghadamnia AA, Rastegar H, Larijani B. Design and synthesis of new imidazo[1,2-b]pyrazole derivatives, in vitro α-glucosidase inhibition, kinetic and docking studies. Mol Divers 2020;24(1):69–80.

[78] Deswal L, Verma V, Kumar D, Kaushik CP, Kumar A, Deswal Y, Punia S. Synthesis and antidiabetic evaluation of benzimidazole-tethered 1,2,3-triazoles. Arch Pharm (Weinheim) 2020;353(9), e2000090.

[79] Agelis G, Resvani A, Durdagi S, Spyridaki K, Tůmová T, Slaninová J, Giannopoulos P, Vlahakos D, Liapakis G, Mavromoustakos T, Matsoukas J. The discovery of new potent non-peptide angiotensin II AT1 receptor blockers: a concise synthesis, molecular docking studies and biological evaluation of N-substituted 5-butylimidazole derivatives. Eur J Med Chem 2012;55:358–74.

[80] Zhang Y, Xu J, Li Y, Yao H, Wu X. Design, synthesis and pharmacological evaluation of novel NO-releasing benzimidazole hybrids as potential antihypertensive candidate. Chem Biol Drug Des 2015;85(5):541–8.

[81] Sharma MC, Sharma S, Sharma P, Kumar A, Bhadoriya KS. Structural insights for substituted acyl sulfonamides and acyl sulfamides derivatives of imidazole as angiotensin II receptor antagonists using molecular modeling approach. J Taiwan Inst Chem Eng 2014;45(1):12–23.

[82] Theivendren P, Subramanian A, Murugan I, Joshi SD, More UA. Graph theoretical analysis, in silico modeling, design, and synthesis of compounds containing benzimidazole skeleton as antidepressant agents. Chem Biol Drug Des 2017;89(5):714–22.

[83] Kenchappa R, Bodke YD, Telkar S, Aruna SM. Antifungal and anthelmintic activity of novel benzofuran derivatives containing thiazolo benzimidazole nucleus: an in vitro evaluation. J Chem Biol 2016;10(1):11–23.

[84] Brusina MA, Nikolaev DN, Fundamenskii VS, Gurzhii VV, Zolotarev AA, Selitrenikov AV, Zevatskii YE, Potapkin AM, Ramsh SM, Piotrovskii LB. Water-soluble form of 1-alkyl (aryl) imidazole-4, 5-dicarboxylic acids. Structure and anticonvulsant activity of the triethanolammonium salt of 1-propylimidazole-4, 5-dicarboxylic acid. Pharm Chem J 2018;52(4):299–303.

[85] Naureen S, Chaudhry F, Asif N, Munawar MA, Ashraf M, Nasim FH, Arshad H, Khan MA. Discovery of indole-based tetraarylimidazoles as potent inhibitors of urease with low antilipoxygenase activity. Eur J Med Chem 2015;102:464–70.

[86] Ren W, Ren Y, Dong M, Gao Y. Design, synthesis, and thrombin inhibitory activity evaluation of some novel benzimidazole derivatives. Helv Chim Acta 2016;99(4):325–32.

[87] Yoon YK, Choon TS. Structural modifications of benzimidazoles via multi-step synthesis and their impact on sirtuin-inhibitory activity. Arch Pharm (Weinheim) 2016;349(1):1–8.

[88] Rajesh R, Manikandan A, Sivakumar A, Ramasubbu C, Nagaraju N. Substituted methoxybenzyl-sulfonyl-1H-benzo[d]imidazoles evaluated as effective H^+/K^+-ATPase inhibitors and anti-ulcer therapeutics. Eur J Med Chem 2017;139:454–60.

[89] Bhalekar SB, Shelke SN. Synthesis, biological and molecular docking study of benzimidazole-clubbed tetrahydrothieno [3, 2-c] pyridine as platelet inhibitors. Chemistry Select 2019;4(41):12170–5.

[90] Im D, Moon H, Kim J, Oh Y, Jang M, Hah JM. Conformational restriction of a type II FMS inhibitor leading to discovery of 5-methyl-*N*-(2-aryl-1*H*-benzo[d]imidazo-5-yl) isoxazole-4-carboxamide analogues as selective FLT3 inhibitors. J Enzyme Inhib Med Chem 2019;34(1):1716–21.

Chapter 8

Imidazole-based drugs and drug discovery: Present and future perspectives

Ayushi Sethiya[a], Jay Soni[a], Dinesh K. Agarwal[b] and Shikha Agarwal[a]
[a]*Synthetic Organic Chemistry Laboratory, Department of Chemistry, Mohanlal Sukhadia University, Udaipur, Rajasthan, India,* [b]*Department of Pharmacy, PAHER University, Udaipur, Rajasthan, India*

1. Introduction

Heterocyclic compounds have gained much attention in the past decades due to their reoccurrence in numerous pharmaceutically active molecules. The literature study showed that most of the biologically active compounds contain nitrogen in the ring system [1]. N-based heterocycles demonstrate fascinating biological activities and are used in several fields of pharmacology as well as industries [2, 3]. The imidazole ring is highly polar, amphoteric, can donate or accept protons easily, readily forms diverse weak interactions, and has several binding sites that permit it to interrelate effortlessly with various bio-molecules of the living systems. The imidazole ring exists as a structural subunit in diverse complex natural products, and these natural compounds are essential for performing a range of physiological actions for decisive biological compounds like vitamin B_{12}, histamine, hemoglobin, deoxyribonucleic acid (DNA), etc. Imidazole-containing compounds have different applications, varying from hormones and herbicides to antifungal, antibacterial, antihistamine, antidiabetic, antitubercular, anticancer agents, antiepileptic, etc. in the whole range of medicinal chemistry [4–7]. Consequently, imidazole-bearing compounds have become part of a vital approach for the design of various pharmaceutical drugs, man-made materials, agrochemicals, artificial acceptors, biomimetic catalysts, and supramolecular ligands [8–12]. The advancement of additional efficient synthetic protocols and armamentariums has been an ongoing quest. To combat severe life-threatening diseases and global escalating drug resistance, imidazole derivatives have been considered as fascinating lead structures to create new hybrid molecules with improved potency, excellent bioavailability, efficacy, good tissue

Imidazole-Based Drug Discovery. https://doi.org/10.1016/B978-0-323-85479-5.00004-6
Copyright © 2022 Elsevier Inc. All rights reserved.

penetrability, pharmacological profile, and permeability coupled with a relatively low incidence of adverse effects and lesser toxicity [13–16]. Due to these abovementioned characteristics, imidazole has been clubbed and incorporated with several other moieties to generate hybrid molecules with enhanced biological profiles [17]. Numerous imidazole-based derivatives have been fabricated and evaluated for their in vitro and in vivo biological activities and have further undergone clinical trials, and several of them have proven to be endowed with tremendous pharmacological characteristics [18]. The several therapeutic applications of imidazole derivatives are shown in Fig. 8.1.

This chapter investigates the progress made in the area of imidazoles in the last decades along with the prospects, sustained challenges, and future viewpoints. The favored pharmacokinetics and pharmacodynamics characteristics of imidazole-based drugs and recommendations for designed composite progression in the diverse phases of drug discovery are also discussed. These studies include identification of hit to preclinical examination, as well as clinical development of imidazole-based drugs. Wherever necessary, important reviews and articles on the particular issues are cited and readers are strongly encouraged to refer to these for in-depth learning. To compile the literature systematically, the chapter is divided into subsections for the easier understanding of readers.

2. Imidazole-based market available drugs

Several imidazole-based drugs are clinically used to cure diverse types of diseases. Table 8.1 summarizes different drugs with their therapeutic efficiency and mechanism of action. This tabular representation is intended to enable easier reading for the vast community of researchers.

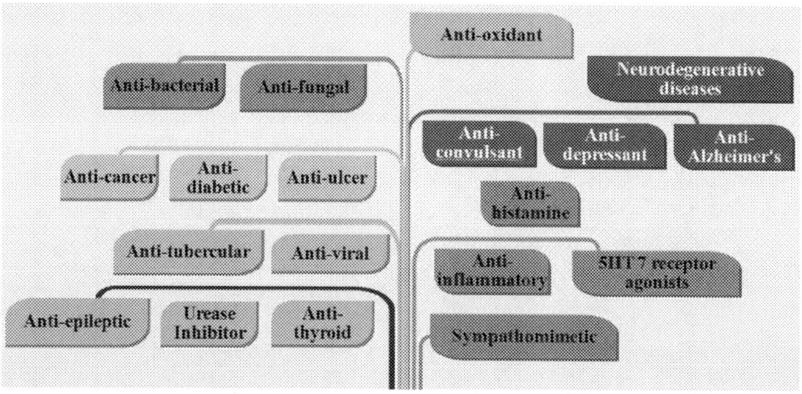

FIG. 8.1 Miscellaneous therapeutic efficiency of imidazole derivatives.

TABLE 8.1 List of imidazole-based drugs, therapeutic indications, and their mechanisms of action.

S. no.	Drug name	Therapeutic indication	Mechanism of actions
1	Dacarbazine	Anticancer	It is used as a chemotherapeutic agent as it is a drug of alkylating category. It sticks to the DNA and damages it
2	Fadrozole	Antineoplastic agents which are used to cure hormone-sensitive tumors like breast, endometrium, prostate, and lymphomas, and certain leukemias	It acts as an aromatase inhibitor, thereby reducing formation of estrogenic steroid hormones
3	Tipifarnib	Acute myeloid leukemia (AML) and other types of cancer	It inhibits the enzyme farnesyl protein transferase, thereby preventing the activation of Ras oncogenes that cause cell growth, provoke apoptosis, and restrain angiogenesis
4	Nilotinib	Chronic myelogenous leukemia, anticoronaviral agent	Tyrosin kinase inhibitor. It binds with the inactive conformation of the kinase domain of the ABL protein of the BCR-ABL fusion protein and causes inhibition of the BCR-ABL-mediated propagation of Philadelphia chromosome-positive (Ph +) chronic myeloid leukemia (CML) cells
5	Etanidazole	Radiosensitizing agent used for hypoxic tumor cells	It diminishes the conc. of glutathione and inhibits glutathione S-transferase that makes tissues more sensitive toward ionizing radiation used in cancer treatment
6	Misonidazole	Radio-sensitizer used in radiation therapy	It causes generation of free radicals and reduces radio-protective thiols, causing sensitization of hypoxic cells to the cytotoxic effects of ionizing radiations
7	Indimitecan	Anticancer	Topoisomerase I inhibitor. It thereby helps in cell apoptosis
8	Idelalisib	Used in treatment of blood cancer	It prevents the activation of the PI3K signaling pathway and reduces cell proliferation of tumors

Continued

TABLE 8.1 List of imidazole-based drugs, therapeutic indications, and their mechanisms of action—cont'd

S. no.	Drug name	Therapeutic indication	Mechanism of actions
9	Nocodazole	Antineoplastic agent	Tubulin-binding agent that stops mitosis and provokes apoptosis in tumor cells
10	Ponatinib	Antiangiogenic, antineoplastic	Tyrosine kinase inhibitor
11	Rodotinib	Chronic myelogenous leukemia	Tyrosine kinase inhibitor
12	Nimorazole	Antibacterial and radio-sensitizer	It induces free radical generation and causes sensitization of hypoxic cells
13	Azathioprine	Immunosuppressive medication. It is used in rheumatoid arthritis, granulomatosis with polyangiitis, Crohn's disease, systemic lupus erythematosus, and ulcerative colitis	It interrupts the formation of RNA and DNA via 6-thioguanine
14	Zoledronic acid	It is used to treat osteoporosis, Paget's disease of bone, and bone disease in cancer patients	It slows bone breakdown, escalating bone density (thickness) and lessening the quantity of calcium excreted from the bones into the blood
15	Secnidazole	Antibiotic	It enters the bacterial cell as an inactive prodrug where the nitro group got reduced to radical anions. These radical anions obstruct the DNA synthesis
16	Metronidazole	Antibiotic and antiprotozoal	It hinders the synthesis of nucleic acid by forming nitroso radicals that interrupt the DNA of microbial cells
17	Ornidazole	Antibiotic	It disrupts the DNA replication and transcription process
18	Tinidazole	Antiprotozoal and antibacterial	The nitro radical generated by its reduction causes antiprotozoal activity
19	Dimetridazole	Antiprotozoal	NA

20	Ridinilazole	Antibiotic (used to cure *Clostridioides difficile* infection)	It is involved in the inhibition of cell division
21	Thiabendazole	Anthelmintic agent	It inhibits the helminth-specific enzyme fumarate reductase
22	Albendazole	Anthelmintic agent	It binds to the colchicine-sensitive site of β-tubulin, thereby preventing its polymerization
23	Enviroxime	Antiviral	Kinase inhibitor. It inhibits the replication of rhinoviruses and poliovirus by targeting protein 3A and 3B
24	Ledipasvir	Antiviral, to cure hepatitis C	It is an inhibitor of NS5A
25	Sertaconazole	Antifungal	It diminishes the growth of fungi
26	Fenticonazole	Antifungal	It destroys the cell membrane of fungi, thereby killing and stopping their growth
27	Luliconazole	Antifungal, used for skin diseases	It inhibits ergosterol synthesis by inhibiting the enzyme lanosterol demethylase, thereby decreasing the growth of fungus
28	Miconazole	Antifungal	It inhibits the synthesis of ergosterol, thereby affecting growth of fungi
29	Ketoconazole	Antifungal	It inhibits synthesis of ergosterol, thereby preventing growth of fungi
30	Clotrimazole	Fungal skin infections	It inhibits the growth of *Candida* or fungal cells by varying the cell wall permeability. It attaches to phospholipids, thereby inhibiting the biosynthesis of ergosterol and diminishing the growth of fungi
31	Pretomanid	Antimycobacterium	Reduction of pretomanid forms an active metabolite that causes greater generation of NO, leading to bactericidal activities

Continued

TABLE 8.1 List of imidazole-based drugs, therapeutic indications, and their mechanisms of action—cont'd

S. no.	Drug name	Therapeutic indication	Mechanism of actions
32	Delamanid	Antimycobacterium	It forms a reactive intermediate metabolite, which hinders the synthesis of the mycobacterial cell wall
33	Losartan	High blood pressure, diabetic kidney disease	Angiotensin II receptor antagonist
34	Eprosartan	High blood pressure	Angiotensin II receptor antagonist
35	Olmesartan	High blood pressure and diabetic kidney disease	Angiotensin II receptor antagonist
36	Etonitazene	Analgesic	NA
37	Astemizole	Antihistaminic (antiallergic)	It competes with histamine to bind with H_1-receptor sites in the GI tract, thereby suppressing the formation of edema, flare, and pruritus
38	Megazole	Antiulcer	It is a combination of two drugs, domperidone and pantoprazole. Domperidone is a prokinetic that works on the upper digestive tract and thereby increases the movement of the stomach and intestine, which helps the food to travel more effortlessly through the stomach. Pantoprazole is a (PPI) proton pump inhibitor that reduces the amount of acid in the stomach
39	Lansoprazole	Antiulcer	Proton pump inhibitor
40	Cimetidine	Gastroesophageal reflux disease (GERD)	Histamine H_2 receptor antagonist
41	Methimazole	Antithyroid	It interferes the iodination of tyrosine residues in thyroglobulin, thereby obstructing the synthesis of thyroxine (T4) and triiodothyronine (T3)
42	Carbimazole	Antithyroid	It converts to methimazole and causes antithyroid action
43	Tolazoline	Vasodilator	Nonselective competitive α-adrenergic receptor antagonist

44	Naphazoline	Decongestant	It works by temporarily narrowing the blood vessels
45	Tetrahydrozoline	Decongestant	It works by temporarily narrowing the blood vessels
46	Etomidate	Anesthetic agent	It works on CNS to motivate gamma-aminobutyric acid (GABA) receptors, thereby decreasing the reticular activating system
47	Pilocarpine	Cholinergic agonist	Muscarinic receptor agonist
48	Pimobendan	Calcium sensitizer and phosphodiesterase inhibitor (veterinary medication)	It increases myocardial contractility. It enhances the binding efficiency of cardiac troponin in the myofibril to calcium ions
49	Fenbendazole	Antihelmintic (veterinary medication)	It binds to tubulin and disrupts the tubulin microtubule equilibrium
50	Dabigatran	Anticoagulant	It is a thrombin inhibitor, and thereby diminishes generation of fibrin and decreases thrombin-encouraged platelet aggregation
51	Dexmedetomidine	Anxiolytic, anesthetic, and analgesic	Selective $\alpha 2$-adrenoceptor agonism
52	Midazolam	Anticonvulsant, anesthesia, sedative, and hypnotic	It augments the inhibitory activity of the amino acid neurotransmitter gamma-aminobutyric acid (GABA)
53	Flumazenil	Benzodiazepine antagonist	It competitively blocks the binding site of benzodiazepine on the GABA/benzodiazepine receptor
54	Nafimidone	Anticonvulsant	It blocks voltage-gated sodium channels (VGSCs) and augments g-aminobutyric acid (GABA)-mediated response

NA, not available.

3. Pharmacokinetics and pharmacodynamic profiles of a drug

As per the pharmacokinetics characteristics, a high-quality drug candidate should have adequate solubility and stability in the GI tract and liver, and be well-absorbed. It should have good permeability, and systematic distribution. A good drug should follow the Lipinski Rule of Five [19]. Inadequate drug absorption via the gut, elimination in the liver and kidney, and early biotransformation can extensively diminish the quantity of orally administered drugs that go into the systemic circulation. Good drugs should also include the following properties: excellent bio-availability, extent of distribution, moderated to high half-life, medium to low clearance, low dose requirement, and the drug in the blood may be protein or lipid-bound or exist in a free soluble form. The drug should be easily taken up by target organs and cells, as well as being easily excreted by the kidneys. To match these characteristics, there is a persuasive requirement to design novel drug candidates with a novel mode of action for different diseases with improved pharmacological characteristics. Imidazole-based drugs are a step in this direction. The worldwide scenario of imidazole-based drugs has noticeably evolved in the last decades. Numerous newer and repurposed drugs are under experimental trials [20–22].

4. Synergy between imidazole-based compounds and drug development

4.1 Rational approaches to find imidazole-based drugs

The amazing success of imidazole-containing drugs discussed in Table 8.1 provides enormous motivation to chemists to invent new pathways for the preparation of imidazole-containing hybrid compounds. The continuous interest in synthetic pathways embedding imidazole with other organic molecules is an all-time elevated, ever presenting incredible ingenuity. It has reached record levels in research work, supported by the number of publications and citations [23–26]. All the renowned journals and publications are flooded with imidazole-containing compounds.

4.1.1 Drug design

Drug design is an ingenious procedure of finding novel compounds having medicinal use based on the comprehension of a biological target. Designing a drug is a prolonged and costly procedure. Approximately 12–15 years, plus a huge amount of expenditure, are required to bring a drug to the market. For roughly every 10,000 complex compounds that are estimated in vivo and in vitro studies, only 10 will undergo human clinical trials to facilitate a single compound on the market. There are three phases in clinical trials before drug approval:

- *Phase I*: To examine the safety, tolerability, dosage levels, toxic effects, pharmacokinetic properties, and pharmacological effects in approximately 100 healthy volunteers (generally a few months to a year and a half).
- *Phase II*: To evaluate the efficiency of the drug, finding out the side effects and other safety aspects, in a few hundred patients who are suffering from that disease (about 1–3 years).
- *Phase III*: This trial is done on a large scale with thousands of patients which are already hospitalized in clinics and hospitals, to establish the effectiveness of the drug and observe adverse reactions from long-term use (about 2–6 years) [27–29]. After completion of successful clinical trials, a new drug application (NDA) is put forward to the Food and Drug Administration (FDA), and it can take a few months to several years before its approval for commercial use. A pictorial representation of drug designing is shown in Fig. 8.2.

Several imidazole-based compounds are under investigation to estimate their therapeutic efficiencies. Combinatorial chemistry plays a key role in synthetic organic chemistry for the advancement of the drug-like molecules [30]. Researchers from academia and industry are carrying out exhaustive hard work to design a facile approach for the synthesis of drug candidates via eco-benign pathways with improved biological and pharmacological profiles. Presently, the main aim of scientists is to design a novel approach for the fabrication of a drug using green aspects of chemistry as well as by minimizing their adverse impact. In vitro, in vivo, and in silico studies are the major contributors for designing a drug. Diverse drug design protocols and virtual screening methods like computational-based drug approach, molecular docking, DFT study, and their structure–activity relationship are the foremost steps for designing a drug. These methodologies are very helpful

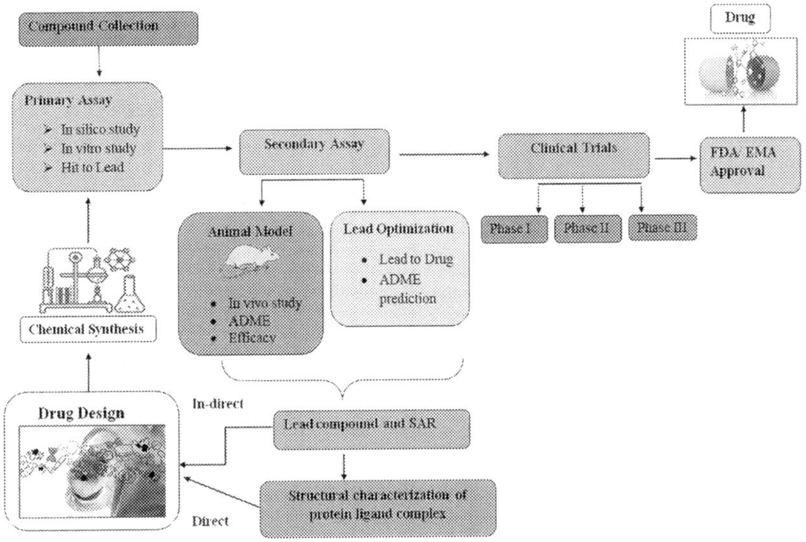

FIG. 8.2 Graphical representation for designing of a drug.

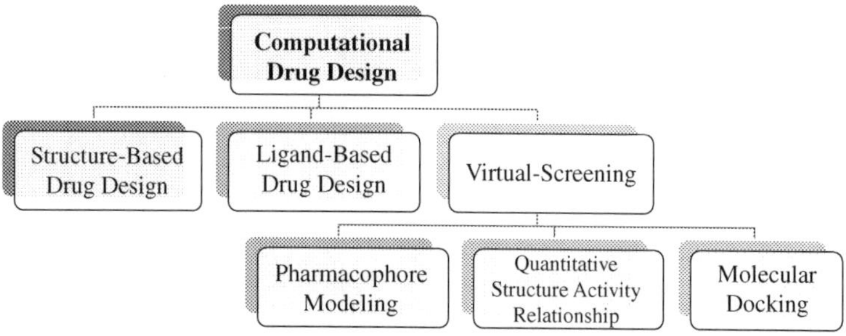

FIG. 8.3 Pictorial representation of computational approach in drug discovery.

for target-based designing [31]. The graphical representation of the computational approach in drug discovery is depicted in Fig. 8.3.

4.2 Imidazole hybrids as a hit for several diseases

The imidazole ring, including either as a substituent or with diverse substitutions on the ring itself, is among the most often observed heterocyclic scaffolds in the structure of various natural products and synthetic materials. The electronic characteristics on the ring may be responsible for better interaction and binding with receptors and enzymes that lead to further modifications in the scaffold to achieve an ideal activity profile. Several drugs containing imidazole moiety are already available on the market, while others are undergoing clinical trials. Several potent drug candidates like YM116, flutrimazole, azanidazole, dimetridazole, fipamezole, irtemazole, ramucirumab, fexinidazole, dazoxiben, AZD5597, etc., contain imidazole scaffolds [32–38].

In Chapters 2–7 of this book, we have thoroughly explained the in vitro and in vivo studies of synthesized compounds depending upon their biological activities, and the literature related to the biological profile of imidazoles has already been cited in these previous chapters. In view of this, Rulhania et al. [39] reviewed the medicinal viewpoint of synthetic hybrids of imidazole derivatives. In 2015, Munagala and coworkers [40] synthesized derivatives of 6-nitro-2,3-dihydroimidazooxazole (NHIO) scaffolds having triazolyl- and isoxazolyl moieties. The authors synthesized 41 novel triazolyl- and isoxazolyl-based NHIO analogs and assessed them against (MTB) H37Rv. From all the synthesized compounds, the potent compounds having MICs in the range of 0.57–0.13 μM were further examined against dormant, and also with resistant strains of MTB. As per the results of in vitro study, five compounds were appraised for in vivo experiment and compound **1** (Fig. 8.4, Entry 1) is the most potent and represented as an alternative to the lead candidate. In continuation of this work, the authors synthesized sulfonyl, uridyl, and thiouridyl derivatives of NHIO and evaluated them against (MTB) H37Rv. Among all the derivatives, the most potent six compounds were further tested against resistant (RifR and MDR) and dormant strains of MTB, and were also tested for cytotoxicity in the HepG2 cell line.

Compound **2** (Fig. 8.4, Entry 2) possessed the highest microsomal stability, solubility, and favorable oral in vivo pharmacokinetics, and efficacy [41]. In 2015,

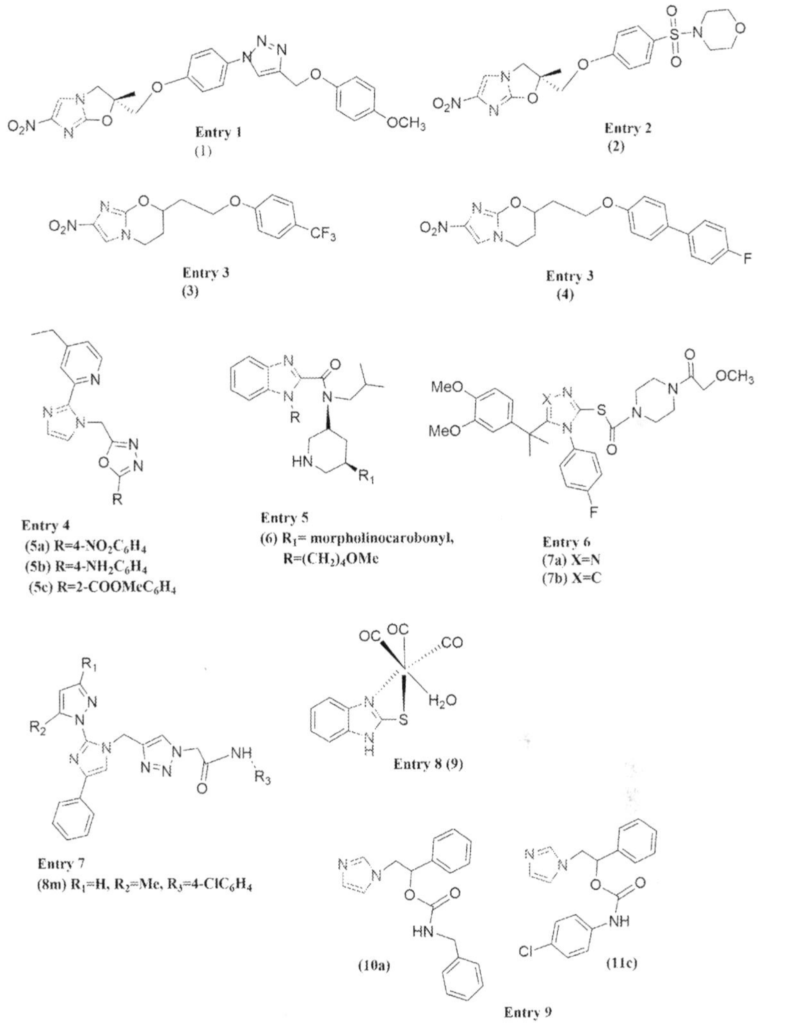

FIG. 8.4 Imidazole-containing scaffolds.

Kang and coauthors [42] structurally modified the potent TB drug PA-824 at the C-7 position by substituent benzyl ether, phenyl ether, benzyl carbonate, and benzyl carbamate, and these derivatives were examined against (MTB) H37Rv by in vitro methods. Among the tested compounds, compound **3 (R)** with the trifluoromethoxybenzyl group was five-fold more potent than PA-824 *in vitro* assay and **4 (R)** analogs having phenyl ether substitution (Fig. 8.4, Entry 3) exhibited the most potent antimycobacterial activity. Wani and coworkers [43] synthesized imidazole-based oxadiazole hybrids and examined their antifungal activity by in vitro assay. Among the tested compounds, three compounds **5a**, **5b**, and **5c** (Fig. 8.4, Entry 4), showed good antimicrobial activity and were also found to inhibit ergosterol biosynthesis. Their mechanism of action is similar to that of the drug fluconazole. A molecular docking study was also performed to find the binding interactions in cytochrome P450 lanosterol 14α-demethylase of *C. albicans* protein with PDB ID: 4K0F. Tokhuhara et al. [44] synthesized pyridine analogs of benzimidazole and studied their renin inhibitory activity. Among the synthesized compounds, derivative **6** (Fig. 8.4, Entry 5) showed high five-fold of plasma renin inhibition compared to aliskiren in cynomolgus monkeys at dose ratio.

Agarwal and coworkers [45] designed a library of new 2-mercapto imidazole and triazole analogs and estimated their TGR5 agonist activity, and compound **7a** showed good activity. Compound **7a** (Fig. 8.4, Entry 6) was capable of enhancing hTGR5 receptors with EC_{50}=25nM in CRE-directed luciferase-based RG assay. Furthermore, compound **7a** bearing an imidazole ring and **7b** bearing a triazole ring have satisfactory values for physicochemical characteristics including lipophilicity and CLogP. In 2020, advancement in the synthesis of the hybrid structure was done and a series of pyrazole-imidazole-triazole hybrids were synthesized by Punia and coworkers [46] using click reactions. The prepared compounds were screened for antimicrobial activity and compound **8m** exhibited exceptional potency against *A. niger* (MIC value 0.0064 μmol/mL); this was better than the reference drug fluconazole. Further compounds **8g** and **8m** (Fig. 8.4, Entry 7) were docked with protein PDB ID: 1KZN and 5TZ1 to determine the binding interactions, and it was observed that crucial interactions were present. In the same year, Rizvi et al. [47] synthesized 2-mercaptobenzimidazole derivatives, radiolabeled them with ^{99m}Tc, and used them in a radiotracer for hypoxic tumor imaging. The ^{99m}Tc-2-MBI complex (**9**) (Fig. 8.4, Entry 8) was found to be stable in saline and serum media with no visible degradation at RT and 37 °C, respectively, over 24 h. ^{99m}Tc-2-MBI was noticeably gathered in hypoxic tumor tissue, signifying that it would be a promising radiotracer for near the beginning stage analysis of tumor hypoxia. Recently, Ammazzalorso and coworkers [48] synthesized a library of imidazole-triazole carbamate derivatives and assessed these compounds against human aromatase by in vitro fluorimetric assay. From the result of this study, derivatives **10a** and **11c** (Fig. 8.4, Entry 9) were further examined via an in vitro method against the MCF-7 cell line by MTT assay.

4.3 Clinical trials on imidazole-based compounds

Drug resistance and coronavirus have put the science and technology on the back foot. Researchers are working day and night to develop effective therapeutic agents to combat severe diseases. In the present scenario, novel drugs with target base delivery are urgently needed. Numerous newer and repurposed drugs are undergoing clinical trials, and a couple of novel drugs have been approved by regulatory authorities in recent times. Different classes of drugs under clinical trials are summarized in Table 8.2.

4.4 Patents on imidazole-based compounds

Diverse molecules bearing imidazole have been examined and several patents on imidazole analogs have been claimed in the past decades. The patents published in the last 10 years are summarized in Table 8.3.

TABLE 8.2 Imidazole-based compounds under clinical trials.

Drug	Title	Mode of action	Clinical trial phase and identifier	Sponsor
TBI223	A phase 1 study to evaluate safety, tolerability, and pharmacokinetics of TBI-223 in healthy adults	It inhibits protein synthesis.	Phase 1 NCT03758612	Global Alliance for TB Drug Development
Contezolid	Contezolid acefosamil versus linezolid for the treatment of acute bacterial skin and skin structure infection	It inhibits protein synthesis	Phase 2 NCT03747497	MicuRx
Delamanid	A placebo-controlled, Phase 2 trial to evaluate OPC 67683 in patients with pulmonary sputum culture-positive, multidrug-resistant tuberculosis (TB)	It forms NO species that hinders synthesis of cell wall	Phase 2 NCT00685360	Otsuka Pharmaceutical Development & Commercialization
	Safety and efficacy trial of delamanid for 6 months in participants with multidrug-resistant tuberculosis	–	Phase 3 NCT01424670	Otsuka Pharmaceutical Development & Commercialization
	A 6-month safety, efficacy, and pharmacokinetic (pk) trial of delamanid in pediatric participants with multidrug resistant tuberculosis (MDR-TB)	–	Phase 2 NCT01859923	Otsuka Pharmaceutical Development & Commercialization
Pretomanid	Assessing PA-824 for tuberculosis	It forms NO species that hinder the growth of bacteria	Phase 2 NCT02256696	Johns Hopkins University
	A Phase 1 trial to evaluate the male reproductive safety of pretomanid in healthy volunteers (PaSEM)	–	Phase 1 NCT04179500	Global Alliance for TB Drug Development

Continued

TABLE 8.2 Imidazole-based compounds under clinical trials—cont'd

Drug	Title	Mode of action	Clinical trial phase and identifier	Sponsor
TBA354	A Phase 1 study to evaluate the safety, tolerability, and pharmacokinetics of TBA-354 in healthy adult subjects	It forms NO species	Phase 1 NCT02288481	Global Alliance for TB Drug Development
Telacebec (Q203)	A Phase 2 study to evaluate early bactericidal activity, safety, tolerability, and pharmacokinetics of multiple oral doses of telacebec (Q203)	It binds to QcrB subunit of cytochrome *bc*1 and inhibits respiration, and is useful in drug-sensitive TB	Phase 2 NCT03563599	Qurient Co., Ltd.
Nirogacestat	Nirogacestat for adults with desmoid tumor/aggressive fibromatosis (DT/AF) (DeFi)	It is a noncompetitive inhibitor of γ-secretase (GS) with a potential antitumor activity	Phase 3 NCT03785964	SpringWorks Therapeutics, Inc.
	A study of a new drug, nirogacestat, for treating desmoid tumors that cannot be removed by surgery	It might prevent the growth of tumor cells by inhibiting a few enzymes needed for cell growth	Phase II NCT04195399	Children's Oncology Group
Telithromycin (HMR3647)	TELI TON—Telithromycin in Tonsillitis	–	Phase 3 NCT00315042	Sanofi
PF-04991532	A 2-week trial of PF-04991532 in patients with type 2 diabetes	–	Phase 1 NCT01469065	Pfizer
	Multiple dose study of PF-04991532 in patients with type 2 diabetes	–	Phase 1 NCT01129258	Pfizer
IMC-1121B	Study of IMC-1121B in patients with advanced solid tumors not responding to standard therapy	It inhibits the growth of cancer cells	Phase 1 NCT00786383	Eli Lilly and Company

IMC-1121B (it is given in combination with other drugs)	A study of IMC-1121B (ramucirumab) in colorectal cancer	–	Phase 2 NCT00862784	Eli Lilly and Company
Dacarbazine + Brentuximab Vedotin + Adriamycin	Brentuximab vedotin plus AD in nonbulky limited stage Hodgkin's lymphoma	It binds particularly to Hodgkin's lymphoma cells. It enters the cells, and then excretes the drug to demolish the cells	Phase 2 NCT02505269	Massachusetts General Hospital
Ketoconazole	The comparative assessment of mycological efficacy, safety, recurrence, and cost-effectiveness of selenium sulfide 1.8% shampoo versus ketoconazole 2% shampoo in pityriasis versicolor: a double-blind randomized controlled trial	It is effective against pityriasis versicolor	Phase 4 NCT04007237	Indonesia University
Fluconazole	Generation of biological samples positive to fluconazole for antidoping control (FLUC)	It inhibits particular cytochrome P-450 enzymes that contribute in the preparation of steroid hormones	Phase 1 NCT04201054	Parc de Salut Mar
Fenobam	Pharmacokinetics and side effects of the mGlu5 antagonist fenobam in adult healthy volunteers	It is a selective negative allosteric modulator of the metabotropic glutamate receptor	Phase 1 NCT01806415	Washington University School of Medicine
Dacarbazine	Adjuvant, combined interleukin 2 (proleukin) and DTIC (dacarbazine) in high-risk melanoma patients (DTIC)	It kills micro-metastatic seed of melanoma cells	Phase 2 NCT00553618	James Graham Brown Cancer Center

Continued

TABLE 8.2 Imidazole-based compounds under clinical trials—cont'd

Drug	Title	Mode of action	Clinical trial phase and identifier	Sponsor
18F-fluoromisonidazole ([18F]FMISO)	Understanding grid radiation therapy effects on human tumor oxygenation and interstitial pressure to increase translation of solid tumor therapy	It is a radio-imaging agent that connects to huge molecules in tumor cells which have low oxygen levels	Early Phase 1	University of Arkansas
	[18F]FMISO PET/CT after transcatheter arterial embolization in imaging tumors in patients with liver cancer	–	Phase 2 NCT02695628	Sanjiv Sam Gambhir
Pimonidazole	Study of intratumoral hypoxia using preoperative administration of pimonidazole	It specifically reduces and binds to intracellular large molecules in areas of hypoxia	NCT01248637	University Health Network, Toronto
	Prostate hypoxia—TIC	–	NCT02095249	University Health Network, Toronto
	Correlation of FAZA PET hypoxia imaging to 3D histology in oral tongue cancer	–	Phase 1/Phase 2 NCT03181035	University Health Network, Toronto
	Novel window of opportunity trial to evaluate the impact of statins to oppose prostate cancer	–	Phase 2 NCT01992042	University Health Network, Toronto
Liarozole	Efficacy and safety of two doses of liarozole vs placebo for the treatment of lamellar ichthyosis	It inhibits degradation retinoic acid, which is the main endogenous controller of growth and differentiation of epithelial tissues in mammals. It is useful in lamellar ichthyosis	Phase 2 Phase 3 NCT00282724	Stiefel, a GSK Company
Ridinilazole (SMT19969)	A study of ridinilazole (SMT19969) compared with fidaxomicin for the treatment of *Clostridioides difficile* infection (CDI)	It is used for treatment of *Clostridioidesdifficile* infection (CDI)	Phase 2 NCT02784002	Summit Therapeutics

AT9283	Aurora kinase inhibitor AT9283 in treating patients with advanced or metastatic solid tumors or non-Hodgkin's lymphoma	It may prevent the development of cancer cells as it blocks some of the enzymes required for cell growth	Phase 1 NCT00443976	NCIC Clinical Trials Group
Veliparib	Veliparib (ABT-888), an oral PARP inhibitor, and VX-970, an ATR inhibitor, in combination with cisplatin in people with refractory solid tumors	It inhibits PARPs, thus obstructing DNA repair and acting as a DNA-damaging agent	Phase 1 NCT02723864	National Cancer Institute (NCI), USA
Arasertaconazole	Arasertaconazole nitrate pessaries—dose finding study for vulvovaginal candidiasis (VVC) treatment	It is a broad-spectrum antifungal drug	Phase 2 NCT01144286	Ferrer Internacional S.A.
Galeterone	1911GCCC: galeterone or galeterone with gemcitabine for patients with metastatic pancreatic adenocarcinoma	It is an androgen receptor inhibitor	Phase 2 NCT04098081	University of Maryland, Baltimore
Rivoglitazone	Randomized, double-blind, active-controlled, study of rivoglitazone in type 2 diabetes mellitus	It is useful for treating type 2 diabetes	Phase 3 NCT00571519	Daiichi Sankyo, Inc.
Mizoribine	A study to evaluate the efficacy and safety of mizoribine in the treatment of lupus nephritis	It causes complete inhibition of guanine nucleotide synthesis without incorporation into nucleotides, and is thereby helpful in renal diseases	Phase 3 NCT02256150	Asahi Kasei Pharma Corporation
Nazartinib	Phase III study of nazartinib (EGF816) versus erlotinib/gefitinib in first-line locally advanced/metastatic NSCLC with EGFR activating mutations	It binds to and hinders the action of mutant forms of EGFR	Phase III NCT03529084	Novartis Pharmaceuticals
	EGF816 and trametinib in patients with nonsmall cell lung cancer harboring activating EGFR mutations (EATON)	It binds to and hinders the action of mutant forms of EGFR	Phase 1 NCT03516214	University of Cologne

Note: The investigation of clinical trials is done using the ClinicalTrials.gov website. All the data summarized in Table 8.2 has been collected from this site.

TABLE 8.3 List of patents published on imidazole analogs in various patent agencies.

Title	Patent no. and patent published date	Invention disclosed
Antimicrobial oxazolidinone, hydantoin and imidazolidinone compositions	WO 2010/054009A1 (14/05/2010)	It includes the antimicrobial activity of N-chlorinated oxazolidinone, hydantoin, and imidazolidinone compounds, and associated compositions [49]
Substituted fused imidazole derivatives, pharmaceutical compositions, and methods of use thereof	US-2011201604A1 (18/02/2010)	The method of preparation and pharmaceutical application of substituted fused imidazole analogs and their use to treat inflammation and control the activity and amount of heme-oxygenase have been investigated [50]
	US-8759535 B2 (24/06/2014)	
	US-10030011 (24/07/2018)	
	US-10287284 (14/05/2019)	
Prodrugs of LXR modulating imidazole derivatives	WO/2012/135082 (4/10/2012)	Imidazole-based compounds were investigated and examined as modulators of the activity of liver X receptors (LXR) [51]
	EP2688871B1 (26/08/2015)	
The prodrug of LXR modulating imidazole derivatives	CN103443082B (25/11/2015)	Imidazole-based compounds were investigated and examined as modulators of the activity of liver X receptors (LXR) [52]
Imidazole prodrug LXR modulators	TW201242953A (1/11/2012)	Imidazole analogs were synthesized and their action as modulator nuclear receptor, liver X receptor (LXR) was examined [53]
	US8901106B2 (02/12/2014)	

1-*N*-phenyl-amino-1H-imidazole derivatives and pharmaceutical compositions containing them	US-8476455-B2 (2/07/2013)	Imidazole analogs were synthesized and their biological applications were studied and related to the area of hormone- and nonhormone-dependent cancer and endocrine disorders [54]
TNF-alpha modulating benzimidazoles	WO2013186229A1 (19/12/2013)	A library of benzimidazole derivatives was synthesized and investigated for TNF-alpha activity, and they were found to be helpful in curing and prevention of human ailments, including autoimmune and inflammatory disorders, pain, neurological and neurodegenerative disorders, nociceptive disorders, etc. [55]
Imidazopyridine derivatives as modulators of TNF activity	WO2014009295A1 16/01/2014 US20150203486A1 23/07/2015	Imidazo[l,2-a]pyridine analogs of formula (I) were synthesized, studied for TNF-alpha activity, and the study revealed that the compounds were beneficial in human ailments, including autoimmune and inflammatory disorders, pain, neurological and neurodegenerative disorders, nociceptive disorders, etc. [56]
Imidazopyridine derivatives as modulators of TNF activity	WO2015086500A1 18/06/2015	A library of substituted 3H-imidazo[4,5-c]pyridine derivatives of formula (I) was synthesized and examined for TNF-alpha activity. As per the outcomes, the compounds were found to be beneficial in the curing of inflammatory and cardiovascular diseases, pain and nociceptive disorders, ocular disorders, etc. [57]
Imidazopyrazine derivatives as modulators of TNF activity	WO2014009296A1 (16/01/2014) CN104470924 (25/03/2015) CA2877543C (22/09/2020)	Different imidazopyrazine derivatives were synthesized and examined for human TNFα activity. They were found to be potent for the treatment of inflammatory diseases and neurological disorders, pain or nociceptive disorders, etc. [58]
Substituted imidazo[1,2-a]pyrazines as TNF activity modulators	US20150191482 09/07/2015	Imidazo[1,2-a]pyrazine derivatives were examined for human TNFα activity. They were found to be potent for the treatment of inflammatory diseases, cardiovascular diseases, autoimmune disease, and neurological disorders, pain or nociceptive disorders, etc. [59].

Continued

TABLE 8.3 List of patents published on imidazole analogs in various patent agencies—cont'd

Title	Patent no. and patent published date	Invention disclosed
Fused tricyclic benzimidazoles derivatives as modulators of TNF activity	WO2015086525A1 (18/06/2015)	Fused analogs of benzimidazole like dihydro-1H-imidazo [1,2-a]benzimidazole, dihydro-1H-pyrazino[1,2-a]benzimidazole, dihydro-1H-pyrrolo [1,2-a] benzimidazole, dihydrothiazolo[3,4-a]benzimidazolem,dihydro-1H-[1,4] oxazino[4,3-a]benzimidazole, and analogs were prepared and found to be potent modulators of human TNF activity. They are helpful in several human ailments, autoimmune and inflammatory disorders, cardiovascular diseases, etc. [60]
Fused pentacyclic imidazole derivatives	WO2016050975A1 07/04/2016	7-dihydro-7,14-methanobenzimidazo[l,2-b][2,5]benzodiazocin-5(14H)-one derivatives and analogs were synthesized and examined for their TNF activity and were found to be useful for the treatment of autoimmune and inflammatory disorders, pain and nociceptive disorders, metabolic disorders, ocular disorders, oncological disorders, etc. [61]
Fused pentacyclic imidazole derivatives as modulators of TNF activity	WO2017167996A1 5/10/2017	Imidazole analogs were synthesized and estimated as modulators of signaling of TNFα and were found to be helpful in inflammatory diseases and neurological disorders [62]
Substituted imidazole derivatives and methods of use thereof	US-9598375B2 21/03/2017	The study described the synthesis of imidazole derivatives and their application in the treatment of Alzheimer's disease [63]
Fused pentacyclic imidazole derivatives as modulators of TNF activity	WO2017167996A1 5/10/2017	Fused pentacyclic imidazole derivatives were synthesized and TNF activity was determined [64]
Fused pentacyclic imidazole derivatives as modulators of TNF activity	AU2018259040 1/11//2018	The derivatives of imidazole were prepared and investigated as modulators of signaling of TNFα, and were found to be helpful in inflammatory and cardiovascular diseases, pain and nociceptive disorders, ocular disorders, etc. [65]
Fused pentacyclic imidazole derivatives as modulators of TNF activity	US-20190100525 04/04/2019	A library of pentacyclic benzimidazole derivatives were synthesized and examined for TNFα activity. These compounds were found to be useful in the treatment and preclusion of several human ailments, and autoimmune and inflammatory disorders [66]
Fused pentacyclic imidazole derivatives as modulators of TNF activity	WO2020084008 30/04/2020	The derivatives of imidazole were prepared and investigated as modulators of signaling of TNFα, and were helpful in inflammatory and cardiovascular diseases, pain and nociceptive disorders, ocular disorders, etc. [67]

5. Conclusion and future prospects

The simple biologically potent heterocyclic scaffolds, imidazole, benzimidazole, and hybrid compounds have played a key role in drug designing and development. Great efforts have been devoted toward the synthesis of imidazole- and benzimidazole-based hybrid molecules and have led to the generation of diverse drugs for the cures of several diseases with immense therapeutic efficacy. Developing cost-effective, biologically potent drugs with improved pharmacological and pharmacokinetic characteristics is a vital and current topic that has encouraged the quest for sustainable approaches. In this chapter, we have highlighted recent findings related to the development of imidazole-based compounds to design new potential candidates. We have given an outline of various drugs available on the market and other important imidazole-based compounds, which are undergoing clinical trials, along with the published patents. This detailed study is intended to provide a single platform for researchers who are investigating this moiety for the overall benefits of humankind. This chapter is complementary to earlier published chapters and review articles, and focuses on the latest developments on imidazole-based compounds.

Having pointed out several unnoticed features of contemporary research, we would like to underline the great significance of imidazole moiety. Though many advancements have been made, it is clear that the progress of all probable applications of imidazole-containing heterocycles as drug candidates, pathologic probes, and diagnostic agents, will continue to be a subject of tremendous interest for a long time. Moreover, numerous characteristics need to be explored and investigated further:

- There is a requirement for information about the toxicity and biodegradability of drugs.
- There is a significant discrepancy in the study that makes it hard to relate and achieve as which conjugation is superior to the other.
- Significant attention should be given to structural alteration of marketed imidazole drugs to ascertain the benefits of these drugs and conquer their shortcomings. The imperative approach is to use various functional groups or structural analogs that are suitable for improving physicochemical characteristics with target sites to alter imidazole-based drugs. The main purpose is to boost their biological potency, widen the active spectrum, and overcome drug resistance.
- Progress should be made on imidazole-based chiral compounds so that they can target chiral sites with minimal side effects and unspecific toxicity.
- Imidazole-containing supramolecular drugs will illustrate escalating concentrations and become more active subjects of research and provide good drug ability.
- There is a requirement to focus on the synthesis of imidazole-based artificial receptors, pathological probes, and fluorescent molecules for diagnosis in biological systems.

- The imidazole ring should be explored as an isostere to swap with simple heterocyclic moieties like triazole, pyrazole, oxazole, amide, tetrazole, thiazole, etc., to modify the clinical drugs.

Conflict of interest

The authors declare no conflict of interest, financial or otherwise.

Acknowledgment

The authors are grateful to the Department of Chemistry, Mohan Lal Sukhadia University, Udaipur, India for providing the necessary library facilities for carrying out the work. A. Sethiya is also grateful to UGC-MANF (201819-MANF-2018-19-RAJ-91971) for providing a Senior Research Fellowship to carry out this work.

Funding source

This work was supported by UGC-MANF (201819-MANF-2018-19-RAJ-91971).

References

[1] Taylor AP, Robinson RP, Fobian YM, Blakemore DC, Jones LH, Fadeyi O. Modern advances in heterocyclic chemistry in drug discovery. Org Biomol Chem 2016;14:6611–37.

[2] Tandon R, Singh I, Luxami V, Tandon N, Paul K. Recent advances and developments of in vitro evaluation of heterocyclic moieties on cancer cell lines. Chem Rec 2019;19:362–93.

[3] Hossain M, Nanda AK. A review on heterocyclic: synthesis and their application in medicinal chemistry of imidazole moiety. Sci J Chem 2018;6(5):83–94. https://doi.org/10.11648/j.sjc.20180605.12.

[4] Vishvakarmaa VK, Reetu SN, Kumari K, Patel R, Singh P. A model to study the inhibition of nsP2B-nsP3 protease of dengue virus with imidazole, oxazole, triazole thiadiazole, and thiazolidine based scaffolds. Heliyon 2019;5(8), e02124.

[5] Rani N, Sharma A, Gupta GK, Sign R. Imidazoles as potential antifungal agents: a review. Mini Rev Med Chem 2013;13(11):1626–55.

[6] Rani N, Sharma A, Sign R. Imidazoles as promising scaffolds for antibacterial activity: a review. Mini Rev Med Chem 2013;13(12):1812–35.

[7] Mishra R, Ganguly S. Imidazole as an anti-epileptic: an overview. Med Chem Res 2012;21:3929–39.

[8] Tahlan S, Kumar S, Narasimhan B. Antimicrobial potential of 1*H*-benzo[*d*]imidazole scaffold: a review. BMC Chem 2019;13:18. https://doi.org/10.1186/s13065-019-0521-y.

[9] Aleksandrova E, Kravchenko A, Kochergin P. Properties of haloimidazoles. Chem Heterocycl Compd 2011;47(3):261–89. https://doi.org/10.1007/s10593-011-0754-8.

[10] Narasimhan B, Sharma D, Kumar P. Biological importance of imidazole nucleus in the new millennium. Med Chem Res 2010;20(8):1119–40. https://doi.org/10.1007/s0.0.044-010-9472-5.

[11] Zheng X, Ma Z, Zhang D. Synthesis of imidazole-based medicinal molecules utilizing the van Leusen imidazole synthesis. Pharmaceuticals 2020;13(3):37. https://doi.org/10.3390/ph13030037.

[12] Siwach A, Verma PK. Synthesis and therapeutic potential of imidazole containing compounds. BMC Chem 2021;15:12. https://doi.org/10.1186/s13065-020-00730-1.

[13] Najjar A, Najjar A, Karaman R. Newly developed prodrugs and prodrugs in development; an insight of the recent years. Molecules 2020;25(4):884. https://doi.org/10.3390/molecules25040884.

[14] Dong J, Chen S, Li R, et al. Imidazole-based pinanamine derivatives: discovery of dual inhibitors of the wild-type and drug-resistant mutant of the influenza A virus. Eur J Med Chem 2016;108:605–15. https://doi.org/10.1016/j.ejmech.2015.12. 013.

[15] Molina P, Tárraga A, Otón F. Imidazole derivatives: a comprehensive survey of their recognition properties. Org Biomol Chem 2012;10(9):1711–24. https://doi.org/10.1039/c2ob06808g.

[16] Claesson A, Minidis A. Systematic approach to organizing structural alerts for reactive metabolite formation from potential drugs. Chem Res Toxicol 2018;31:389–411.

[17] Shaveta MS, Singh P. Hybrid molecules: the privileged scaffolds for various pharmaceuticals. Eur J Med Chem 2016;124:500–36.

[18] Zhang L, Peng X-M, Damu GLV, Geng R-X, Zhou C-H. Comprehensive review in current developments of imidazole-based medicinal chemistry. Med Res Rev 2013;34(2):340–437.

[19] Lipinski CA, Lombardo F, Dominy BW, Feeney PJ. Experimental and computational approaches to estimate solubility and permeability in drug discovery and development settings. Adv Drug Deliv Rev 2001;46(1–3):3–26.

[20] Chopra PN, Sa JK. Biological significance of imidazole-based analogues in new drug development. Curr Drug Discov Technol 2020;17(5):574–84. https://doi.org/10.2174/1570163816666190320123340.

[21] Shalmali N, Ali MR, Bawa S. Imidazole: an essential edifice for the identification of new lead compounds and drug development. Mini Rev Med Chem 2018;18(2):142–63. https://doi.org/10.2174/1389557517666170228113656.

[22] Swikriti BR, Arora S. A comprehensive review on therapeutic potential of benzimidazole: a miracle scaffold. J Pharm Technol Res Manag 2020;8(1):23–9. https://doi.org/10.15415/jptrm.2020.81004.

[23] Tabassum K, Ekta P, Kumar KP. Imidazole and pyrazole: privileged scaffolds for anti-infective activity. Mini Rev Med Chem 2018;15(6):459–75. https://doi.org/10.2174/157019 3X15666171211170100.

[24] Gaba M, Mohan C. Development of drugs based on imidazole and benzimidazole bioactive heterocycles: recent advances and future directions. Med Chem Res 2016;25:173–210.

[25] Rani N, Kumar P, Singh R, De Sousa DP, Sharma P. Current and future prospective of a versatile moiety: imidazole. Curr Drug Targets 2020;21(11):1130–55. https://doi.org/10.2174/1389450121666200530203247.

[26] Ali I, Lone MN, Aboul-Enein HY. Imidazoles as potential anticancer agents. Med Chem Commun 2017;8(9):1742–73.

[27] Duelen BR, CorvelynI M, Tortorella I, Sampaolesi M, Sampaolesi M. Medicinal biotechnology for disease modeling, clinical therapy, and drug discovery and development. In: Introduction to biotech entrepreneurship: From idea to business. Springer Cham; 2019. https://doi.org/10.1007/978-3-030-22141-6_5.

[28] Zhou S-F, Zhong W-Z. Drug design and discovery: principles and applications. Molecules 2017;22(2):279. https://doi.org/10.3390/molecules22020279.

[29] Copeland RA. The drug–target residence time model: a 10-year retrospective. Nat Rev Drug Discov 2016;87–95. https://doi.org/10.1038/nrd.2015.18.

[30] Sliwoski G, Kothiwale SK, Meiler J, Lowe EW, Eric Jr L. Computational methods in drug discovery. Barker Pharmacol Rev 2014;66(1):334–95. https://doi.org/10.1124/pr.112.007336.

[31] Lin X, Li X, Lin X. A review on applications of computational methods, in drug screening and design. Molecules 2020;25(6):1375. https://doi.org/10.3390/molecules25061375.

[32] Ideyama Y, Kudoh M, Tanimoto K, Susaki Y, Nanya T, Nakahara T, Ishikawa H, Fujikura T, Akaza H, Shikama H. YM116, 2-(1H-imidazol-4-ylmethyl)-9H-carbazole, decreases adrenal androgen synthesis by inhibiting C17-20 Lyase activity in NCI-H295 human adrenocortical carcinoma cells. Jpn J Pharmacol 1999;79(2):213–20. https://doi.org/10.1254/jjp.79.213.

[33] Savola JM, Hill M, Engstrom M, Merivuori H, Wurster S, McGuire SG, Fox SH, Crossman AR, Brotchie JM. Fipamezole (JP-1730) is a potent alpha2 adrenergic receptor antagonist that reduces levodopa-induced dyskinesia in the MPTP-lesioned primate model of Parkinson's disease. Mov Disord 2003;18(8):872–83. https://doi.org/10.1002/mds.10464.

[34] Cantelli-Forti G, Paolini M, Hrelia P, Sapigni E, Biagi GL. Effects of metronidazole, azanidazole, and azathioprine on cytochrome P450 and various mono oxygenase activities in hepatic microsomes from control and induced mice. Toxicol Mech Methods 1987. https://doi.org/10.1007/978-3-642-72558-6_48. Springer, Berlin, Heidelberg.

[35] Fuchs CS, Tomasek J, Yong CJ, Dumitru F, Passalacqua R, Goswami C, et al. Ramucirumab monotherapy for previously treated advanced gastric or gastro-oesophageal junction adenocarcinoma (REGARD): an international, randomised, multicentre, placebo-controlled, phase 3 trial. Lancet 2014;383(9911):31–9.

[36] Belch JJ, Cormie J, Newman P, McLaren M, Barbenel J, Capell H, Leiberman P, Forbes CD, CR Prentice CR. Dazoxiben, a thromboxane synthetase inhibitor, in the treatment of Raynaud's syndrome: a double-blind trial. Br J Clin Pharmacol 1983;15(S1):113S–6S. https://doi.org/10.1111/j.1365-2125.1983.tb02119.x.

[37] Kaiser M, Bray MA, Cal M, Trunz BB, Torreele E, Brun R. Antitrypanosomal activity of fexinidazole, a new oral nitroimidazole drug candidate for treatment of sleeping sickness. Antimicrob Agents Chemother 2011;55(12):5602–8. https://doi.org/10.1128/AAC.00246-11.

[38] Jones CD, Andrews DM, Barkera AJ, Blades K, et al. The discovery of AZD5597, a potent imidazole pyrimidine amide CDK inhibitor suitable for intravenous dosing. Bioorg Med Chem Lett 2008;18(24):6369–73.

[39] Rulhania S, Kumar S, Nehra B, Gupta GD, Monga V. An insight into the medicinal perspective of synthetic analogs of imidazole. J Mol Struct 2021;1232:129982.

[40] Munagala G, Yempalla KR, Singh S, Sharma S, Kali NP, Rajput VS, Kumar S, Sawant SD, Khan IA, Vishwakarma RA, Singh PP. Synthesis of new generation triazolyl- and isoxazolyl-containing 6-nitro-2,3-dihydroimidazooxazoles as anti-TB agents: in vitro, structure–activity relationship, pharmacokinetics and in vivo evaluation. Org Biomol Chem 2015;13:3610–24.

[41] Yempalla KR, Munagala G, Singh S, Kour G, Sharma S, Chib R, Kumar S, Wazir P, Singh GD, Raina S, Bharate SS, Khan IA, Vishwakarma RA, Singh PP. Synthesis and biological evaluation of polar functionalities containing nitrodihydroimidazooxazoles as anti-TB agents. ACS Med Chem Lett 2015;6:1059–64. https://doi.org/10.1021/acsmedchemlett.5b00202.

[42] Kang YG, Park C-Y, Shin H, Singh R, Arora G, Yu C-M, Lee Y. Synthesis and anti-tubercular activity of 2-nitroimidazooxazines with modification at the C-7 position as PA-824 analogs. Bioorg Med Chem Lett 2015;25(17):3650–3. https://doi.org/10.1016/j.bmcl.2015.06.060.

[43] Wani MY, Ahmad A, Shiekh RA, Al-Ghamdi KJ, Sobral AJF. Imidazole clubbed 1,3,4-oxadiazole derivatives as potential antifungal agents. Bioorg Med Chem 2015;23(15):4172–80. https://doi.org/10.1016/j.bmc.2015.06.053.

[44] Tokuhara H, Imaeda Y, Fukase Y, Iwanaga K, Taya N, Watanabe K, Kanagawa R, Matsuda K, Kajimoto Y, Keiji Kusumoto K, Kondo M, Snell G, Behnke CA, Kuroita T. Discovery of benzimidazole derivatives as orally active renin inhibitors: optimization of 3,5-disubstituted piperidine to improve pharmacokinetic profile. Bioorg Med Chem 2018;26:3261–86.

[45] Agarwal S, Sasane S, Kumar J, Darji B, Bhayani H, Soman S, Kulkarni N, Jain M. Novel 2-mercapto imidazole and triazole derivatives as potent TGR5 receptor agonists. Med Drug Discov 2019;1:100002.

[46] Punia S, Verma V, Kumar D, Kumar A, Deswal L. Facile synthesis, antimicrobial evaluation and molecular docking studies of pyrazole-imidazole-triazole hybrids. J Mol Struct 2021;1223:129216.

[47] Rizvi SFR, Zhang H, Mehmood S, Sanad M. Synthesis of 99mTc-labeled 2-mercaptobenzimidazole as a novel radiotracer to diagnose tumor hypoxia. Transl Oncol 2020;13:100854.

[48] Ammazzalorso A, Gallorini M, Fantacuzzi M, Gambacorta N, Filippis BD, Giampietro L, Cristina Maccallini C, Nicolotti O, Cataldi A, Amoroso R. Design, synthesis and biological evaluation of imidazole and triazolebased carbamates as novel aromatase inhibitors. Eur J Med Chem 2021;211:113115.

[49] Jain RK, Low E, Francavilla C, Shiau TP, Kim B, Nair SK. Antimicrobial oxazolidinone, hydantoin and imidazolidinone compositions; May 14, 2010. WO 2010/054009A1.

[50] Mjalli AM, Rao PD, Nabeel KJ, Matthew KJ, Mustafa G. Substituted fused imidazole derivatives, pharmaceutical compositions, and methods of use thereof; Feb 18, 2010. US-2011201604-A1; (2a) Mjalli AMM, Polisetti DR, Kassis JN, Kostura MJ, Guzel M, Attucks OC, Andrews RC, Victory S, Gupta S. Substituted fused imidazole derivatives, pharmaceutical compositions, and methods of use thereof. US-8759535 B2; June 24, 2014; (2b) Mjalli AMM, Polisetti DR, Kassis JN, Kostura MJ, Guzel M, Attucks OC, Andrews RC, Victory S, Gupta S. Substituted fused imidazole derivatives, pharmaceutical compositions, and methods of use thereof. US-10030011; July 24, 2018; (2c) Mjalli MM, Polisetti DR, Kassis JN, Kostura MJ, Guzel M, Attucks OC, Andrews RC, Victory S, Gupta S. Substituted fused imidazole derivatives, pharmaceutical compositions, and methods of use thereof. US-10287284; May 14, 2019.

[51] Kick EK, Hageman MJ, Guarino VR, Su C-C, Wei C, Warrier JS, Nair SK. Prodrugs of LXR modulating imidazole derivatives; Oct 4, 2012. WO/2012/135082; (3a) Kick EK, Hageman MJ, Guarino VR, Su C-C, Wei C, Warrier JS, Nair SK. Prodrugs of LXR modulating imidazole derivatives. EP2688871B1, WO/2012/135082; August 26, 2015.

[52] Kick EK, Hageman MJ, Guarino VR, Su C-C, Wei C, Warrier JS, Nair SK. The prodrugs of LXR modulating imidazole derivatives; November 25, 2015. CN103443082B.

[53] Kick EK, Hageman MJ, Guarino VR, Su C-C, Wei C, Warrier JS, Nair SK. Imidazole prodrug LXR modulators; November 1, 2012. TW201242953A; (5a) Kick EK, Hageman MJ, Guarino VR, Su C-C, Wei C, Warrier JS, Nair SK. Imidazole prodrug LXR modulators. US8901106B2, December 12, 2014.

[54] Lafay J, Rondot B, Bonnet P, Clerc T, Shields J, Duc I, Eric D, Puccio F, Blot C, Maillos P. 1-N-phenyl-amino-1h-imidazole derivatives and pharmaceutical compositions containing them. US-8476455-B2; July 2, 2013.

[55] Brookings DC, Calmiano MD, Gallimore EO, Horsley HT, Hutchings MC, Johnson JA, Kroeplien B, Lecomte FC, Lowe MA, Norman TJ, Porter JR, Quincey JR, Reuberson JT, Selby MD, Shaw MA, Zhu Z, Foley AM. TNF-alpha modulating benzimidazoles; December 19, 2013. WO2013186229A1.

[56] Bentley JM, Brookings DC, Brown JA, Cain TP, Chovatia PT, et al. Imidazopyridine derivatives as modulators of TNF activity; January 16, 2014. WO2014009295A1; (8a) Bentley JM, Brookings DC, Brown JA, Cain TP, Chovatia PT, et al. Imidazopyridine derivatives as modulators of TNF activity. US20150203486A1; July 23, 2015.

[57] Ve J, Kroeplien B, Porter JR. Imidazopyridine derivatives as modulators of TNF activity; June 18, 2015.

[58] Bentley JM, Brookings DC, Brown JA, Cain TP, et al. Imidazopyrazine derivatives as modulators of TNF activity; January 16, 2014. WO2014009296A1; (10a) Bentley JM, Brookings DC, Brown JA, Cain TP, et al. Imidazopyrazine derivatives as modulators of TNF activity. CN104470924; March 25, 2015; (10b) Bentley JM, Brookings DC, Brown JA, Cain TP, et al. Imidazopyrazine derivatives as modulators of TNF activity. CA2877543C; September 22, 2020.

[59] Bentley JM, Brookings DC, Brown JA, Cain TP, et al. Substituted imidazo[1,2-a]pyrazines as TNF activity modulators; July 9, 2015. US20150191482.

[60] Alexander RP, Calmiano MD, Sabine Defays S, Véronique Durieu V, Michael Deligny M, Heer JP, et al. Fused tricyclic benzimidazoles derivatives as modulators of TNF activity; June 18, 2015. WO2015086525A1.

[61] Garcia TD, Deligny M, Heer JP, Quincey JR, Xuan M, Zhu Z, et al. Fused pentacyclic imidazole derivatives; 7 April 2016. WO2016050975A1.

[62] Deligny M, Louis R, Heer JP. Fused pentacyclic imidazole derivatives; 10 May 2017. WO2017167996A1.

[63] Jones D, Gowda RB, Xie R. Substituted imidazole derivatives and methods of use thereof; March 21, 2017. US-9598375B2.

[64] Heer JP, Keyaerts J. Fused pentacyclic imidazole derivatives as modulators of TNF activity; Oct 5, 2017. WO2017167996A1.

[65] Brookings CD, Garcia DH, Foricher T, Horsley Y, Hutchings HT, Johnson MC, Maccoss JA, Xuan M, Zhaoning MZ. Fused pentacyclic imidazole derivatives as modulators of TNF activity; November 1, 2018. AU2018259040.

[66] Deligny RML, Heer JP, Keyaerts J, Lepissier LE, Lowe MA. Fused pentacyclic imidazole derivatives as modulators of TNF activity; April 4, 2019.

[67] Johnson JA, Gallimore EO, Xuan M. Fused pentacyclic imidazole derivatives as modulators of TNF activity. WO2020084008; April 30, 2020.

Further reading

Other important links

[1] https://go.drugbank.com/drug.

[2] https://www.clinicaltrials.gov.

[3] https://pubchem.ncbi.nlm.nih.gov/compound.

Index

Note: Page numbers followed by *t* indicate tables.

O

P

Q

R

S

T

V

Z

Printed in the United States
by Baker & Taylor Publisher Services